Lecture Notes in Computer Scien

Edited by G. Goos, J. Hartmanis, and J. van L

T0238148

# Springer

*Berlin*
*Heidelberg*
*New York*
*Barcelona*
*Hong Kong*
*London*
*Milan*
*Paris*
*Tokyo*

Zoltán Ésik   Zoltán Fülöp (Eds.)

# Developments in Language Theory

7th International Conference, DLT 2003
Szeged, Hungary, July 7-11, 2003
Proceedings

 Springer

Series Editors

Gerhard Goos, Karlsruhe University, Germany
Juris Hartmanis, Cornell University, NY, USA
Jan van Leeuwen, Utrecht University, The Netherlands

Volume Editors

Zoltán Ésik
Zoltán Fülöp
University of Szeged, Department of Informatics
Árpád tér 2, 6720 Szeged, Hungary
E-mail: {ze,fulop}@inf.u-szeged.hu

Cataloging-in-Publication Data applied for

Bibliographic information published by Die Deutsche Bibliothek
 Die Deutsche Bibliothek lists this publication in the Deutsche Nationalbibliografie;

detailed bibliographic data is available in the Internet at <http://dnb.ddb.de>.


CR Subject Classification (1998): F.4.3, F.4.2, F.4, F.3, F.1, G.2

ISSN 0302-9743
ISBN 3-540-40434-1 Springer-Verlag Berlin Heidelberg New York

Springer-Verlag Berlin Heidelberg New York
a member of BertelsmannSpringer Science+Business Media GmbH

http://www.springer.de

© Springer-Verlag Berlin Heidelberg 2003
Printed in Germany

Typesetting: Camera-ready by author, data conversion by PTP-Berlin GmbH
Printed on acid-free paper      SPIN: 10928752      06/3142      5 4 3 2 1 0

# Preface

DLT 2003 was the 7th Conference on Developments in Language Theory. It was intended to cover all important areas of language theory, such as grammars, acceptors and transducers for strings, trees, graphs, arrays, etc., algebraic theories for automata and languages, combinatorial properties of words and languages, formal power series, decision problems, efficient algorithms for automata and languages, relations to complexity theory and logic, picture description and analysis, DNA computing, quantum computing, cryptography, and concurrency.

DLT 2003 was held at the University of Szeged, Hungary, from July 7 to 11, 2003. Previous DLTs were held in Turku (1993), Magdeburg (1995), Thessaloniki (1997), Aachen (1999), Vienna (2001), and Kyoto (2002). Since 2001, a DLT conference takes place in every odd year in Europe, and in every even year in Asia.

Altogether 57 papers were submitted. Each submission was evaluated by at least four members of the Program Committee, who were often assisted by their referees. This volume contains the 27 selected papers and the texts of the seven invited presentations.

We would like to thank the members of the Program Committee for the evaluation of the submissions and the numerous referees who assisted in this work. The list of referees is as complete as we could achieve and we apologize for any omissions and errors. We are grateful to the contributors of DLT 2003, in particular to the invited speakers for their willingness to present interesting new developments. We also thank the members of the Organizing Committee and all those whose work behind the scenes contributed to this volume.

DLT 2003 was sponsored by the Department of Informatics, University of Szeged, the Ministry of Education of Hungary, the Fund for Szeged, and Hewlett-Packard Hungary, Ltd.

April 2003                                     Zoltán Ésik and Zoltán Fülöp

# Organization

DLT 2003 was organized by the Department of Computer Science, University of Szeged, Hungary.

## Invited Lecturers

A. Bertoni (Milan)
J. Esparza (Stuttgart)
F. Gécseg (Szeged)
F. Otto (Kassel)
J.-É. Pin (Paris)
W. Plandowski (Warsaw)
H. Vollmer (Hannover)

## Program Committee

S. Bozapalidis (Thessaloniki)
Ch. Choffrut (Paris)
V. Diekert (Stuttgart)
Z. Ésik (Szeged, co-chair)
Z. Fülöp (Szeged, co-chair)
H.J. Hoogeboom (Leiden)
M. Ito (Kyoto)
J. Karhumäki (Turku)
W. Kuich (Vienna)
A. Restivo (Palermo)
B. Rovan (Bratislava)
A. Salomaa (Turku)
W. Thomas (Aachen)
Sheng Yu (London, ON, Canada)

## Organizing Committee

Z. Alexin (Szeged)
Z. Ésik (Szeged, co-chair)
Z. Fülöp (Szeged, co-chair)
Zs. Gazdag (Szeged)
É. Gombás (Szeged)
Z.L. Németh (Szeged)

# Referees

# Table of Contents

## Invited Presentations

## Contributions

# Quantum Computing: 1-Way Quantum Automata*

Alberto Bertoni, Carlo Mereghetti, and Beatrice Palano

Dipartimento di Scienze dell'Informazione
Università degli Studi di Milano
via Comelico 39/41, 20135 Milano, Italy
{bertoni, mereghetti, palano}@dsi.unimi.it

**Abstract.** In this paper we analyze several models of 1-way quantum finite automata, in the light of formal power series theory. In this general context, we recall two well known constructions, by proving:

1. Languages generated with isolated cut-point by a class of bounded rational formal series are regular.
2. If a class of formal series is closed under f-complement, Hadamard product and convex linear combination, then the class of languages generated with isolated cut-point is closed under boolean operations.

We introduce a general model of 1-way quantum automata and we compare their behaviors with those of measure-once, measure-many and reversible 1-way quantum automata.

**Keywords:** formal power series, quantum finite automata

## 1 Introduction

Quantum computing is a research area, halfway between computer science and physics [14]. In the early 1980's, Feynman suggests that the computational power of quantum mechanical processes might be beyond that of traditional computation models [11]. Almost at the same time, Benioff already determines that such processes are at least as powerful as Turing machines [4]. Discussing the notion of "quantum computer", Deutsch [10] develops the model of quantum Turing machine as a physically realizable model for a quantum computer. From the point of view of structural complexity [9], it is introduced the class **BQP** of problems solvable in polynomial time on quantum Turing machines, to be compared with the corresponding deterministic class **P** or probabilistic class **BPP**.

A well known result witnessing quantum power is Shor's algorithm for integer factorization which runs in polynomial time on a quantum computer [26]. Other relevant progress is made by Grover [13], who proposes a quantum algorithm for searching an item in an unsorted database containing $n$ items in time $O(\sqrt{n})$.

The power of quantum paradigm crucially relies on the features of quantum systems: superposition, interference and observation. The state of a quantum

---

* Partially supported by MURST, under the project "Linguaggi formali: teoria ed applicazioni".

Z. Ésik and Z. Fülöp (Eds.): DLT 2003, LNCS 2710, pp. 1–20, 2003.

machine can be seen as a linear combination of classical states (superposition). A unitary transformation makes the machine evolving from superposition to superposition. Superposition can transfer the complexity of the problem from a large number of sequential steps to a large number of coherently superposed quantum states. Entanglement is used to create complicated correlations that permit interference between the parallel "computations" performed by the machine. Furthermore, the machine can be observed: this process makes the superposition collapse to a particular state.

Some efforts have been made for constructing quantum devices, and their realizations seems to be a difficult task. For this reason it can be useful to study the computational characteristics of simple devices such as quantum finite automata. Some models have been introduced: measure-once [6,8,21], measure-many [1,17], "enhanced" [22], reversible [12]. In this paper we introduce a more general model — 1-way quantum automata with control language — and we compare its computational power with that of the other considered models.

In Section 2 we briefly recall elements of quantum physics and we discuss, in relation to the so called Moore law, the interest of quantum computing in the development of computing technologies. After introducing basic concepts and notations on quantum models, Section 4 is devoted to recall elements of rational power series theory. Since stochastic behaviors of 1-way quantum automata are rational power series, it seems a correct abstraction level to study properties of automata in the light of rational series theory. In the case of bounded series we rephrase Rabin's theorem for stochastic automata (Theorem 3); moreover, we show how closure properties of classes of series reflect on closure properties of the class of languages generated with isolated cut-point by the series (Theorem 4).

In Section 5, we recall some models of 1-way quantum automata and we introduce the notion of 1-way quantum automata with control language (Definition 2). In Section 6, it is proved that languages recognized by quantum automata with control language are a class of regular languages closed under boolean operations (Theorem 6 and Theorem 8) and in Section 7, the computational power of the considered models is compared (Theorem 9 and Theorem 10).

## 2   Laws of Nature and Technological Developments

### 2.1   Quantum Mechanics

Quantum physics is an elegant theory that describes with precision a large spectrum of the phenomena of nature. Many physicists cite 1900 as the date of birth of quantum physics. In such a year, Max Planck publishes his work on blackbody radiation, postulating that energy consists of discrete quantities called "energy quanta". De Broglie makes another important step in 1924. He suggests that the dichotomy between the physics of particles and that of waves be abolished to produce a single theory of "matter-waves" that could behave either like waves or like particles, depending on the experiments performed on them.

The formulation known as "quantum mechanics" was given by Schrödinger and Heisenberg who independently developed two mathematically equivalent versions in 1925 and 1926.

To introduce Schrödinger's description of quantum systems, let us restrict to a very simple system consisting of a single particle (e.g., an electron) moving on a straight line. The state of this system at any given time $t$ is represented by a complex function $\Psi(x,t)$, such that $\int_{+\infty}^{-\infty} |\Psi(x,t)|^2 dx = 1$; in the Born's interpretation $|\Psi(x,t)|^2 dx$ represents the probability to find an electron between $x$ and $x + dx$. $\Psi(x,t)$ verifies the Schrödinger's equation

$$i\hbar \frac{\partial}{\partial t}\Psi(x,t) = H\Psi(x,t), \tag{1}$$

where $H$ is an Hermitian operator which denotes system energy.

If $\Psi(x,0)$ is the state at time 0, by solving (1) we obtain the state $\Psi(x,t)$ at time $t$:

$$\Psi(x,t) = \Psi(x,0)e^{-\frac{i}{\hbar}Ht}.$$

Observe that the linear operator $e^{-\frac{i}{\hbar}Ht}$ is unitary and its inverse is $e^{\frac{i}{\hbar}Ht}$; the dynamics of a quantum system is reversible: starting from $\Psi(x,t)$ and $t$, we can recollect $\Psi(x,0) = \Psi(x,t)e^{\frac{i}{\hbar}Ht}$.

On the contrary, the process of observation (i.e., the measure of a physical observable) turns out to be irreversible. For instance, suppose we want to determine the position at time $t$ of the electron whose dynamics is described by Equation (1). If our measure apparatus gives the position $x'$, we are sure that our observation process has fixed the state of the system, at time $t^+$, in $\delta(x - x')$, where $\delta$ is Dirac's distribution. This means that:

- the observation has modified the state of the system,
- from the state at time $t^+$ (after measurement) we cannot reconstruct the state $\Psi(x,t)$ of the system at time $t$ (before measurement).

## 2.2   Quantum Computing and Moore's Law

Moore's law is a well-known statement describing performance increasing of computing systems along years [20]. In 1975, on the basis of actual observations on a period of 16 years, Moore emphasizes the growth by a factor of 2 per year of the number of transistors within a chip. This leads him to state such an exponential increase as a law for microelectronics development. Up to now, such a law has demonstrated his validity, although with slight tuning, and there is no reasons why it could not describe the near future [3].

The importance of such an exponential growth from an economical point of view is straightforward, and so a lot of research and technological efforts are devoted to support this trend. However, the exponential growth of transistor integration density, put the accent on some limits reducing the advantages brought by this evolution. For instance, a deep level of miniaturization yields delay in signal transmission and increase of thermal dissipation within chips. Moreover, there are also several physical limits for microelectronics developments, among others nonscalability of transistors, or limits imposed by the technology of lithography in chip design.

Given such technological and physical limitations, it is natural to search for new technologies able to overcome them. Quantum computation could represent a possible candidate. This direction is also motivated by the actual ability of operating at a microscopic level and hence with systems exhibiting typical quantum features (superposition, interference, ...). For instance, some memory devices has been built where the transition on/off is determined by a single electron instead of $10^5$ electrons as in conventional devices.

At the moment, among all possible quantum devices, it seems more realistic the construction of "small" quantum components (i.e., with memory consisting of few quantum bits) embedded into classical computers. A useful mathematical model to study the computational power of such small component is represented by quantum finite automata.

## 3   Preliminaries

### 3.1   Linear Algebra

We quickly resume some notations of linear algebra. For more details, we refer the reader to, e.g., [18,19].

We denote by $\mathbf{C}$ the field of complex numbers and by $\mathbf{C}^{n \times m}$ the set $n \times m$ matrices having entries in $\mathbf{C}$. Given a complex number $z \in \mathbf{C}$, its *conjugate* is denoted by $\bar{z}$, and its *modulus* is $|z| = \sqrt{z\bar{z}}$. The *adjoint* of a matrix $M \in \mathbf{C}^{n \times m}$ is the matrix $M^\dagger \in \mathbf{C}^{m \times n}$, where $M_{ij}^\dagger = \overline{M_{ji}}$. For matrices $A, B \in \mathbf{C}^{n \times m}$, their sum is the matrix $(A + B)_{ij} = A_{ij} + B_{ij}$. For matrices $C \in \mathbf{C}^{n \times m}$ and $D \in \mathbf{C}^{m \times r}$, their product is the matrix $(CD)_{ij} = \sum_{k=1}^{m} C_{ik} D_{kj}$ in $\mathbf{C}^{n \times r}$. For matrices $A \in \mathbf{C}^{n \times m}$ and $B \in \mathbf{C}^{p \times q}$, their direct sum and Kronecker (or direct) product are the $(m+p) \times (n+q)$ and $mp \times nq$ matrices defined, respectively, as:

$$A \oplus B = \begin{pmatrix} A & \mathbf{0} \\ \mathbf{0} & B \end{pmatrix}, \quad A \otimes B = \begin{pmatrix} A_{11}B & \cdots & A_{1m}B \\ \vdots & \ddots & \vdots \\ A_{n1}B & \cdots & A_{nm}B \end{pmatrix}.$$

When operations can be performed, we get $(A \otimes B) \cdot (C \otimes D) = AC \otimes BD$.

An Hilbert space of dimension $n$ is the linear space $\mathbf{C}^{1 \times n}$ equipped with sum and product by elements in $\mathbf{C}$, in which the *inner product* $(\pi, \xi) = \pi \xi^\dagger$ is defined. If $(\pi, \xi) = 0$ we say that $\pi$ is *orthogonal* to $\xi$. The norm of *vector* $\pi$, is defined as $\|\pi\| = \sqrt{(\pi, \pi)}$. Two subspaces $X, Y$ are orthogonal if each vector in $X$ is orthogonal to any vector in $Y$; in this case, the linear space generated by $X \cup Y$ is denoted by $X \oplus Y$.

A matrix $M \in \mathbf{C}^{n \times n}$ can be view as the automorphism $\pi \mapsto \pi M$ of the Hilbert space $\mathbf{C}^{1 \times n}$ in itself. $M$ is said to be *unitary* whenever $MM^\dagger = I = M^\dagger M$, where $I$ is the identity matrix; moreover, a matrix is unitary if and only if it preserves the norm, i.e., $\| xM \| = \| x \|$ for each vector $x \in \mathbf{C}^{1 \times n}$. The eigenvalues of unitary matrices are complex numbers of modulus 1, i.e., they are in the form $e^{i\vartheta}$, for some real $\vartheta$. $M$ is said to be *Hermitian* whenever $M = M^\dagger$. Given an Hermitian matrix $\mathcal{O}$, let $c_1, \ldots, c_s$ be its eigenvalues and $E_1, \ldots E_s$ the corresponding eigenspaces. It is well known that each eigenvalue $c_k$ is real, that $E_i$ is orthogonal to $E_j$, for any $i \neq j$ and that $E_1 \oplus \ldots \oplus E_s = \mathbf{C}^{1 \times n}$. Each vector $\pi$ can be uniquely decomposed as $\pi = \pi_1 + \cdots + \pi_s$, where $\pi_j \in E_j$; the linear transformation $\pi \mapsto \pi_j$ is called projector $P_j$ on the subspace $E_j$. An Hermitian matrix $\mathcal{O}$ is biunivocally determined by its eigenvalues and its eigenspaces (or, equivalently, by its projectors); it holds that $\mathcal{O} = c_1 P_1 + \ldots + c_s P_s$.

### 3.2 Axiomatic for Quantum Mechanics in Short

In the following, we use the previous formalism as follows.

Given a set $Q = \{q_1, \ldots, q_m\}$, every $q_i$ can be represented by its characteristic vector $e_i = (0, \ldots, 1, \ldots, 0)$. A *quantum state* on $Q$ is a superposition $\pi = \sum_{k=1}^{m} \pi_k e_k$, where the coefficients $\pi_k$ are complex *amplitudes* and $\| \pi \| = 1$. Every $e_k$ is called *pure state*. Given alphabet $\Sigma = \{\sigma_1, \ldots, \sigma_l\}$, with every symbol $\sigma_i$ we associate a unitary transformation $U(\sigma_k) : \mathbf{C}^{1 \times m} \to \mathbf{C}^{1 \times m}$. An *observable* is described by an Hermitian matrix $\mathcal{O} = c_1 P_1 + \ldots + c_s P_s$. Suppose that a quantum system is described by the quantum state $\pi$. Then, we can operate

1. *Evolution $U(\sigma_j)$.* In this case, the new state $\xi = \pi U(\sigma_j)$ is reached; this dynamics is reversible, since $\pi = \xi U^\dagger(\sigma_j)$.
2. *Measurement of $\mathcal{O}$.* In this case, every result in $\{c_1, \ldots, c_s\}$ can be obtained; $c_j$ is obtained with probability $\| \pi P_j \|^2$ and the state after such a measurement is $\pi P_j / \| \pi P_j \|$. The state transformation induced by a measurement is typically irreversible.

## 4 Formal Power Series

In this section, we recall basics on formal power series in noncommutative variables with coefficients in $\mathbf{C}$. In particular, we show Theorem 3 and Theorem 4. Suitable specializations of these results will enable us to investigate properties of the classes of languages accepted by several models of quantum automata. For more details on formal power series, the reader is referred to, e.g., [5,28].

### 4.1 Rational Power Series

Let $\Sigma^*$ be the free monoid generated by the finite alphabet $\Sigma$. A *formal power series (in noncommutative variables)* with coefficients in the field $\mathbf{C}$ is any function $\varphi : \Sigma^* \to \mathbf{C}$, usually expressed by the formal sum $\varphi = \sum_{w \in \Sigma^*} \varphi(w)\, w$. For instance, the series $\mathbf{1}(w) = 1$, for every $w \in \Sigma^*$, writes as $\mathbf{1} = \sum_{w \in \Sigma^*} w$.

The class of all formal power series $\varphi : \Sigma^* \to \mathbf{C}$ writes as $\mathbf{C}\langle\langle \Sigma \rangle\rangle$. The *support* of $\varphi$ is the language $\mathrm{supp}(\varphi) = \{w \in \Sigma^* \mid \varphi(w) \neq 0\}$. A *polynomial* is a series with finite support, and the class of polynomials is denoted by $\mathbf{C}\langle\Sigma\rangle$. For the sake of simplicity, for any $\alpha \in \mathbf{C}$, we write $\alpha$ for the polynomial $\alpha\,\varepsilon$.

A structure of ring is defined on $\mathbf{C}\langle\langle\Sigma\rangle\rangle$ by introducing the operations of sum and Cauchy product:

*Sum:* $(\varphi + \psi)(w) = \varphi(w) + \psi(w)$.
*Cauchy product:* $(\varphi \cdot \psi)(w) = \sum_{xy=w} \varphi(x)\psi(y)$.

A series $\varphi$ is *proper* if $\varphi(\varepsilon) = 0$. In this case, the *star* operation can be defined as

$$\varphi^* = \sum_{i\geq 0} \varphi^{(i)}, \text{ where } \varphi^{(0)} = 1\,\varepsilon \text{ and } \varphi^{(i)} = \varphi \cdot \varphi^{(i-1)}, \text{ for } i > 0.$$

An important subclass of $\mathbf{C}\langle\langle\Sigma\rangle\rangle$ is the class of rational series:

**Definition 1.** *The class* $\mathbf{C}^{\mathrm{Rat}}\langle\langle\Sigma\rangle\rangle$ *of rational power series is the smallest subclass of* $\mathbf{C}\langle\langle\Sigma\rangle\rangle$ *containing* $\mathbf{C}\langle\Sigma\rangle$ *and closed under the rational operations of sum, Cauchy product and star.*

A well known result due to Schützenberger [27] proposes an alternative characterization of $\mathbf{C}^{\mathrm{Rat}}\langle\langle\Sigma\rangle\rangle$ by the notion of linear representation of series. Such a result can be seen as an extension of Kleene's Theorem. A *linear representation of dimension* $m$ of a series $\varphi : \Sigma^* \to \mathcal{C}$ is a triple $(\pi, \{A(\sigma)\}_{\sigma\in\Sigma}, \eta)$, with $\pi, \eta \in \mathbf{C}^{1\times m}$ and $A(\sigma) \in \mathbf{C}^{m\times m}$, such that, for any $w = x_1 x_2 \cdots x_n \in \Sigma^*$ we get

$$\varphi(w) = \pi A(w)\eta^\dagger = \pi \left(\prod_{i=1}^n A(x_i)\right)\eta^\dagger.$$

**Theorem 1.** [27] *A formal power series is rational if and only if it has a linear representation (of finite dimension).*

Further useful operations can be defined on $\mathbf{C}\langle\langle\Sigma\rangle\rangle$:

*Hadamard product:* $(\varphi \odot \psi)(w) = \varphi(w) \cdot \psi(w)$.
*Linear combination with coefficients* $\alpha, \beta \in \mathbf{C}$: $(\alpha\varphi + \beta\psi)(w) = \alpha\varphi(w) + \beta\psi(w)$.
*Conjugation:* $\overline{\varphi}(w) = \overline{\varphi(w)}$.
*f-complement:* $(\mathbf{1} - \varphi)(w) = 1 - \varphi(w)$.

Furthermore, a linear combination with coefficients $\alpha, \beta \in [0,1]$ satisfying $\alpha + \beta = 1$ is said to be a *convex linear combination*. It is not hard to prove that the class $\mathbf{C}^{\mathrm{Rat}}\langle\langle\Sigma\rangle\rangle$ is closed under all these operations. In the following Example, we prove this closure properties by also evaluating the dimensions of the resulting linear representations.

*Example 1.* Let us consider $\varphi_1, \varphi_2 \in \mathbf{C}^{\mathrm{Rat}}\langle\langle\Sigma\rangle\rangle$ with associated linear representations $(\pi_1, \{A_1(\sigma)\}_{\sigma\in\Sigma}, \eta_1)$ and $(\pi_2, \{A_2(\sigma)\}_{\sigma\in\Sigma}, \eta_2)$ of dimensions $m_1$ and $m_2$, respectively.

(i) A linear representation of $\varphi_1 \odot \varphi_2$ is given by

$$(\pi_1 \otimes \pi_2, \{A_1(\sigma) \otimes A_2(\sigma)\}_{\sigma \in \Sigma}, \eta_1 \otimes \eta_2).$$

Such a representation has dimension $m_1 \cdot m_2$.

(ii) Given $\alpha, \beta \in \mathbf{C}$, a linear representation of $\alpha\varphi_1 + \beta\varphi_2$ is given by

$$(\alpha\pi_1 \oplus \beta\pi_2, \{A_1(\sigma) \oplus A_2(\sigma)\}_{\sigma \in \Sigma}, \eta_1 \oplus \eta_2).$$

Such a representation has dimension $m_1 + m_2$.

(iii) A linear representation of $\overline{\varphi_1}$ is given by

$$(\overline{\pi_1}, \{\overline{A_1(\sigma)}\}_{\sigma \in \Sigma}, \overline{\eta_1}).$$

Such a representation has dimension $m_1$.

## 4.2   Formal Series and Languages

In this section, we will be focusing on those series in $\mathbf{C}\langle\langle \Sigma \rangle\rangle$ having real coefficients, i.e., $\varphi(w) \in \mathbf{R}$, for every $w \in \Sigma^*$. As an example of such kind of series, we recall the notion of stochastic event induced by a probabilistic automaton:

*Example 2.* A probabilistic automaton [25,23] (1pfa, for short) on $\Sigma$ with $m$ control-states is represented by $A = \langle \pi, \{M(\sigma)\}_{\sigma \in \Sigma}, \eta \rangle$, where $\pi$ is a $1 \times m$ stochastic vector (probability distribution of initial state), $M(\sigma)$ is a $m \times m$ stochastic matrix (transition matrix), for every $\sigma \in \Sigma$, and $\eta \in \{0,1\}^m$ is the characteristic vector of final states; $A$ is the linear representation of the formal series $\mathcal{P}_A : \Sigma^* \to [0,1]$, where the stochastic event $\mathcal{P}_A(w)$ is the probability of accepting the word $w$.

Given a formal power series $\varphi$ and a real $\lambda$, *the language $L_{\varphi,\lambda}$ defined by $\varphi$ with cut-point $\lambda$* writes as

$$L_{\varphi,\lambda} = \{w \in \Sigma^* \mid \varphi(w) > \lambda\}.$$

The cut-point is said to be *isolated* if there exists a positive real $\delta$ such that $|\varphi(w) - \lambda| > \delta$, for any $w \in \Sigma^*$. Given a class $\mathcal{S} \subset \mathbf{C}\langle\langle \Sigma \rangle\rangle$ of real valued series, we let

$$\Lambda(\mathcal{S}) = \{L_{\varphi,\lambda} \subseteq \Sigma^* \mid \varphi \in \mathcal{C} \text{ and } \lambda \in \mathbf{R}\},$$
$$\Lambda^{\mathrm{Is}}(\mathcal{S}) = \{L_{\varphi,\lambda} \subseteq \Sigma^* \mid \varphi \in \mathcal{C} \text{ and isolated } \lambda \in \mathbf{R}\}.$$

In what follows, we derive properties of $\Lambda(\mathcal{S})$ (or $\Lambda^{\mathrm{Is}}(\mathcal{S})$) by using simple algebraic (or topological) properties of the class $\mathcal{S}$. The first result is basically due to Turakainen [29], and states that with every real $\lambda$ and real valued series $\varphi$, a real $\lambda'$ and a probabilistic automaton $\langle \pi, \{A(\sigma)\}_{\sigma \in \Sigma}, \eta \rangle$ can be associated such that $\varphi(w) > \lambda$ if and only if $\pi A(w)\eta^\dagger > \lambda'$. Therefore:

**Theorem 2.** *If $\mathcal{S} \subset \mathbf{C}^{\mathrm{Rat}}\langle\langle \Sigma \rangle\rangle$ is a class of real valued series, then $\Lambda(\mathcal{S})$ consists of stochastic languages.*

We call *bounded* a rational series $\varphi \in \mathbf{C}\langle\langle \Sigma \rangle\rangle$ that admits a linear representation $(\pi, \{A(\sigma)\}_{\sigma \in \Sigma}, \eta)$ such that

$$\| \pi A(w) \| \leq K,$$

for a fixed positive constant $K$ and every $w \in \Sigma^*$. We prove a nice property of bounded series, in the style of the well known result shown by Rabin [25] in the realm of stochastic automata.

**Theorem 3.** *If $\mathcal{S} \subset \mathbf{C}^{\mathrm{Rat}}\langle\langle \Sigma \rangle\rangle$ is a class of real valued bounded series, then $\Lambda^{\mathrm{Is}}(\mathcal{S})$ consists of regular languages.*

*Proof.* Let $(\pi, \{A(\sigma)\}_{\sigma \in \Sigma}, \eta)$ be the linear representation of a series $\varphi \in \mathcal{S}$. Then, there exists a positive constant $K$ such that, for every $w \in \Sigma^*$, we have $\| \pi A(w) \| \leq K$ and $|\varphi(w) - \lambda| \geq \delta$, for given $\lambda$ and $\delta > 0$. We are going to prove that $L_{\varphi, \lambda}$ is regular.

First, we prove that there exists a positive constant $D$ such that, for each vector $\nu$ in the vector space $\mathcal{V}$ spanned by vectors $\{\pi A(w) \mid w \in \Sigma^*\}$, we have

$$\| \nu A(w) \| \leq D \, \| \nu \|. \tag{2}$$

Let the vectors $a_1 = \pi A(w_1), a_2 = \pi A(w_2), \ldots, a_m = \pi A(w_m)$ be a basis for $\mathcal{V}$, for some $w_1, w_2, \ldots, w_m \in \Sigma^*$. For each $1 \leq i \leq m$, we let $b_i$ be a norm 1 vector orthogonal to $a_1, \ldots, a_{i-1}, a_{i+1}, \ldots, a_m$. Then, for $1 \leq i \neq j \leq m$, we have $a_i b_i^{\dagger} = 0$ and $a_i b_j^{\dagger} \neq 0$. Clearly, there exist $\alpha_1, \alpha_2, \ldots, \alpha_m \in \mathbf{C}$ such that $\nu = \sum_{i=1}^{m} \alpha_i a_i$, and

$$\| \nu \| \geq |\nu b_i^{\dagger}| = |\alpha_i| \cdot |a_i b_i^{\dagger}|. \tag{3}$$

Then, we can write

$$\| \nu A(x) \| = \| (\sum_{i=1}^{m} \alpha_i a_i) A(x) \|$$

$$\leq \sum_{i=1}^{m} |\alpha_i| \, \| \pi A(w_i x) \| \quad \text{(since } a_i = \pi A(w_i))$$

$$\leq \sum_{i=1}^{m} |\alpha_i| K \quad \text{(by hypothesis } \| \pi A(w_i x) \| \leq K)$$

$$\leq \| \nu \| \sum_{i=1}^{m} K / |a_i b_i^{\dagger}| \quad \text{(by inequality (3))}$$

$$= D \, \| \nu \|.$$

Now, suppose by contradiction that $L_{\varphi, \lambda}$ is not a regular language. Then, its syntactic congruence must be of infinite index. For $k \geq 0$, we let $x_k$ be the string representing the $k$th class of the congruence. So, for $i \neq j$, there exists $x \in \Sigma^*$

such that $x_i x \in L_{\varphi,\lambda}$ if and only if $x_j x \notin L_{\varphi,\lambda}$. Hence, by inequality (2), we can write

$$2\delta \leq |\varphi(x_i x) - \varphi(x_j x)| = \| (\pi A(x_i) - \pi A(x_j))A(x)\eta \|$$
$$\leq \| (\pi A(x_i) - \pi A(x_j))A(x) \| \, \| \eta \| \leq D \, \| \pi A(x_i) - \pi A(x_j) \| \, \| \eta \| .$$

This is a contradiction since, for each $i$, $\pi A(x_i)$ is in the sphere of radius $K$, which is a compact.  $\square$

In the following theorem, we single out properties of classes of series $\mathcal{S} \subseteq \mathbf{C}\langle\langle \Sigma \rangle\rangle$ guaranteeing that the corresponding classes of languages $\Lambda^{\mathrm{Is}}(\mathcal{S})$ are closed under union, intersection and complement. We denote by $\mathbf{0}$ the series $\mathbf{0}(w) = 0$, for every $w \in \Sigma^*$, and let $\varphi^k = \varphi \odot \cdots \odot \varphi$, $k$ times.

**Theorem 4.** *Let $\mathcal{S} \subseteq \mathbf{C}^{\mathrm{Rat}}\langle\langle \Sigma \rangle\rangle$ be a class of formal power series $\varphi : \Sigma^* \to [0,1]$ containing the series $\mathbf{0}$ and closed under f-complement, Hadamard product, and convex linear combination. Then $\Lambda^{\mathrm{Is}}(\mathcal{S})$ is closed under union, intersection and complement.*

*Proof.*

(i) If $L$ is defined by $\varphi \in \mathcal{S}$ with isolated cut-point $\lambda$, then its complement $L^c$ can be defined by $\mathbf{1} - \varphi$ with isolated cut-point $1 - \lambda$. This shows that $\Lambda^{\mathrm{Is}}(\mathcal{S})$ is closed under complement.

(ii) It is enough to prove that $\Lambda^{\mathrm{Is}}(\mathcal{S})$ is closed under intersection. Again, let $L$ be defined by $\varphi \in \mathcal{S}$ with isolated cut-point $\lambda$. For a fixed positive $M$, define the series

$$\psi = \sum_{0 \leq k \leq \lambda M} \frac{\binom{M}{k}}{2^M} \varphi^k \odot (1 - \varphi)^{M-k}.$$

Such a series is easily seen to belong to $\mathcal{S}$ since it is a convex linear combination of series in $\mathcal{S}$. Observe that, for any $w \in \Sigma^*$,

$$\psi(w) = \frac{1}{2^M} \, \mathrm{prob}\left\{ \frac{\sum_{i=1}^M X_i}{M} \geq \lambda \right\},$$

where $X_i$'s are i.i.d. random variables in $\{0,1\}$ with $\mathrm{prob}\{X_i = 1\} = \varphi(w)$. If $w \notin L$, we know that $\varphi(w) \leq \lambda - \delta$, for suitable $\lambda, \delta > 0$. Hence, by Höffdings' inequality [16], we can write

$$\mathrm{prob}\left\{ \frac{\sum_{i=1}^M X_i}{M} \geq \lambda \right\} \leq \mathrm{prob}\left\{ \frac{\sum_{i=1}^M X_i}{M} - \varphi(w) \geq \delta \right\} \leq e^{-2\delta^2 M},$$

and hence $\psi(w) \leq \frac{1}{2^M} e^{-2\delta^2 M}$. If $w \in L$, by the same reasoning we obtain $\psi(w) \geq \frac{1}{2^M}(1 - e^{-2\delta^2 M})$. In conclusion, fixed an arbitrary $\varepsilon > 0$, for every

$M$ such that $\varepsilon \geq e^{-2\delta^2 M}$ we get $|\frac{1}{2^M}\chi_L(w) - \psi(w)| \leq \frac{\varepsilon}{2^M}$, where $\chi_L$ is the characteristic function of $L$.

Now, consider languages $L_1, L_2 \in \Lambda^{\text{Is}}(\mathcal{S})$. By the previous reasoning, fixed $\varepsilon = \frac{1}{10}$, we can find $M$ and two series $\psi_1, \psi_2 \in \mathcal{S}$ such that $|\frac{1}{2^M}\chi_{L_1}(w) - \psi_1(w)| \leq \frac{1}{10}\frac{1}{2^M}$ and $|\frac{1}{2^M}\chi_{L_2}(w) - \psi_2(w)| \leq \frac{1}{10}\frac{1}{2^M}$. Let us consider the series $\psi = \frac{1}{2}\psi_1 + \frac{1}{2}\psi_2$. This series belongs to $\mathcal{S}$, moreover it is easy to verify that $\lambda = \frac{8}{10}\frac{1}{2^M}$ is an isolated cut-point for $\psi$ and $\psi(w) > \lambda$ if and only if $w \in L_1 \cap L_2$.

□

If, in addition, we have that $\varphi \in \mathcal{S}$ implies $\sum_{0 \leq k \leq \lambda M} \binom{M}{k}\varphi^k \odot (\mathbf{1} - \varphi)^{M-k} \in \mathcal{S}$ then the characteristic functions of languages in $\Lambda^{\text{Is}}(\mathcal{S})$ can be arbitrarily approximated by series in $\mathcal{S}$. In fact, by the previous construction, with every language $L \in \Lambda^{\text{Is}}(\mathcal{S})$ and $\varepsilon > 0$, a series $\psi \in \mathcal{S}$ can be associated so that $|\chi_L(w) - \psi(w)| \leq \varepsilon$, for every $w \in \Sigma^*$.

## 5 Models of 1-Way Quantum Automata

1-way quantum finite automata (1qfa, for short) are computational devices particularly interesting because of the simplicity of their realization. Moreover, their analysis provides a good insight into the nature of quantum computation, since 1qfa's are a theoretical model for a quantum computer with finite memory.

From the point of view of computational capabilities, quantum models of 1-way finite automata present both advantages and disadvantages with respect to their classical (deterministic or probabilistic) counterpart. Essentially, quantum superposition offers some computational advantages with respect to probabilistic superposition. On the other hand, quantum dynamics are reversible: because of limitation of memory, it is generally impossible to simulate classical automata by quantum automata. Limitations due to reversibility can be partially attenuated by systematically introducing measurements of suitable observables as computational steps.

Several models of quantum automata have been proposed in the literature. Basically, they differ in measurement policy [15]. In this section we recall the following models:

- Measure-Once Quantum Finite Automata,
- Measure-Many Quantum Finite Automata,
- Reversible Quantum Finite Automata.

Yet, in the spirit of the so called "enhanced quantum automata" [22], we introduce an hybrid model, with a classical control on the results of measurements. We call such a model *Quantum Finite Automaton with control language*.

**Measure-Once 1qfa** (MO-1qfa): The "measure-once" model is the simplest 1qfa [6,8,21]. In this case, the transformation on a symbol of the input alphabet is realized by a unitary operator. A unique measurement is performed at the end of computation.

More formally, a measure-once automaton with $m$ control states on the alphabet $\Sigma$ is a system $A = \langle \pi, \{U(\sigma)\}_{\sigma \in \Sigma}, \mathcal{O} \rangle$, where $\pi \in \mathbf{C}^{1 \times m}$, for any $\sigma \in \Sigma$, $U(\sigma) \in \mathbf{C}^{m \times m}$ is a unitary transformation, and $\mathcal{O}$ is an observable with results in $\{a, r\}$, completely described by the projectors $P(a)$ and $P(r)$. The behavior of $A$ is the stochastic event $\mathcal{P}_A : \Sigma^* \to [0,1]$ defined by the formal series

$$\mathcal{P}_A(x_1 \cdots x_n) = \| \pi \left( \prod_{i=1}^{n} U(x_i) \right) P(a) \|^2 .$$

In the following, we denote by $\mathbf{C_{MO}}$ the class of stochastic events induced by MO-1qfa's, and by $\mathbf{BMO}$ the class of languages $\Lambda^{\mathrm{ls}}(\mathbf{C_{MO}})$.

**Measure-Many 1qfa** (MM-1qfa): The main difference between the "measure-many" model [17,1] and the previous one, is that a MO-1qfa performs a unique measurement at the end of computation, while a MM-1qfa performs a measurement at each step.

Formally a MM-1qfa with $m$ control states on the alphabet $\Sigma$ and $\sharp \notin \Sigma$ is a system $A = \langle \pi, \{U(\sigma)\}_{\sigma \in \Sigma \cup \{\sharp\}}, \mathcal{O} \rangle$, where $\pi \in \mathbf{C}^{1 \times m}$, for any $\sigma \in \Sigma \cup \{\sharp\}$, $U(\sigma) \in \mathbf{C}^{m \times m}$ is a unitary transformation, and $\mathcal{O}$ is an observable with results in $\{a, r, g\}$, completely described by the projectors $P(a)$, $P(r)$ and $P(g)$.

With every word $x_1 \cdots x_n \sharp$, where $x_1 \cdots x_n \in \Sigma^n$, we associate the following computation: starting from $\pi$, $U(x_1)$ is applied and a measurement of $\mathcal{O}$ is performed obtaining a new current state. Then, $U(x_2)$ is applied and a new measurement of $\mathcal{O}$ is performed. This process continues as far as measurements yields the result $g$. As soon as the result of measurement is $a$, the computation halts and the word is accepted. Hence, the stochastic event $\mathcal{P}_A : \Sigma^* \to [0,1]$ induced by $A$ is the formal series

$$\mathcal{P}_A(x_1 \cdots x_n) = \sum_{k=1}^{n+1} \| \pi \left( \prod_{i=1}^{k-1} U(x_i) P(g) \right) U(x_k) P(a) \|^2 ,$$

where, for simplicity, we let $x_{n+1} = \sharp$.

In the following, we denote by $\mathbf{C_{MM}}$ the class of stochastic events induced by MM-1qfa's, and by $\mathbf{BMM}$ the class of languages $\Lambda^{\mathrm{ls}}(\mathbf{C_{MM}})$.

**Reversible 1qfa:** If we consider observables whose eigenspaces are monodimensional, the result of a measurement leads to a simple configuration, not to a superposition. If measurements are performed after each step, the automaton is actually probabilistic. In fact, for the sake of simplicity, suppose that the eigenspaces of the observable $\mathcal{O}$ are defined by vectors $e_1, \ldots, e_m$, where $e_i \in \mathbf{C}^{1 \times m}$ is a vector having 1 at the $i$-th component, and 0 elsewhere. Suppose also that $t$ steps of computation lead the system to the state $e_j$. After an evolution described by a unitary transformation $U$ and a measurement of $\mathcal{O}$, the new state at step $t+1$ is $e_k$ with probability $|e_j U e_k^\dagger|^2 = |U_{jk}|^2$. This may give evidence of the interest of analyzing particular probabilistic automata called quantum reversible, or their extension called probabilistic reversible [12].

The probabilistic automaton $\mathcal{A} = \langle \pi, \{M(\sigma)\}_{\sigma \in \Sigma}, \eta \rangle$ is said to be probabilistic reversible (*quantum reversible*) if, for all $\sigma \in \Sigma$, the matrix $M(\sigma)$ is double stochastic (*orthostochastic*, respectively). A matrix $M$ is said to be double stochastic if both $M$ and $M^\dagger$ are stochastic, while $M$ is orthostochastic if there exists a unitary matrix $U$ such that $M_{ij} = |U_{ij}|^2$. Since $U^\dagger$ is unitary, every orthostochastic matrix is double stochastic, while the vice versa generally does not hold: for instance, the matrix

$$\begin{pmatrix} \frac{1}{2} & \frac{1}{2} & 0 \\ \frac{1}{2} & 0 & \frac{1}{2} \\ 0 & \frac{1}{2} & \frac{1}{2} \end{pmatrix}$$

is double stochastic but not orthostochastic [19].

In the following, we denote by $\mathbf{C_{RQ}}$ the class of stochastic events induced by reversible 1qfa's, and by $\mathbf{BRQ}$ the class of languages $\Lambda^{ls}(\mathbf{C_{RQ}})$.

**1qfa with control language** (1qfc): Here, we introduce a model where, as in "enhanced 1qfa's" proposed in [22], the state of the system can be observed after each symbol is processed. An observable $\mathcal{O}$ with a fixed, but arbitrary, set of possible results $\mathcal{C} = \{c_1, \ldots, c_s\}$ is considered. On any given input word $x$, the computation displays a sequence $y \in \mathcal{C}^*$ of results of measurements of $\mathcal{O}$ with a certain probability $p(y; x)$: the computation is "accepting" if and only if $y$ belongs to a fixed regular control language $\mathcal{L} \subseteq \mathcal{C}^*$. More formally:

**Definition 2.** *Given an alphabet $\Sigma$ and an endmarker symbol $\sharp \notin \Sigma$, an $m$-state 1-way quantum finite automata with control language (1qfc, for short) is a system $\mathcal{A} = \langle \pi, \{U(\sigma)\}_{\sigma \in \Gamma}, \mathcal{O}, \mathcal{L} \rangle$, for $\Gamma = \Sigma \cup \{\sharp\}$, where*

- *$\pi \in \mathbf{C}^{1 \times m}$ is the initial amplitude vector satisfying $\| \pi \| = 1$; for all $\sigma \in \Gamma$, $U(\sigma) \in \mathbf{C}^{m \times m}$ is a unitary matrix;*
- *$\mathcal{O}$ is an observable on $\mathbf{C}^{1 \times m}$; if $\mathcal{C} = \{c_1, \ldots, c_s\}$ is the class of all possible results of measurements of $\mathcal{O}$, $P(c_i)$ denotes the projector on the eigenspace corresponding to $c_i$, for all $c_i \in \mathcal{C}$;*
- *$\mathcal{L} \subseteq \mathcal{C}^*$ is a regular language (control language).*

Now, we define the behavior of $\mathcal{A}$ on any word $x_1 \cdots x_n \in \Gamma^*$.

At any time, the state of $\mathcal{A}$ is a vector $\xi \in \mathbf{C}^{1 \times m}$ with $\| \xi \| = 1$. The computation starts in the state $\pi$, then transformations associated with the symbols in the word $x_1 \cdots x_n$ are applied in succession. The transformation corresponding to a symbol $\sigma \in \Gamma$ consists of two steps:

1. First, $U(\sigma)$ is applied to the current state $\xi$ of the automaton yielding the new state $\xi'$.
2. Then, the observable $\mathcal{O}$ is measured on $\xi'$. According to quantum mechanics principles (see Section 2.1), the result of measurement is $c_k$, with probability $\| \xi' P(c_k) \|^2$, and the state of automaton "collapses" to $\xi' P(c_k) / \| \xi' P(c_k) \|$.

Thus, a computation on $x_1 \cdots x_n$ leads to a sequence $y_1 \cdots y_n$ of results of the measurements of $\mathcal{O}$ with probability $p(y_1 \cdots y_n; x_1 \cdots x_n)$ given by

$$p(y_1 \cdots y_n; x_1 \cdots x_n) = \| \pi \prod_{i=1}^{n} U(x_i) P(y_i) \|^2 .$$

A computation leading to the word $y_1 \cdots y_n$ is *accepting* if $y_1 \cdots y_n \in \mathcal{L}$, otherwise it is rejecting. Hence, the probability that, on input $x_1 \cdots x_n$, the automaton generates an accepting computation is

$$\psi_{\mathcal{A}}(x_1 \cdots x_n) = \sum_{y_1 \cdots y_n \in \mathcal{L}} p(y_1 \cdots y_n; x_1 \cdots x_n). \tag{4}$$

In what follows, we are interested in the behavior of $\mathcal{A}$ on words in $\Sigma^* \sharp$, where $\sharp$ is the endmarker symbol. Therefore, the stochastic event $\mathcal{P}_{\mathcal{A}} : \Sigma^* \to [0,1]$ induced by $\mathcal{A}$ is

$$\mathcal{P}_{\mathcal{A}}(x_1 \cdots x_n) = \psi_{\mathcal{A}}(x_1 \cdots x_n \sharp),$$

i.e., the probability that $x_1 \cdots x_n \sharp$ generates an accepting computation.

In the following, we denote by $\mathbf{C_{QC}}$ the class of stochastic events induced by 1qfc's, and by $\mathbf{C_{QC}}(\mathcal{C}, \mathcal{L})$ the class of stochastic events induced by 1qfc's of the form $\mathcal{A} = \langle \pi, \{U(\sigma)\}_{\sigma \in \Gamma}, \mathcal{O}, \mathcal{L} \rangle$, where results of measurements of $\mathcal{O}$ are in the fixed set $\mathcal{C}$ and $\mathcal{L} \subseteq \mathcal{C}^*$. Analogously, we denote by $\mathbf{BQC}$ the class of languages $\Lambda^{\mathrm{ls}}(\mathbf{C_{QC}})$ and by $\mathbf{BQC}(\mathcal{C}, \mathcal{L})$ the class of languages $\Lambda^{\mathrm{ls}}(\mathbf{C_{QC}}(\mathcal{C}, \mathcal{L}))$.

## 6 Languages Recognized by 1qfc's and Closure Properties

To provide some indications on the computational power of 1-way quantum automata with control language, we show that the series in the class $\mathbf{C_{QC}}$ are bounded rational, and that $\mathbf{C_{QC}}$ is closed under f-complement, Hadamard product and convex linear combination. From Theorem 3 and Theorem 4, we get that $\mathbf{BQC}$ is a subclass of regular languages closed under boolean operations.

**Theorem 5.** *The formal series in $\mathbf{C_{QC}}$ are bounded rational.*

*Proof.* For $\mathcal{P}_{\mathcal{A}} \in \mathbf{C_{QC}}$, let $\psi_{\mathcal{A}}$ the series associated with the 1qfc $\mathcal{A} = \langle \pi, \{U(\sigma)\}_{\sigma \in \Sigma \cup \{\sharp\}}, \mathcal{O}, \mathcal{L} \rangle$, such that $\mathcal{P}_{\mathcal{A}}(w) = \psi_{\mathcal{A}}(w\sharp)$, for any $w \in \Sigma^*$. Since the control language $\mathcal{L} \subseteq \mathcal{C}^*$ is regular, there exists a deterministic automaton recognizing $\mathcal{L}$. We associate with such an automaton the linear representation $\langle \alpha, \{M(c)\}_{c \in \mathcal{C}}, \beta \rangle$ realizing the characteristic function $\chi_{\mathcal{L}}$ of $\mathcal{L}$ and such that, for any word $y \in \mathcal{C}^*$, $\alpha M(y)$ is a vector with a unique nonzero component of value 1. Therefore, we get that $\forall y \in \mathcal{C}^*, \ \| \alpha M(y) \| = 1$.

Now, consider the linear representation $\langle \hat{\pi}, \{\hat{U}(\sigma)\}_{\sigma \in \Sigma \cup \{\sharp\}}, \hat{\eta} \rangle$, where:

- $S = \sum_{c \in \mathcal{C}} P(c) \otimes P(c) \otimes M(c)$,
- $\hat{\pi} = (\pi \otimes \overline{\pi} \otimes \alpha)$,
- $\hat{U}(\sigma) = (U(\sigma) \otimes \overline{U(\sigma)} \otimes I) \cdot S$,

- $\hat{\eta} = \sum_{k=1}^{m} e_k \otimes e_k \otimes \beta$, where $e_k$ is the boolean vector having 1 only at the $k$-th component.

we have the following properties:

1. $\psi_A$ *is rational.* In fact $\langle \hat{\pi}, \{\hat{U}(\sigma)\}_{\sigma \in \Sigma}, \hat{\eta} \rangle$ is a linear representation of $\psi_A$. To this regard, let $x_1 \cdots x_n \in \Sigma^n$. Then

$$\hat{\pi}\hat{U}(x_1) \cdots \hat{U}(x_n)\hat{\eta}^\dagger =$$
$$\sum_{k=1}^{m} \sum_{y=y_1 \ldots y_n \in \mathcal{C}^n} (\pi \prod_{j=1}^{n} U(x_j)P(y_j))_k (\overline{\pi} \prod_{j=1}^{n} \overline{U(x_j)}P(y_j))_k \alpha M(y)\beta^\dagger =$$
$$\sum_{y=y_1 \ldots y_n \in \mathcal{C}^n} \chi_{\mathcal{L}}(y) \sum_{k=1}^{m} |(\pi \prod_{j=1}^{n} U(x_j)P(y_j))_k|^2 =$$
$$\sum_{y_1 \ldots y_n \in \mathcal{L}} \| \pi \prod_{j=1}^{n} U(x_j)P(y_j) \|^2 = \psi_A(x_1 \cdots x_n).$$

2. $\psi_A$ *is bounded.* In fact, by recalling that for complex vectors $\nu, \nu'$, we have $\| \nu \otimes \nu' \| = \| \nu \| \| \nu' \|$, and using Lemma 1, Section 7, we can write

$$\| \hat{\pi}\hat{U}(x_1) \cdots \hat{U}(x_n) \| =$$
$$\| \sum_{y=y_1 \ldots y_n \in \mathcal{C}^n} ((\pi \prod_{j=1}^{n} U(x_j)P(y_j)) \otimes (\overline{\pi} \prod_{j=1}^{n} \overline{U(x_j)}P(y_j)) \otimes \alpha M(y)) \| \leq$$
$$\sum_{y=y_1 \ldots y_n \in \mathcal{C}^n} \| ((\pi \prod_{j=1}^{n} U(x_j)P(y_j)) \otimes \overline{\pi} \prod_{j=1}^{n} \overline{U(x_j)}P(y_j) \otimes \alpha M(y)) \| =$$
$$\sum_{y=y_1 \ldots y_n \in \mathcal{C}^n} \| \pi \prod_{j=1}^{n} U(x_j)P(y_j) \|^2 \| \alpha M(y) \| = \| \pi \|^2 = 1.$$
$\square$

Then, we can conclude that

**Theorem 6.** *A language accepted with isolated cut point by a 1-way quantum finite automaton with control language is regular.*

*Proof.* Immediate from Theorem 5 and Theorem 3.    $\square$

**Theorem 7.** *The class* $\mathbf{C_{QC}}$ *is closed under f-complement, Hadamard product and convex linear combination.*

*Proof.* Let $\mathcal{E}_1, \mathcal{E}_2$ be the stochastic events realized by $\mathcal{A}_1, \mathcal{A}_2$, respectively, where $\mathcal{A}_1 = \langle \pi_1, \{U_1(\sigma)\}_{\sigma \in \Sigma \cup \{\sharp\}}, \mathcal{O}_1, \mathcal{L}_1 \rangle$ with results of measurements of $\mathcal{O}_1$ in $\mathcal{C}_1$ and $\mathcal{A}_2 = \langle \pi_2, \{U_2(\sigma)\}_{\sigma \in \Sigma \cup \{\sharp\}}, \mathcal{O}_2, \mathcal{L}_2 \rangle$, with results of measurements of $\mathcal{O}_2$ in $\mathcal{C}_2$. Then:

1. $1 - \mathcal{E}_1$ is realized by $\langle \pi_1, \{U_1(\sigma)\}_{\sigma \in \Sigma \cup \{\sharp\}}, \mathcal{O}_1, \mathcal{L}_1^c \rangle$, where $\mathcal{L}_1^c$ is the complement of $\mathcal{L}_1$.
2. $\mathcal{E}_1 \odot \mathcal{E}_2$ is realized by $\mathcal{A}_3 = \langle \pi_3, \{U_3(\sigma)\}_{\sigma \in \Sigma \cup \{\sharp\}}, \mathcal{O}_3, \mathcal{L}_3 \rangle$, where:
    - $\pi_3 = \pi_1 \otimes \pi_2$,
    - $U_3(\sigma) = U_1(\sigma) \otimes U_2(\sigma)$,
    - $\mathcal{O}_3$ is an observable with results in $\mathcal{C}_1 \times \mathcal{C}_2$, and $P_3(c_1, c_2) = P(c_1) \otimes P(c_2)$,
    - $\mathcal{L}_3 = \{(y_1, z_1) \cdots (y_n, z_n) \mid y_1 \cdots y_n \in \mathcal{L}_1 \text{ and } z_1 \cdots z_n \in \mathcal{L}_2\}$.
3. $\alpha \mathcal{E}_1 + \beta \mathcal{E}_2$ is realized by $\mathcal{A}_4 = \langle \pi_4, \{U_4(\sigma)\}_{\sigma \in \Sigma \cup \{\sharp\}}, \mathcal{O}_4, \mathcal{L}_4 \rangle$, where:
    - $\pi_4 = \sqrt{\alpha}\pi_1 \oplus \sqrt{\beta}\pi_2$,
    - $U_4(\sigma) = U_1(\sigma) \oplus U_2(\sigma)$,
    - $\mathcal{O}_4$ is an observable with results in $\mathcal{C}_1 \cup \mathcal{C}_2$, where for $c \in \mathcal{C}_1$ then $P_4(c) = P_1(c) \oplus \mathbf{0}$, for $c \in \mathcal{C}_2$ then $P_4(c) = \mathbf{0} \oplus P_2(c)$,
    - $\mathcal{L}_4 = \mathcal{L}_1 \cup \mathcal{L}_2$.

We verify, for instance, that $\mathcal{E}_3 = \mathcal{E}_1 \odot \mathcal{E}_2$. Let $x_1 \cdots x_n$ be an input string for $\mathcal{A}_3$, then

$$
\begin{aligned}
\mathcal{E}_3(x_1 \cdots x_n) &= \sum_{(y_1,z_1)\cdots(y_n,z_n)\in\mathcal{L}_3} \left\| \pi_3 \prod_{i=1}^{n} U_3(x_i)P_3(y_i,z_i) \right\|^2 \\
&= \sum_{(y_1,z_1)\cdots(y_n,z_n)\in\mathcal{L}_3} \left\| (\pi_1 \prod_{i=1}^{n} U_1(x_i)P_1(y_i)) \otimes (\pi_2 \prod_{i=1}^{n} U_2(x_i)P_2(z_i)) \right\|^2 \\
&= \sum_{y_1\cdots y_n\in\mathcal{L}_1} \sum_{z_1\cdots z_n\in\mathcal{L}_2} \left\| \pi_1 \prod_{i=1}^{n} U_1(x_i)P_1(y_i) \right\|^2 \left\| \pi_2 \prod_{i=1}^{n} U_2(x_i)P_2(z_i) \right\|^2 \\
&= \mathcal{E}_1(x_1 \cdots x_n) \cdot \mathcal{E}_2(x_1 \cdots x_n).
\end{aligned}
$$

$\square$

By Theorem 7 and Theorem 4, we can state

**Theorem 8.** *The class **BQC** of languages accepted by 1qfc's with isolated cut point is closed under boolean operations.*

## 7    Relations among Models

Aim of this section is to compare the 1qfa models previously introduced. To this purpose, we survey some properties of the classes **BMO**, **BMM** and **BRQ** and compare them with **BQC**$(\{a,r\}, a^*)$, **BQC**$(\{a,r,g\}, g^*a\{a,r,g\}^*)$ and **BQC**, the classes of languages recognized with isolated cut-point by 1-way quantum automata with control language.

First of all, it is easy to prove that the stochastic events realized by M0-1qfa's, reversible 1qfa's and 1qfc's are closed under f-complement, Hadamard product and convex linear combination. This implies

**Theorem 9. BMO, BRQ** *and* **BQC** *are closed under boolean operations.*

This enables us to prove the following

**Theorem 10.**

- **BMO** $\subset$ **BQC**$(\{a,r\}, a^*)$ $\subseteq$ **BQC**$(\{a,r,g\}, g^*a\{a,r,g\}^*)$ = **BMM** $\subset$ **BQC**
- **BRQ** $\subseteq$ **BQC**

*Proof.*

- **BMO** $\subset$ **BQC**$(\{a,r\}, a^*)$: In fact, for any M0-1qfa $A = \langle \pi, \{U(\sigma)\}_{\sigma\in\Sigma}, \mathcal{O} \rangle$, it is easy to construct a 1qfc $\hat{A} = \langle \hat{\pi}, \{\hat{U}(\sigma)\}_{\sigma\in\Sigma\cup\{\sharp\}}, \hat{\mathcal{O}}, \hat{\mathcal{L}} \rangle$, such that $\mathcal{P}_A = \mathcal{P}_{\hat{A}}$. It is enough to let:

  - $\hat{\pi} = \pi \oplus \mathbf{0}$, $\hat{U}(\sigma) = U(\sigma) \oplus I$, for any $\sigma \in \Sigma$, and $\hat{U}(\sharp) = \begin{pmatrix} \mathbf{0} & I \\ I & \mathbf{0} \end{pmatrix}$,

- the observable $\hat{\mathcal{O}}$ with results $\{a, r\}$ described by the projectors $\hat{P}(a) = I \oplus P(a)$ and $\hat{P}(r) = \mathbf{0} \oplus P(r)$,
- the control language $\hat{\mathcal{L}} = a^*$.

It is easy to verify that $\{\varepsilon\} \in \mathbf{BQC}(\{a, r\}, a^*)$, while $\{\varepsilon\} \notin \mathbf{BMO}$ since every non empty language in $\mathbf{BMO}$ must be infinite [7].

- $\mathbf{BQC}(\{a, r\}, a^*) \subseteq \mathbf{BQC}(\{a, r, g\}, g^* a \{a, r, g\}^*)$: Construction similar to the previous point.
- $\mathbf{BQC}(\{a, r, g\}, g^* a \{a, r, g\}^*) = \mathbf{BMM}$: Let $A = \langle \pi, \{U(\sigma)\}_{\sigma \in \Sigma \cup \{\sharp\}}, \mathcal{O}\rangle$ a MM-1qfa. Then, $\mathcal{A} = \langle \pi, \{U(\sigma)\}_{\sigma \in \Sigma \cup \{\sharp\}}, \mathcal{O}, g^* a \{a, r, g\}^*\rangle$ is a 1qfc with the same behavior of $A$. The key of the proof is the following Lemma, easy to prove by induction:

**Lemma 1.** Let $U(\sigma)$ be a unitary matrix, for $\sigma \in \Sigma$, and $\mathcal{O}$ an observable with results in $\mathcal{C}$ described by projectors $P(c)$, for $c \in \mathcal{C}$. For any complex vector $\alpha$ and word $x_1 \cdots x_s \in \Sigma^s$, we get

$$\sum_{y_1 \cdots y_s \in \mathcal{C}^s} \left\| \alpha \prod_{i=1}^s U(x_i) P(y_i) \right\|^2 = \| \alpha \|^2.$$

Now, we associate with $\mathcal{A}$ the formal series $\psi_{\mathcal{A}}$:

$$\psi_{\mathcal{A}}(x_1 \cdots x_n) = \sum_{y_1 \cdots y_n \in g^* a \{a, r, g\}^*} \left\| \pi \prod_{i=1}^n U(x_i) P(y_i) \right\|^2$$

$$= \sum_{k=0}^{n-1} \sum_{y_{k+2} \cdots y_n} \left\| \pi \prod_{i=1}^k (U(x_i) P(g)) U(x_{k+1}) P(a) \prod_{j=k+2}^n U(x_j) P(y_j) \right\|^2$$

$$= \sum_{k=0}^{n-1} \left\| \pi \prod_{i=1}^k (U(x_i) P(g)) U(x_{k+1}) P(a) \right\|^2 \quad \text{(by Lemma 1)}.$$

On words in $\Sigma^* \sharp$, $\psi_{\mathcal{A}}$ is the behavior $\mathcal{P}_A$ of the MM-1qfa $A = \langle \pi, \{U(\sigma)\}_{\sigma \in \Gamma}, \mathcal{O}\rangle$, with $\Gamma = \Sigma \cup \{\sharp\}$. The converse follows by the same reasoning.

- $\mathbf{BQC}(\{a, r, g\}, g^* a \{a, r, g\}^*) \subset \mathbf{BQC}$: $\mathbf{BQC}(\{a, r, g\}, g^* a \{a, r, g\}^*)$ coincides with $\mathbf{BMM}$ which is not closed under boolean operations. On the contrary, $\mathbf{BQC}$ is closed.
- $\mathbf{BRQ} \subseteq \mathbf{BQC}$: Suppose $\psi$ is induced by an $m$-state reversible 1qfa $\langle \alpha, \{M(\sigma)\}_{\sigma \in \Sigma}, \eta\rangle$. Let $\pi \in \mathbf{C}^m$ with $\pi_i = \sqrt{\alpha_i}$, for any $\sigma \in \Sigma$, let $U(\sigma)$ be a unitary transformation such that $M_{ij}(\sigma) = |U_{ij}(\sigma)|^2$, and $U(\sharp) = I$. Moreover, let $\mathcal{O}$ be the observable with possible results $\{c_1, \ldots, c_m\}$, where $P(c_k)$ projects onto the monodimensional subspaces generated by $e_k$, and let $\mathcal{L} = \{c_1, \ldots, c_m\}^* F$, where $F = \{c_i \mid \eta_i = 1\}$. Then, $\mathcal{A} = \langle \pi, \{U(\sigma)\}_{\sigma \in \Sigma \cup \{\sharp\}}, \mathcal{O}, \mathcal{L}\rangle$ is a 1qfc such that $\mathcal{P}_{\mathcal{A}} = \psi$.

$\square$

A remarkable characterization [6,8] of $\mathbf{BMO}$ uses the notion of group language [24]. A language $L \subset \Sigma^*$ is a *group language* if it is recognized by a deterministic automaton $A = \langle Q, \Sigma, \delta, q_0, F \rangle$ where, for every $\sigma \in \Sigma$, $\delta(\sigma, -)$ is a permutation. We denote by **GFA** the class of group languages.

Moreover, in [12] it is proved that for $L \in$ **BRQ**, $\chi_L \in$ **C$_{RQ}$** if and only if $L \in$ **BMO**. Hence:

**Theorem 11.**

- **BMO = GFA.**
- $\chi_L \in$ **C$_{RQ}$** *if and only if* $L \in$ **BMO.**

On the contrary, the class of stochastic events realized by MM-1qfa's is closed under f-complement and convex combination, but not for Hadamard product. So, it seems reasonable to expect that **BMM** is not closed under boolean operations and that, for certain languages $L \in$ **BMM**, the characteristic series $\chi_L$ cannot be arbitrarily approximated by series in **C$_{MM}$**.

This two problem have been studied in [1]. In particular, it has been observed that MM-1qfa's can simulate reversible automata. A deterministic automaton $A = \langle Q, \Sigma, \delta, q_0, F \rangle$ is said to be *reversible* if, for any $q \in Q$ and $\sigma \in \Sigma$, there exists at most one state $q' \in Q$ such that $\delta(q', \sigma) = q$ or, if there exist distinct states $q_1, q_2 \in Q$ and symbol $\sigma \in \Sigma$ yielding $\delta(q_1, \sigma) = q = \delta(q_2, \sigma)$, then $\delta(q, \Sigma) = q$ (i.e., $q$ is a spin state) [1,8]. We denote by **RFA** the class of languages accepted by reversible automata.

To have just an idea on the features of MM-1qfa's useful in simulating reversible automata, consider the problem of simulating a final spin state. To this purpose, we could build a MM-1qfa in which the acceptance probability exceeds the cutpoint during the computation: every continuation will be accepted.

MM-1qfa's can accept even non reversible languages, for instance $0^*1^*$ or all unary regular languages. In [1], a 4-state MM-1qfa is exhibited, which realizes a stochastic event $\psi$ satisfying $|\chi_{0^*1^*}(w) - \psi(w)| < 0.32$. Indeed, it can be proved that $\chi_{0^*1^*}$ cannot be arbitrarily approximated by series in **C$_{MM}$**, and this is not an isolated example, as shown in

**Theorem 12.** [1] *Let $L \subseteq \Sigma^*$ be a language in **BMM** such that there exists a series $\psi \in$ **C$_{MM}$** satisfying $|\psi(w) - \chi_L(w)| < 2/9 - \varepsilon$, $\varepsilon > 0$, for any $w \in \Sigma^*$. Then, $L \in$ **RFA**. Vice versa, for every language $L \in$ **RFA**, there exists a series $\psi$ such that $|\psi(w) - \chi_L(w)| < \frac{2}{9} - \varepsilon$, for a given $\varepsilon > 0$.*

Consider now the following reversible 1qfa on input alphabet $\{0,1\}$ [12]:

$$\mathcal{A} = \langle (1,0,0), M(0) = \begin{pmatrix} 1 & 0 & 0 \\ 0 & \frac{1}{2} & \frac{1}{2} \\ 0 & \frac{1}{2} & \frac{1}{2} \end{pmatrix}, M(1) = \begin{pmatrix} \frac{1}{2} & \frac{1}{2} & 0 \\ \frac{1}{2} & \frac{1}{2} & 0 \\ 0 & 0 & 1 \end{pmatrix}, \mathcal{O} = \{P_1, P_2\} \rangle,$$

where $P_1$ projects onto the first component, and $P_2$ is the orthogonal projector. It is not hard to see that $\mathcal{A}$ induces the event $\mathcal{P}_\mathcal{A}$ where: $\mathcal{P}_\mathcal{A}(0^i1^j) = 1$, while $\mathcal{P}_\mathcal{A}(0^i1^j0w) \leq 3/4$, for $w \in \{0,1\}^*$. By Theorem 9 and Theorem 12 we can conclude that:

**Theorem 13.** *For any $\varepsilon > 0$, there is $\psi \in \mathbf{C_{RQ}}$ such that $|\psi(w) - \chi_{0^*1^*}(w)| \leq \varepsilon$, for any $w \in \{0, 1\}^*$. On the contrary, since $0^*1^*$ is not a reversible language, for any $\varphi \in \mathbf{C_{MM}}$ we have $|\varphi(w) - \chi_{0^*1^*}(w)| \geq 2/9$, for some $w \in \{0, 1\}^*$.*

At the moment, a full characterization of the class **BMM** has not yet been discovered and appears to be a really difficult problem. However, several necessary conditions for membership in **BMM** have been devised. Basically, such conditions state that a given regular language cannot belong to **BMM** if its minimal deterministic automaton contains certain "forbidden figures". More precisely, in [8,12] it is shown that

**Theorem 14.** *Given a regular language $L \subseteq \Sigma^*$, let $M$ be its minimal (deterministic) automaton containing the construction in Figure 1, for some states $p, q$ and strings $x, y \in \Sigma^*$. Then, $L$ belongs neither to **BMM** nor to **BRQ**.*

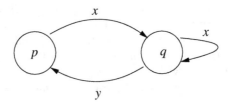

**Fig. 1.** If the minimal automaton of a regular language contains this pattern, then the language is neither in **BMM** nor in **BRQ**.

As an easy consequence of this theorem, one immediately gets that the language $\{0, 1\}^*0$ cannot belong neither to **BMM** nor to **BRQ**. In fact, it is immediate that the minimal automaton for $\{0, 1\}^*0$ contains the forbidden construction in Figure 1.

More general forbidden constructions have been exhibited in the literature. For instance, in [2], the following result is proved:

**Theorem 15.** *Given a regular language $L \subseteq \Sigma^*$, let $M$ be its minimal (deterministic) automaton containing the construction in Figure 2, for some states $p, q, t$ and strings $x, y, z, z' \in \Sigma^*$. Then, $L$ cannot belong to **BMM**.*

Among others, this theorem enables to show the following [2]

**Theorem 16.** *The class **BMM** is not closed under union.*

*Proof.* Let the regular languages

- $L_1 = (0^2)^*1(0^2)^*0$,
- $L_2 = 0(0^2)^*1(0^2)^*0$.

In [2], 8-states MM-1qfa's are exhibited accepting these languages. Now, consider their union $L_1 \cup L_2 = 0^*1(0^2)^*0$. It is not hard to verify that the minimal automaton for $L_1 \cup L_2$ contains the forbidden construction in Figure 2.   □

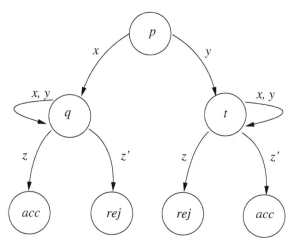

**Fig. 2.** If the minimal automaton of a regular language contains this pattern, then the language is not in **BMM**.

# References

1. A. Ambainis and R. Freivalds. 1-way quantum finite automata: strengths, weaknesses and generalizations. In *Proc. 39th Symposium on Foundations of Computer Science*, pp. 332–342, 1998.
2. A. Ambainis, A. Kikusts, and M. Valdats. On the class of languages recognizable by 1-way quantum finite automata. In *Proc. 18th Annual Symposium on Theoretical Aspects of Computer Science*, LNCS 2010, Springer, pp. 305–316, 2001. Also as Technical Report `quant-ph/0001005`.
3. L. Baldi and G. Cerofolini. La Legge di Moore e lo sviluppo dei circuiti integrati. *Mondo Digitale*, 3:3–15, 2002 (in Italian).
4. P. Benioff. Quantum mechanical Hamiltonian models of Turing machines. *J. Stat. Phys.*, 29:515–546, 1982.
5. J. Berstel and C. Reutenauer. *Rational series and their languages*. EATCS Monographs on Theoretical Computer Science, vol. 12, Springer-Verlag, 1988.
6. A. Bertoni and M. Carpentieri. Regular languages accepted by quantum automata. *Information and Computation*, 165:174–182, 2001.
7. A. Bertoni and M. Carpentieri. Analogies and differences between quantum and stochastic automata. *Theoretical Computer Science*, 262:69–81, 2001.
8. A. Brodsky and N. Pippenger. Characterizations of 1-way quantum finite automata. Technical Report TR-99-03, Department of Computer Science, University of British Columbia, 2000.
9. E. Bernstein and U. Vazirani. Quantum complexity theory. *SIAM J. Comput.*, 26:1411–1473, 1997. A preliminary version appeared in *Proc. 25th ACM Symp. on Theory of Computation*, pp. 11–20, 1993.
10. D. Deutsch. Quantum theory, the Church-Turing principle and the universal quantum computer. *Proc. Roy. Soc. London Ser. A*, 400:97–117, 1985.
11. R. Feynman. Simulating physics with computers. *Int. J. Theoretical Physics*, 21:467–488, 1982.

12. M. Golovkins and M. Kravtsev. Probabilistic Reversible Automata and Quantum Automata. In *Proc. 8th International Computing and Combinatorics Conference*, LNCS 2387, Springer, pp. 574–583, 2002

13. L. Grover. A fast quantum mechanical algorithm for database search. In *Proc. 28th ACM Symposium on Theory of Computing*, pp. 212–219, 1996.

14. J. Gruska. *Quantum Computing.* McGraw-Hill, 1999.

15. J. Gruska. Descriptional complexity issues in quantum computing. *J. Automata, Languages and Combinatorics*, 5:191–218, 2000.

16. W. Höffdings. Probability inequalities for sums of bounded random variables. *J. American Statistical Association*, 58:13–30, 1963.

17. A. Kondacs and J. Watrous. On the power of quantum finite state automata. In *38th Symposium on Foundations of Computer Science*, pp. 66–75, 1997.

18. M. Marcus and H. Minc. *Introduction to Linear Algebra.* The Macmillan Company, 1965. Reprinted by Dover, 1988.

19. M. Marcus and H. Minc. *A Survey of Matrix Theory and Matrix Inequalities.* Prindle, Weber & Schmidt, 1964. Reprinted by Dover, 1992.

20. G.E. Moore. Progress in Digital Integrated Electronics. In *Digest of the 1975 International Electron Devices Meeting*, IEEE, pp. 11–13, 1975.

21. C. Moore and J. Crutchfield. Quantum automata and quantum grammars. *Theoretical Computer Science*, 237:275–306, 2000. A preliminary version of this work appears as Technical Report in 1997.

22. A. Nayak. Optimal lower bounds for quantum automata and random access codes. In *Proc. 40th Symposium on Foundations of Computer Science.* pp. 369–376, 1999.

23. A. Paz. *Introduction to Probabilistic Automata.* Academic Press, 1971.

24. J.E. Pin. On languages accepted by finite reversible automata. In *Proc. 14th International Colloquium on Automata, Languages and Programming*, LNCS 267, pp. 237–249. Springer-Verlag, 1987.

25. M. Rabin. Probabilistic automata. *Information and Control*, 6:230–245, 1963.

26. P. Shor. Polynomial-time algorithms for prime factorization and discrete logarithms on a quantum computer. *SIAM Journal on Computing*, 26:1484–1509, 1997. A preliminary version appeared in *Proc. 35th IEEE Symp. on Foundations of Computer Science*, pp. 20–22, 1994.

27. M.P. Schützenberger. On the definition of a family of automata. *Information and Control*, 4:245–270, 1961.

28. A. Salomaa and M. Soittola. *Automata-theoretic aspects of formal power series.* Texts and Monographs in Computer Science, Springer-Verlag, 1978.

29. P. Turakainen. Generalized automata and stochastic languages. *Proc. Amer. Math. Soc.*, 21:303-309, 1969.

# An Automata-Theoretic Approach to Software Verification

Javier Esparza

Institute for Formal Methods in Computer Science
Software Reliability and Security Group
University of Stuttgart, Germany

During the last two decades, significant achievements have been obtained on the *automatic* verification of *finite-state systems* (i.e., systems with finitely many configurations). Many companies have already created verification groups that use these techniques.

However, while the finite-state framework is well suited for verifying systems like logical circuits, it cannot deal with essential aspects of software systems like communication protocols. For this we must be able to reason about models with *infinitely* many configurations.

There are different "sources of infinity". A simple analysis permits to identify at least the following five:

- **Data manipulation.** Software systems manipulate data structures over infinite domains, e.g., integer counters.
- **Control structures.** Procedures lead to potentially unbounded call stacks, while process creation leads to unboundedly many concurrent processes.
- **Asynchronous communication.** This kind of communication is modelled using unbounded queues for messages sent but not yet delivered.
- **Parameterisation.** Many distributed systems should be correct whatever the value of some parameter. The problem is to validate the system for all possible values. (Strictly speaking, we deal here with an infinite family of systems, but the family can usually be seen as one single infinite-state system.)
- **Real-time constraints.** Models of real-time systems use clocks to measure delays between events. Since time domains are infinite (even dense), these models have infinitely many configurations.

In the last few years automata-theoretic techniques have been developed to deal with all these infinities, many of them within the IST-FET project ADVANCE, funded by the European Commission and coordinated by Ahmed Bouajjani. The system to be verified is modelled as an *extended automaton*, i.e., an automaton whose transitions are labelled with actions over some data structures (stacks, integers, queues ...). The property that the system should satisfy is also modelled as a suitable automaton (often an automaton over infinite words or trees). The verification problem is reduced to checking emptiness for the product of the two automata.

In the talk I will present our work on applying this automata-theoretic approach to the infinities caused by control structures. Time permitting I will also sketch some of the techniques for other sources of infinity.

Z. Ésik and Z. Fülöp (Eds.): DLT 2003, LNCS 2710, p. 21, 2003.

# Comments on Complete Sets of Tree Automata*

Ferenc Gécseg

University of Szeged, Szeged, Hungary,
gecseg@inf.u-szeged.hu

**Abstract.** Products of tree automata do not preserve the basic properties of homomorphically and metrically complete systems of finite state
automata. To remedy it, we have introduced the concept of the quasi-
product of tree automata which is only a slightly more general than
the product. In this paper we present the main properties of the quasi-
product concerning homomorphic and metric representation of tree automata, and compare the representing powers of special quasi-products.

## 1 Introduction

In [12] Magnus Steinby introduced the concept of products of tree automata as
a generalization of the Gluškov-type product of finite state automata. In the
same paper he shows that the characterization of isomorphically complete sets
of ordinary automata can be carried over to tree automata in a natural way.
This is not so in the case of homomorphic completeness. For example, in the
classical case every homomorphically complete set of automata contains a single automaton which itself is homomorphically complete. On the other hand,
depending on the rank type of the considered tree automata, for every natural
number $k$, there exists a $k$-element set of tree automata which is homomorphically complete and minimal (see [7]). One reason for this behavior is that by the
definition of the product of tree automata no operation symbol can be replaced
by another one if they have different ranks even in the case when both of them
depend on the same variables. To remedy it, in [8] we introduced the concept
of the quasi-product which is slightly more general than the product given in
[12]. In a quasi-product the inputs of the component automata are operational
symbols in which permutation and unification of variables are allowed. It will
be seen that in sets of tree automata which are homomorphically complete with
respect to the quasi-product the essentially unary operations play the basic role
among all operations with nonzero ranks. Furthermore, we give a characterization of homomorphically complete sets which is similar to the classical one. The
next section of this study contains partial results for permutation products and
products. Finally, we show that the basic results concerning metric representation of sequential machine maps can be carried over to quasi-products of tree
automata.

* Supported by the Hungarian National Foundation for Scientific Research under
Grant T 037258

We note that tree automata to be considered here are the so-called frontier-to-root tree automata processing a tree from the frontier towards the root. In this paper we shall not deal with systems working in the opposite direction which are called root-to-frontier tree automata. Results concerning completeness of root-to-frontier tree automata can be found in [13].

## 2    Notions and Notations

Sets of operational symbols will be denoted by $\Sigma$ and $\Omega$ with or without superscripts. If $\Sigma$ is finite, then it is called a *ranked alphabet*. For the subset of $\Sigma$ consisting of all $l$-ary operational symbols from $\Sigma$ we shall use the notation $\Sigma_l$ ($l \geq 0$). By a $\Sigma$-*algebra* we mean a pair $\mathcal{A} = (A, \{\sigma^{\mathcal{A}} | \sigma \in \Sigma\})$, where $A$ is a nonempty set and $\sigma^{\mathcal{A}}$ is an $l$-ary operation on $A$ if $\sigma \in \Sigma_l$. If there will be no danger of confusion then we omit the superscript $\mathcal{A}$ in $\sigma^{\mathcal{A}}$ and simply write $\mathcal{A} = (A, \Sigma)$. Finally, all algebras considered in this paper will be finite, i.e. $A$ is finite and $\Sigma$ is a ranked alphabet.

A *rank type* is a non void set $R$ of nonnegative integers. A ranked alphabet $\Sigma$ is of *rank type* $R$ if $\{l \mid \Sigma_l \neq \emptyset\} = R$. An algebra $\mathcal{A} = (A, \Sigma)$ is of *rank type* $R$ if $\Sigma$ is of rank type $R$. To avoid trivial cases, we shall suppose that each rank type considered in this paper contains an $l$ with $l > 0$.

Let $\Xi$ be a set of variables. The set $T_\Sigma(\Xi)$ of $\Sigma\Xi$-*trees* is defined as follows:

(i)   $\Xi \bigcup \Sigma_0 \subseteq T_\Sigma(\Xi)$,
(ii)  $\sigma(p_1, \ldots, p_l) \in T_\Sigma(\Xi)$ whenever $l > 0$, $\sigma \in \Sigma_l$ and $p_1, \ldots, p_l \in T_\Sigma(\Xi)$, and
(iii) every $\Sigma\Xi$-tree can be obtained by applying the rules (i) and (ii) a finite number of times.

For a nonnegative integer $k$, $T_\Sigma^k(\Xi)$ denotes the subset of $T_\Sigma(\Xi)$ consisting of all trees from $T_\Sigma(\Xi)$ with height less than or equal to $k$.

In the sequel $\Xi$ will stand for the countable set $\{\xi_1, \xi_2, \ldots\}$ and for every $m \geq 0$, $\Xi_m$ will denote the subset $\{\xi_1, \ldots, \xi_m\}$ of $\Xi$.

If $p \in T_\Sigma(\Xi_l)$ and $p_1, \ldots, p_l \in T_\Sigma(\Xi_m)$ are trees, then $p(p_1, \ldots, p_l) \in T_\Sigma(\Xi_m)$ is the tree obtained by replacing each occurrence of $\xi_i$ ($i = 1, \ldots, l$) in $p$ by $p_i$.

A tree $p \in T_\Sigma(\Xi_1)$ is *completely balanced* if

1) all paths leading from the root of $p$ to leaves of $p$ are of the same length, and
2) if $p_1$ and $p_2$ are subtrees of $p$ with the same height, then $p_1 = p_2$.

The set of all completely balanced trees from $T_\Sigma(\Xi_1)$ will be denoted by $\hat{T}_\Sigma(\Xi_1)$.

In the sequel we shall suppose that all ranked alphabets considered in this paper are disjoint with $\Xi$.

Next we recall the two basic models of tree automata theory. For this, two more countable sets of variables $X = \{x_1, x_2, \ldots\}$ and $Y = \{y_1, y_2, \ldots\}$ are needed. Moreover, we set $X_m = \{x_1, \ldots, x_m\}$ and $Y_m = \{y_1, \ldots, y_m\}$ for any positive integer $m$.

A $\Sigma X_m$-*recognizer* is a system $\mathbf{A} = (\mathcal{A}, \mathbf{a}, X_m, A')$, where

1) $\mathcal{A} = (A, \Sigma)$ is a finite $\Sigma$-algebra,
2) $\mathbf{a} = (a^{(1)}, \dots, a^{(m)}) \in A^m$ is the *initial vector*,
3) $A' \subseteq A$ is the set of *final states*.

The *tree language* $T(\mathbf{A})$ *recognized by* $\mathbf{A}$ is given by

$$T(\mathbf{A}) = \{p \in T_\Sigma(X_m) \mid p(\mathbf{a}) \in A'\}.$$

A *tree transducer* is a system $\mathfrak{A} = (\Sigma, X_m, A, \Omega, Y_n, A', P)$, where

1) $\Sigma$ and $\Omega$ are ranked alphabets.
2) $X_m$ and $Y_n$ are the *frontier alphabets*.
3) $A$ is a ranked alphabet consisting of unary operators, the *state set* of $\mathfrak{A}$. It is assumed that $A$ is disjoint with all other sets in the definition of $\mathfrak{A}$, except $A'$.
4) $A' \subseteq A$ is the set of *final states*.
5) $P$ is a finite set of *productions* of the following two types:
   i) $x \to a(q)$ $(x \in X_m, \ a \in A, \ q \in T_\Omega(Y_n))$,
   ii) $\sigma(a_1, \dots, a_l) \to a(q(\xi_1, \dots, \xi_l))$ $(\sigma \in \Sigma_l, \ l > 0, \ a_1, \dots, a_l, a \in A,$
   $q(\xi_1, \dots, \xi_l) \in T_\Omega(Y_n \cup \Xi_l))$.

If $a \in A$ is a state and $p$ is a tree, then we generally write $ap$ for $a(p)$. Similarly, if $T$ is a tree language, then $AT$ will denote the tree language $\{ap \mid a \in A, \ p \in T\}$.

It is said that a tree transducer $\mathfrak{A}$ is *deterministic* if different productions in $P$ have different left hand sides. Moreover, it is *totally defined* if for each $x \in X_m$, there is a production $x \to aq$ in $P$, and for all $\sigma \in \Sigma_l$ and $a_1, \dots, a_l \in A$, $\sigma(a_1, \dots, a_l)$ is the left side of a production in $P$.

For every $p \in T_\Sigma(X_m)$ let $\tau^*_\mathfrak{A}(p)$ be the subset of $AT_\Omega(Y_n)$ given as follows:

(i) if $p \in X_m \bigcup \Sigma_0$ then $aq \in \tau^*_\mathfrak{A}(p)$ for all $p \to aq \in P$,
(ii) if $p = \sigma(p_1, \dots, p_l)$ $(\sigma \in \Sigma_l, \ l > 0)$ then $aq(q_1, \dots, q_l) \in \tau^*_\mathfrak{A}(p)$ for all $\sigma(a_1, \dots, a_m) \to aq \in P$ and $a_i q_i \in \tau^*_\mathfrak{A}(p_i)$ $(i = 1, \dots, l)$.

Then the relation

$$\tau_\mathfrak{A} = \{(p, q) \mid p \in T_\Sigma(X_m), \ q \in T_\Omega(Y_n), \ aq \in \tau^*_\mathfrak{A}(p) \text{ for some } a \in A'\}$$

is called the *tree transformation induced by* $\mathfrak{A}$.

Deterministic totally defined tree transducers will be called DT-*transducers*, too. One can easily show that for every deterministic tree transducer $\mathfrak{A}$ there is a DT-transducer $\mathfrak{B}$ with $\tau_\mathfrak{A} = \tau_\mathfrak{B}$. Accordingly, in this paper we deal with tree transformations induced by DT-transducers.

Other models of tree transducers can be found in [2].

To a DT-transducer $\mathfrak{A} = (\Sigma, X_m, A, \Omega, Y_n, A', P)$ we can correspond a $\Sigma X_m$-recognizer $\mathbf{A} = (\mathcal{A}, \mathbf{a}, X_m, A')$ with $\mathcal{A} = (A, \Sigma)$ and $\mathbf{a} = (a^{(1)}, \dots, a^{(m)})$, where

2) for any $l \geq 0$, $\sigma \in \Sigma_l$ and $a_1, \dots, a_l \in A$, $\sigma^\mathcal{A}(a_1, \dots, a_l) = a$ if the production $\sigma(a_1, \dots, a_l) \to aq$ is in $P$ for a $q$, and

1) $a^{(i)} = a$ if the production $x_i \to aq$ is in $P$ for some $q$ $(i = 1, \ldots, m)$.

This uniquely defined recognizer will be denoted by $\mathrm{rec}(\mathfrak{A})$.

Now take a $\Sigma X_m$-recognizer $\mathbf{A} = (\mathcal{A}, \mathbf{a}, X_m, A')$ with $\mathcal{A} = (A, \Sigma)$ and $\mathbf{a} = (a^{(1)}, \ldots, a^{(m)})$. Define a tree transducer $\mathfrak{A} = (\Sigma, X_m, A, \Omega, Y_n, A', P)$ by

$$P = \{x_i \to a^{(i)}q^{(i)} \mid q^{(i)} \in T_\Omega(Y_n), \ i = 1, \ldots, m\} \bigcup$$
$$\{\sigma(a_1, \ldots, a_l) \to \sigma^{\mathcal{A}}(a_1, \ldots, a_l)q^{(\sigma, a_1, \ldots, a_l)} \mid$$
$$\sigma \in \Sigma_l, \ l \geq 0, \ a_1, \ldots, a_l \in A, \ q^{(\sigma, a_1, \ldots, a_l)} \in T_\Omega(Y_n \cup \Xi_l)\},$$

where the ranked alphabet $\Omega$, the integer $n$ and the trees in the right sides of the productions in $P$ are fixed arbitrarily. Obviously, $\mathfrak{A}$ is a DT-transducer. Denote by $\mathrm{tr}(\mathbf{A})$ the class of all DT-transducers obtained in the above way. It is easy to see that for arbitrary DT-transducer $\mathfrak{A}$ the inclusion $\mathfrak{A} \in \mathrm{tr}(\mathrm{rec}(\mathfrak{A}))$ holds. Therefore, for every DT-transducer $\mathfrak{A}$ there exists a recognizer $\mathbf{A}$ such that $\mathfrak{A} \in \mathrm{tr}(\mathbf{A})$.

The basic part of a tree recognizer or (deterministic) tree transducer is an algebra: both types of systems are built on algebras. From the point of view of the structure theory of tree automata, important classes of algebras are those which are complete under one of the following definitions.

A class $\mathcal{K}$ of algebras of rank type $R$ is *isomorphically complete* if for every algebra $\mathcal{A}$ of rank type $R$ there is an algebra $\mathcal{B}$ in $\mathcal{K}$ such that a subalgebra of $\mathcal{B}$ is isomorphic to $\mathcal{A}$.

A class $\mathcal{K}$ of algebras of rank type $R$ is *homomorphically complete* if for every algebra $\mathcal{A}$ of rank type $R$ there is an algebra $\mathcal{B}$ in $\mathcal{K}$ such that $\mathcal{A}$ is a homomorphic image of a subalgebra of $\mathcal{B}$.

A class $\mathcal{K}$ of algebras of rank type $R$ is *language complete* if for every tree language $T$ recognizable by a tree recognizer built on an algebra of rank type $R$ there is an algebra $\mathcal{A}$ in $\mathcal{K}$ such that $T$ can be recognized by a tree recognizer built on $\mathcal{A}$.

A class $\mathcal{K}$ of algebras of rank type $R$ is *transformation complete* if for every tree transformation $\tau$ induced by a DT-transducer built on an algebra of rank type $R$ there is an algebra $\mathcal{A}$ in $\mathcal{K}$ such that $\tau$ can be induced by a tree transducer built on $\mathcal{A}$.

Let $\mathfrak{A} = (\Sigma, X_m, A, \Omega, Y_n, A', P)$ and $\mathfrak{B} = (\Sigma, X_m, B, \Omega, Y_n, B', P)$ be two DT-transducers and $k \geq 0$ an integer. We write $\tau_{\mathfrak{A}} \overset{k}{=} \tau_{\mathfrak{B}}$ if $\tau_{\mathfrak{A}}(p) = \tau_{\mathfrak{B}}(p)$ for all $p \in T_\Sigma^k(X_m)$.

A class $\mathcal{K}$ of algebras of rank type $R$ is *metrically complete* (*m-complete*, for short) if for any DT-transducer $\mathfrak{A} = (\Sigma, X_m, A, \Omega, Y_n, P, A')$ with $\Sigma$ of rank type $R$ and integer $k \geq 0$ there exist an algebra $\mathcal{B} = (B, \Sigma)$ in $\mathcal{K}$, a vector $\mathbf{b} \in B^m$ and a subset $B' \subseteq B$ such that $\tau_{\mathfrak{A}} \overset{k}{=} \tau_{\mathfrak{B}}$ for some $\mathfrak{B} \in \mathrm{tr}(\mathbf{B})$, where $\mathbf{B} = (\mathcal{B}, \mathbf{b}, X_m, B')$. (The name metric completeness comes from the fact that such systems are tree automata theoretic generalizations of metrically complete systems of finite automata introduced in [3].)

The aim of the structure theory is to represent tree recognizers and tree transducers by means of products of other tree recognizers and tree transducers.

Since both tree recognizers and tree transducers are built on algebras, we define products of algebras only.

Let $R$ be a rank type and take the algebras $\mathcal{A}_i = (A_i, \Sigma^{(i)})$ $(i = 1, \ldots, k > 0)$ with rank type $R$ and let

$$\varphi = \{\varphi_l : (A_1 \times \ldots \times A_k)^l \times \Sigma_l \to$$
$$T_{\Sigma^{(1)}}(\Xi_l) \times \ldots \times T_{\Sigma^{(k)}}(\Xi_l) \mid l \in R\}$$

be a family of mappings, where $\Sigma$ is an arbitrary ranked alphabet (not necessarily of rank type $R$). Then by the *generalized product* of $\mathcal{A}_1, \ldots, \mathcal{A}_k$ with respect to $\Sigma$ and $\varphi$ we mean the algebra $\mathcal{A} = (A, \Sigma)$ with $A = A_1 \times \ldots \times A_k$ such that for any $\sigma \in \Sigma_l$ $(l \in R)$ and $(a_{11}, \ldots, a_{1k}), \ldots, (a_{l1}, \ldots, a_{lk}) \in A$,

$$\sigma^{\mathcal{A}}((a_{11}, \ldots, a_{1k}), \ldots, (a_{l1}, \ldots, a_{lk})) =$$
$$(p_1^{\mathcal{A}_1}(a_{11}, \ldots, a_{l1}), \ldots, p_k^{\mathcal{A}_k}(a_{1k}, \ldots, a_{lk})),$$

where $(p_1, \ldots, p_k) = \varphi_l((a_{11}, \ldots, a_{1k}), \ldots, (a_{l1}, \ldots, a_{lk}), \sigma)$. For this generalized product we use the notation $\prod_{i=1}^{k} \mathcal{A}_i[\Sigma, \varphi]$. If $\mathcal{A}_1 = \ldots = \mathcal{A}_k = \mathcal{B}$ then we speak of a *generalized power* of $\mathcal{B}$, and write $\mathcal{B}^n[\Sigma, \varphi]$.

Obviously, for every $\varphi_l$ there are mappings

$$\varphi_l^{(i)} : (A_1 \times \ldots \times A_k)^l \times \Sigma_l \to T_{\Sigma^{(i)}}(\Xi_l) \quad (i = 1, \ldots, k)$$

such that

$$\varphi_l((a_{11}, \ldots, a_{1k}), \ldots, (a_{l1}, \ldots, a_{lk}), \sigma) =$$
$$(\varphi_l^{(1)}((a_{11}, \ldots, a_{1k}), \ldots, (a_{l1}, \ldots, a_{lk}), \sigma), \ldots,$$
$$\varphi_l^{(k)}((a_{11}, \ldots, a_{1k}), \ldots, (a_{l1}, \ldots, a_{lk}), \sigma)).$$

If every $\varphi_l^{(i)}$ $(i = 1, \ldots, k)$ may depend on $a_{11}, \ldots, a_{1i-1}, \ldots, a_{l1}, \ldots, a_{li-1}$ and $\sigma$ only, then we speak of a *cascade generalized product*.

A similar generalization of the cascade product has been introduced in [1].

Now let us extend $\varphi_l$ to arbitrary nonnegative integer $m$ and mapping $\varphi_m$ from $(A_1 \times \ldots \times A_k)^m \times T_\Sigma(\Xi_m)$ to $T_{\Sigma^{(1)}}(\Xi_m) \times \ldots \times T_{\Sigma^{(k)}}(\Xi_m)$ in the following way: for any $\mathbf{a} \in A^m$ and $p \in T_\Sigma(\Xi_m)$

(i) if $p = \xi_j$ $(1 \leq j \leq m)$, then $\varphi_m(\mathbf{a}, \xi_j) = (\xi_j, \ldots, \xi_j)$,
(ii) if $p = \sigma(p_1, \ldots, p_l)$ $(\sigma \in \Sigma_l, l \geq 0)$, then

$$\varphi_m(\mathbf{a}, p) =$$

$$(q_1(\pi_1(\varphi_m(\mathbf{a}, p_1)), \ldots, \pi_1(\varphi_m(\mathbf{a}, p_l))), \ldots,$$
$$q_k(\pi_k(\varphi_m(\mathbf{a}, p_1)), \ldots, \pi_k(\varphi_m(\mathbf{a}, p_l))))$$

where $(q_1, \ldots, q_k) = \varphi_l(p_1(\mathbf{a}), \ldots, p_l(\mathbf{a}), \sigma)$ and $\pi_j$ is the $j$th projection.

A generalized product $\mathcal{A} = (A, \Sigma) = \prod_{i=1}^{k} \mathcal{A}_i[\Sigma, \varphi]$, where $\mathcal{A}_i = (A_i, \Sigma^{(i)})$ $(i = 1, \ldots, k)$, is called a *quasi-product* if for $\varphi_l(\mathbf{a}_1, \ldots, \mathbf{a}_l, \sigma) = (p_1, \ldots, p_k)$ $(\mathbf{a}_1, \ldots, \mathbf{a}_l \in A_1 \times \ldots \times A_k$, $\sigma \in \Sigma_l$, $l \geq 0)$ we have $p_i = \sigma_i(\xi_{i_1}, \ldots, \xi_{i_{l_i}})$ $(1 \leq i_1, \ldots, i_l \leq l)$ and $\sigma_i \in \Sigma_{l_i}^{(i)}$ $(i = 1, \ldots, k)$. Let us note that the above quasi-product is called a *general product* if $l_i = l$ and $\xi_{i_j} = \xi_j$ for all $i = 1, \ldots, k$ and $j = 1, \ldots, l$. Observe that if $\mathcal{A}$ is a general product then it is of rank type $R$.

Powers for special generalized products are defined in the natural way.

The cascade product introduced by G. Ricci in [11] is a special quasi-product.

# 3  Preliminaries

Let *product* mean any of the products introduced above. Take two ranked alphabets $R$ and $R'$. We call a set $\mathcal{K}$ of algebras of rank type $R$ *homomorphically complete (isomorphically complete, language complete, transformation complete) for $R'$ with respect to the product* if the class of all products of algebras from $\mathcal{K}$ with rank type $R'$ is homorphically complete (isomorphically complete, language complete, transformation complete).

We say that a class $\mathcal{K}$ of algebras is *closed under deletion of operations*, if for all $\mathcal{A} = (A, \Sigma) \in \mathcal{K}$, $l$ with $|\Sigma_l| \geq 2$, and $\sigma \in \Sigma_l$, the algebra $\mathcal{A}' = (A, \Sigma')$, where $\Sigma'_l = \Sigma_l \setminus \{\sigma\}$, $\Sigma'_k = \Sigma_k$ if $k \neq l$, and $\bar{\sigma}^{\mathcal{A}'}(a_1, \ldots, a_m) = \bar{\sigma}^{\mathcal{A}}(a_1, \ldots, a_m)$ $(\bar{\sigma} \in \Sigma'_m$, $m \geq 0$, $a_1, \ldots, a_m \in A)$, is also in $\mathcal{K}$.

It can be seen from the proof of Theorem 7 in [7] that the next result holds.

**Theorem 1.** *Let $\mathcal{K}$ be a class of algebras of rank type $R$ closed under deletion of operations. Then the following statements are equivalent.*

1) *$\mathcal{K}$ is homomorphically complete.*
2) *$\mathcal{K}$ is language complete.*
3) *$\mathcal{K}$ is transformation complete.*

It is obvious that if $R$ and $R'$ are arbitrary rank types and $\mathcal{K}$ is a set of algebras of rank type $R$ then the class of all algebras with rank type $R'$ obtained by an arbitrarily fixed type of our products from algebras in $\mathcal{K}$ is closed under deletion of operations. Thus, from Theorem 1 we obtain

**Corollary 1.** *Let $R$ and $R'$ be rank types and let product mean any of the products introduced in this paper. For arbitrary set $\mathcal{K}$ of algebras of rank type $R$ the following statements are equivalent.*

1) *$\mathcal{K}$ is homomorphically complete for $R'$ with respect to the product.*
2) *$\mathcal{K}$ is language complete for $R'$ with respect to the product.*
3) *$\mathcal{K}$ is transformation complete for $R'$ with respect to the product.*

Therefore, in studying these three types of completeness, for any of our products we may confine ourselves to homomorphic completeness.

In the classical case, i.e. when $R = R' = \{1\}$ we have the following result which is from [10] (see, also [6]).

**Theorem 2.** *Let* $R = \{1\}$. *A set* $\mathcal{K}$ *of algebras of rank type* $R$ *is homomorphically complete for* $R$ *with respect to the general product if and only if there are an* $\mathcal{A} = (A, \Sigma) \in \mathcal{K}$, *an element* $a_0 \in A$, *two operational symbols* $\sigma_1, \sigma_2 \in \Sigma (= \Sigma_1)$, *and two trees* $p_1, p_2 \in T_\Sigma(\Xi_1)$ *such that* $\sigma_1(a_0) \neq \sigma_2(a_0)$ *and* $p_1(\sigma_1(a_0)) = p_2(\sigma_2(a_0)) = a_0$.

This theorem directly implies

**Corollary 2.** *Let* $R$ *be a rank type with* $R = \{1\}$. *If a set* $\mathcal{K}$ *of algebras of rank type* $R$ *is homomorphically complete for* $R$ *with respect to the general product, then* $\mathcal{K}$ *contains a single algebra which itself is homomorphically complete for* $R$ *with respect to the general product.*

The following result, which is from [7], shows that this is not true for tree automata.

**Theorem 3.** *Let* $R$ *be a rank type with* $|R| = t$. *Then there is a* $t$-*element set of algebras of rank type* $R$ *which is homomorphically complete for* $R$ *with respect to the general product and minimal.*

The use of generalized product remedies this problem as it is shown by the next result (see, [9]).

**Theorem 4.** *Let* $R$ *and* $R'$ *be arbitrary rank types and* $\mathcal{K}$ *a set of algebras of rank type* $R$. *Then the following statements are equivalent.*

 *i)* $\mathcal{K}$ *is isomorphically complete for* $R'$ *with respect to the generalized product,*
 *ii)* $\mathcal{K}$ *is homomorphically complete for* $R'$ *with respect to the generalized product,*
 *iii)* $\mathcal{K}$ *is language complete for* $R'$ *with respect to the generalized product,*
 *iv)* $\mathcal{K}$ *is transformation complete for* $R'$ *with respect to the generalized product.*

As this result shows, the generalized product is too general since isomorphic and homomorphic representations coincide for it. Thus, we restricted it to such an extent that the resultant product, the quasi-product, has the fundamental property of the classical case and it is only a slightly more general than the general product.

## 4    Homomorphic Representation by Quasi-products

In the rest of this paper, if it is not stated otherwise, we shall suppose that rank types do not contain 0.

The next theorem gives necessary and sufficient conditions for a set of algebras to be homomorphically complete with respect to the quasi-product (see, [8]).

**Theorem 5.** *Let* $R$ *and* $R'$ *be rank types. A set* $\mathcal{K}$ *of algebras of rank type* $R$ *is homomorphically complete for* $R'$ *with respect to the quasi-product if and only if there exist an algebra* $\mathcal{A} = (A, \Sigma) \in \mathcal{K}$, *an element* $a_0 \in A$, *two natural numbers* $l_1, l_2 \in R$, *two operational symbols* $\sigma_1 \in \Sigma_{l_1}$, $\sigma_2 \in \Sigma_{l_2}$, *and two completely balanced trees* $p, q \in \hat{T}_\Sigma(\Xi_1)$ *such that* $\sigma_1(a_0, \ldots, a_0) \neq \sigma_2(a_0, \ldots, a_0)$ *and* $p(\sigma_1(a_0, \ldots, a_0)) = q(\sigma_2(a_0, \ldots, a_0)) = a_0$.

From this theorem we directly obtain

**Corollary 3.** *Let $\mathcal{K}$ be an arbitrary set of algebras of any rank type. If $\mathcal{K}$ is homomorphically complete for a rank type with respect to the quasi-product then it contains an algebra which itself is homomorphically complete for arbitrary rank type with respect to the quasi-product.*

The following example shows that this is not true if 0-ary operational symbols are allowed.

*Example 1.* Let $R = \{0, 1\}$ be a rank type. Take the algebra

$$\mathcal{B} = (\{0, 1\}, \Sigma)$$

with $\Sigma = \Sigma_0 \cup \Sigma_1$, $\Sigma_0 = \{\sigma_0, \sigma_0'\}$, and $\Sigma_1 = \{\sigma, \sigma'\}$ defined by $\sigma_0^{\mathcal{B}} = 0$, $\sigma_0'^{\mathcal{B}} = 1$,

$$\sigma(a) = \begin{cases} 0, \text{ if } a = 0, \\ 1, \text{ if } a = 1 \end{cases}$$

and

$$\sigma'(a) = \begin{cases} 1, \text{ if } a = 0, \\ 0, \text{ if } a = 1. \end{cases}$$

It can be shown in a similar way as in the proof of Lemma 4 in [8] that for every algebra $\mathcal{A} = (A, \Sigma')$ of any rank type $R'$ there is a quasi-power $\mathcal{C} = (C, \Sigma')$ of $\mathcal{B}$ such that $\mathcal{A}$ is isomorphic to a subalgebra of $\mathcal{C}$. Therefore, $\mathcal{B}$ is isomorphically complete for $R$ with respect to the quasi-product.

Consider the algebras $\mathcal{A}_1 = (\{0, 1\}, \Sigma^{(1)})$ and $\mathcal{A}_2 = (\{0, 1, 2\}, \Sigma^{(2)})$ with $\Sigma^{(1)} = \Sigma_0^{(1)} \cup \Sigma_1^{(1)}$, $\Sigma_0^{(1)} = \{\sigma_0, \sigma_0'\}$, $\Sigma_1^{(1)} = \{\sigma\}$, and $\Sigma^{(2)} = \Sigma_0^{(2)} \cup \Sigma_1^{(2)}$, $\Sigma_0^{(2)} = \{\sigma_0, \sigma_0'\}$, $\Sigma_1^{(2)} = \{\sigma, \sigma'\}$. Moreover, $\sigma_0^{\mathcal{A}_1} = 0$, $\sigma_0'^{\mathcal{A}_1} = 1$, $\sigma^{\mathcal{A}_1}(0) = 0$, $\sigma^{\mathcal{A}_1}(1) = 1$ and $\sigma_0^{\mathcal{A}_2} = \sigma_0'^{\mathcal{A}_2} = 2$, $\sigma'^{\mathcal{A}_2}(2) = 0$, $\sigma^{\mathcal{A}_2}(2) = 1$, $\sigma'^{\mathcal{A}_2}(0) = 1$, $\sigma'^{\mathcal{A}_2}(1) = 0$, $\sigma^{\mathcal{A}_2}(0) = 0$, $\sigma^{\mathcal{A}_2}(1) = 1$.

Take the product $\mathcal{A} = (A, \Sigma) = \mathcal{A}_1 \times \mathcal{A}_2[\Sigma, \varphi]$, where $\varphi$ is given as follows: $\varphi_0(\sigma_0) = (\sigma_0, \sigma_0)$, $\varphi_0(\sigma_0') = (\sigma_0', \sigma_0')$, $\varphi_1(0, 2, \sigma) = (\sigma, \sigma')$, $\varphi_1(0, 0, \sigma) = (\sigma, \sigma)$, $\varphi_1(1, 0, \sigma) = (\sigma, \sigma)$, $\varphi_1(1, 2, \sigma) = (\sigma, \sigma)$, $\varphi_1(1, 1, \sigma) = (\sigma, \sigma)$, $\varphi_1(0, 1, \sigma) = (\sigma, \sigma)$, $\varphi_1(0, 2, \sigma') = (\sigma, \sigma)$, $\varphi_1(0, 0, \sigma') = (\sigma, \sigma')$, $\varphi_1(1, 0, \sigma') = (\sigma, \sigma')$, $\varphi_1(1, 2, \sigma') = (\sigma, \sigma')$, $\varphi_1(1, 1, \sigma') = (\sigma, \sigma')$, $\varphi_1(0, 1, \sigma') = (\sigma, \sigma')$.

Finally, consider the mapping $\tau : A \to \{0, 1\}$ given by $\tau((0, 2)) = \tau((0, 0)) = \tau((1, 0)) = 0$ and $\tau((1, 2)) = \tau((1, 1)) = \tau((0, 1)) = 1$. A straightforward computation shows that the algebra $\mathcal{B}$ is a homomorphic image of $\mathcal{A}$. Therefore, $\{\mathcal{A}_1, \mathcal{A}_2\}$ is homomorphically complete for $R$ with respect to the quasi-product.

The algebra $\mathcal{A}_1$ itself does not form a homomorphically complete set for $R$ with respect to the quasi-product since each unary operation in every quasi-power of $\mathcal{A}_1$ is idempotent. The algebra $\mathcal{A}_2$ is not homomorphically complete for $R$ with respect to the quasi-product either, since different 0-ary operational symbols have the same realization in each quasi-power of $\mathcal{A}_2$. Therefore, $\{\mathcal{A}_1, \mathcal{A}_2\}$ is minimal.

## 5  Comparison of Product Types

In this section we define special quasi-products and compare them with each other and with the general product.

Let us call a quasi-product $\mathcal{A} = \prod_{i=1}^{k} \mathcal{A}_i[\Sigma, \varphi]$ a *unification product* if for $\varphi_l(\mathbf{a}_1, \ldots, \mathbf{a}_l, \sigma) = (p_1, \ldots, p_k)$ $(\mathbf{a}_1, \ldots, \mathbf{a}_l \in A_1 \times \ldots \times A_k,\ \sigma \in \Sigma_l,\ l > 0)$ with $p_i = \sigma_i(\xi_{i_1}, \ldots, \xi_{i_{l_i}})$ $(\sigma_i \in \Sigma_{l_i}^{(i)},\ i = 1, \ldots, k)$ we have $\xi_{i_1} = \ldots = \xi_{i_{l_i}}$ $(1 \leq i_1, \ldots, i_{l_i} \leq l)$. Moreover, let us say that a quasi-product is a *permutation product* if for $\varphi_l(\mathbf{a}_1, \ldots, \mathbf{a}_l, \sigma) = (p_1, \ldots, p_k)$ $(\mathbf{a}_1, \ldots, \mathbf{a}_l \in A_1 \times \ldots \times A_k,\ \sigma \in \Sigma_l,\ l > 0)$ with $p_i = \sigma_i(\xi_{i_1}, \ldots, \xi_{i_{l_i}})$ $(\sigma_i \in \Sigma_{l_i}^{(i)},\ i = 1, \ldots, k)$, $l_i = l$ and $i_1, \ldots, i_l$ is a permutation of $1, \ldots, l$. A rank type $R$ is *homogeneous* if $R = \{l\}$ for some $l > 0$. Let $Q_1$-product and $Q_2$-product mean any of the quasi-product, unification product, permutation product or general product. We say that the $Q_1$-product is *more general* than the $Q_2$-product with respect to the homomorphic completeness, if the following two conditions are satisfied:

(1) For any two rank types $R$ and $R'$, if a set $\mathcal{K}$ of algebras of rank type $R$ is homomorphically complete for $R'$ with respect to the $Q_2$-product then $\mathcal{K}$ is homomorphically complete for $R'$ with respect to the $Q_1$-product.

(2) There are two rank types $R$ and $R'$ and a set $\mathcal{K}$ of algebras of rank type $R$ which is homomorphically complete for $R'$ with respect to the $Q_1$-product and $\mathcal{K}$ is not homomorphically complete for $R'$ with respect to the $Q_2$-product.

The $Q_1$-product is *equivalent* to the $Q_2$-product with respect to the homomorphic completeness if for any two rank types $R$ and $R'$ and any set $\mathcal{K}$ of algebras of rank type $R$, $\mathcal{K}$ is homomorphically complete for $R'$ with respect to the $Q_1$-product if and only if it is homomorphically complete for $R'$ with respect to the $Q_2$-product.

**Theorem 6.** *The quasi-product is equivalent to the unification product with respect to the homomorphic completeness.*

*Proof.* Let $R$ and $R'$ be two rank types and $\mathcal{K}$ a set of algebras of rank type $R$ which is homomorphically complete for $R'$ with respect to the quasi-product. Then $\mathcal{K}$ satisfies the conditions of Theorem 5. Therefore, by Theorem 2, the algebra $\mathcal{B}$ given in Example 1 is a homomorphic image of a subalgebra of a unification power of an algebra in $\mathcal{K}$. As it is shown in the proof of Lemma 4 in [8], every algebra of any rank type is isomorphic to a subalgebra of a unification power of $\mathcal{B}$. (Observe, that a quasi-power of $\mathcal{B}$ is always a unification power.) Since the formation of the unification product is transitive, $\mathcal{K}$ is homomorphically complete for $R'$ with respect to the unification product.

It is obvious that for inhomogeneous rank types the quasi-product is more powerful than the permutation product or the general product. Now we show that, with respect to the homomorphic completeness, even for homogeneous rank types the quasi-product is more general than the permutation product which is more general than the general product.

**Theorem 7.** *With respect to the homomorphic completeness the quasi-product is more general than the permutation product.*

*Proof.* Let $R = \{2\}$. Take the algebra $\mathcal{A} = (A, \Sigma)$ with $A = \{0, 1\}$, $\Sigma = \Sigma_2 = \{\sigma_1, \sigma_2\}$, $\sigma_1(0, 0) = \sigma_1(1, 1) = 0$, $\sigma_2(0, 0) = 1$, $\sigma_2(1, 1) = 0$ and $\sigma_1(x, y) = \sigma_2(x, y) = 1$ if $x \neq y$. By Theorem 5, $\{\mathcal{A}\}$ is homomorphically complete for $R$ with respect to the quasi-product. Moreover, let $\mathcal{I} = (I, \Sigma)$ be the algebra, where $I = \{0, 1\}$, $\sigma_1(0, 0) = 0$, $\sigma_1(1, 1) = 1$, $\sigma_2(0, 0) = 1$, $\sigma_2(1, 1) = 0$ and $\sigma_1(0, 1) = 0$, $\sigma_2(0, 1) = 1$, $\sigma_1(1, 0) = 0$ and $\sigma_2(1, 0) = 1$. Obviously, $\mathcal{I}$ has the property that for all pairs $((x, y), z)$ $(x, y, z \in I)$ there is an operation $\sigma$ with $\sigma(x, y) = z$. Assume that a subalgebra $\mathcal{C} = (C, \Sigma)$ of a permutation-power $\mathcal{B} = \mathcal{A}^n[\Sigma, \varphi]$ can be mapped homomorphically onto $\mathcal{I}$. Let $\mathbf{a} \in C$ be an arbitrary element with maximal, say $k$, numbers of occurrences of 1. It is clear that $k > 0$. By the choice of $\mathcal{I}$, $\sigma_1^{\mathcal{C}}(\sigma_1^{\mathcal{C}}(\mathbf{a}, \mathbf{a}), \mathbf{a}) \neq \sigma_2^{\mathcal{C}}(\sigma_1^{\mathcal{C}}(\mathbf{a}, \mathbf{a}), \mathbf{a})$. Therefore, since $\sigma_1^{\mathcal{A}}(1, 1) = \sigma_2^{\mathcal{A}}(1, 1) = 0$ and $\sigma_1^{\mathcal{A}}(0, 1) = \sigma_1^{\mathcal{A}}(1, 0) = \sigma_2^{\mathcal{A}}(0, 1) = \sigma_2^{\mathcal{A}}(1, 0) = 1$, thus $\sigma_1^{\mathcal{C}}(\sigma_1^{\mathcal{C}}(\mathbf{a}, \mathbf{a}), \mathbf{a})$ or $\sigma_2^{\mathcal{C}}(\sigma_1^{\mathcal{C}}(\mathbf{a}, \mathbf{a}), \mathbf{a})$ must have at least $k+1$ occurrences of 1, which is a contradiction. Consequently, $\mathcal{A}$ is not homomorphically complete for $R$ with respect to the permutation product. $\quad\blacksquare$

**Theorem 8.** *With respect to the homomorphic completeness the permutation product is more general than the product.*

*Proof.* Let $R = \{2\}$. Take the algebra $\mathcal{A} = (A, \Sigma)$ with $A = \{0, 1\}$, $\Sigma = \Sigma_2 = \{\sigma_1, \sigma_2\}$, $\sigma_1(0, 0) = 0$, $\sigma_1(1, 1) = 1$, $\sigma_2(0, 0) = 1$, $\sigma_2(1, 1) = 0$ and $\sigma_1(x, y) = \sigma_2(x, y) = x$ if $x \neq y$. Similarly as in the proof of Lemma 4 in [8], it can be shown that every algebra of rank type $R$ is isomorphic to a subalgebra of a permutation power of $\mathcal{A}$. Therefore, $\{\mathcal{A}\}$ is homomorphically complete for $R$ with respect to the permutation product. Assume that a subalgebra $\mathcal{C} = (C, \Sigma)$ of a power $\mathcal{B} = \mathcal{A}^n[\Sigma, \varphi]$ can be mapped homomorphically onto the algebra $\mathcal{I}$ given in the proof of the previous theorem. Let $\mathbf{a}, \mathbf{b} \in C$ be a pair such that they are different in maximal, say $k$, numbers of their components. By the construction of $\mathcal{I}$, $\sigma_1^{\mathcal{C}}(\mathbf{a}, \mathbf{b})$ and $\sigma_2^{\mathcal{C}}(\mathbf{a}, \mathbf{b})$ are different. Moreover, they are equal to $\mathbf{a}$ in those components in which $\mathbf{a}$ and $\mathbf{b}$ differ. Therefore, $\mathbf{b}$ and $\sigma_1^{\mathcal{C}}(\mathbf{a}, \mathbf{b})$ or $\mathbf{b}$ and $\sigma_2^{\mathcal{C}}(\mathbf{a}, \mathbf{b})$ are different at least in $k + 1$ components, which is a contradiction. Therefore, $\mathcal{A}$ is not homomorphically complete for $R$ with respect to the product. $\quad\blacksquare$

## 6  Metrical Representation by Quasi-products

Let $R$ and $R'$ be rank types. A set $\mathcal{K}$ of algebras of rank type $R$ is *metrically complete* (*m-complete*, for short) *with respect to the quasi-product* (*cascade quasi-product*) if the class of all quasi-products (cascade quasi-products) of algebras from $\mathcal{K}$ with rank type $R'$ is metrically complete.

Let $\mathcal{A} = (A, \Sigma)$ be an algebra, $m > 0$ an integer and $\mathbf{a} \in A^m$ a vector. For an integer $k \geq 0$, set $A_{\mathbf{a}}^{(k)} = \{p^{\mathcal{A}}(\mathbf{a}) \mid p \in T_{\Sigma}^k(X_m)\}$. The system $(\mathcal{A}, \mathbf{a})$ is called *k-free* if $|A_{\mathbf{a}}^{(k)}| = |T_{\Sigma}^k(X_m)|$, i.e. $p \neq q$ implies $p(\mathbf{a}) \neq q(\mathbf{a})$ whenever $p, q \in T_{\Sigma}^k(X_m)$.

We recall the following result from [5].

**Theorem 9.** *A class $\mathcal{K}$ of algebras of any rank type $R$ is m-complete if and only if for arbitrary integers $m, k > 0$ there is a k-free system $(\mathcal{A}, \mathbf{a})$ with $\mathcal{A} \in \mathcal{K}$ and $\mathbf{a} \in A^m$.*

In this section we give necessary and sufficient conditions for a set of algebras to be m-complete with respect to the quasi-product. It will turn out that, regarding metric completeness, the cascade quasi-product is as powerful as the quasi-product.

**Theorem 10.** *Let $R$ and $R'$ be rank types. Assume that a set $\mathcal{K}$ of algebras of rank type $R$ is m-complete for $R'$ with respect to the quasi-product. Then for every integer $k \geq 0$, there exist an algebra $\mathcal{A} \in \mathcal{K}$, an element $a_0 \in A$, a completely balanced tree $p \in \hat{T}_\Sigma(\Xi_1)$ with $h(p) = k$ and two operational symbols $\sigma_1, \sigma_2 \in \Sigma$ such that $\sigma_1(p(a_0), \ldots, p(a_0)) \neq \sigma_2(p(a_0), \ldots, p(a_0))$.*

*Proof.* Suppose that the conditions of Theorem 10 are satisfied. Take a $\Sigma'$ of rank type $R'$ with $|\Sigma'| > 1$, a tree $p' \in \hat{T}_{\Sigma'}(\Xi_1)$ with $h(p') = k$ and two different operational symbols $\sigma_1' \in \Sigma_{l_1}'$ and $\sigma_2' \in \Sigma_{l_2}'$. By Theorem 9, there are a quasi-product $\mathcal{B} = (B, \Sigma') = \prod_{i=1}^t \mathcal{B}_i[\Sigma', \varphi]$ of algebras $\mathcal{B}_1, \ldots, \mathcal{B}_t$ from $\mathcal{K}$ and an element $\mathbf{b} \in B$ such that $(\mathcal{B}, \mathbf{b})$ is $(k+1)$-free. Let $\mathbf{b} = (b_1, \ldots, b_t)$. Then for an $i$ $(1 \leq i \leq t)$, $\mathcal{B}_i$, $b_i$, $\varphi_1^{(i)}(\mathbf{b}, p')$, $\pi_i(\varphi_{l_1}((p'(\mathbf{b}), \ldots, p'(\mathbf{b})), \sigma_1'))$ and $\pi_i(\varphi_{l_2}((p'(\mathbf{b}), \ldots, p'(\mathbf{b})), \sigma_2'))$ should satisfy the conditions of Theorem 10.

The following result is from [4] (see, also [6]).

**Theorem 11.** *Let $\mathcal{K}$ be a set of algebras of rank type $\{1\}$ such that for every integer $k \geq 0$, there exist an algebra $\mathcal{A} = (A, \Sigma) \in \mathcal{K}$, an element $a_0 \in A$, a tree $p \in \hat{T}_\Sigma(\Xi_1)$ with $h(p) = k$ and two operational symbols $\sigma_1, \sigma_2 \in \Sigma$ satisfying $\sigma_1(p(a_0)) \neq \sigma_2(p(a_0))$. Then $\mathcal{K}$ is m-complete for $\{1\}$ with respect to the cascade quasi-product.*

The next result shows that Theorem 11 is true for any rank type.

**Theorem 12.** *Let $\mathcal{K}$ be a set of algebras of rank type $\{1\}$ satisfying the conditions of Theorem 11. Then for arbitrary rank type $R$ and integers $m > 0$, $k \geq 0$ there exists a cascade quasi-product $\mathcal{B} = (B, \Sigma')$ of algebras from $\mathcal{K}$ with rank type $R$ and a vector $\mathbf{b} \in B^m$ such that $(\mathcal{B}, \mathbf{b})$ is k-free.*

*Proof.* We shall proceed by induction on $k$. Since $\mathcal{A}$ has at least two elements and the direct power is a cascade quasi-product, our claim is true for $k = 0$. Assume that Theorem 12 has been proved up to a $k \geq 0$. Let $\mathcal{C} = (C, \Sigma')$ be a cascade quasi-product of algebras from $\mathcal{K}$ of rank type $R$ such that $(\mathcal{C}, \mathbf{c})$ is $k$-free for a $\mathbf{c} = (c_1, \ldots, c_m) \in C^m$. Let $t$ be the number of trees from $T_{\Sigma'}(X_m)$ with height $k + 1$. Moreover, let $\bar{\Sigma}$ be a unary rank alphabet with cardinality $t + 1$. Take a fixed element $\bar{\sigma}_0 \in \bar{\Sigma}$, and let $\chi$ be a one-to-one mapping of $\{\sigma(p_1, \ldots, p_l) \mid p_1, \ldots, p_l \in T_{\Sigma'}(X_m), h(\sigma(p_1, \ldots, p_l)) = k + 1\}$ onto $\bar{\Sigma} \setminus \{\bar{\sigma}_0\}$. Let $(\mathcal{D}, d_0)$ be a

$(k+1)$-free system, where $\mathcal{D} = (D, \bar{\Sigma})$ is a cascade quasi-product of algebras from $\mathcal{K}$ and $d_0 \in D$. Form the cascade quasi-product $\mathcal{B} = (B, \Sigma') = (\mathcal{C} \times \mathcal{D})[\Sigma', \varphi]$, where for any $\sigma \in \Sigma'_l$ and $p_1, \ldots, p_l \in T_{\Sigma'}(X_m)$, $\varphi$ is given in the following way: $\varphi_l^{(1)}(\sigma) = \sigma$, and

$$\varphi_l^{(2)}(p_1(\mathbf{c}), \ldots, p_l(\mathbf{c}), \sigma) = \begin{cases} \bar{\sigma}_0(\xi_1), \text{if } \max\{h(p_i) \mid i = 1, \ldots, l\} < k, \\ \chi(\sigma(p_1, \ldots, p_l))(\xi_1) \text{if } \max\{h(p_1) \ldots, h(p_l)\} = k, \\ \text{an arbitrary element of } \bar{\Sigma} \text{ in all other cases.} \end{cases}$$

It is easy to show that the system $(\mathcal{B}, \mathbf{b})$ with $\mathbf{b} = ((c^{(1)}, d_0), \ldots, (c^{(m)}, d_0))$ is $(k + 1)$-free. Since the formation of the cascade quasi-product is transitive, this ends the proof of the Theorem.

Using Theorem 9, we obtain

**Corollary 4.** *Let $R$ and $R'$ be rank types. A set $\mathcal{K}$ of algebras of rank type $R$ is m-complete for $R'$ with respect to the quasi-product if and only if $\mathcal{K}$ is m-complete for $R'$ with respect to the cascade quasi-product.*

**Corollary 5.** *Let $\mathcal{K}$ be a set of algebras of any rank type. If $\mathcal{K}$ is m-complete for a rank type with respect to the quasi-product then it is m-complete for arbitrary rank type with respect to the quasi-product.*

Thus, in case of quasi-products we may speak simply of m-completeness without mentioning the rank type.

*Example 2.* Let $R = \{1, 2\}$. Take the ranked alphabet $\Sigma = \Sigma_1 \cup \Sigma_2$ with $\Sigma_1 = \{\sigma_1, \sigma'_1\}$ and $\Sigma_2 = \{\sigma_2, \sigma'_2\}$. Consider the algebras $\mathcal{A} = (\{a_1, a_2\}, \Sigma)$ and $\mathcal{B} = (\{b_1, b_2\}, \Sigma)$ , where $\sigma_1^{\mathcal{A}}(c_1) = a_1$, $\sigma'_1{}^{\mathcal{A}}(c_1) = a_2$, and $\sigma_2^{\mathcal{A}}(c_1, c_2) = \sigma'_2{}^{\mathcal{A}}(c_1, c_2) = a_1$ for all $c_1, c_2 \in \{a_1, a_2\}$. Moreover, $\sigma_1^{\mathcal{B}}(c_1) = \sigma'_1{}^{\mathcal{B}}(c_1) = b_1$, $\sigma_2^{\mathcal{B}}(c_1, c_2) = b_1$ and $\sigma'_2{}^{\mathcal{B}}(c_1, c_2) = b_2$ for all $c_1, c_2 \in \{b_1, b_2\}$. By Lemma 3 of [5], the set $\{\mathcal{A}, \mathcal{B}\}$ is m-complete (for $R$) with respect to the general product. On the other hand, none of the single sets $\{\mathcal{A}\}$ and $\{\mathcal{B}\}$ satisfies the conditions of Theorem 2 in [5]. Therefore, none of $\{\mathcal{A}\}$ and $\{\mathcal{B}\}$ is m-complete with respect to the general product.

It follows from Theorem 1 in [4] that in the classical case if a finite set of finite automata is m-complete with respect to the general product, then it contains an automaton which itself forms an m-complete set. Example 2 shows that this is not true for the general products of tree automata. Now we show, that quasi-products of tree automata have this property.

**Theorem 13.** *If a finite set $\mathcal{K}$ of algebras of rank type $R$ is m-complete with respect to the quasi-product then there is an algebra $\mathcal{A} \in \mathcal{K}$ such that $\{\mathcal{A}\}$ is m-complete with respect to the quasi-product.*

*Proof.* Let $t$ be a positive integer which is greater than the number of elements of algebras in $\mathcal{K}$. By Theorem 10, there exist an algebra $\mathcal{A} = (A, \Sigma) \in \mathcal{K}$, an element $a_0 \in A$, a tree $p(\xi_1) = \sigma_t(\ldots(\sigma_1(\xi_1, \ldots, \xi_1))\ldots) \in \hat{T}_\Sigma(\Xi_1)$ of height $t$ and two operational symbols $\sigma$ and $\sigma'$ in $\Sigma$ such that $\sigma(p(a_0), \ldots, p(a_0)) \neq \sigma'(p(a_0), \ldots, p(a_0))$. Moreover, there are integers $k$ and $l$ with $1 \leq k < l \leq t$ under which

$$\sigma_k(\ldots(\sigma_1(a_0, \ldots, a_0))\ldots) = \sigma_l(\ldots(\sigma_1(a_0, \ldots, a_0))\ldots)$$

Therefore, $\{\mathcal{A}\}$ satisfies the conditions of Theorem 10, and thus, by Theorems 9, 11 and 12, it is m-complete.

From the proof of the previous Theorem we obtain

**Corollary 6.** *There exists an algorithm to decide for a finite set of algebras of any rank type if it is m-complete with respect to the quasi-product.*

Finally, we mention that there is no finite set of algebras which is homomorphically complete with respect to the cascade generalized product (see, [9]).

# References

1. Ésik, Z.: Definite tree automata and their cascade compositions. Publ. Math. **48** (1996), 243–261.
2. Fülöp, Z., Vogler, H.: Syntax-Directed Semantics: Formal Models Based on Tree Transducers. Springer-Verlag, 1998.
3. Gécseg, F.: On R-products of automata I. Studia Sci. Math. Hungar. **1** (1966), 437–441.
4. Gécseg, F.: Metrically complete systems of automata (Russian). Kibernetika (Kiev) **3** (1968), 96–98.
5. Gécseg, F.: On a representation of deterministic frontier-to-root tree transformations. Acta Sci. Math. **45** (1983), 177–187.
6. Gécseg, F.: Products of automata. Springer-Verlag, 1986.
7. Gécseg, F.: Homomorphic representations by products of tree automata. In: Results and Trends in Theoretical Computer Science (Proceedings, Colloquium in Honour of Arto Salomaa, Graz, 1994), Lecture Notes in Computer Science, Vol. 812, Springer-Verlag, 1994, 131–139.
8. Gécseg, F.: On quasi-products of tree automata. J.UCS **2** (2002), 184–192.
9. Gécseg, F., Imreh, B.: On complete sets of tree automata. In: Third International Conference Developments in Language Theory (Thessaloniki, 1997), 37–47.
10. Letičevskiǐ, A. A.: A condition for the completeness of finite automata (Russian). Žurn. vyčisl. matem. i matem. fiz. **1** (1961), 702–710.
11. Ricci, W.: Cascades of tree automata and computations in universal algebra. Math. Systems Theory **7** (1973), 201–218.
12. Steinby, M.: On the structure and realization of tree automata. In: Second Coll. sur les Arbres en Algèbre et en Programmation (Lille, 1977), 235–248.
13. Virágh, J.: Deterministic ascending tree automata II. Acta Cybernet. **6** (1983), 291–301.

# On a Conjecture of Schnoebelen

Antonio Cano Gómez[1] and Jean-Éric Pin[2]

[1] Departamento de Sistemas Informáticos y Computación, Universidad Politénica de Valencia, Camino de Vera s/n, P.O. Box: 22012, E-46020, Valencia
acano@dsic.upv.es
[2] LIAFA, Université Paris VII and CNRS, Case 7014, 2 Place Jussieu, 75251 Paris Cedex 05, France. Work supported by INTAS project 1224
Jean-Eric.Pin@liafa.jussieu.fr
http://liafa.jussieu.fr/~jep

**Abstract.** The notion of sequential and parallel decomposition of a language over a set of languages was introduced by Schnoebelen. A language is decomposable if it belongs to a finite set of languages $S$ such that each member of $S$ admits a sequential and parallel decomposition over $S$. We disprove a conjecture of Schnoebelen concerning decomposable languages and establish some new properties of these languages.

## 1 Introduction

The shuffle product is a standard tool for modeling process algebras [1]. This motivates the study of "robust" classes of recognizable languages which are closed under shuffle product.

In [7], Schnoebelen introduced the notion of *sequential and parallel decomposition* (a precise definition is given in Section 3). A language is decomposable if it belongs to a finite set of languages $S$ such that each member of $S$ admits a sequential and parallel decomposition over $S$. Schnoebelen proved that the class of decomposable languages is a quite robust class of rational languages, since it is closed under finite union, product and shuffle.

It is a challenging question to find an effective characterization of this class of languages. In [7], Schnoebelen proved that a finite union of products of commutative languages are decomposable and conjectured there are no other decomposable languages.

In this paper, we disprove this conjecture and establish some new properties of decomposable languages: they are closed under left and right quotients, but are not closed under intersection. They are also closed under inverse length preserving morphisms, but not under inverse morphisms. Finally, we propose a new conjecture for the description of decomposable languages.

## 2 Preliminaries

In this section, we recall a few important results about recognizable and rational sets. Let $M$ be a monoid. A subset $P$ of $M$ is *recognizable* if there exists a finite monoid $F$, and a monoid morphism $\varphi : M \to F$ such that $P = \varphi^{-1}(\varphi(P))$.

Z. Ésik and Z. Fülöp (Eds.): DLT 2003, LNCS 2710, pp. 35–54, 2003.

The class Rat($M$) of *rational* subsets of $M$ is the smallest set $\mathcal{R}$ of subsets of $M$ satisfying the following properties:

(1) For each $m \in M$, $\{m\} \in \mathcal{R}$
(2) The empty set belongs to $\mathcal{R}$, and if $X$, $Y$ are in $\mathcal{R}$, then $X \cup Y$ and $XY$ are also in $\mathcal{R}$.
(3) If $X \in \mathcal{R}$, the submonoid $X^*$ generated by $X$ is also in $\mathcal{R}$.

Recognizable sets are closed under boolean operations, quotients and inverse morphisms. Rational sets are closed under rational operations (union, product and star) and under morphisms. Kleene theorem states that for every finite alphabet $A$, Rec($A^*$) = Rat($A^*$). We conclude by three standard results.

**Theorem 1.** (McKnight) *The intersection of a rational set and of a recognizable set is rational.*

**Theorem 2.** (Mezei) *Let $M_1, \ldots, M_n$ be monoids. A subset of $M_1 \times \cdots \times M_n$ is recognizable if and only if it is a finite union of subsets of the form $R_1 \times \cdots \times R_n$, where $R_i \in$ Rec($M_i$).*

**Proposition 1.** *Let $A_1, \ldots, A_n$ be finite alphabets. Then Rec($A_1^* \times A_2^* \times \cdots \times A_n^*$) is closed under product.*

A *substitution* from $A^*$ into a monoid $M$ is a monoid morphism from $A^*$ into $\mathcal{P}(M)$, the monoid of subsets of $M$.

## 3    Decompositions of Languages

Schnoebelen [7] introduced the following definitions. Consider the transductions $\tau$ and $\sigma$ from $A^*$ into $A^* \times A^*$ defined as follows:

$$\tau(w) = \{(u, v) \in A^* \times A^* \mid w = uv\}$$
$$\sigma(w) = \{(u, v) \in A^* \times A^* \mid w \in u \amalg v\}$$

Therefore, if $K$ is a language

$$\tau(K) = \{(u, v) \in A^* \times A^* \mid uv \in K\}$$
$$\sigma(K) = \{(u, v) \in A^* \times A^* \mid (u \amalg v) \cap K \neq \emptyset\}$$

**Proposition 2.** *The transduction $\sigma$ is a substitution.*

*Proof.* We claim that $\sigma(x_1 x_2) = \sigma(x_1)\sigma(x_2)$. First, if $(u_1, v_1) \in \sigma(x_1)$ and $(u_2, v_2) \in \sigma(x_2)$, then $x_1 \in u_1 \amalg v_1$ and $x_2 \in u_2 \amalg v_2$. It follows that $x_1 x_2 \in u_1 u_2 \amalg u_2 v_2$ and thus $(u_1 u_2, v_1 v_2) \in \sigma(x_1 x_2)$. Conversely, if $(u, v) \in \sigma(x_1 x_2)$, then $x_1 x_2 \in u \amalg v$. Therefore, $u$ and $v$ can be decomposed as $u = u_1 u_2$, $v = v_1 v_2$ in such a way that $x_1 \in u_1 \amalg v_1$ and $x_2 \in u_2 \amalg v_2$. It follows that $(u, v) \in \sigma(x_1)\sigma(x_2)$.

Let $S$ be a set of languages. A language $K$ admits a *sequential decomposition* over $S$ if $\tau(K)$ is a finite union of sets of the form $L \times R$, where $L, R \in S$. Sequential decompositions of recognizable languages where considered by Conway [2]. Actually, Conway proved that every recognizable language $K$ admits a *maximal* sequential decomposition

$$\tau(K) = \bigcup_{1 \le i \le n} L_i \times R_i$$

in the following sense. If $LR \subseteq K$, then for some $i$, $L \subseteq L_i$ and $R \subseteq R_i$.

A language $K$ admits a *parallel decomposition* over $S$ if $\sigma(K)$ is a finite union of sets of the form $L \times R$, where $L, R \in S$.

A *sequential (resp. parallel) system* is a finite set $S$ of languages such that each member of $S$ admits a sequential (resp. parallel) decomposition over $S$. A language is *sequentially decomposable* if it belongs to some sequential system. It is *decomposable* if it belongs to a system which is both sequential and parallel. Thus, for each decomposable language $L$, one can find a sequential and parallel system $S(L)$ containing $L$.

**Theorem 3.** *Let $K$ be a language of $A^*$. The following conditions are equivalent:*

(1) *$K$ is rational,*

(2) *$\tau(K)$ is recognizable,*

(3) *$K$ is sequentially decomposable.*

*Proof.* (1) implies (3). Let $K$ be a rational language and let $\mathcal{A}$ be the minimal automaton of $K$. For each state $p$, $q$ of $\mathcal{A}$, let $K_{p,q}$ be the language accepted by $\mathcal{A}$ with $p$ as initial state and $q$ as unique final state. Let $S$ be the set of finite unions of languages of the form $K_{p,q}$. Since $\mathcal{A}$ is finite, $S$ is a finite set. We claim that

$$\tau(K_{p,q}) = \bigcup_{r \in Q} K_{p,r} \times K_{r,q} \tag{1}$$

Indeed, if, for some state $r$, $u \in K_{p,r}$ and $v \in K_{r,q}$, then $p \cdot u = r$ and $r \cdot v = q$ in $\mathcal{A}$. It follows that $uv \in K_{p,q}$ and $(u, v) \in \tau(K_{p,q})$. Conversely, let $(u, v) \in \tau(K_{p,q})$ and let $r = p \cdot u$. Since $uv \in K_{p,q}$, $p \cdot uv = q$, and thus $r \cdot v = q$, whence $v \in K_{r,q}$. Thus $(u, v) \in K_{p,r} \times K_{r,q}$.

It follows from (1) that each language of $S$ admits a sequential decomposition on $S$. Thus $S$ is a sequential system and $K$ itself is sequentially decomposable.

(3) implies (1) (Arnold, Carton). Let $S$ be a sequential system containing $K$, and let $\sim_S$ be the equivalence on $A^*$ defined by $u \sim_S v$ if, for every $F \in S$, $u \in F$ is equivalent to $v \in F$. Clearly, $\sim_S$ is an equivalence of finite index, which saturates $K$ by definition. Therefore, it suffices to show that $\sim_S$ is a congruence. Suppose that $u \sim_S v$ and let $F$ be a language of $S$. Let $w \in A^*$ and suppose that $uw \in F$. Since $F$ is sequentially decomposable over $S$, there exist two languages $L, R \in S$ such that $u \in L$, $w \in R$ and $LR \subset F$. It follows that $v \in L$, since

$u \sim_S v$, and thus $vw \in F$. Similarly, $vw \in F$ implies $uw \in F$ and thus $uw \sim_S vw$. A dual argument would show that $wu \sim_S wv$, which concludes the proof.

(2) implies (1). Suppose that $\tau(K)$ is recognizable. Observing that

$$\tau(K) \cap \{1\} \times A^* = \{1\} \times K$$

it follows by McKnight's theorem, that $\{1\} \times K$ is rational. Now $K = \pi(\{1\} \times K)$, where $\pi$ denotes the second projection from $A^* \times A^*$ onto $A^*$. It follows that $K$ is rational.

(3) implies (2). If $K$ is sequentially decomposable, it belongs to some sequential system $\mathcal{S}$. By definition, each element of $\mathcal{S}$ is sequentially decomposable and thus is rational, since (3) implies (1). Furthermore, $\tau(K)$ is a finite union of languages of the form $L \times R$, where $L, R \in \mathcal{S}$. It follows, by Mezei's theorem, that $\tau(K)$ is recognizable.

It follows that every decomposable language is rational. Another important consequence is the following:

**Proposition 3.** *For each decomposable language $K$, $\sigma(K)$ is recognizable.*

*Proof.* Indeed, if $\mathcal{S}$ is a parallel and sequential system for $K$, every language of $\mathcal{S}$ is decomposable and hence rational, by Theorem 3. Since is sequentially decomposable over $\mathcal{S}$, $\sigma(K)$ is a finite union of languages of the form $L \times R$, where $L, R \in \mathcal{S}$. It follows, by Mezei's theorem, that $\sigma(K)$ is recognizable.

We shall see later that the converse does not hold. That is, there exist a rational, indecomposable language $K$ such that $\sigma(K)$ is recognizable. However, Proposition 3 remains a powerful tool for proving that a language is not decomposable. The most important example was already given in [7].

**Proposition 4.** (Schnoebelen) *The set $\sigma((ab)^*)$ is not recognizable. In particular, the language $(ab)^*$ is not decomposable.*

*Proof.* Let $K = (ab)^*$. Suppose that

$$\sigma(K) = \bigcup_{1 \leq i \leq n} L_i \times R_i$$

Then, for every $k \geq 0$, $(a^k, b^k) \in \sigma(K)$ since $(ab)^k \in a^k \ {III}\ b^k$. It follows that one of the blocks $L_i \times R_i$ contains two distinct pairs $(a^r, b^r)$ and $(a^s, b^s)$ with $r \neq s$. It follows that $(a^r, b^s)$ also belongs to this block, and hence $a^r \ {III}\ b^s$ should contain a word of $(ab)^*$, a contradiction.

## 4   Closure Properties of Decomposable Languages

The following closure properties were established in [7], where the closure under shuffle is credited to Arnold. For the convenience of the reader, we give a self-contained proof.

**Proposition 5.** *Decomposable languages are closed under finite union, product and shuffle.*

*Proof.* Let $K$ and $K'$ be decomposable languages. Let $\mathcal{S}$ (resp. $\mathcal{S}'$) be a sequential and parallel system for $K$ (resp. $K'$). Then $\mathcal{S} \cup \mathcal{S}' \cup \{K \cup K'\}$ is a sequential and parallel system for $K \cup K'$. Let us show that

$$\mathcal{R} = \mathcal{S} \cup \mathcal{S}' \cup \{XX' \mid X \in \mathcal{S} \text{ and } X' \in \mathcal{S}'\}$$

is a sequential and parallel system for $KK'$. Assume that

$$\sigma(K) = \bigcup_{1 \le i \le n} L_i \times R_i \quad \text{and} \quad \sigma(K') = \bigcup_{1 \le j \le n'} L'_j \times R'_j$$

Since $\sigma$ is a substitution by Proposition 2,

$$\sigma(KK') = \bigcup_{\substack{1 \le i \le n \\ 1 \le j \le n'}} L_i L'_j \times R_i R'_j$$

Furthermore, if

$$\tau(K) = \bigcup_{1 \le i \le n} L_i \times R_i \quad \text{and} \quad \tau(K') = \bigcup_{1 \le j \le n'} L'_j \times R'_j$$

then

$$\tau(KK') = \bigcup_{1 \le i \le n} (L_i \times R_i K') \cup \bigcup_{1 \le j \le n} (KL'_i \times R'_i)$$

It follows that $KK'$ is decomposable. Finally, the system

$$\mathcal{T} = \{X \amalg X' \mid X \in \mathcal{S} \text{ and } X' \in \mathcal{S}'\}$$

is a sequential and parallel system for $K \amalg K'$. Indeed, we have, with the previous notation,

$$\tau(K \amalg K') = \bigcup_{\substack{1 \le i \le n \\ 1 \le j \le n'}} (L_i \amalg L'_j) \times (R_i \amalg R'_j)$$

$$\sigma(K \amalg K') = \bigcup_{\substack{1 \le i \le n \\ 1 \le j \le n'}} (L_i \amalg R'_i) \times (L_j \amalg R'_j)$$

Therefore $K \amalg K'$ is decomposable.

We establish some new closure properties. We first consider inverse morphisms. We shall see later (Proposition 19) that decomposable languages are not closed under inverse morphisms. However, closure under inverse morphisms holds for a restricted class of morphisms. Recall that a morphism $\varphi : A^* \to B^*$ is *length preserving* (or *litteral*, or *strictly alphabetic*) if each letter of $A$ is mapped onto a letter of $B$.

**Lemma 1.** *Let* $\varphi : A^* \to B^*$ *be a morphism and let* $K$ *be a language of* $B^*$. *If*

$$\tau(K) = \bigcup_{1 \leq i \leq n} L_i \times R_i$$

*then*

$$\tau(\varphi^{-1}(K)) = \bigcup_{1 \leq i \leq n} \varphi^{-1}(L_i) \times \varphi^{-1}(R_i)$$

*Furthermore, if* $\varphi$ *is length preserving, and if*

$$\sigma(K) = \bigcup_{1 \leq i \leq n} L_i \times R_i$$

*then*

$$\sigma(\varphi^{-1}(K)) = \bigcup_{1 \leq i \leq n} \varphi^{-1}(L_i) \times \varphi^{-1}(R_i)$$

*Proof.* The first result follows from the following sequence of equivalences

$$
\begin{aligned}
(u,v) \in \tau(\varphi^{-1}(K)) &\iff uv \in \varphi^{-1}(K) \\
&\iff \varphi(uv) \in K \\
&\iff \varphi(u)\varphi(v) \in K \\
&\iff \text{for some } i \in \{1,..,n\},\ (\varphi(u),\varphi(v)) \in L_i \times R_i \\
&\iff \text{for some } i \in \{1,..,n\},\ (u,v) \in \varphi^{-1}(L_i) \times \varphi^{-1}(R_i)
\end{aligned}
$$

If $\varphi$ is length preserving, then a word $w$ belongs to $u \amalg v$ if and only if $\varphi(w)$ belongs to $\varphi(u) \amalg \varphi(v)$. Therefore, the following sequence of equivalences holds:

$$
\begin{aligned}
(u,v) \in \sigma(\varphi^{-1}(K)) &\iff (u \amalg v) \cap \varphi^{-1}(K) \neq \emptyset \\
&\iff (\varphi(u) \amalg \varphi(v)) \cap K \neq \emptyset \\
&\iff (\varphi(u),\varphi(v)) \in \sigma(K) \\
&\iff \text{for some } i \in \{1,..,n\},\ (\varphi(u),\varphi(v)) \in L_i \times R_i \\
&\iff \text{for some } i \in \{1,..,n\},\ (u,v) \in \varphi^{-1}(L_i) \times \varphi^{-1}(R_i)
\end{aligned}
$$

This proves the second part of the lemma.

**Proposition 6.** *Decomposable languages are closed under inverse of length preserving morphisms.*

*Proof.* Let $\varphi : A^* \to B^*$ be a length preserving (or letter to letter) morphism. Let $K$ be a decomposable language of $B^*$ and let $\mathcal{S}$ be a sequential and parallel system for $K$. By Lemma 1, the set $\{\varphi^{-1}(L) \mid L \in \mathcal{S}\}$ is a sequential and parallel system for $\varphi^{-1}(K)$.

**Proposition 7.** *Decomposable languages are closed under left and right quotients.*

*Proof.* Let $\mathcal{S}$ be a sequential and parallel system for $L$ and let $a \in A$. Let

$$\tau(L) = \bigcup_{1 \le i \le n} L_i \times R_i$$

Observing that $\{a\} \times a^{-1}L \subseteq \tau(L)$, we set

$$J = \{j \mid a \in L_j \text{ and } a^{-1}L \cap R_j \ne \emptyset\}$$

We claim that $a^{-1}L = \bigcup_{j \in J} R_j$. Indeed, if $u \in a^{-1}L$, then $au \in L$ and $(a, u) \in \tau(L)$. Therefore, $(a, u) \in L_j \times R_j$ for some $j$. But since $a \in L_j$ and $u \in a^{-1}L \cap R_j$, $j$ belongs to $J$. Thus $u \in \bigcup_{j \in J} R_j$. Conversely, if $u \in R_j$ for some $j \in J$, then $a \in L_j$ and hence $(a, u) \in \tau(L)$, whence $au \in L$ and $u \in a^{-1}L$.

It follows from the claim that $a^{-1}L$ is a finite union of elements of $\mathcal{S}$. Since the elements of $\mathcal{S}$ are decomposable, $a^{-1}L$ is decomposable.

A symmetrical argument would show that $La^{-1}$ is decomposable.

It is tempting to conjecture that a language is decomposable if and only if $\sigma(L)$ is recognizable. However, Proposition 7 can be used to give a counterexample. Let $L = (ab)^* \cup bA^* \cup aaA^*$. Then

$$\sigma(L) = (aA^* \times aA^*) \cup (bA^* \times A^*) \cup (aaA^* \times A^*)$$
$$\cup (A^* \times aaA^*) \cup (1 \times L) \cup (L \times 1)$$

However, $L$ is not decomposable, since $(ab)^{-1}L = (ab)^*$.

The next result was proved by in [7].

**Proposition 8.** (Schnoebelen) *Every commutative rational language is decomposable.*

*Proof.* Let $L$ be a commutative rational language. We claim that $\tau(L) = \sigma(L)$. Indeed, if $uv \in L$, then $u \text{ III } v \cap L \ne \emptyset$. Conversely, if $u \text{ III } v$ meets $L$, there exist two facorizations $u = u_0 u_1 \cdots u_k$ and $v = v_1 v_2 \cdots v_k$ such that $u_0 v_1 u_1 \cdots v_k u_k \in L$. It follows, since $L$ is commutative, that $uv \in L$. It follows from the claim that any sequential decomposition of $L$ is also a parallel decomposition. Now, by Theorem 3, $L$ has a sequential decomposition. Therefore, $L$ is decomposable.

Recall that the *polynomial closure* of a class of languages $\mathcal{C}$ is the class of languages which are union of products of languages of $\mathcal{C}$. Proposition 5 shows that decomposable languages are closed under polynomial closure. In [7], Schnoebelen proposed the following conjecture.

*Conjecture 1.* A language is decomposable if and only if it belongs to the polynomial closure of the commutative languages.

Denote by Pol($\mathcal{C}om$) the *polynomial closure* of the commutative languages. This class contains in particular the finite and the cofinite languages. One direction of the conjecture is an immediate consequence of Propositions 8 and 5.

**Theorem 4.** (Schnoebelen) *Every language of Pol(Com) is decomposable.*

We shall give a counterexample to Schnoebelen's conjecture in the next section. Let us first give a last closure property of the class of decomposable languages, due to Arnold. It is based on the following property.

**Proposition 9.** *Let $K$ and $L$ be languages. Then $\tau(K \cap L) = \tau(K) \cap \tau(L)$. Furthermore, if one of the languages is commutative, $\sigma(K \cap L) = \sigma(K) \cap \sigma(L)$.*

*Proof.* The first formula follows from the following sequence of equivalent statements:

$$(u, v) \in \tau(K \cap L) \Longleftrightarrow uv \in K \cap L \Longleftrightarrow uv \in K \text{ and } uv \in L$$
$$\Longleftrightarrow (u, v) \in \tau(L) \text{ and } (u, v) \in \tau(K)$$

The inclusion $\sigma(K \cap L) \subseteq \sigma(K) \cap \sigma(L)$ is trivial. Assume that $L$ is commutative and let $(u, v) \in \sigma(K) \cap \sigma(L)$. Since $L$ is commutative, $\sigma(L) = \tau(L)$ and hence $uv \in L$. Furthermore there exist a word $w$ of $K$ such that $w = u_0 v_1 u_1 \cdots v_n u_n$ for some factorizations $u = u_0 u_1 \cdots u_n$ and $v = v_1 v_2 \cdots v_n$ of $u$ and $v$. Now $uv$ and $w$ are commutatively equivalent, and thus $w \in K \cap L$. It follows that $(u, v)$ belongs to $\sigma(K \cap L)$ and hence $\sigma(K \cap L) = \sigma(K) \cap \sigma(L)$.

**Corollary 1.** (Arnold) *The intersection of a decomposable language with a commutative recognizable language is decomposable*

Proposition 9 can be used to give non-trivial examples of indecomposable language.

**Proposition 10.** *Let $A = \{a, b, c\}$. The language $(ab)^* c A^*$ is not decomposable.*

*Proof.* Let $L = (ab)^* c A^*$. If $L$ is decomposable, the language

$$Lc^{-1} = (ab)^* \cup (ab)^* c A^*$$

is decomposable by Proposition 7. The intersection of this language with the recognizable commutative language $\{a, b\}^*$ is equal to $(ab)^*$, and thus by Corollary 1, $(ab)^*$ should also be decomposable. But this contradicts Proposition 4 and thus $L$ is not decomposable.

We shall see later that decomposable languages are not closed under intersection.

## 5   Schnoebelen's Conjecture

In this section, we disprove Schoebelen's conjecture by giving an example of a decomposable language which is not a finite union of products of commutative languages.

An algebraic characterization of the languages of Pol($Com$) was given in [6], but a much weaker property will suffice for our purpose: Every group in the syntactic monoid of a language of Pol($Com$) is commutative.

Let $u$ and $v$ be two words of $A$. A word $u = a_1 a_2 \cdots a_k$ is said to be a *subword* of $v$ if $v$ can be factorized as $v = v_0 a_1 v_1 \cdots a_k v_k$ where $v_0, v_1, \ldots, v_k \in A^*$. Following Eilenberg [3], we set

$$\binom{v}{u} = \mathrm{Card}\{(v_0, v_1, \ldots, v_k) \in A^* \times A^* \times \cdots \times A^* \mid v_0 a_1 v_1 \cdots a_k v_k = v\}$$

Thus $\binom{v}{u}$ is the number of distinct ways to write u as a subword of v. For instance, $\binom{aabbaa}{aba} = 8$ and $\binom{a^n}{a^m} = \binom{n}{m}$. The basic properties of these *binomial coefficients* are summarized by the following formulae

(1) For every word $u \in A^*$, $\binom{u}{1} = 1$.

(2) For every non-empty word $u \in A^*$, $\binom{1}{u} = 0$.

(3) If $w = uv$, then $\binom{w}{x} = \sum_{x_1 x_2 = x} \binom{u}{x_1}\binom{v}{x_2}$.

Let $A = \{a, b\}$ be an alphabet, and let $K$ be the language of words $x$ over the alphabet $A$ such that $\binom{x}{ab} \equiv 1 \mod 2$. The minimal automaton of this language is represented in Figure 1.

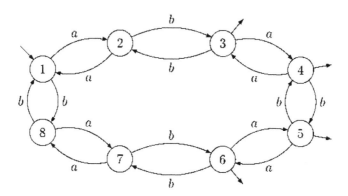

**Fig. 1.** An automaton for $M$.

The syntactic monoid of $K$ is a non-abelian group with eight elements. Therefore $K$ does not belong to Pol($Com$). We now prove that $K$ is decomposable. For $i, j, k \in \{0, 1\}$ and $c \in A$, let us set

$$M_k^{i,j} = \left\{x \in A^* \mid |x|_a \equiv i \mod 2, \ |x|_b \equiv j \mod 2 \text{ and } \binom{x}{ab} \equiv k \mod 2\right\}$$

$$M^{i,j} = \left\{x \in A^* \mid |x|_a \equiv i \mod 2, \ |x|_b \equiv j \mod 2\right\}$$

$$M_c^{i,j} = M^{i,j} \cap A^* c A^*$$

Let $\mathcal{F}$ be the set of languages that are finite union of languages of the form $M_k^{i,j}$, $M_c^{i,j}$ or $\{1\}$. Observe that, since

$$M^{i,j} = M_0^{i,j} \cup M_1^{i,j}$$

$M^{i,j}$ belongs to $\mathcal{F}$. We claim that $\mathcal{F}$ is a sequential and parallel system. We first show it is a sequential system by proving that the languages $M_k^{i,j}$, $M_c^{i,j}$ or $\{1\}$ have a sequential decomposition over $\mathcal{F}$. This is the purpose of the next proposition.

**Proposition 11.**

(a) $\tau(1) = \{1\} \times \{1\}$

(b) *For* $i, j, k \in \{0, 1\}$,

$$\tau(M_h^{i,j}) = \bigcup_{\substack{k+m \equiv i \bmod 2 \\ \ell+n \equiv j \bmod 2 \\ h \equiv p+q+kn \bmod 2}} (M_p^{k,\ell} \times M_q^{m,n}) \tag{2}$$

(c) *For* $i, j \in \{0, 1\}$, *and any* $c \in A$,

$$\tau(M_c^{i,j}) = \bigcup_{\substack{k+m \equiv i \bmod 2 \\ \ell+n \equiv j \bmod 2}} (M_c^{k,\ell} \times M^{m,n}) \bigcup_{\substack{k+m \equiv i \bmod 2 \\ \ell+n \equiv j \bmod 2}} (M^{m,n} \times M_c^{k,\ell}) \tag{3}$$

*Proof.* (a) is trivial.

(b) Let $(u, v) \in \tau(M_h^{i,j})$, that is, $uv \in M_h^{i,j}$. Define $k, \ell, m, n, p, q \in \{0, 1\}$ by the conditions $|u|_a \equiv k \bmod 2$, $|u|_b \equiv \ell \bmod 2$, $\binom{u}{ab} \equiv p \bmod 2$, $|v|_a \equiv m \bmod 2$, $|v|_b \equiv n \bmod 2$ and $\binom{v}{ab} \equiv q \bmod 2$. By definition, $u \in M_p^{k,\ell}$ and $v \in M_q^{m,n}$. Furthermore, since $|uv|_a \equiv i \bmod 2$ and $|uv|_b \equiv j \bmod 2$, $k + m \equiv i \bmod 2$ and $\ell + n \equiv j \bmod 2$. Finally, since $\binom{u}{ab} \equiv h \bmod 2$, the formula

$$\binom{uv}{ab} \equiv \binom{u}{ab} + \binom{v}{ab} + |u|_a|v|_b \equiv p + q + kn \quad \bmod 2 \tag{4}$$

shows that $p + q + kn \equiv h \bmod 2$. This proves that $(u, v)$ belongs to the right hand side of (2).

In the opposite direction, let $u \in M_p^{k,\ell}$ and $v \in M_q^{m,n}$, with $k + m \equiv i \bmod 2$, $\ell + n \equiv j \bmod 2$ and $h \equiv p + q + kn \bmod 2$. Then $|uv|_a \equiv k + m \equiv i \bmod 2$, $|uv|_b \equiv \ell + n \equiv j \bmod 2$, and, by Formula (4), $\binom{u}{ab} \equiv p + q + kn \equiv h \bmod 2$.

(c) Let $(u, v) \in \tau(M_c^{i,j})$, that is, $uv \in M_c^{i,j}$. Define $k, \ell, m, n \in \{0, 1\}$ by the conditions $|u|_a \equiv k \bmod 2$, $|u|_b \equiv \ell \bmod 2$, $|v|_a \equiv m \bmod 2$ and $|v|_b \equiv n \bmod 2$. Since $|uv|_a \equiv i \bmod 2$ and $|uv|_b \equiv j \bmod 2$, the relations $k + m \equiv i \bmod 2$ and $\ell + n \equiv j \bmod 2$ hold. Furthermore, since $uv \in M_c^{i,j}$, the letter $c$ occurs at least once in $uv$. Thus it occurs in $u$ or in $v$ and hence $(u, v) \in (M_c^{k,\ell} \times M^{m,n}) \cup (M^{m,n} \times M_c^{k,\ell})$.

In the opposite direction, let $u \in M_c^{k,\ell}$ and $v \in M^{m,n}$, with $k + m \equiv i \bmod 2$ and $\ell + n \equiv j \bmod 2$ (the proof would be similar for $u \in M^{k,\ell}$ and $v \in M_c^{m,n}$). Then $|uv|_a \equiv k + m \equiv i \bmod 2$, $|uv|_b \equiv \ell + n \equiv j \bmod 2$, and since $|u|_c > 0$, $|uv|_c > 0$ and $uv \in M_c^{i,j}$, that is, $(u, v) \in \tau(M_c^{i,j})$.

We now prove that $\mathcal{F}$ is a parallel system. The proof relies on a simple, but useful observation.

**Lemma 2.** *For any words* $x, y \in A^*$, $\binom{xaby}{ab} = \binom{xbay}{ab} + 1$.

*Proof.* We have on the one hand

$$\binom{xaby}{ab} = \binom{y}{ab} + \binom{xa}{a}\binom{by}{b} + \binom{x}{ab} = \binom{y}{ab} + (|x|_a + 1)(|y|_b + 1) + \binom{x}{ab}$$

and, on the other hand

$$\binom{xbay}{ab} = \binom{ay}{ab} + \binom{xb}{a}\binom{ay}{b} + \binom{xb}{ab} = \binom{y}{ab} + |y|_b + |x|_a|y|_b + |x|_a + \binom{x}{ab}$$

which proves the lemma.

**Proposition 12.**

(a) $\sigma(1) = \{1\} \times \{1\}$

(b)

$$\text{For any } i, j \in \{0, 1\}, \sigma(M_0^{i,j}) = \bigcup_{\substack{k+m \equiv i \bmod 2 \\ \ell+n \equiv j \bmod 2}} (M^{k,\ell} \times M^{m,n}) \tag{5}$$

(c) *For any* $i, j \in \{0, 1\}$,

$$\sigma(M_1^{i,j}) = \bigcup_{\substack{k+m \equiv i \bmod 2 \\ \ell+n \equiv j \bmod 2}} \left( (M_a^{k,\ell} \times M_b^{m,n}) \cup (M_b^{k,\ell} \times M_a^{m,n}) \right)$$

$$\cup (\{1\} \times M_1^{i,j}) \cup (M_1^{i,j} \times \{1\}) \tag{6}$$

(d) *For any* $i, j \in \{0, 1\}$, *for any* $c \in A$, $\sigma(M_c^{i,j}) = \tau(M_c^{i,j})$.

*Proof.* (a) is trivial.

(b) Let $(u, v) \in \sigma(M_0^{i,j})$, and let $w \in (u \; \text{III} \; v) \cap M^{i,j}$. Define $k, \ell, m, n \in \{0, 1\}$ by the conditions $|u|_a \equiv k \bmod 2$, $|u|_b \equiv \ell \bmod 2$, $|v|_a \equiv m \bmod 2$ and $|v|_b \equiv n \bmod 2$. Then $(u, v) \in M^{k,\ell} \times M^{m,n}$. Furthermore, since $|w|_a \equiv i \bmod 2$ and $|w|_b \equiv j \bmod 2$, the relations $k + m \equiv i \bmod 2$ and $\ell + n \equiv j \bmod 2$ hold. Thus $(u, v)$ belongs to the right hand side of (5).

Conversely, let $(u, v) \in M^{k,\ell} \times M^{m,n}$ with $k + m \equiv i \bmod 2$ and $\ell + n \equiv j \bmod 2$. First, if $(u, v) \in (A^*aA^* \times A^*bA^*) \cup (A^*bA^* \times A^*aA^*)$, then $\{u, v\} = \{(x_1ay_1, x_2by_2)\}$ for some words $x_1, x_2, y_1, y_2 \in A^*$. Setting $x = x_1x_2$ and $y = y_1y_2$, the set $u \; \text{III} \; v$ contains the words $xaby$ and $xbay$. Note that $|xaby|_a = |xbay|_a = |uv|_a \equiv i \bmod 2$ and $|xaby|_b = |xbay|_b = |uv|_b \equiv i \bmod 2$. It follows by Lemma 2 that one of these words is in $M_0^{i,j}$ and thus $(u, v) \in \sigma(M_0^{i,j})$. If now, $(u, v) \notin (A^*aA^* \times A^*bA^*) \cup (A^*bA^* \times A^*aA^*)$, then $(u, v) \in (a^* \times a^*) \cup (b^* \times b^*)$ and thus $\binom{uv}{ab} = 0$. Therefore $uv \in M_0^{i,j}$ and $(u, v) \in \sigma(M_0^{i,j})$.

(c) The proof of is quite similar to that of (b). Let $(u, v) \in \sigma(M_1^{i,j})$, and let $w \in (u \; \text{III} \; v) \cap M^{i,j}$. Since $\binom{w}{ab} \equiv 1 \bmod 2$, $|w|_a > 0$ and $|w|_b > 0$.

(d) holds, since $M_c^{i,j}$ is a commutative language.

**Theorem 5.** *The language K is decomposable.*

*Proof.* Indeed, $K$ belongs to the parallel and sequential system $\mathcal{F}$, since $K = \bigcup_{0 \leq i,j \leq 1} M_1^{i,j}$.

# 6   Non Decomposable Languages

In this section, we give several examples of non decomposable languages. We first generalize Proposition 4 to a larger class of languages.

**Proposition 13.** *Let $u$ and $v$ be two words of $A^*$ such that $uv \neq vu$. Then $\sigma((uv)^*)$ is not recognizable. In particular, the language $(uv)^*$ is not recognizable.*

*Proof.* We first remind the reader with a classical result of combinatorics on words (see [4]): two words commute if and only if they are powers of the same word. Next we claim that $\sigma((uv)^*)$ is recognizable if and only if $\sigma((vu)^*)$ is recognizable. Indeed, $(vu)^* = \{1\} \cup v(uv)^*u$ and since $\sigma$ is a substitution,

$$\sigma((vu)^*) = \sigma(1) \cup \sigma(v)\sigma((uv)^*)\sigma(u)$$

Now $\sigma(1)$, $\sigma(u)$ and $\sigma(v)$ are finite and, by Mezei's theorem, belong to $\mathrm{Rec}(A^* \times A^*)$. Furthermore, $\mathrm{Rec}(A^* \times A^*)$ is closed under union and product. It follows that if $\sigma((uv)^*)$ is recognizable, so is $\sigma((vu)^*)$. The converse also holds by duality, which proves the claim.

We can now assume, without loss of generality, that $|u| \leq |v|$. Let $K = (uv)^*$. If $\sigma(K)$ is recognizable, it can be written as a finite union of blocks $L \times R$, where $L, R \in \mathrm{Rec}(A^*)$. For each $n \geq 0$, $(uv)^n \in u^n$ III $v^n$, and hence $(u^n, v^n) \in \sigma(K)$. Therefore, there exists a block $L \times R$ such that the set

$$S = \{n \in \mathbb{N} \mid (u^n, v^n) \in L \times R\}$$

is infinite. Let $s = \min S$. By assumption, we have $u^s \in L$ and $v^n \in R$ for each $n \in S$ and thus $(u^s$ III $v^n) \cap (uv)^* \neq \emptyset$.

Let us fix $n \geq 4s|u|$ and let $k$ be such that $(uv)^k \in u^s$ III $v^n$. There exist two factorizations $u^s = x_0 \cdots x_{\ell+1}$ and $v^n = y_1 \cdots y_\ell$ such that $x_0, x_{\ell+1} \in A^*$, $x_1, y_1, \ldots, x_\ell, y_\ell \in A^+$ and $x_0 y_1 x_1 \cdots x_\ell y_\ell x_{\ell+1} = (uv)^k$. It follows, by the choice of $n$

$$|y_1 \cdots y_\ell| = n|v| \geq 4s|u||v|$$

Now, since $\ell \leq |x_1 \cdots x_\ell| \leq s|u|$, one of the words $y_i$ has length $\geq 4|v|$. Now, since $y_i$ is a factor of $(uv)^k$ and since $|u| \leq |v|$, $y_i$ contains $vuv$ as a factor. At the same time, $y_i$ is a factor of $v^n$ of length $\geq 4|v|$, that is, a factor of $v^5$. It follows that $vuv$ is a factor of $v^5$. Let us write $v$ as the power of some primitive word, say $v = t^p$. Then $tut$ is a factor of $vuv$, which is in turn a factor of $t^{5p}$. Now, a primitive word cannot have any conjugate, which means that if $t^{5p} = xtuty$, then $x$ and $y$ are necessarily powers of $t$. But this forces $u$ itself to be a power of $t$, a contradiction, since $u$ and $v$ are not powers of the same word. Thus $\sigma(K)$ is not recognizable.

## 7   Inverse Morphisms

In this section, we show that decomposable languages are not closed under inverse morphisms.

Let $L$ be the language defined over the alphabet $A = \{a, b\}$ by the following regular expression:

$$L = (aab)^* \cup A^*b(aa)^*abA^*$$

We claim that this language is decomposable. We now show that $L$ is decomposable by constructing, step by step, a parallel and sequential system containing $L$. First, consider the automaton $\mathcal{A}$ represented in Figure 2.

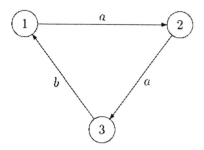

**Fig. 2.** The automaton $\mathcal{A}$.

Let $K_{i,j}$ be the language accepted by $\mathcal{A}$ with $i$ as initial state and $j$ as the only final state. These languages are $K_{1,1} = (aab)^*$, $K_{1,2} = (aab)^*a$, $K_{1,3} = (aab)^*aa$, $K_{2,1} = (aba)^*ab$, $K_{2,2} = (aba)^*$, $K_{2,3} = (aba)^*a$, $K_{3,1} = (baa)^*b$, $K_{3,2} = (baa)^*ba$ and $K_{3,3} = (baa)^*$. In the next proposition, we compute the image by $\tau$ of these languages.

**Proposition 14.** *For* $1 \leq i, j \leq 3$,

$$\tau(K_{i,j}) = \bigcup_{1 \leq k \leq 3} (K_{i,k} \times K_{k,j})$$

*Proof.* If $(u, v) \in \tau(K_{i,j})$, then $uv \in K_{i,j}$, and thus $i \cdot uv = j$ in $\mathcal{A}$. Setting $k = i \cdot u$, we have $k \cdot v = j$ and thus $u \in K_{i,k}$ and $v \in K_{k,j}$. In the opposite direction, if $u \in K_{i,k}$ and $v \in K_{k,j}$, then $i \cdot u = k$ and $k \cdot v = j$, whence $i \cdot uv = j$ and $uv \in K_{i,j}$.

Let $L_0 = A^*b(aa)^*abA^*$. We first compute the image of $L_0$ under $\sigma$ and $\tau$.

**Proposition 15.**

$$\tau(L_0) = (A^* \times L_0) \cup (L_0 \times A^*) \cup (A^*b(aa)^* \times a(aa)^*bA^*) \cup$$
$$(A^*b(aa)^*a \times (aa)^*bA^*)$$

*Proof.* Let $(u, v) \in \tau(L_0)$, that is, $uv \in L_0$. First suppose that $u \in L_0$ or $v \in L_0$, this implies that $(u, v) \in (A^* \times L_0)$ or $(u, v) \in (L_0 \times A^*)$. Otherwise we can write $uv$ as $u_1 b(aa)^n abv_1$ where $n \geq 0$, $u_1$ is a prefix of $u$ and $v_1$ is a suffix of $v$, and then $(u, v) \in (A^* b(aa)^* \times a(aa)^* bA^*)$ or $(u, v) \in (A^* b(aa)^* a \times (aa)^* bA^*)$. In the opposite direction, if $u \in L_0$ its clear that for any $v \in A^*$, $uv, vu \in L_0$. Now suppose that $u = u_1 b(aa)^n$ and $v = a(aa)^m bv_1$ (resp. $u = u_1(aa)^n a$ and $v = (aa)^m bv_1$) for some $n, m \geq 0$, then $uv \in L_0$.

**Proposition 16.**

$$\sigma(L_0) = (L_0 \times A^*) \cup (A^* \times L_0) \cup (A^* aA^* bA^* \times A^* bA^*)$$
$$\cup (A^* bA^* aA^* \times A^* bA^*) \cup (A^* bA^* bA^* \times A^* aA^*)$$
$$\cup (A^* bA^* \times A^* aA^* bA^*) \cup (A^* bA^* \times A^* bA^* aA^*)$$
$$\cup (A^* aA^* \times A^* bA^* bA^*)$$

*Proof.* If $(u, v) \in \sigma(L_0)$, then by definition $(u \; \mathrm{III} \; v) \cap L_0 \neq \emptyset$. Let us take some $w \in (u \; \mathrm{III} \; v) \cap L_0$. By definition of $L_0$, $w$ can be decomposed as $w = w_1 w_2 w_3$, where $w_1, w_3 \in A^*$ and $w_2 = ba^n b$ with $n$ odd. Since $w \in u \; \mathrm{III} \; v$, $w_2$ can be written as $w_2 = u_1 v_1 \cdots u_n v_n$ with $u_i, v_i \in A^*$ for $1 \leq i \leq n$, $u = u_0 u_1 \cdots u_n u_{n+1}$ and $v = v_0 v_1 \cdots v_n v_{n+1}$ for some $u_0, v_0, \dots, u_{n+1}, v_{n+1} \in A^*$. Let us consider the words $\bar{u} = u_1 \cdots u_n$ and $\bar{v} = v_1 \cdots v_n$. Since $w_2 = ba^n b$, the words $\bar{u}$ and $\bar{v}$ can take the following values, where $n = n_1 + n_2$:

(1) $\bar{u} = ba^{n_1} b$ and $\bar{v} = a^{n_2}$
(2) $\bar{u} = a^{n_1}$ and $\bar{v} = ba^{n_2} b$
(3) $\bar{u} = ba^{n_1}$ and $\bar{v} = a^{n_2} b$
(4) $\bar{u} = a^{n_1} b$ and $\bar{v} = ba^{n_2}$

Let $R$ be the right hand side of the equality to be proved. We show that $\sigma(L_0)$ is a subset of $R$ by considering the four cases separately.

Case (1). If $n_1$ is odd, then $u \in L_0$, and so $(u, v) \in (L_0 \times A^*)$, otherwise, if $n_1$ is even, and $n_1 + n_2$ is odd, we have $n_2 > 0$, and then $(u, v) \in A^* bA^* bA^* \times A^* aA^*$.

Case (2). A proof similiar to that of case (1) shows that $(u, v)$ belongs either to $A^* \times L_0$ or to $A^* aA^* \times A^* bA^* bA^*$.

Case (3). If $n_1 > 0$, $(u, v) \in A^* bA^* aA^* \times A^* bA^*$. Otherwise, since $n_1 + n_2$ is odd, $n_2 > 0$ and then $(u, v) \in A^* bA^* \times A^* aA^* bA^*$.

Case (4). A proof similiar to that of case (3) shows that $(u, v)$ belongs either to $A^* bA^* \times A^* bA^* aA^*$ or to $A^* aA^* bA^* \times A^* bA^*$.

We now prove that $R$ is a subset of $\sigma(L_0)$. If $(u, v) \in (L_0 \times A^*) \cup (A^* \times L_0)$, it is clear that $(u \; \mathrm{III} \; v) \cap L_0 \neq \emptyset$. Suppose that $(u, v) \in (A^* aA^* bA^* \times A^* bA^*)$. Observing that $A^* aA^* bA^* = A^* abA^*$, the word $u$ can be decomposed as $u = u_1 abu_2$ with $u_1, u_2 \in A^*$, and $v$ can be decomposed as $v = v_1 bv_2$ with $v_1, v_2 \in A^*$. Now $v_1 u_1 babv_2 \in (u \; \mathrm{III} \; v) \cap L_0$ and thus $(u, v) \in \sigma(L_0)$. The proof is similar if $(u, v)$ belongs to $A^* bA^* aA^* \times A^* bA^*$, $A^* bA^* \times A^* bA^* aA^*$ or $A^* bA^* \times A^* aA^* bA^*$.

Finally, if $(u, v) \in A^* bA^* bA^* \times A^* aA^*$, then $u$ can be decomposed as $u = u_1 ba^n bu_2$ with $n \geq 0$, $u_1, u_2 \in A^*$, and $v$ can be decomposed as $v = v_1 av_2$ with

$v_1, v_2 \in A^*$. If $n$ is even, $u$ and $v$ can be shuffled as $u_1 v_1 b a^{(n+1)} b v_2 u_2$, a word of $L_0$, otherwise, $u$ and $v$ can be shuffled as $u_1 v_1 b a^n b a v_2 u_2$, another word of $L_0$. Thus $(u, v) \in \sigma(L_0)$ in both cases. The proof for $A^* a A^* \times A^* b A^* b A^*$ is similar.

For $1 \leq i, j \leq 3$, define the languages $L_{i,j}$ as the union of $L_0$ with the language $K_{i,j}$. We first compute the image of these languages under $\tau$.

**Proposition 17.** *For* $1 \leq i, j \leq 3$,

$$\tau(L_{i,j}) = \tau(L_0) \cup \bigcup_{1 \leq k \leq 3} (L_{i,k} \times L_{k,j})$$

*Proof.* Let $R$ be the right handside of the relation to be proved. Since $L_{i,j} = L_0 \cup K_{i,j}$, $\tau(L_{i,j}) = \tau(L_0) \cup \tau(K_{i,j})$. Now, by Proposition 14, $\tau(K_{i,j}) = \bigcup_{1 \leq k \leq 3}(K_{i,k} \times K_{k,j})$, and since each $K_{i,j}$ is a subset of $L_{i,j}$, $\tau(K_{i,j})$ is a subset of $R$. It follows that $\tau(L_{i,j})$ is a subset of $R$.

We now prove the opposite inclusion. Since $L_0$ is a subset of $L_{i,j}$, $\tau(L_0)$ is a subset of $\tau(L_{i,j})$. Furthermore, if $(u, v) \in L_{i,k} \times L_{k,j}$ for some $k$, two possibilities arise. First, if $u \in L_0$ or $v \in L_0$ then $(u, v) \in \tau(L_0)$ and hence $(u, v) \in \tau(L_{i,j})$. Otherwise, $u \in K_{i,k}$, $v \in K_{k,j}$, $uv \in K_{i,j}$ and thus $(u, v) \in \tau(L_{i,j})$.

In order to compute the image of the languages $L_{i,j}$ under $\sigma$, we introduce the languages

$$M_{i,j} = K_{i,j} \cap a^*(b \cup 1)a^*$$

These languages are clearly finite and thus decomposable, and they belong to some parallel and sequential systems $\mathcal{S}(M_{i,j})$.

**Proposition 18.** *For* $1 \leq i, j \leq 3$,

$$\sigma(L_{i,j}) = \sigma(L_0) \cup \sigma(M_{i,j}) \cup (L_{i,j} \times \{1\}) \cup (\{1\} \times L_{i,j})$$

*Proof.* If $(u, v) \in \sigma(L_{i,j})$, then $(u \amalg v) \cap L_{i,j} \neq \emptyset$. First, if $u = 1$ or $v = 1$ then $(u, v) \in (L_{i,j} \times \{1\}) \cup (\{1\} \times L_{i,j})$. Now, suppose that $u$ and $v$ are non-empty words.

If $|u|_b + |v|_b < 2$ then necessarily $(u, v) \in \sigma(M_{i,j})$. Suppose that $|u|_b + |b|_v \geq 2$. We claim that $(u, v) \in (A^* a A^* b A^* \times A^* b A^*) \cup (A^* b A^* a A^* \times A^* b A^*) \cup (A^* b A^* b A^* \times A^* a A^*) \cup (A^* b A^* \times A^* a A^* b A^*) \cup (A^* b A^* \times A^* b A^* a A^*) \cup (A^* a A^* \times A^* b A^* b A^*) \subseteq \sigma(L_0)$. Since $u$ and $v$ are both non-empty, they contain at least one letter. If $u$ and $v$ are equal to $b$, $(u \amalg v) \cap L_0 = \emptyset$, and so $u$ or $v$ contains the letter $a$ and since $|u|_b + |v|_b < 2$ the result holds. Conversely, if $(u, v) \in \sigma(L_0) \cup \sigma(M_{i,j}) \cup (L_{i,j} \times \{1\}) \cup (\{1\} \times L_{i,j})$ by the definition of $M_{i,j}$, Proposition 16 and the definition of $L_{i,j}$, $(u, v) \in \sigma(L_{i,j})$.

We are now ready to show that $L$ is decomposable.

**Theorem 6.** *The language* $L$ *is decomposable.*

*Proof.* By Theorem 4, all languages which are products of commutative languages are decomposable. In particular, the following languages are decomposable:

$$A^*b(aa)^*, (aa)^*abA^*, A^*b(aa)^*a, (aa)^*bA^*, A^*aA^*bA^*,$$
$$A^*bA^*aA^*, A^*bA^*bA^*, A^*aA^*, A^*bA^*$$

Let us define a system $\mathcal{S}$ consisting of the unions of the following languages: $L_0$, $\{1\}$, $A^*$, $L_{i,j}$, for $1 \leq i, j \leq 3$, the languages of the systems associated with the languages in display and the languages of $\mathcal{S}(M_{i,j})$ for $1 \leq i, j \leq 3$. Note that $L$ belongs to $\mathcal{S}$ since $L = L_{1,1}$.

It remains to show that for any language $L'$ of $\mathcal{S}$, $\tau(L')$ and $\sigma(L')$ can be written as the union of languages $(M \times N)$ where $M$ and $N$ belong to $\mathcal{S}$. For $L_0$, the result follows from Propositions 15 and 16. For $L_{i,j}$, the result follows from Propositions 17 and 18. Clearly, $\tau(\{1\}) = \sigma(\{1\}) = (\{1\} \times \{1\})$ and $\sigma(A^*) = \tau(A^*) = (A^* \times A^*)$. For any language of the systems described above, its image under $\tau$ and $\sigma$ can be obtained from the languages of these systems, because they are parallel and sequential systems. Finally, since for any language $N_1$, $N_2$, the formulas $\tau(N_1 \cup N_2) = \tau(N_1) \cup \tau(N_2)$ and $\sigma(N_1 \cup N_2) = \sigma(N_1) \cup \sigma(N_2)$ hold, the system $\mathcal{S}$ is a parallel and sequential system.

**Proposition 19.** *Decomposable languages are not closed under inverse morphism.*

*Proof.* Let $A = \{a, b\}$ and let $\varphi : A^* \to A^*$ be the morphism of monoids defined by $\psi(a) = aa$ and $\psi(b) = b$. If $L = (aab)^* \cup A^*b(aa)^*abA^*$, then $\varphi^{-1}(L) = (ab)^*$. Now $L$ is decomposable by Theorem 6 but, by Proposition 4, $\varphi^{-1}(L)$ is not.

**Corollary 2.** *Decomposable languages do not form a positive variety of languages.*

## 8    Intersection

Let $L_1$ and $L_2$ be the languages defined over the alphabet $A = \{a, b\}$ by the following regular expressions:

$$L_1 = (ab)^+ \cup (ab)^*bA^*$$
$$L_2 = (ab)^+ \cup (ab)^*aaA^*$$

One can show (proof omitted) that these two languages are decomposable but that their intersection is not decomposable. Therefore, decomposable languages are not closed under intersection.

## 9    More on Group Languages

Recall that a *group language* is a language recognized by a finite group. Equivalently, this is a language whose minimal automaton is a permutation automaton. The aim of this section is to prove the following result

**Theorem 7.** *If $L$ is a group language, then $\sigma(L)$ is recognizable.*

The proof relies on several lemmas of independent interest. Let $G$ be a finite group, let $\pi : A^* \to G$ be a surjective morphism.

**Lemma 3.** *Let $g_1, g_2, \ldots, g_n$ be a sequence of elements of $G$, with $n \geq |G|$. Then there exist two indices $i, j$, with $i \leq j$ such that $g_i \cdots g_j = 1$.*

*Proof.* Consider the sequence $g_1, g_1 g_2, \ldots, g_1 g_2 \cdots g_n$. Either one of these elements is equal to 1, or, since $n > |G| - 1$, two of them are equal, say $g_1 \cdots g_{i-1} = g_1 \cdots g_j$. In this case, $g_i \cdots g_j = 1$.

Let $L = \pi^{-1}(1)$. The next lemma states that sufficiently long words contain a factor in $L$. More precisely

**Lemma 4.** *Every word of $A^*$ of length $\geq |G|$ contains a non-empty factor in $L$.*

*Proof.* Let $a_1 \cdots a_n$ be a word of length $n \geq |G|$. By Lemma 3, there exist two indices $i, j$, with $i \leq j$ such that $\pi(a_i) \cdots \pi(a_j) = 1$. It follows that $\pi(a_i \cdots a_j) = 1$ and hence $a_i \cdots a_j \in L$.

**Lemma 5.** *Let $x$ be a word of $L$ and let $u$ and $v$ be two words. Then $uv \in L$ if and only if $uxv \in L$.*

*Proof.* Indeed, if $x \in L$, then $\pi(x) = 1$. It follows that

$$\pi(uxv) = \pi(u)\pi(x)\pi(v) = \pi(u)\pi(v) = \pi(uv)$$

Therefore $\pi(uxv) = 1$ if and only if $\pi(uv) = 1$, that is if and only if $uxv \in L$.

**Lemma 6.** *Let $a_1, \ldots, a_r$ be letters, let $x$ be a word of $L$ and let $u$ and $v$ be two words. If $uv \in La_1 La_2 L \cdots La_r L$, then $uxv \in La_1 La_2 L \cdots La_r L$.*

*Proof.* Indeed, if $uv \in La_1 La_2 L \cdots La_r L$, then there exist an index $I$ and two words $x', x'' \in A^*$ such that $u \in La_1 L \cdots La_i x'$ and $v \in x'' a_{i+1} L \cdots La_r L$. It follows immediately that $uxv \in La_1 La_2 L \cdots La_r L$.

**Proposition 20.** *If the language $L$ is decomposable, then every language recognized by $\pi$ is decomposable.*

*Proof.* Let $P$ be a subset of $G$. Since decomposable languages are closed under union and since

$$\pi^{-1}(P) = \bigcup_{g \in P} \pi^{-1}(g)$$

it suffices to prove that each language $\pi^{-1}(g)$ is decomposable. Now, let $g$ be a fixed element of $G$ and let $u$ be a word such that $\pi(u) = g^{-1}$. We claim that $\pi^{-1}(g) = Lu^{-1}$. Indeed if $x \in \pi^{-1}(g)$, then $\pi(xu) = \pi(x)\pi(u) = gg^{-1} = 1$. It follows that $xu \in L$ and $x \in Lu^{-1}$. Conversely, if $x \in Lu^{-1}$, then $xu \in L$, that is, $\pi(xu) = 1$. Therefore $\pi(x) = \pi(u)^{-1} = g$ and $x \in \pi^{-1}(g)$, which proves the claim. Now, Proposition 7 shows that decomposable languages are closed under quotients, and thus $\pi^{-1}(g)$ is decomposable.

Theorem 7 now follows from the next proposition, which gives an explicit formula for $\sigma(L)$.

**Proposition 21.** *The following formula holds, with* $N = 2|G|^4$:

$$\sigma(L) = \bigcup_{\substack{r,s \leq N \\ (a_1 \cdots a_r \text{ III } b_1 \cdots b_s) \cap L \neq \emptyset}} (La_1 La_2 L \cdots La_r L) \text{ III } (Lb_1 Lb_2 L \cdots Lb_s L)$$

*Proof.* Let $K$ be the right member of the equality to be proved. Let $a_1 \cdots a_r$ and $b_1 \cdots b_s$ be two words such that $r, s \leq N$ and

$$(a_1 \cdots a_r \text{ III } b_1 \cdots b_s) \cap L \neq \emptyset$$

Then there exists in $L$ a word $w = c_1 \cdots c_{r+s}$, a partition $(I, J)$ of $\{1, \cdots, r+s\}$ and two bijections $\alpha : \{1, \cdots, r\} \to I$ and $\beta : \{1, \cdots, s\} \to J$ such that, for $1 \leq i \leq r+s$,

$$c_i = \begin{cases} a_{\alpha^{-1}(i)} & \text{if } i \in I \\ b_{\beta^{-1}(i)} & \text{if } j \in I \end{cases}$$

Suppose that $(u, v) \in (La_1 La_2 L \cdots La_r L)$ III $(Lb_1 Lb_2 L \cdots Lb_s L)$. Then $u = u_0 a_1 u_1 \cdots a_r u_r$ and $v = v_0 b_1 v_1 \cdots b_s v_s$ for some words $u_0, u_1, \ldots, u_r, v_0, \ldots, v_r \in L$. Let, for $1 \leq i \leq r+s$,

$$x_i = \begin{cases} c_i u_{\alpha^{-1}(i)} & \text{if } i \in I \\ c_i v_{\beta^{-1}(i)} & \text{if } i \in J \end{cases}$$

and let $x = u_0 x_1 \cdots x_{r+s}$. For instance, if $w = a_1 a_2 b_1 a_3 b_2 b_3 a_4$, then $x = u_0 a_1 u_1 a_2 u_2 b_1 v_1 a_3 u_3 b_2 v_2 b_3 v_3 a_4 u_4$. Then $x$ belongs to $u$ III $v$ par construction. Furthermore, since for $1 \leq i \leq r$, $\pi(u_i) = 1$ and for $1 \leq i \leq s$, $\pi(v_i) = 1$, we have, for $1 \leq i \leq r+s$, $\pi(x_i) = \pi(c_i)$. It follows that $\pi(x) = \pi(c_1 \cdots c_{r+s}) = \pi(w)$ and hence $\pi(x) = 1$, since $w \in L$. Therefore $x \in (u$ III $v) \cap L$ and $(u, v) \in \sigma(L)$.

In the opposite direction, consider a pair $(u, v) \in \sigma(L)$. We prove by induction on $|u| + |v|$ that $(u, v) \in K$. First assume that $|u| + |v| \leq N$. Let $u = a_1 \cdots a_r$ and $v = b_1 \cdots b_s$ with $r, s \leq N$. Since $1 \in L$, one has $u \in La_1 La_2 L \cdots La_r L$

and $v \in Lb_1 Lb_2 L \cdots Lb_s L$. Furthermore, $(a_1 \cdots a_r \ \text{III} \ b_1 \cdots b_s) \cap L \neq \emptyset$, since $(u, v) \in \sigma(L)$. Therefore $(u, v) \in K$.

We may now assume that $|u| + |v| > N$. By assumption, there are two factorizations $u = u_1 u_2 \cdots u_n$ and $v = v_1 v_2 \cdots v_n$, with $u_1 \in A^*$, $v_1, u_2, v_2, \ldots$, $v_{n-1}, u_n$ in $A^+$, $v_n \in A^*$ such that $u_1 v_1 \cdots u_n v_n \in L$.

Suppose first that one of the words $u_1, \ldots, u_n, v_1, \ldots, v_n$ has length $\geq |G|$. We may assume for instance, that this word is some $u_i$ (a symmetrical argument works for $v_i$). Then by Lemma 4, this word contains a non-empty factor in $L$, that is $u_i = u_i' x u_i''$ with $x \in L \cap A^+$. It follows by Lemma 5 that $u_i' u_i'' \in L$. Let $u' = u_1 \cdots u_{i-1} u_i'$ and $u'' = u_i'' u_{i+1} \cdots u_n$. Then $u'u''$ is the word obtained from $u$ by deleting the factor $x$ in $u_i$. By Lemma 5, the word $u_1 v_1 \cdots u_{i-1} v_{i-1} u_i' u_i'' v_i \cdots u_n v_n$ belongs to $L$ and thus $(u'u'', v) \in \sigma(L)$. It follows by the induction hypothesis that $(u'u'', v) \in K$. Therefore, there exist two words $a_1 \cdots a_r$ and $b_1 \cdots b_s$ such that $(a_1 \cdots a_r \ \text{III} \ b_1 \cdots b_s) \cap L \neq \emptyset$, $u'u'' \in La_1 La_2 L \cdots La_r L$ and $v \in Lb_1 Lb_2 L \cdots Lb_s L$. Now by Lemma 6, $u'xu'' \in La_1 La_2 L \cdots La_r L$, and hence $(u, v) \in K$.

Suppose now that the length of each $u_i$ and $v_i$ is strictly smaller that $|G|$. Since $|u_1 v_1 \cdots u_n v_n| = |u| + |v| \geq 2|G|^4$, we have $n \geq |G|^3$. By Lemma 3, applied to the group $G^3$, there exist two indices $i$ and $j$, with $i \leq j$ such that $(\pi(u_i \cdots u_j), \pi(u_i v_i \cdots u_j v_j), \pi(v_i \cdots v_j)) = (1, 1, 1)$, which means that $u_i \cdots u_j$, $u_i v_i \cdots u_j v_j$ and $v_i \cdots v_j$ are in $L$. Now, since $u_1 v_1 \cdots u_n v_n \in L$, it follows by Lemma 5 that $u_1 v_1 \cdots u_{i-1} v_{i-1} u_{j+1} v_{j+1} \cdots u_n v_n \in L$. Therefore

$$(u_1 \cdots u_{i-1} u_{j+1} \cdots u_n, v_1 \cdots v_{i-1} v_{j+1} \cdots v_n) \in \sigma(L)$$

and by the induction hypothesis, $(u_1 \cdots u_{i-1} u_{j+1} \cdots u_n, v_1 \cdots v_{i-1} v_{j+1} \cdots v_n) \in K$. It follows by Lemma 6 that $(u, v) \in K$. Therefore $\sigma(L) = K$.

## 10    Conclusion

We have given a counterexample to the original Schnoebelen's Conjecture, and we have studied the closure properties of the class of decomposable languages. This class is closed under union, product, shuffle, inverse of length preserving morphisms and residuals, but it is not closed under intersection and inverse morphisms.

However the main problem remains open: is it decidable whether a given recognizable language is decomposable?

An example of a non-decomposable language $L$ such that $\sigma(L)$ is recognizable was given in Section 4. Nevertheless, we think that decomposable languages might be characterized by some recognizability property of $\sigma$, and we propose the following precise formulation:

*Conjecture 2.* Let $K$ be a recognizable language and let $\bigcup_{1 \leq i \leq n} L_i \times R_i$ be the Conway canonical form of $\tau(K)$. Then $L$ is decomposable if and only if all the sets $\sigma(L_i)$ and $\sigma(R_i)$ are recognizable.

If this conjecture is true, Theorem 7 implies that every group language is decomposable. Actually, the conjecture would imply more generally that every language of the polynomial closure of the group languages (languages that are finite union of products of group languages) is decomposable. These languages form a positive variety of languages and admit a nice algebraic characterization: in their ordered syntactic monoid, every idempotent $e$ satisfies $e \leq 1$. They are also the rational open sets of the pro-group topology [5]. Proving directly that these sets are decomposable seems to be an accessible challenge and a good test for our conjecture.

However it would not be a characterization of the class of decomposable languages, since decomposable languages do not form a positive variety.

# References

1. J.C.M. Baeten and W.P. Weijland. *Process algebra*, volume 18 of *Cambridge Tract in Theoretical Computer Science*. Cambridge University Press, Cambridge UK, 1990.
2. John H. Conway. *Regular Algebra and Finite Machines*. Chapman and Hall, London, 1971.
3. Samuel Eilenberg. *Automata, Languages and Machines*, volume B. Academic Press, New York, 1976.
4. Lothaire. *Combinatorics on Words*, volume 17 of *Encyclopedia of Mathematics and its Applications*. aw, reading, 1983.
5. Jean-Éric Pin. Polynomial closure of group languages and open sets of the hall topology. *Theoretical Computer Science*, 169:185–200, 1996.
6. Jean-Eric Pin and Pascal Weil. Polynomial closure and unambiguous product. *Theory Comput. Systems*, 30:1–39, 1997.
7. Ph. Schnoebelen. Decomposable regular languages and the shuffle operator. *EATCS Bull.*, (67):283–289, 1999.

# Restarting Automata and Their Relations to the Chomsky Hierarchy

Friedrich Otto

Fachbereich Mathematik/Informatik, Universität Kassel
34109 Kassel, Germany
otto@theory.informatik.uni-kassel.de
http://www.theory.informatik.uni-kassel.de

**Abstract.** The restarting automaton, introduced by Jančar et al in 1995, is motivated by the so-called 'analysis by reduction,' a technique from linguistics. By now there are many different models of restarting automata, and their investigation has proved very fruitful in that they offer an opportunity to study the influence of various kinds of resources on their expressive power. Here a survey on the various models and their properties is given, their relationships to the language classes of the Chomsky hierarchy are described, and some open problems are presented.

## 1 Introduction

A *restarting automaton* as described by Jančar, Mráz, Plátek and Vogel [15] is a nondeterministic machine model that processes strings which are stored in a list (or a 'rubber' tape). It has a finite control, and it has a read/write-head with a finite look-ahead working on the list of symbols. As such it can be seen as a modification of the *list automaton* presented in [6] and the *forgetting automaton* [12,13]. However, the restarting automaton can only perform two kinds of operations: move-right transitions, which shift the read/write-head one position to the right changing the actual state, and restart transitions, which delete some symbols from the read/write window, place this window over the left end of the list, and put the automaton back into its initial state. Hence, after performing a restart transition a restarting automaton has no way to remember that it has already performed some steps of a computation. Further, by each application of a restart transition the tape is shortened. In this aspect the restarting automaton is similar to the *contraction automaton* [44]. It follows that restarting automata are linearly time-bounded.

As illustrated by Jančar et al (cf., e.g., [19]) these automata were invented to model the *analysis by reduction* of sentences in a natural language [45,46]. This analysis consists of a stepwise simplification of a sentence in such a way that the syntactical correctness or incorrectness of the sentence is not affected. After a finite number of steps either a correct simple sentence is obtained, or an error is detected. Generative devices capable of simulating this process to some extent include the *contextual grammars* (see, e.g., [8,22]) and the *g-systems* [43].

Z. Ésik and Z. Fülöp (Eds.): DLT 2003, LNCS 2710, pp. 55–74, 2003.
© Springer-Verlag Berlin Heidelberg 2003

Subsequently Jančar et al extended their model in various ways. Instead of simply deleting some symbols from the actual contents of the read/write window, this contents is replaced by a shorter string [16]. Further, the use of auxiliary symbols was added in [18], which yields the so-called RWW-automata, and various notions of *monotonicity* have been discussed for these automata. It turned out that the monotone RWW-automata accept the context-free languages, and that the various forms of monotone and deterministic RWW-automata accept the deterministic context-free languages. In [18] also the restart transition was separated from the rewrite transition so that, after performing a rewrite step, the automaton can still read the remaining part of the tape before performing a restart transition. This gives the so-called RRWW-automata.

Here we present a survey on the restarting automaton and its many variants. In particular, we will address the relationships between these automata and the classes of the Chomsky hierarchy. The paper is structured as follows. In the next section the basic variants of the restarting automaton are defined. The definitions given here differ slightly from the ones found in the literature (cf., e.g., [19,36]), but they are easily seen to be equivalent to them. In Section 3 the monotone restarting automata are considered, and the relationships between the various language classes thus obtained and the (deterministic) context-free languages are described. Then we turn to the various types of deterministic restarting automata (Section 4), and we discuss their relationships to the Church-Rosser languages [24,29]. In Section 5 the weakly monotone restarting automata are introduced, and it is shown that they correspond to the so-called growing context-sensitive languages [4,7,23]. In the last section we then present some further variants of the restarting automaton in short. Throughout the paper many problems will be mentioned that are still open.

Due to the page limit this paper cannot possibly contain all the many facets of restarting automata in full detail, not to mention all the predecessor models and related topics. Also no proofs are given, but for all results presented the appropriate references are cited. Naturally a presentation like this reflects the author's interests and his view of the topic discussed. I apologize to all who feel that important aspects of the theory are missing.

Thoughout the paper we will use the following notation. All alphabets considered are finite. For an alphabet $\Sigma$, $\Sigma^*$ denotes the set of all strings over $\Sigma$ including the empty string $\varepsilon$, and $\Sigma^+$ is the set of all non-empty strings. For a string $w$, $|w|$ denotes its *length*, and $w^n$ is defined by $w^0 := \varepsilon$ and $w^{n+1} := w^n w$, where $uv$ simply denotes the concatenation of the strings $u$ and $v$. Finally, $w^R$ will denote the mirror image of $w$. Further, by DCFL, CFL, and CSL we denote the classes of deterministic context-free languages, context-free languages, and context-sensitive languages, respectively.

## 2    Restarting Automata

Here we give the definition of the restarting automaton and its main variants. In doing so we will not follow the historical development outlined above, but we will first present the most general model, the RRWW-automaton, and then describe the other variants as restrictions thereof.

A *restarting automaton*, RRWW-automaton for short, is a one-tape machine that is described by an 8-tuple $M = (Q, \Sigma, \Gamma, \text{\textcent}, \$, q_0, k, \delta)$, where $Q$ is the finite set of states, $\Sigma$ is the finite input alphabet, $\Gamma$ is the finite tape alphabet containing $\Sigma$, $\text{\textcent}, \$ \notin \Gamma$ are the markers for the left and right border of the work space, respectively, $q_0 \in Q$ is the initial state, $k \geq 1$ is the size of the *read/write window*, and

$$\delta : Q \times \mathcal{PC}^{\leq k} \to \mathfrak{p}((Q \times (\{\mathsf{MVR}\} \cup \mathcal{PC}^{\leq k-1})) \cup \{\mathsf{Restart}, \mathsf{Accept}\})$$

is the *transition relation*. Here $\mathfrak{p}(S)$ denotes the powerset of the set $S$, and

$$\mathcal{PC}^{\leq k} := (\text{\textcent} \cdot \Gamma^{k-1}) \cup \Gamma^k \cup (\Gamma^{\leq k-1} \cdot \$) \cup (\text{\textcent} \cdot \Gamma^{\leq k-2} \cdot \$)$$

is the set of *possible contents* of the read/write window of $M$, where $\Gamma^{\leq n} := \bigcup_{i=0}^{n} \Gamma^i$, and $\mathcal{PC}^{\leq k-1}$ is defined analogously.

The transition relation describes four different types of transition steps:

1. A *move-right step* is of the form $(q', \mathsf{MVR}) \in \delta(q, u)$, where $q, q' \in Q$ and $u \in \mathcal{PC}^{\leq k}$, $u \neq \$$. If $M$ is in state $q$ and sees the string $u$ in its read/write window, then this move-right step causes $M$ to shift the read/write window one position to the right and to enter state $q'$.
2. A *rewrite step* is of the form $(q', v) \in \delta(q, u)$, where $q, q' \in Q$, $u \in \mathcal{PC}^{\leq k}$, $u \neq \$$, and $v \in \mathcal{PC}^{\leq k-1}$ such that $|v| < |u|$. It causes $M$ to replace the contents $u$ of the read/write window by the string $v$, and to enter state $q'$. Further, the read/write window is placed immediately to the right of the string $v$. However, some additional restrictions apply in that the border markers $\text{\textcent}$ and $\$ $ must not disappear from the tape nor that new occurrences of these markers are created. Further, the read/write window must not move across the right border marker $\$ $.
3. A *restart step* is of the form $\mathsf{Restart} \in \delta(q, u)$, where $q \in Q$ and $u \in \mathcal{PC}^{\leq k}$. It causes $M$ to move its read/write window to the left end of the tape, so that the first symbol it sees is the left border marker $\text{\textcent}$, and to reenter the initial state $q_0$.
4. An *accept step* is of the form $\mathsf{Accept} \in \delta(q, u)$, where $q \in Q$ and $u \in \mathcal{PC}^{\leq k}$. It causes $M$ to halt and accept.

Obviously, each computation of $M$ proceeds in cycles. Starting from an initial configuration $q_0\text{\textcent}w\$$, the head moves right, while move-right and rewrite steps are executed until finally a restart step takes $M$ back into a configuration of the form $q_0\text{\textcent}w_1\$$. It is required that in each such cycle *exactly one* rewrite step is executed. By $\vdash_M^c$ we denote the execution of a complete cycle. As by a rewrite step the contents of the tape is shortened, only a linear number of cycles can be executed within any computation. That part of a computation of $M$ that follows after the execution of the last restart is called the *tail* of the computation. It contains at most one application of a rewrite step.

An input $w \in \Sigma^*$ is accepted by $M$, if there exists a computation of $M$ which starts with the initial configuration $q_0\text{\textcent}w\$$, and which finally ends with executing an accept step. By $L(M)$ we denote the *language accepted by* $M$, and $\mathcal{L}(\mathsf{RRWW})$ will denote the class of languages that are accepted by RRWW-automata.

The following lemma can easily be proved by using standard techniques from automata theory.

**Lemma 1.** *Each* RRWW-*automaton* $M$ *is equivalent to an* RRWW-*automaton* $M'$ *that satisfies the following additional restriction:*

(*) $M'$ *makes an accept or a restart step only when it sees the right border marker* $ *in its read/write window.*

This lemma means that in each cycle and also during the tail of a computation the read/write window moves all the way to the right before a restart is made, respectively, before the machine halts and accepts.

Based on this fact each cycle (and also the tail) of a computation of an RRWW-automaton $M$ consists of three phases. Accordingly, the transition relation of an RRWW-automaton can be described through a sequence of so-called *meta-instructions* [35] of the form $(R_1, u \to v, R_2)$, where $R_1$ and $R_2$ are regular languages, called the *regular constraints* of this instruction, and $u$ and $v$ are strings such that $|u| > |v|$. The rule $u \to v$ stands for a rewrite step of the RRWW-automaton $M$ considered. On trying to execute this meta-instruction $M$ will get stuck (and so reject) starting from the configuration $q_0 \, ¢ \, w \, $$, if $w$ does not admit a factorization of the form $w = w_1 u w_2$ such that $¢ w_1 \in R_1$ and $w_2 $ \in R_2$. On the other hand, if $w$ does have a factorization of this form, then one such factorization is chosen nondeterministically, and $q_0 \, ¢ \, w \, $$ is transformed into $q_0 \, ¢ \, w_1 v w_2 \, $$. In order to describe the tails of accepting computations accept instructions may be used within meta-instructions.

This description of the RRWW-automaton $M$ corresponds to the characterization of the class $\mathcal{L}(\mathsf{RRWW})$ by certain infinite prefix-rewriting systems as given in [36], Corollary 6.4.

A restarting automaton is called an RWW-*automaton* if it makes a restart immediately after performing a rewrite operation. In particular, this means that it cannot perform a rewrite step during the tail of a computation.

A cycle of a computation of an RWW-automaton $M$ consists of two phases only. Accordingly, the transition relation of an RWW-automaton can be described by a finite sequence of *meta-instructions* of the form $(R, u \to v)$, where $R$ is a regular language, and $u$ and $v$ are strings such that $|u| > |v|$. This description corresponds to the characterization of the class $\mathcal{L}(\mathsf{RWW})$ by certain infinite prefix-rewriting systems as given in [36], Corollary 6.2.

An R(R)WW-automaton is *deterministic* if its transition relation is a function. It is called an R(R)W-*automaton* if it is an R(R)WW-automaton for which the tape alphabet $\Gamma$ coincides with the input alphabet $\Sigma$, that is, if no auxiliary symbols are available. Finally, it is an R(R)-*automaton* if it is an R(R)W-automaton for which the right-hand side $v$ of each rewrite step $(q', v) \in \delta(q, u)$ is a scattered subword of the left-hand side $u$.

Figure 1 summarizes the obvious inclusions between the language classes defined by the various types of restarting automata. Here P and NP denote the well-known complexity classes. In the following we are concerned with two main topics:

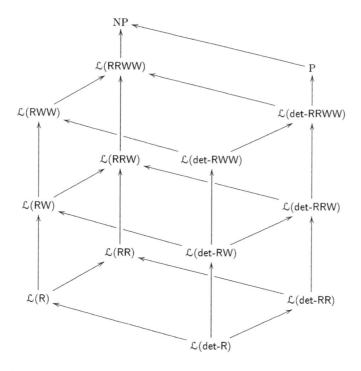

**Fig. 1.** Inclusions between the language classes defined by the basic types of restarting automata

(a) Which of the inclusions in Figure 1 are proper?
(b) Where do the language classes defined by the various types of restarting automata lie in relation to the Chomsky hierarchy?

To attack these questions additional characterizations of some of these classes will be derived, and some closure properties will be considered. The role that the auxiliary symbols play in restarting automata can be qualified as follows.

**Theorem 1.** [36]
*A language L is accepted by a (deterministic)* R(R)WW-*automaton if and only if there exist a (deterministic)* R(R)W-*automaton $M_1$ and a regular language R such that $L = L(M_1) \cap R$ holds.*

This characterization shows that $\mathcal{L}$(det-R(R)WW) and $\mathcal{L}$(R(R)WW) are closed under the operation of intersection with regular languages.

In the literature further restrictions of R-automata have been considered. In [14] the *normal* R-automata are considered, and it is shown that the class of languages accepted by them can be characterized by a certain type of *contextual grammars*. In [28] these investigations are continued by considering *restarting automata with a delete operation*, the so-called DR-automata. For these automata the rewrite transitions are replaced by *delete* transitions that simply delete a factor from the current contents of the read/write window. However, in contrast to the other models, a DR-automaton is allowed to perform several rewrite (that is, delete) steps during a cycle.

# 3  Monotone Restarting Automata

In [16] the notion of monotonicity is introduced for restarting automata. Let $M$ be an RRWW-automaton. Each computation of $M$ can be described by a sequence of cycles $C_1, C_2 \ldots, C_n$, where $C_n$ is the last cycle, which is followed by the tail of the computation. Each cycle $C_i$ of this computation contains a unique configuration of the form $\text{¢}xuy\$$ such that $u \to v$ is the rewrite step applied during this cycle. By $D_r(C_i)$ we denote the $r$-*distance* $|y\$|$ of this cycle. The sequence of cycles $C_1, C_2, \ldots, C_n$ is called *monotone* if $D_r(C_1) \geq D_r(C_2) \geq \ldots \geq D_r(C_n)$ holds. The RRWW-automaton $M$ is called *monotone* if all its computations are monotone.

**Proposition 1.** [19] *It is decidable whether an* RRWW-*automaton is monotone.*

**Proof idea.** From a given RRWW-automaton $M$, a nondeterministic finite-state acceptor $A_M$ can be constructed such that the language $L(A_M)$ is non-empty if and only if $M$ admits a non-monotone computation.                                    □

If an RRWW-automaton $M$ is monotone, then this means that after a restart, $M$ moves right across the actual tape contents until it has completely seen the string that was written by the preceding rewrite operation. Thus, it is intuitively clear that each monotone RWW-automaton can be simulated by a pushdown automaton $P_M$, and that $P_M$ can be made to be deterministic, if $M$ is. In fact, it turns out that the same is true for RRWW-automata.

**Proposition 2.** [19] $\mathcal{L}$(det-mon-RRWW) $\subseteq$ DCFL *and* $\mathcal{L}$(mon-RRWW) $\subseteq$ CFL.

On the other hand, based on the characterization of deterministic context-free languages by LR(0)-grammars (see, e.g., [11]) the following strong result is obtained.

**Proposition 3.** [19] DCFL $\subseteq \mathcal{L}$(det-mon-R).

It follows that all variants of deterministic monotone restarting automata coincide in their expressive power.

**Theorem 2.** [16,17,18,19]
*For all types* $X \in \{R, RR, RW, RRW, RWW, RRWW\}$, $\mathcal{L}$(det-mon-X) = DCFL.

For nondeterministic restarting automata it turns out that the use of auxiliary symbols is necessary to obtain a characterization of the class CFL of context-free languages.

**Theorem 3.** [16,17,18,19] $\mathcal{L}$(mon-RWW) = $\mathcal{L}$(mon-RRWW) = CFL.

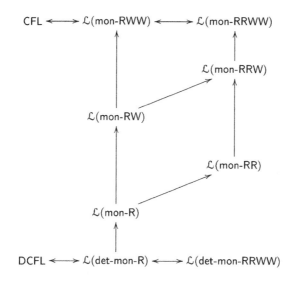

**Fig. 2.** The taxonomy of monotone restarting automata

And indeed, for the monotone restarting automata we have the taxonomy given in Figure 2, where an arrow indicates a proper inclusion, a two-sided arrow expresses equality, and where two language classes are incomparable under inclusion, if there does not exist a directed path between them in this diagram. The separation results are obtained from the following example languages:

$$L_1 := \{\, a^n b^n c \mid n \geq 0 \,\} \cup \{\, a^n b^{2n} d \mid n \geq 0 \,\},$$
$$L_2 := \{\, a^n b^n \mid n \geq 0 \,\} \cup \{\, a^n b^m \mid m > 2n \geq 0 \,\},$$
$$L_3 := \{f, ee\} \cdot \{\, a^n b^n \mid n \geq 0 \,\} \cup \{g, ee\} \cdot \{\, a^n b^m \mid m > 2n \geq 0 \,\},$$
$$L_4 := \{\, a^n b^m \mid 0 \leq n \leq m \leq 2n \,\}.$$

It is shown in [19] that $L_1 \in \mathcal{L}(\text{mon-RR}) \smallsetminus \mathcal{L}(\text{RW})$, $L_2 \in \text{CFL} \smallsetminus \mathcal{L}(\text{RRW})$, $L_3 \in \mathcal{L}(\text{mon-RW}) \smallsetminus \mathcal{L}(\text{RR})$, and $L_4 \in \mathcal{L}(\text{mon-R}) \smallsetminus \mathcal{L}(\text{det-RW})$. Together with the above results these facts yield all the relations displayed in Figure 2.

Also some other notions of monotonicity have been considered in the literature. A restarting automaton $M$ is called *a-monotone* if all its accepting computations are monotone, and it is called *g-monotone* if there exists an accepting monotone computation for each string $w \in L(M)$. In the deterministic case these notions still give the class DCFL, while in the nondeterministic case additional intermediate classes are obtained [18].

## 4    Deterministic Restarting Automata

There is a close correspondence between the deterministic restarting automata and the Church-Rosser languages. Therefore, we begin by restating in short the main definitions regarding the various classes of Church-Rosser languages. For additional information concerning the notions introduced the reader is asked to consult the literature, where [3] serves as our main reference concerning the theory of string-rewriting systems.

Let $\Sigma$ be a finite alphabet. A function $\varphi : \Sigma \to \mathbb{N}_+$ is called a *weight-function*. Its extension to $\Sigma^*$, which we will also denote by $\varphi$, is defined inductively through $\varphi(\varepsilon) := 0$ and $\varphi(wa) := \varphi(w) + \varphi(a)$ for all $w \in \Sigma^*$ and $a \in \Sigma$. A particular weight-function is the *length-function* $|.| : \Sigma \to \mathbb{N}_+$, which assigns each symbol the weight (length) 1.

A *string-rewriting system* $R$ on $\Sigma$ is a subset of $\Sigma^* \times \Sigma^*$. An element $(\ell, r) \in R$ is called a *rewrite rule* or simply a *rule*, and it will usually be written as $(\ell \to r)$.

The string-rewriting system $R$ induces several binary relations on $\Sigma^*$, the most basic of which is the *single-step reduction relation*

$$\to_R := \{(u\ell v, urv) \mid u, v \in \Sigma^*, (\ell \to r) \in R\}.$$

Its reflexive and transitive closure is the *reduction relation* $\to_R^*$ induced by $R$. If $u \to_R^* v$, then $u$ is an *ancestor* of $v$, and $v$ is a *descendant* of $u$. If there is no $v \in \Sigma^*$ such that $u \to_R v$ holds, then the string $u$ is called *irreducible* (mod $R$). By $\mathrm{IRR}(R)$ we denote the set of all irreducible strings. If $R$ is finite, then $\mathrm{IRR}(R)$ is obviously a regular language.

The string-rewriting system $R$ is called

- *length-reducing* if $|\ell| > |r|$ holds for each rule $(\ell \to r) \in R$,
- *weight-reducing* if there exists a weight-function $\varphi$ such that $\varphi(\ell) > \varphi(r)$ holds for each rule $(\ell \to r) \in R$,
- *confluent* if, for all $u, v, w \in \Sigma^*$, $u \to_R^* v$ and $u \to_R^* w$ imply that $v$ and $w$ have a common descendant.

If a string-rewriting system $R$ is weight-reducing, then it allows no infinite reduction sequence; indeed, if $w_0 \to_R w_1 \to_R \ldots \to_R w_m$, then $m \leq \varphi(w_0)$. If, in addition, $R$ is confluent, then each string $w \in \Sigma^*$ has a unique irreducible descendant $\hat{w} \in \mathrm{IRR}(R)$. Observe that $\hat{w}$ can be determined from $w$ in linear time [3]. This observation was one of the main motives to introduce the Church-Rosser languages.

**Definition 1.** *A language $L \subseteq \Sigma^*$ is a* Church-Rosser language *if there exist an alphabet $\Gamma \supsetneq \Sigma$, a finite, length-reducing, confluent string-rewriting system $R$ on $\Gamma$, two strings $t_1, t_2 \in (\Gamma \setminus \Sigma)^* \cap \mathrm{IRR}(R)$, and a symbol $Y \in (\Gamma \setminus \Sigma) \cap \mathrm{IRR}(R)$ such that, for all $w \in \Sigma^*$, $t_1 w t_2 \to_R^* Y$ if and only if $w \in L$ [24,29]. By* CRL *we denote the class of Church-Rosser languages.*

By admitting weight-reducing instead of length-reducing string-rewriting systems in the above definition, we obtain the class GCRL of *generalized Church-Rosser languages* [5]. From [24,29] we obtain the sequence of inclusions

$$\mathsf{DCFL} \subset \mathsf{CRL} \subseteq \mathsf{GCRL} \subset \mathsf{CSL},$$

where the first and the last one are known to be proper. Also it has been shown in [24] that CRL is not contained in the class CFL, as the non-context-free language $L_5 := \{ a^{2^n} \mid n \geq 0 \}$ is Church-Rosser.

The Church-Rosser languages have been characterized by certain models of automata. The following definition is a variant of the one given in [5].

**Definition 2.** *A* two-pushdown automaton (TPDA) *with* pushdown windows *of size* $k$ *is a nondeterministic automaton with two pushdown stores. Formally, it is defined as a 10-tuple* $P = (Q, \Sigma, \Gamma, \delta, k, q_0, \perp, t_1, t_2, F)$, *where*

- $Q$ *is the finite set of states,*
- $\Sigma$ *is the finite input alphabet,*
- $\Gamma$ *is the finite tape alphabet with* $\Gamma \supsetneq \Sigma$ *and* $\Gamma \cap Q = \emptyset$,
- $q_0 \in Q$ *is the initial state,*
- $\perp \in \Gamma \setminus \Sigma$ *is the bottom marker of the pushdown stores,*
- $t_1, t_2 \in (\Gamma \setminus \Sigma)^*$ *are the preassigned contents of the left/right pushdown store, respectively,*
- $F \subseteq Q$ *is the set of final (or halting) states, and*
- $\delta : Q \times {}_\perp\Gamma^{\leq k} \times \Gamma_\perp^{\leq k} \to \mathfrak{P}_{fin}(Q \times \Gamma^* \times \Gamma^*)$ *is the transition relation, where* ${}_\perp\Gamma^{\leq k} := \Gamma^k \cup \{ \perp u \mid |u| \leq k - 1 \}$, $\Gamma_\perp^{\leq k} := \Gamma^k \cup \{ v\perp \mid |v| \leq k - 1 \}$, *and* $\mathfrak{P}_{fin}(Q \times \Gamma^* \times \Gamma^*)$ *denotes the set of finite subsets of* $Q \times \Gamma^* \times \Gamma^*$.

*The automaton* $P$ *is a* deterministic two-pushdown automaton (DTPDA), *if* $\delta$ *is a (partial) function from* $Q \times {}_\perp\Gamma^{\leq k} \times \Gamma_\perp^{\leq k}$ *into* $Q \times \Gamma^* \times \Gamma^*$.

A configuration of a (D)TPDA is described as $uqv$, where $q \in Q$ is the actual state, $u \in \Gamma^*$ is the contents of the first pushdown store with the first symbol of $u$ at the bottom and the last symbol of $u$ at the top, and $v \in \Gamma^*$ is the contents of the second pushdown store with the last symbol of $v$ at the bottom and the first symbol of $v$ at the top. For an input string $w \in \Sigma^*$, the corresponding initial configuration is $\perp t_1 q_0 w t_2 \perp$. The computation relation of a (D)TPDA $P$ is denoted by $\vdash_P^*$, where $\vdash_P$ is the single-step computation relation. The (D)TPDA $P$ accepts with empty pushdown stores, that is,

$$L(P) := \{ w \in \Sigma^* \mid \perp t_1 q_0 w t_2 \perp \vdash_P^* q \text{ for some } q \in F \}$$

is the *language accepted by* $P$.

**Definition 3.** *A (D)TPDA is called* shrinking *if there exists a weight-function* $\varphi : Q \cup \Gamma \to \mathbb{N}_+$ *such that, for all* $q \in Q$, $u \in {}_\perp\Gamma^{\leq k}$, *and* $v \in \Gamma_\perp^{\leq k}$, $(p, u', v') \in \delta(q, u, v)$ *implies that* $\varphi(pu'v') < \varphi(quv)$. *By* sTPDA *and* sDTPDA *we denote the corresponding classes of shrinking automata.*
*A (D)TPDA is called* length-reducing *if, for all* $q \in Q$, $u \in {}_\perp\Gamma^{\leq k}$, *and* $v \in \Gamma_\perp^{\leq k}$, $(p, u', v') \in \delta(q, u, v)$ *implies* $|u'v'| < |uv|$. *We denote the corresponding classes of length-reducing automata by* lrTPDA *and* lrDTPDA, *respectively.*

Obviously, the length-reducing TPDA is a special case of the shrinking TPDA. Observe that the input is provided to a TPDA as the initial contents of its second pushdown store, and that in order to accept a TPDA is required to empty its pushdown stores. Thus, it is forced to consume its input completely. Using standard techniques from automata theory it can be shown that, for a (shrinking/length-reducing) (deterministic) TPDA $P$, we may require that the special symbol $\perp$ can only occur at the bottom of a pushdown store, and that no other symbol can occur at that place. In addition, we can enforce that all the halting and accepting configurations of $P$ are of the form $q$, where $q \in F$, and all the halting and rejecting configurations of $P$ are of the form $\perp q$, where $q \in F$. In addition, we can assume that $P$ has a single halting state only.

The definition of the (D)TPDA given here differs from that given by Buntrock and Otto [5], as in the original definition the preassigned contents $t_1$ and $t_2$ of the pushdown stores are always empty, and the (D)TPDA only sees the topmost symbol on each of its pushdown stores. However, for shrinking (D)TPDA's this model is equivalent to the original model [31].

In [5] it is shown that the shrinking DTPDA characterize the class GCRL, and that the shrinking TPDA accept exactly the growing context-sensitive languages. Here a language is called *growing context-sensitive* if it is generated by a phrase-structure grammar $G = (N, T, S, P)$ such that the start symbol $S$ does not appear on the right-hand side of any production of $G$, and $|\alpha| < |\beta|$ holds for all productions $(\alpha \to \beta) \in P$ satisfying $\alpha \neq S$. By GCSL we denote the class of *growing context-sensitive languages*.

From the definitions it is easily seen that a language that is accepted by a length-reducing DTPDA is Church-Rosser. The following result, which can be proved by a direct, but technically quite involved simulation, establishes the converse inclusion.

**Proposition 4.** [31] *For each shrinking TPDA $P$, there exists a length-reducing TPDA $P'$ that accepts the same language. If $P$ is deterministic, then so is $P'$.*

Hence, we have the following characterizations.

**Theorem 4.** [31,32] CRL = GCRL.

**Corollary 1.** [31]
(a) *A language is Church-Rosser, if and only if if it is accepted by a length-reducing DTPDA, if and only if it is accepted by a shrinking DTPDA.*
(b) *A language is growing context-sensitive, if and only if it is accepted by a length-reducing TPDA, if and only if it is accepted by a shrinking TPDA.*

A (deterministic) TPDA $P$ that is length-reducing can be simulated by a (deterministic) RWW-automaton $M$ [34]. A configuration $\perp uqv \perp$ of $P$, where $q$ is a state, is encoded by the tape contents $\mathcal{c} \hat{u} q v \$$ of $M$, where $\hat{u}$ is a copy of $u$ that consists of marked symbols. $M$ simply moves its read/write window to the right until it encounters the left-hand side of a transition of $P$, which it then simulates by applying a rewrite step. There is, however, a slight technical problem in that $M$ starts with a tape contents of the form $\mathcal{c} w \$$ that does not contain a state symbol of $P$, but this problem can be overcome.

Conversely, each deterministic RRWW-automaton $M$ can be simulated by a shrinking DTPDA $P$ [36]. The key observation is the following. If $M$ rewrites $\mathcal{c} x u y \$$ into $\mathcal{c} x v y \$$, then in the next cycle $M$ must read all of $x$ and at least the first symbol of $vy\$$ before it can execute another rewrite step. However, while reading the prefix $x$, $M$ will undergo the same state transitions as in the previous cycle. Thus, instead of reading $x$ again, $P$ simply stores the information on the states of $M$ together with the symbols of $x$ on its left-hand pushdown store. Hence, it can simply reconstruct the state of $M$. Further, after performing the rewrite step $\text{¢}xuy\$ \vdash \text{¢}xvy\$$, $M$ will continue with MVR-steps until it reaches the right delimiter \$, and then it either restarts, halts and accepts, or halts without accepting. Hence, with the string $y$, we can associate two subsets $Q_+(y)$ and $Q_{rs}(y)$ of the state set $Q$ of $M$ as follows:

A state $q \in Q$ is contained in $Q_+(y)$ ($Q_{rs}(y)$) if, starting from the configuration $\text{¢}qy\$$, $M$ makes only MVR-steps until it scans the \$-symbol, and halts and accepts (respectively, restarts) then.

The DTPDA $P$ stores the string $y$ on its right-hand pushdown. In addition, for each suffix $w$ of $y$, its stores descriptions of the sets $Q_+(w)$ and $Q_{rs}(w)$ under the first symbol of $w$. Using this information it does not need to simulate the MVR-steps of $M$ step-by-step, but it can immediately simulate a restart or halt and accept (or reject). As $P$ starts with the input on its right-hand pushdown, it needs a preprocessing stage to determine and to store the sets $Q_+(w)$ and $Q_{rs}(w)$, but this is easily realized even by a shrinking DTPDA.

Thus, we obtain the following results.

**Theorem 5.** [34,36] (a) CRL $=$ $\mathcal{L}$(det-RWW) $=$ $\mathcal{L}$(det-RRWW).
(b) GCSL $\subseteq$ $\mathcal{L}$(RWW) $\subseteq$ $\mathcal{L}$(RRWW).

The Gladkij language $L_{Gl} := \{ w\#w^R\#w \mid w \in \{a,b\}^* \}$ is not growing context-sensitive [2,4,10]. However, its complement $L_{Gl}^c := \{ w \in \{a, b, \#\}^* \mid w \notin L_{Gl} \}$ is a linear language. Thus, we see the following.

**Corollary 2.** [5] *The classes* CFL *and* CRL *are incomparable under inclusion.*

In [24] it was conjectured that the unambiguous context-free language $L_{2pal} = \{ ww^R \mid w \in \{0,1\}^* \}$ is not Church-Rosser. Some arguments supporting this conjecture were presented, but it was proved only recently by using Kolmogorov complexity. In this way the following result is obtained.

**Corollary 3.** [20] *The classes* UCFL *of unambiguous context-free languages and* CRL *are incomparable under inclusion.*

In [47] an interesting normal form result for presentations of Church-Rosser languages is derived. Essentially it says that for each Church-Rosser language, there is a finite, weight-reducing, and confluent string-rewriting system $R$ presenting this language in the sense of Definition 1 such that each rule of $R$ is of the form $(uvw \to uxw)$ for some strings $u, v, x, w$ satisfying $|v| \geq |x|$ and $|x| \leq 1$. In [48] this characterization is used to give a construction for the languages of prefixes of some Church-Rosser languages. Observe, however, that in general the language of prefixes of a Church-Rosser language is not itself Church-Rosser. In fact, the class CRL has the so-called *basis property* [39], that is, each recursively enumerable language $L \subseteq \Sigma^*$ can be presented as $\pi_\Sigma(C)$, where $C \subseteq \Gamma^*$ is a Church-Rosser language on $\Gamma \supsetneq \Sigma$, and $\pi_\Sigma : \Gamma^* \to \Sigma^*$ is the natural projection.

Concerning the closure properties of CRL the following results are known.

**Theorem 6.** [31]

(a) *The class* CRL *is closed under the following operations: complementation, intersection with regular languages, reversal, left and right quotients with single strings, and inverse morphisms.*

(b) *The class* CRL *is not closed under the following operations: union, intersection, product, Kleene star, projections, $\varepsilon$-free morphisms, or the power operation.*

To conclude the discussion of CRL it should be mentioned that in [1] the so-called *McNaughton families of languages* have been defined by considering other classes of string-rewriting systems in Definition 1. As string-rewriting systems are as powerful as Turing machines, it is not surprising that in this way almost all the language classes from the Chomsky hierarchy are obtained. However, there is one exception: the regular languages. No characterization as a McNaughton family of languages is currently known for the class REG, but some other language classes below DCFL are obtained in this way.

Finally we turn to the interrelation between the language classes that are defined by the various types of deterministic restarting automata. Here we have the strict inclusions given in Figure 3, where the interpretation is as in Figure 2. The separation results are obtained from the following example languages:

$$
\begin{aligned}
L_6 &:= L_{6,1} \cup L_{6,2} \cup L_{6,3}, \\
L_{6,1} &:= \{ (ab)^{2^n - i} c(ab)^i \mid n \geq 1, 0 \leq i \leq 2^n \}, \\
L_{6,2} &:= \{ (ab)^{2^n - 2i} (abb)^i \mid n \geq 1, 0 \leq i \leq 2^{n-1} \}, \\
L_{6,3} &:= \{ (abb)^{2^n - i} (ab)^i \mid n \geq 1, 0 \leq i \leq 2^n \}, \\
L_7 &:= L_{7,1} \cup L_{7,2}, \\
L_{7,1} &:= \{ a^{2^n - 2i} c a^i \mid n \geq 1, 0 \leq 2i < 2^n \}, \\
L_{7,2} &:= \{ a^i d a^{2^n - 2i} \mid n \geq 1, 0 \leq 2i < 2^n \}.
\end{aligned}
$$

It can be shown that $L_6 \in \mathcal{L}(\text{det-RR}) \smallsetminus \mathcal{L}(\text{RW})$, and $L_7 \in \mathcal{L}(\text{det-RW}) \smallsetminus \mathcal{L}(\text{RR})$. Further, $L_1 \in \text{CRL} \smallsetminus \mathcal{L}(\text{det-RRW})$ [36], and based on $L_5$ a non-context-free language is described in [16] that belongs to $\mathcal{L}(\text{det-R})$. Together with the above results these facts yield the proper inclusions displayed in Figure 3.

Before closing this section we shortly consider the Church-Rosser congruential languages, which are the strict variants of the Church-Rosser languages.

**Definition 4.** [24,29] *A language $L \subseteq \Sigma^*$ is a Church-Rosser congruential language (CRCL), if there exist a finite, length-reducing, and confluent string-rewriting system $R$ on $\Sigma$ and a finite set of irreducible strings $\{w_1, \ldots, w_n\}$ such that $L = \bigcup_{i=1}^{n} [w_i]_R$.*

Each Church-Rosser congruential language is accepted by some deterministic RW-automaton, but CRCL contains languages that are not in $\mathcal{L}(\text{det-RR})$ [34]. On the other hand, the deterministic context-free language $L_8 := \{ a^n b^n \mid n \geq 0 \} \cup \{a\}^*$ is not congruential. This yields the results concerning CRCL in Figure 3.

It is further shown in [33,34] that CRCL and $\mathcal{L}(\text{det-R})$ each form a *quotient basis* for the recursively enumerable languages. Actually, this result is shown

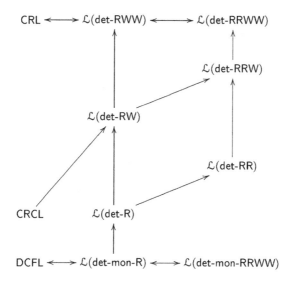

**Fig. 3.** The taxonomy of deterministic restarting automata

for the class of *confluent internal contextual languages*, which is contained in the intersection of CRCL and $\mathcal{L}$(det-R), and which is a restricted version of the *internal contextual languages*, for which a corresponding result is proved in [9]. Thus, CRCL is quite an expressive language class. Nevertheless the question of whether each regular language is contained in CRCL, stated already in [24], is still open. In [30] it is shown that all regular languages of polynomial density are Church-Rosser congruential, and in [37] it is shown that also some regular languages of exponential density, among them all languages from level 1 of the Straubing-Thérien hierarchy, are Church-Rosser congruential.

## 5 Weakly Monotone Restarting Automata

Dahlhaus and Warmuth [7] proved that the membership problem for each growing context-sensitive language is solvable in polynomial time. A detailed discussion of the class GCSL, which has many nice closure properties, can be found in [4].

As seen in Theorem 5, GCSL is contained in $\mathcal{L}$(RWW). On the other hand, in [21] an RWW-automaton $M$ is presented that accepts the Gladkij language $L_{\mathrm{Gl}}$. This RWW-automaton works in two major steps:

(1.) It first transforms an input $u\#v\#w$, where $u, v, w \in \{a, b\}^*$, into a string of the form $u_1\#v_1\#w_1$ using auxiliary symbols that allow to encode two input symbols into one auxiliary symbol. This transformation succeeds only if $w$ is a subsequence of $v^R$, and $v^R$ is a subsequence of $u$.
(2.) Then $M$ checks whether or not $|u_1| = |v_1| = |w_1|$ holds. If this is the case, then also $|u| = |v| = |w|$ implying that actually $u = v^R$ and $w = v^R$, that is, the input belongs to $L_{\mathrm{Gl}}$.

As $L_{\mathrm{GI}}$ is not growing context-sensitive, this yields the following result.

**Theorem 7.** *The class* GCSL *is properly contained in* $\mathcal{L}(\mathsf{RWW})$.

In fact, $\mathcal{L}(\mathsf{RWW})$ even contains some NP-complete languages [21]. In order to derive a characterization of GCSL in terms of restarting automata, we consider a weaker notion of monotonicity. Let $M$ be an R(R)WW-automaton. We say that $M$ is *weakly monotone* if there is a constant $c \in \mathbb{N}$ such that, for each computation of $M$, the corresponding sequence $C_1, C_2, \ldots, C_n$ of cycles satisfies $D_r(C_{i+1}) \leq D_r(C_i) + c$ for all $i = 1, \ldots, n-1$. By using the prefix wmon-, we denote the corresponding classes of restarting automata.

Let $M$ be a deterministic R(R)WW-automaton, and let $C_1, C_2, \ldots, C_n$ be the sequence of cycles of a computation of $M$. If $C_i$ contains the rewrite step $\text{\textcent} xuy\$ \to \text{\textcent} xvy\$$, then $D_r(C_i) = |y| + 1$. In the next cycle, $C_{i+1}$, $M$ cannot perform a rewrite step before it sees at least the first symbol of $vy\$$ in its read/write window. Thus, $D_r(C_{i+1}) \leq |vy\$| - 1 = D_r(C_i) + |v| - 1$. By taking the constant $c := \max(\{\,|v| - 1 \mid u \to v \text{ is a rewrite step of } M\,\} \cup \{0\})$, we see that $M$ is necessarily weakly monotone. Hence, it is only for the various nondeterministic restarting automata that these additional restrictions can (and will) make a difference.

For each language $L \in$ GCSL, there exists a length-reducing TPDA $P$ such that $L = L(P)$ (Corollary 1). This TPDA can be simulated by an RWW-automaton $M$ that encodes a configuration $\perp uqv \perp$ of $P$ by the tape contents $\text{\textcent} \hat{u}qv\$$ (see the discussion following Corollary 1). The automaton $M$ simply moves its read/write window from left to right across its tape until it discovers the left-hand side of a transition of $P$, which it then simulates by a rewrite step. As this rewrite step includes the unique state symbol of $P$ contained on the tape, we see that $M$ is indeed weakly monotone.

On the other hand, the simulation of a deterministic RRWW-automaton $M$ by a shrinking deterministic TPDA carries over to nondeterministic RRWW-automata that are weakly monotone. Just notice the following two facts:

(i) While performing MVR steps $M$ behaves like a finite-state acceptor. Hence, the first part of each cycle can be simulated deterministically. Nondeterminism comes in as soon as a rewrite step is enabled.
(ii) As pointed out in Section 4, with each string $w \in \Gamma^*$, two subsets $Q_+(w)$ and $Q_{\mathrm{rs}}(w)$ can be associated. For the deterministic case these sets are necessarily disjoint for each string $w$. If $M$ is nondeterministic, then this is not true anymore. However, if $Q_+(w)$ is nonempty, then $M$ will accept, and so the simulation simply accepts, and if $Q_+(w)$ is empty, but $Q_{\mathrm{rs}}(w)$ is nonempty, then a restart step must be simulated.

As the shrinking TPDA yields another characterization of GCSL (Corollary 1), this shows that $L(M)$ is growing context-sensitive as required. Hence, we obtain the following characterization.

**Theorem 8.** GCSL $= \mathcal{L}(\mathsf{wmon\text{-}RWW}) = \mathcal{L}(\mathsf{wmon\text{-}RRWW})$.

Instead of requiring that the R(R)WW-automaton considered is weakly monotone, we can consider *left-most computations* of R(R)WW-automata, which gives another characterization of GCSL. To complete our discussion of GCSL we present still another characterization that is in terms of grammars.

A grammar $G = (N, T, S, P)$ is called *(strictly) monotone* if the start symbol does not occur on the right-hand side of any production, and if $|\alpha| \leq |\beta|$ ($|\alpha| < |\beta|$) holds for each production $\alpha \to \beta$ of $P$ satisfying $\alpha \neq S$. Recall from Section 4 that the class GCSL is defined in terms of *strictly monotone* grammars. A grammar $G = (N, T, S, P)$ is called *weight-increasing* if there is a weight-function $\varphi : N \cup T \to \mathbb{N}$ such that $\varphi(\alpha) < \varphi(\beta)$ holds for each production $\alpha \to \beta$ of $P$ satisfying $(\alpha \to \beta) \neq (S \to \varepsilon)$. In [4] GCSL is characterized as the class of languages that are generated by *weight-increasing* grammars.

A grammar $G = (N, T, S, P)$ is *context-sensitive* if the start symbol does not occur on the right-hand side of any production, and if each production $\alpha \to \beta$ of $P$ satisfying $(\alpha \to \beta) \neq (S \to \varepsilon)$ is of the form $(xAy \to xry)$ for some $x, y, r \in (N \cup T)^*$, $A \in N$, and $r \neq \varepsilon$. It is well-known that the context-sensitive grammars define the same class of languages as the monotone grammars: the class of *context-sensitive languages* CSL.

A context-free grammar $G = (N, T, S, P)$ is called *acyclic* if there does not exist a nonterminal $A \in N$ such that $A \Rightarrow_G^+ A$ holds. For a context-sensitive grammar $G = (N, T, S, P)$, the *context-free kernel* $G' = (N, T, S, P')$ is defined by $P' := \{ (A \to r) \mid \exists x, y \in (N \cup T)^* : (xAy \to xry) \in P \}$. Now a context-sensitive grammar is called *acyclic* if its context-free kernel is acyclic. By ACSL we denote the class of *acyclic context-sensitive languages*, that is, the class of languages that are generated by acyclic context-sensitive grammars. Further, a context-sensitive grammar is called *growing acyclic* if it is strictly monotone. The class of languages generated by these grammars is denoted by GACSL. These definitions date back to [40]. It is known that CFL is properly contained in GACSL [4], and it is easily seen that GACSL $\subseteq$ ACSL $\subseteq$ GCSL holds. However, it was open until recently whether any of these two inclusions is proper.

Based on the techniques used in [47] to develop a normal form for presentations of Church-Rosser languages, it can be shown that, for each weight-increasing grammar, a weight-increasing context-sensitive grammar can be constructed that generates the same language. As a weight-increasing context-sensitive grammar is necessarily acyclic, this gives the following result.

**Theorem 9.** [38] ACSL = GCSL.

However, it remains open whether or not the inclusion GACSL $\subseteq$ ACSL is proper as well.

For the classes of languages characterized by the various types of nondeterministic restarting automata we have the taxonomy presented in Figure 4.

The languages $L_6$ and $L_7$ given in the previous section show that among the classes $\mathcal{L}(R)$, $\mathcal{L}(RR)$, $\mathcal{L}(RW)$, and $\mathcal{L}(RRW)$ only the trivial inclusions hold, and that they are proper. The language $L_2$ is context-free, but not in $\mathcal{L}(RRW)$, and the language $L_{6,2} \cup L_{6,3}$ is not context-free, but it belongs to $\mathcal{L}(R)$ [18]. This shows that CFL is incomparable to all the above classes, and that the inclusion of $\mathcal{L}(R(R)W)$ in $\mathcal{L}(R(R)WW)$ is proper. By Theorem 1 this means that

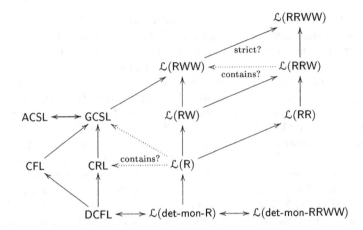

**Fig. 4.** The taxonomy of nondeterministic restarting automata

the language classes $\mathcal{L}(RW)$ and $\mathcal{L}(RRW)$ are not closed under intersection with regular languages. As GCSL is closed under this operation, this theorem also implies that $\mathcal{L}(RW)$ is not contained in GCSL. On the other hand, the language $L_2$ is Church-Rosser, as it can be accepted by a shrinking DTPDA, and so CRL is not contained in $\mathcal{L}(RRW)$, that is, CRL and GCSL are incomparable to $\mathcal{L}(R(R)W)$. However, it remains open whether $\mathcal{L}(R(R))$ is contained in CRL or in GCSL.

Further, it is still open whether the inclusion $\mathcal{L}(RWW) \subseteq \mathcal{L}(RRWW)$ is strict, and whether $\mathcal{L}(RRW)$ is contained in $\mathcal{L}(RWW)$. If $\mathcal{L}(RRW) \subseteq \mathcal{L}(RWW)$, then by Theorem 1 we have $\mathcal{L}(RWW) = \mathcal{L}(RRWW)$, and if $\mathcal{L}(RRW) \not\subset \mathcal{L}(RWW)$, then the inclusion $\mathcal{L}(RWW) \subseteq \mathcal{L}(RRWW)$ is obviously strict, that is, these two open problems are closely related.

## 6  Further Developments and Open Problems

Up to now we have considered the variants of the restarting automaton that are obtained through the following restrictions:

– deterministic versus nondeterministic automata,
– monotone, weakly monotone, or nonmonotone automata,
– automata with or without auxiliary symbols.

Below we address in short some recent results that are obtained by either looking at more fine-grained versions of the above restrictions, or by different variations of the underlying basic model.

### 6.1  Look-Ahead Hierarchies

We start by looking at the size $k$ of the read/write window of the restarting automaton. The influence of the parameter $k$ on the expressive power of restarting automata has been studied in [26] for the various types of restarting automata without auxiliary symbols. The results obtained show that by increasing the size of the read/write window also the expressive power of these types of restarting automata is increased. It remains open, however, whether corresponding results can also be obtained for restarting automata with auxiliary symbols.

## 6.2 Degrees of Weak Monotonicity

Another parameter that allows a quantitative analysis is related to weak monotonicity.

Let $c \in \mathbb{N}_+$, and let $M$ be a restarting automaton. A sequence of cycles $C_1, C_2, \ldots, C_n$ of $M$ is called *weakly c-monotone* if $D_r(C_{i+1}) \leq D_r(C_i) + c$ holds for all $i = 1, \ldots, n-1$. A restarting automaton $M$ is called *weakly c-monotone* if all its computations are weakly $c$-monotone. According to the definition given in Section 5 a restarting automaton $M$ is weakly monotone if it is weakly $c$-monotone for some constant $c$. Now it is possible to distinguish between the different degrees of weak $c$-monotonicity.

For restarting automata without auxiliary symbols it is shown in [27] that the degree of weak monotonicity yields infinite hierarchies for the deterministic as well as for the nondeterministic R-, RR-, RW-, and RRW-automata. However, it is still open whether corresponding results can also be obtained for restarting automata with auxiliary symbols.

## 6.3 Degrees of Monotonicity

Another generalization of monotonicity is discussed in [42]. Let $j \in \mathbb{N}_+$ be a constant, and let $M$ be a restarting automaton. A sequence of cycles $C_1, C_2, \ldots, C_n$ of $M$ is called *j-monotone* if there is a partition of this sequence into at most $j$ subsequences such that each of these subsequences is monotone. Obviously, $C_1, C_2, \ldots, C_n$ is not $j$-monotone if and only if there exist indices $1 \leq i_1 < i_2 < \cdots < i_{j+1} \leq n$ such that $D_r(C_{i_1}) < D_r(C_{i_2}) < \cdots < D_r(C_{i_{j+1}})$ holds. A computation of $M$ is *j-monotone* if the corresponding sequence of cycles is $j$-monotone, and $M$ is called *j-monotone* if all its computations are $j$-monotone. It is shown in [42] that by increasing the value of the parameter $j$, the expressive power of RR- and RRW-automata is increased.

Many variants of the $j$-monotone restarting automaton remain to be studied. Further, it has been conjectured in [42] that, for each $j \geq 1$, it is decidable whether or not a given restarting automaton is $j$-monotone. For 1-monotonicity this is Proposition 1.

## 6.4 Generalizations of Restarting Automata

Here three further generalizations are presented in short. The first one addresses the way in which a restarting automaton reads its tape. Instead of simply scanning the tape from left to right, a *two-way restarting automaton* has move-right and *move-left* transitions [41]. Thus, it has a two-way tape, but as for the models considered before, it is restricted to perform only a single rewrite step between any two restart steps.

It turned out that for the nondeterministic case the two-way tape does not increase the expressive power of restarting automata. For the deterministic variants, however, the two-way tape helps.

The second generalization deals with the number of rewrite steps that a restarting automaton may perform within a single cycle. Motivated by the 'analysis by reduction' each model of restarting automaton considered so far performs exactly one rewrite step in each cycle of each computation. However, as a restarting automaton can be seen as a special type of linear-bounded automaton, it is only natural to also consider variants that perform up to $c$ rewrite steps in each cycle for some constant $c > 1$. How will an increase in the number $c$ influence the expressive power of the various models of restarting automata?

Length-reducing TPDAs are as powerful as shrinking TPDAs (Proposition 4). As the rewrite transitions of a restarting automaton are length-reducing, one might also consider restarting automata for which the rewrite transitions are weight-reducing. For the deterministic case it can be shown that these automata still characterize the class CRL (in the presence of auxiliary symbols), but for the nondeterministic case this variant may actually increase the expressive power.

**Acknowledgement.** The author wants to thank the program committee for inviting him to present this talk. Further, he wants to express his gratitude to Tomasz Jurdziński, František Mráz, Gundula Niemann, and Martin Plátek for many interesting and fruitful discussions regarding restarting automata and related topics. In particular, František Mráz's PhD dissertation [25] has been a very valuable source during the preparation of this paper.

# References

1. Beaudry, M., Holzer, M., Niemann, G., Otto, F.: McNaughton families of languages. Theor. Comput. Sci. **290** (2003) 1581–1628
2. Book, R.V.: Grammars with Time Functions. PhD dissertation. Harvard University, Cambridge, Massachusetts (1969)
3. Book, R.V., Otto, F.: String-Rewriting Systems. Springer, New York (1993)
4. Buntrock, G.: Wachsende kontext-sensitive Sprachen. Habilitationsschrift, Fakultät für Mathematik und Informatik, Universität Würzburg (1996)
5. Buntrock, G., Otto, F.: Growing context-sensitive languages and Church-Rosser languages. Infor. Comput. **141** (1998) 1–36
6. Chytil, M.P., Plátek, M., Vogel, J.: A note on the Chomsky hierarchy. Bulletin of the EATCS **27** (1985) 23–30
7. Dahlhaus, E., Warmuth, M.: Membership for growing context-sensitive grammars is polynomial. J. Comput. Syst. Sci. **33** (1986) 456–472
8. Ehrenfeucht, A., Păun, G., Rozenberg, G.: Contextual grammars and formal languages. In: Rozenberg, G., Salomaa, A. (eds.): Handbook of Formal Languages. Vol. 2. Springer, Berlin Heidelberg New York (1997) 237–293
9. Ehrenfeucht, A., Păun, G., Rozenberg, G.: On representing recursively enumerable languages by internal contextual languages. Theor. Comput. Sci. **205** (1998) 61–83

10. Gladkij, A.W.: On the complexity of derivations for context-sensitive grammars. Algebra i Logika **3** (1964) 29–44 (In Russian)
11. Hopcroft, J.E., Ullman, J.D.: Introduction to Automata Theory, Languages, and Computation. Addison-Wesley, Reading, M.A. (1979)
12. Jančar, P., Mráz, F., Plátek, M.: A taxonomy of forgetting automata. In: Borzyszkowski, A.M., Sokolowski, St. (eds.): MFCS'93, Proc., LNCS 711, Springer, Berlin (1993) 527–536
13. Jančar, P., Mráz, F., Plátek, M.: Forgetting automata and context-free languages. *Acta Infor.* **33** (1996) 409–420
14. Jančar, P., Mráz, F., Plátek, M., Procházka, M., Vogel, J.: Restarting automata, Marcus grammars and context-free languages. In: Dassow, J. Rozenberg, G. and Salomaa, A. (eds.): DLT II, Proc., World Scientific, Singapore (1996) 102–111
15. Jančar, P., Mráz, F., Plátek, M., Vogel, J.: Restarting automata. In: Reichel, H. (ed.): FCT'95, Proc., LNCS 965, Springer, Berlin (1995) 283–292
16. Jančar, P., Mráz, F., Plátek, M., Vogel, J.: On restarting automata with rewriting. In: Păun, G., Salomaa, A. (eds.): New Trends in Formal Languages, LNCS 1218, Springer, Berlin (1997) 119–136
17. Jančar, P., Mráz, F., Plátek, M., Vogel, J.: Monotonic rewriting automata with a restart operation. In: Plášil, F., Jeffery, K.G. (eds.): SOFSEM'97, Proc., LNCS 1338, Springer, Berlin (1997) 505–512
18. Jančar, P., Mráz, F., Plátek, M., Vogel, J.: Different types of monotonicity for restarting automata. In: Arvind, V., Ramanujam, R. (eds.): FSTTCS'98, Proc., LNCS 1530, Springer, Berlin (1998) 343–354
19. Jančar, P., Mráz, F., Plátek, M., Vogel, J.: On monotonic automata with a restart operation. J. Autom. Lang. and Comb. **4** (1999) 287–311
20. Jurdziński, T., Loryś, K.: Church-Rosser languages vs. UCFL. In: Widmayer, P., Triguero, F., Morales, R., Hennessy, M., Eidenbenz, S., Conejo, R. (eds.): ICALP'02, Proc., LNCS 2380, Springer, Berlin (2002) 147–158
21. Jurdziński, T., Loryś, K., Niemann, G., Otto, F.: Some results on RRW- and RRWW-automata and their relationship to the class of growing context-sensitive languages. Mathematische Schriften Kassel, no. 14/01, Fachbereich Mathematik/Informatik, Universität Kassel (2001)
22. Marcus, S.: Contextual grammars and natural languages. In: Rozenberg, G., Salomaa, A. (eds.): Handbook of Formal Languages. Vol. 2. Springer, Berlin Heidelberg New York (1997) 215–235
23. McNaughton, R.: An insertion into the Chomsky hierarchy? In: Karhumäki, J., Maurer, H., Păun, G., Rozenberg, G. (eds.): Jewels are Forever. Springer, Berlin Heidelberg New York (1999) 204–212
24. McNaughton, R., Narendran, P., Otto, F.: Church-Rosser Thue systems and formal languages. J. Assoc. Comput. Mach. **35** (1988) 324–344
25. Mráz, F.: Forgetting and Restarting Automata. PhD thesis, Charles University, Prague (2001)
26. Mráz, F.: Lookahead hierarchies of restarting automata. J. Autom. Lang. and Comb. **6** (2001) 493–506
27. Mráz, F., Otto, F.: Hierarchies of weakly monotone restarting automata. In preparation.
28. Mráz, F., Plátek, M., Procházka, M.: Restarting automata, deleting, and Marcus grammars. In: Martín-Vide, C., Păun, G. (eds.): Recent Topics in Mathem. and Comput. Linguistics. Editura Academiei Române, Bukarest (2000) 218–233
29. Narendran, P.: Church-Rosser and related Thue systems. PhD thesis, Rensselaer Polytechnic Institute, Troy, New York (1984)

30. Niemann, G.: Regular Languages and Church-Rosser congruential languages. In: Freund, R., Kelemenová, A. (eds.): Grammar Systems 2000, Proc., Silesian University, Opava (2000) 359–370

31. Niemann, G.: Church-Rosser languages and related classes. Dissertation, Universität Kassel (2002)

32. Niemann, G., Otto, F.: The Church-Rosser languages are the deterministic variants of the growing context-sensitive languages. In: Nivat, M. (ed.): FoSSaCS'98, Proc., LNCS 1378, Springer, Berlin (1998) 243–257

33. Niemann, G., Otto, F.: Confluent internal contextual languages. In: Martin-Vide, C., Păun, G. (eds.): Recent Topics in Mathematical and Computational Linguistics. The Publishing House of the Romanian Academy, Bucharest (2000) 234–244

34. Niemann, G., Otto, F.: Restarting automata, Church-Rosser languages, and representations of r.e. languages. In: Rozenberg, G., Thomas, W. (eds.): DLT'99, Proc., World Scientific, Singapore (2000) 103–114

35. Niemann, G., Otto, F.: On the power of RRWW-automata. In: Ito, M., Păun, G., Yu, S. (eds.): Words, Semigroups, and Transductions. World Scientific, Singapore (2001) 341–355

36. Niemann, G., Otto, F.: Further results on restarting automata. In: Ito, M., Imaoka, T. (eds.): Words, Languages and Combinatorics III, Proc., World Scientific, Singapore (2003), to appear

37. Niemann, G., Waldmann, J.: Some regular languages that are Church-Rosser congruential. In: Kuich, W., Rozenberg, G., Salomaa, A. (eds.): DLT'01, Proc., LNCS 2295, Springer, Berlin (2002) 330–339

38. Niemann, G., Woinowski, J.R.: The growing context-sensitive languages are the acyclic context-sensitive languages. In: Kuich, W., Rozenberg, G., Salomaa, A. (eds.): DLT'01, Proc., LNCS 2295, Springer, Berlin (2002) 197–205

39. Otto, F., Katsura, M., Kobayashi, Y.: Infinite convergent string-rewriting systems and cross-sections for finitely presented monoids. J. Symbol. Comput. **26** (1998) 621–648

40. Parikh, R.J.: On context-free languages. J. Assoc. Comput. Mach. **13** (1966) 570–581

41. Plátek, M.: Two-way restarting automata and j-monotonicity. In: Pacholski, L., Ružička, P. (eds.): SOFSEM'01, Proc., LNCS 2234, Springer, Berlin (2001) 316–325

42. Plátek, M., Mráz, F.: Degrees of (non)monotonicity of RWW-automata. In: Dassow, J., Wotschke, D. (eds.): Preproceedings of the 3rd Workshop on Descriptional Complexity of Automata, Grammars and Related Structures. Report No. 16, Fakultät für Informatik, Universität Magdeburg (2001) 159–165

43. Rovan, B.: A framework for studying grammars. In: Gruska, I., Chytil, M. (eds.): MFCS'81, Proc., LNCS 118, Springer, Berlin (1981) 473–482

44. Solms, S.H. von: The characterization by automata of certain classes of languages in the context sensitive area. Infor. Control **27** (1975) 262–271

45. Straňáková, M.: Selected types of pg-ambiguity. The Prague Bulletin of Mathematical Linguistics **72** (1999) 29–57

46. Straňáková, M.: Selected types of pg-ambiguity: Processing based on analysis by reduction. In: Sojka, P., Kopeček, I., Pala, K. (eds.): Text, Speech and Dialogue, 3rd Int. Workshop, TSD 2000, Proc., LNCS 1902, Springer, Berlin (2000) 139–144

47. Woinowski, J.: A normal form for Church-Rosser language systems. In: Middeldorp, A. (ed.): RTA'01, Proc., LNCS 2051, Springer, Berlin (2001) 322–337

48. Woinowski, J.: Church-Rosser languages and their application to parsing problems. Dissertation, Technische Universität Darmstadt (2001)

# Test Sets for Large Families of Languages

Wojciech Plandowski

Institute of Informatics
University of Warsaw
Banacha 2, 02-097 Warszawa, Poland
`wojtekpl@mimuw.edu.pl`.

**Abstract.** We study the lower and upper bounds for sizes of test sets for the families of all languages, of commutative languages, of regular languages and of context-free languages.

## 1 Introduction

A *morphism* is a function $f : \Sigma^* \to \Delta^*$ such that $f(uv) = f(u)f(v)$ for each $u$, $v \in \Sigma^*$. A subset $T$ of a language $L \subseteq \Sigma^*$ is a *test set* for $L$ iff for each two morphisms $f, g : \Sigma^* \to \Delta^*$

$$\forall u \in T \ f(u) = g(u) \Rightarrow \forall u \in L \ f(u) = g(u).$$

A famous Ehrenfeucht's Conjecture states that

**Theorem 1 (Ehrenfeucht's Conjecture).** *Each language over a finite alphabet possesses a finite test set.*

The conjecture was proved in 1985 independently by Albert and Lawrence [2] and by Guba [7].

    The Ehrenfeucht's Conjecture can be stated in other monoids than free. The only thing we have to generalize to all monoids is the notion of a morphism. A morphism is now such a function $f : \Sigma^* \to M$, where $M$ is a monoid, that satisfies $f(uv) = f(u)f(v)$ for each $u, v \in \Sigma^*$. The conjecture for some monoids holds true and for other monoids do not hold true [4,12,13,18,21]. In this paper we are interested in free monoids and free groups. The conjecture is then true. However the question of a size of a test set is not completely answered. The aim of this survey is to present existing nontrivial constructions which lead to lower and upper bounds for size of test sets for large families of languages. The families we consider are: the family of all languages, the family of commutative languages, the family of regular languages and the family of context-free languages. In the litterature there is also a construction for a large subfamily of context-sensitive languages called $Q$-rational languages [30]. Unfortunatelly, the construction (not a proof of its correctness) is trivial: a test set for a $Q$-rational language $L$ is a subset of $L$ consisting of words whose length is bounded by polynomial function of parameters of the description of $L$. The sequence of $Q$-rational languages which leads to best known lower bound for the size of a test set for a $Q$-rational

Z. Ésik and Z. Fülöp (Eds.): DLT 2003, LNCS 2710, pp. 75–94, 2003.

languages is the same as the one for general languages. This is why we omit the family of $Q$-rational languages in our considerations.

The author would like to point out that although the paper is a survey it contains new considerations, i.e. the tight bound for size of a test set for regular languages, and the considerations on building large independent systems of equations in free monoid.

## 2   Preliminaries

Denote by 1 the neutral element of any monoid. By a *word* we mean any element of a free monoid. In particular, 1 is the empty word. Denote by $|w|$ the length of a word $w$.

A language $L$ is *commutative* iff if $w \in L$, then each permutation of letters of $w$ is in $L$, too.

We say that two morphisms $f, g : \Sigma^* \to \Delta^*$ are *equivalent* on a language $L \subseteq \Sigma^*$ iff $\forall u \in L$, $f(u) = g(u)$.

Let $\Theta$ be an alphabet of variables. An *word equation* is a pair $(u, v) \in \Theta^* \times \Theta^*$ usually denoted as $u = v$. A *solution* of a word equation in a monoid $M$ is a morphism $h : \Theta^* \to M$ such that $h(u) = h(v)$. A *system of word equations* is any set of word equations. A *solution of a system of word equations* is a morphism which is a solution of all equations in the system. The set of solutions of a system $\mathcal{S}$ is denoted by $Sol(\mathcal{S})$. Two systems $\mathcal{S}_1$ and $\mathcal{S}_2$ are *equivalent* iff $Sol(\mathcal{S}_1) = Sol(\mathcal{S}_2)$. An *independent* system of word equations is such a system $\mathcal{S}$ that for each equation $e \in \mathcal{S}$, $Sol(\mathcal{S}) \neq Sol(\mathcal{S} - \{e\})$. In other words no proper subsystem of $\mathcal{S}$ is equivalent to the whole system.

For introduction to theory of automata we refer to [14]. For introduction to combinatorics of words we refer to [5,22,23].

## 3   Test Sets in the Family of All Languages

We are interested in sizes of minimal test sets for a language. We say that a language is a *test set language* if no proper subset of it is a test set for it. As an easy consequence of Ehrenfeucht's Conjecture we have.

**Corollary 1.**   *1. Each test set language is finite.*
  *2. Each test set for a language $L$ contains a test set language which is a test set for $L$.*
  *3. Each language contains a test set which a test set language.*

Corollary 1 says that the family all test sets for a given language is fully defined by test sets which are test set languages.

Let $EC(n)$ be the the size of a maximal test set language over $n$-letter alphabet. It can happen that $EC(n)$ is unbounded because there is a family of bigger and bigger test set languages over $n$-letter alphabet.

Let $GEC(n)$ be the size of a maximal independent system of equations in $n$ variables. Our next lemma gives the connection between functions $EC$ and $GEC$.

**Lemma 1 ([6]).**

(i) $GEC(n) \leq EC(2n)$
(ii) $EC(n) \leq GEC(2n)$

*Proof.* (i) Let $\mathcal{S}_n = \{u_1 = v_1, u_2 = v_2, \ldots, u_k = v_k\}$ be a maximal independent system of $GEC(n)$ equations in $n$ variables $X = \{x_1, \ldots, x_n\}$. Let $\Sigma = \{x_1, \ldots, x_n, \bar{x}_1, \ldots, \bar{x}_n\}$ be an alphabet of $2n$ symbols. We extend the operator $\bar{\ }$ onto words over $\{x_1, \ldots, x_n\}$ in a morphic manner $\overline{uv} = \bar{u} \cdot \bar{v}$.

*Claim.* The language

$$TS = \{u_1 \bar{v}_1, u_2 \bar{v}_2, \ldots, u_k \bar{v}_k\}$$

is a test set language.

*Proof.* Suppose that $TS - \{u_i \bar{v}_i\}$ is a test set for $TS$. We will prove that, then, $\mathcal{S}_n$ is not independent by proving that if the system $\mathcal{S}_n - \{u_i = v_i\}$ is satisfied then the equation $u_i = v_i$ is satisfied, too. Let $f : X^* \to \Delta^*$ be a solution of $\mathcal{S}_n - \{u_i = v_i\}$. Define two morphisms $h, g : \Sigma^* \to \Delta^*$ in the following way

$$h(x) = \begin{cases} f(x) & \text{if } x \in X \\ 1 & \text{otherwise} \end{cases}, \quad g(x) = \begin{cases} 1 & \text{if } x \in X \\ f(x) & \text{otherwise} \end{cases}$$

Then, $h(u_j \bar{v}_j) = f(u_j) = f(v_j) = g(u_j \bar{v}_j)$ for all $j \neq i$. Hence, the morphisms $h$, and $g$ are equivalent on $TS - \{u_i = v_i\}$. Since $TS - \{u_i = v_i\}$ is a test set for $TS$ they have to be equivalent on $u_i = v_i$, ie. $f(u_i) = h(u_i \bar{v}_i) = g(u_i \bar{v}_i) = f(v_i)$.

The rest of the proof is simple. The language $TS$ is a test set language consiting of $GEC(n)$ words. Hence, $GEC(n) \leq EC(2n)$.

(ii) Let $TS_n = \{u_1, u_2, \ldots, u_k\}$ be a test set language containing $EC(n)$ words over $n$-letter alphabet $\Sigma = \{a_1, \ldots, a_n\}$. Let $X = \{a_1, \ldots, a_n, \bar{a}_1, \ldots, \bar{a}_n\}$ be a set of $2n$ word variables. Now, we extend the operator $\bar{\ }$ onto words over $\{a_1, \ldots, a_n\}$ in a morphic manner $\overline{uv} = \bar{u} \cdot \bar{v}$.

*Claim.* The system of equations

$$\mathcal{S} = \{u_1 = \bar{u}_1, u_2 = \bar{u}_2, \ldots, u_k = \bar{u}_k\}$$

is independent.

*Proof.* Suppose that $\mathcal{S}$ is not independent. Then, for some $i$, if the system $\mathcal{S} - \{u_i = \bar{u}_i\}$ is satisfied then the equation $u_i = \bar{u}_i$ is satisfied, too. We will prove that then $TS_n$ is not a test set language or more precisely that $TS_n - \{u_i\}$ is a test set for $TS_n$. Let $h, g : \Sigma^* \to \Delta^*$ be two morphisms that are equivalent on $TS_n - \{u_i\}$. Define a substitution $f : X \to \Delta^*$ in the following way

$$f(x) = \begin{cases} h(x) & x \in \Sigma \\ g(y) \text{ where } x = \bar{y} & \text{otherwise} \end{cases}$$

Then $f(u_j) = h(u_j) = g(u_j) = f(\bar{u}_j)$ for $i \neq j$. Hence, $f$ is a solution of $\mathcal{S} - \{u_i = \bar{u}_i\}$ so, by the assumption, $f(u_i) = f(\bar{u}_i)$. Hence, $h(u_i) = f(u_i) = f(\bar{u}_i) = g(u_i)$, ie. $h$ and $g$ are equivalent on $u_i$.

Now the rest of the proof is simple. $\mathcal{S}$ is the system of $EC(n)$ independent equations over $2n$ variables. Hence, $EC(n) \leq GEC(2n)$.

## 3.1   Upper Bound

Our proof of the upper bound which is given by the Ehrenfeucht's Conjecture follows the lines of the proof by Victor Guba [7].

By Lemma 1 to prove the Ehrenfeucht's Conjecture it is enough to prove the following theorem.

**Theorem 2 (Generalized Ehrenfeucht's Conjecture).** *Each infinite system of word equations over finite set of variables contains a finite subsystem which is equivalent to the whole system.*

All existing proofs of Theorem 2 [2,7,27,28] use the following Hilbert's Basis Theorem whose proof can be found in most standard books on algebra.

**Theorem 3 (Hilbert's Basis Theorem).** *Let $f_i$ be a family of polynomials in finite number of variables and with integer coefficients. The family contains a finite subfamily $g_1, \ldots g_k$ such that for each $i$ there are polynomials $h_1, \ldots, h_k$ with integer coefficients such that*

$$f_i = h_1 \cdot g_1 + \ldots + h_k \cdot g_k.$$

As a simple corollary we have

**Corollary 2.** *Let $f_i = 0$ be a system of polynomial equations over finite set of variables. There is a finite subsystem of it (e.g. $g_1 = 0$, $g_2 = 0$, $\ldots g_k = 0$) which is equivalent to the whole system.*

Guba's proof of the Generalized Ehrefeucht's Conjecture uses an embedding of free monoids into a monoid $M$ of $2 \times 2$ matrices over integers. Let $\Sigma = \{0, 1\}$. Such an embedding is generated by the following morphism $\mu : \Sigma^* \to M$:

$$\mu(0) = \begin{pmatrix} 2 & 0 \\ 0 & 1 \end{pmatrix}, \ \mu(1) = \begin{pmatrix} 2 & 1 \\ 0 & 1 \end{pmatrix}$$

Denote by $[w]_2$ an integer whose binary expansion is $w$. Recall that $w^R$ is a reverse of a word $w$. Easy induction on $|w|$ proves the following lemma.

**Lemma 2.** *Let $w \in \Sigma^*$. Then*

$$\mu(w) = \begin{pmatrix} 2^{|w|} & [w^R]_2 \\ 0 & 1 \end{pmatrix}$$

Lemma 2 proves that the morphism $\mu$ is injective so it is an embedding. Now the procedure of finding a finite subsystem equivalent to a system of word eqautions is simple:

- Take a system of word equations $\mathcal{S} = \{e_i\}_{i \in I}$.
- Translate each word equation $e_i$ into an integer equation in the following way:
  - Replace each variable $x$ by $\begin{pmatrix} x_1 & x_2 \\ 0 & 1 \end{pmatrix}$ where $x_1$ and $x_2$ are integer variables corresponding to word variable $x$.

- Perform multiplication of $2 \times 2$ matrices on the left and right hand sides of $e_i$.
- Equalize the coefficients of matrices from left and right hand sides. In this way you obtain two nontrivial integer polynomial equations $P_i = P'_i$ and $Q_i = Q'_i$ and two trivial ones, namely $0 = 0$ and $1 = 1$.

– In the previous step you obtained a system of integer polynomial equations $\mathcal{T} = \{P_i - P'_i = 0, Q_i - Q'_i = 0\}_{i \in I}$. Using corollary from Hilbert's Basis Theorem find in this system a finite equivalent subsystem

$$\mathcal{T}' = \{P_{i_1} - P'_{i_1} = 0, \ldots, P_{i_k} - P'_{i_k} = 0, Q_{j_1} - Q'_{j_1} = 0, Q_{j_l} - Q'_{j_l} = 0\}.$$

– The finite system equivalent to $\mathcal{S}$ is

$$\mathcal{S}' = \{e_{i_1}, \ldots, e_{i_k}, e_{j_1}, \ldots, e_{j_l}\}.$$

To prove that $\mathcal{S}'$ is equivalent to $\mathcal{S}$ it is enough to prove that each solution of $\mathcal{S}'$ is a solution of $\mathcal{S}$. Take a solution $h$ of $\mathcal{S}'$. Then the substitution $x_1 = 2^{|h(x)|}$, $x_2 = [h(x)]_2$, for each word variable $x$, is a solution of $\mathcal{T}'$ so it is also a solution of equivalent system $\mathcal{T}$. Suppose that $h$ is not a solution of a word equation $e_i : u_i = v_i$ of $\mathcal{S}$. This means that $h(u_i) \neq h(v_i)$ and since $\mu$ is an embedding it means that $\mu(h(u_i)) \neq \mu(h(v_i))$. Hence, the substitution $x_1 = 2^{|h(x)|}$, $x_2 = [h(x)]_2$ is not a solution of $P_i - P'_i = 0$ or $Q_i - Q'_i = 0$ being the equations in $\mathcal{T}$. A contradiction.

The proof of Generalized Ehrenfeucht's Conjecture for free groups is similar but the proof that a morphism is an embedding is more complicated. According to Sanov Theorem [17] subgroup of the group of invertible $2 \times 2$ integer matrices generated by matrices

$$\begin{pmatrix} 1 & 2 \\ 0 & 1 \end{pmatrix}, \begin{pmatrix} 1 & 0 \\ 2 & 1 \end{pmatrix}$$

is free. This gives the necessary embedding.

An extension of Generalized Ehrenfeucht's Conjecture is considered in [29].

## 3.2  Lower Bound

We start from lower bound in free monoids. We omit here the proof of the following easy lemma.

**Lemma 3.** *There are four words $s, p_1, p_2, p_3$ such that*

$$p_1 p_2 p_3 = s$$
$$p_1^2 p_2^2 p_3^2 = s^2$$
$$p_1^3 p_2^3 p_3^3 \neq s^3$$

**Theorem 4 ([18]).**

(i)  $EC(n) = \Omega(n^4)$.
(ii)  $GEC(n) = \Omega(n^4)$.

*Proof.* Point (i) is a consequence of point (ii) and Lemma 1. It remains to prove point (ii). We will show that there is a family of independent systems of equations $\mathcal{S}_n$ over $\Theta(n)$ variables such that the number of equations in $\mathcal{S}_n$ is $\Theta(n^4)$. The system $\mathcal{S}_n$ is over $10n$ variables $x_i$, $y_i$, $\bar{y}_i$, $\tilde{y}_i$, $u_i$, $\bar{u}_i$, $\tilde{u}_i$, $z_i$, $\bar{z}_i$, $\tilde{z}_i$, for $i = 1..n$ and consists of $n^4$ equations

$$e_{i,j,k,l} : x_i y_j u_k z_l \bar{y}_j \bar{u}_k \bar{z}_l \tilde{y}_j \tilde{u}_k \tilde{z}_l = y_j u_k z_l \bar{y}_j \bar{u}_k \bar{z}_l \tilde{y}_j \tilde{u}_k \tilde{z}_l x_i, \text{ for } i = 1..n.$$

The system $\mathcal{S}_n$ is independent since the substitution

$$x_i = \begin{cases} \varepsilon \text{ if } i \neq i' \\ s \text{ if } i = i' \end{cases}$$

$$y_i = \bar{y}_i = \tilde{y}_i = \begin{cases} \varepsilon & \text{if } i \neq j' \\ p_1 & \text{if } i = j' \end{cases}$$

$$u_i = \bar{u}_i = \tilde{u}_i = \begin{cases} \varepsilon & \text{if } i \neq k' \\ p_2 & \text{if } i = k' \end{cases}$$

$$z_i = \bar{z}_i = \tilde{z}_i = \begin{cases} \varepsilon & \text{if } i \neq l' \\ p_3 & \text{if } i = l' \end{cases}$$

where $s$, $p_1$, $p_2$, $p_3$ are as in Lemma 3, is a solution of $\mathcal{S}_n - \{e_{i',j',k',l'}\}$ but not a solution of $\mathcal{S}_n$. Indeed, all equations in $\mathcal{S}_n - \{e_{i',j',k',l'}\}$ under the considered substitution are of the form

$$s p_1^i p_2^i p_3^i = p_1^i p_2^i p_3^i s$$

for some $i \leq 2$ and by the property of words $s$ and $p_1^i p_2^i p_3^i$ (where $i \leq 2$) they are satisfied. The equation $e_{i',j',k',l'}$ under this substitution is in form

$$s p_1^3 p_2^3 p_3^3 = p_1^3 p_2^3 p_3^3 s$$

i.e. $p_1^3 p_2^3 p_3^3$ commutes with $s$, i.e. $p_1^3 p_2^3 p_3^3$ and $s$ are powers of the same word. Since the length of $p_1^3 p_2^3 p_3^3$ is $3|s|$, it would mean $p_1^3 p_2^3 p_3^3 = s^3$ which is not true.

Note, that the above construction gives a simple method of generating bigger and bigger independent systems of equations. The procedure is simple:

– Find words $s$, $p_i$, for $i = 1..k$, such that

$$p_1^i p_2^i \ldots p_k^i = s^i, \text{for } i < t \tag{1}$$
$$p_1^t p_2^t \ldots p_k^t \neq s^t \tag{2}$$

– Construct an independent system $\mathcal{S}_n$ of $n^{t+1}$ equations over $(k \cdot t + 1) \cdot n$ variables $x_i$, for $i = 1..n$, $y_{q,r}^p$, for $p = 1..k$, $q = 1..t$, $r = 1..n$:

$$e_{i,j_1,\ldots,j_t} : x_i y_{1,j_1}^1 y_{2,j_2}^1 \cdots y_{t,j_t}^1 y_{1,j_1}^2 y_{2,j_2}^2 \cdots y_{t,j_t}^2 \cdots y_{1,j_1}^k y_{2,j_2}^k \cdots y_{t,j_t}^k =$$
$$y_{1,j_1}^1 y_{2,j_2}^1 \cdots y_{t,j_t}^1 y_{1,j_1}^2 y_{2,j_2}^2 \cdots y_{t,j_t}^2 \cdots y_{1,j_1}^k y_{2,j_2}^k \cdots y_{t,j_t}^k x_i$$

for $i, j_1, j_2, \ldots, j_t = 1..n$.

The only obstacle is that the author believes in the following conjecture.

*Conjecture 1.* Let $s$ and $p_i$, for $i = 1..k$ be words. If

$$p_1^i p_2^i \ldots p_k^i = s^i, \text{ for } i \le 3,$$

then

$$p_1^i p_2^i \ldots p_k^i = s^i, \text{ for } i \ge 4.$$

The Conjecture has been almost proved.

**Theorem 5 ([15]).** *If*

$$p_1^i p_2^i \ldots p_k^i = s^i, \text{ for } i \le 4,$$

*then*

$$p_1^i p_2^i \ldots p_k^i = s^i, \text{ for } i \ge 5.$$

This means that the procedure of generating bigger and bigger systems of independent equations cannnot be performed for $t \ge 4$ so this way of bulding independent systems of equations is not capable to build independent systems of size $w(n^5)$ in particular of exponential size.

The procedure of generating polynomial size systems of independent equations can be based on some other systems of equations. For instance, in the litterature [1,9,10,20], there are several results on a system containing equations of the form

$$x_0 u_1^i x_1 u_1^i \ldots u_n^i x_n = y_0 v_1^i y_1 \ldots v_m^i y_m.$$

Suppose we are able to find $k$ such that

$$x_0 u_1^i x_1 u_1^i \ldots u_s^i x_s = y_0 v_1^i y_1 \ldots v_t^i y_t, \text{ for } i < k$$

and

$$x_0 u_1^k x_1 u_1^k \ldots u_s^k x_s \ne y_0 v_1^k y_1 \ldots v_t^k y_t.$$

Then the following system of $n^k$ equations over $(ks + kt)n + (s + t + 2)$ variables

$$\{x_i : i = 0..s\} \cup \{u_{p,q}^r : p = 1..k, q = 1..n, r = 1..s\} \cup$$

$$\{y_i : 0..t\} \cup \{v_{p,q}^r : p = 1..k, q = 1..n, r = 1..t\}$$

is independent

$$e_{j_1,\ldots,j_k} : x_0 u_{1,j_1}^1 u_{2,j_2}^1 \ldots u_{k,j_k}^1 x_1 \ldots x_{s-1} u_{1,j_1}^s u_{2,j_2}^s \ldots u_{k,j_k}^s x_s =$$
$$y_0 v_{1,j_1}^1 v_{2,j_2}^1 \ldots v_{k,j_k}^1 y_1 \ldots y_{t-1} v_{1,j_1}^t v_{2,j_2}^t \ldots v_{k,j_k}^1,$$

for $j_1, \ldots, j_k = 1..n$.

There are extensions and generalizations of Theorem 5 to some families of context-sensitive languages [8,30].

The lower bound for the problem in free groups matches the upper bound, namely $GEC(n)$ for $n \geq 6$ is unbounded. Denote $[a, b] = a^{-1}b^{-1}ab$ and

$$[a_1, a_2, \ldots, a_k] = [a_1, [a_2, [\ldots, [a_{k-1}, a_k]].$$

Of course, using formula for $[a, b]$ we can write $[a_1, a_2, \ldots, a_k]$ as a product of $a_i$ and $a_i^{-1}$. An easy induction on $k$ shows, that in this product two neighbouring symbols are different although one or both of them can be inversed. Note also that $[a, 1] = [1, a] = 1$ for any $a$ and consequently $[a_1, \ldots, a_{i-1}, 1, a_{i+1}, \ldots, a_k] = 1$, for any $a_j$.

**Theorem 6 ([3]).**

*(i)* $GEC(n)$ *is unbounded, for* $n \geq 6$.
*(ii)* $EC(n)$ *is unbounded, for* $n \geq 12$.

*Proof.* Point (ii) is a consequence of point (i) and Lemma 1. It remains to show point (i).

Consider the following system $\mathcal{S}_k$ of $k$ equations over three variables $x$, $y$, $z$

$$e_i : [v_1, \ldots, v_{i-1}, v_{i+1}, \ldots, v_k] = 1,$$

where $v_i = x^i z y^i x^{-i}$. The system $\mathcal{S}_k$ is independent since the substitution $x = b$, $y = a$, $z = a^i$ is a solution of $\mathcal{S}_k - e_i$ but not $\mathcal{S}_k$. Indeed, each of $e_j$ for $j \neq i$ contains $v_i = b^i a^{-i} a^i b^{-i} = 1$ so it is satisfied. The left hand side of the equation $e_i$ consists of $v_j$ and $v_j^{-1}$, for $j \neq i$. By our remark neighbouring symbols in the expression for $[v_1, \ldots, v_{i-1}, v_{i+1}, \ldots, v_k]$ are different so they are in one of the forms: $v_j^{e_1} v_l^{e_2}$ where $j \neq l$ and $e_1, e_2 \in \{1, -1\}$. Under the considered substitution the blocks of $a$ in $v_l$ and $v_k$ cannot by reduced so the whole expression cannot reduce to 1 so $e_i$ is not satisfied.

The system $\mathcal{S}_k$ over three variables consists of $k$ equations and is independent. This would suggest that already $GEC(3)$ is unbounded. Remember however that we used additional operation - inversion. To omit this problem we add three new variables $\bar{x}$, $\bar{y}$, $\bar{z}$ and in $\mathcal{S}_k$ occurrences of $x^{-1}$ we replace by $\bar{x}$, occurrences of $y^{-1}$ by $\bar{y}$ and occurrences of $z^{-1}$ by $\bar{z}$. In this way we obtain an independent system $\mathcal{S}'_k$ over six variables which does not contain the inverse.

## 4    Test Sets for Commutative Languages

### 4.1    Upper Bound

Let $w$ be a word over $\Sigma = \{a_1, \ldots, a_n\}$. Denote by $|w|_a$ the number of occurrences of $a$ in $w$. A *Parikh vector* of $w$ is a vector $\Phi(w) = (|w|_{a_1}, |w|_{a_2}, \ldots, |w|_{a_n})$. Let $L$ be a language. Denote by $\Phi(L)$ the set of vectors $\{\Phi(w) : w \in L\}$. A *basis* of the language $L$ is the set of words $B \subseteq L$ such that

- each vector in $\Phi(L)$ is a linear combination of vectors in $\Phi(B)$,
- vectors in $\Phi(B)$ are linearly independent.

We say that two morphisms $f$ and $g$ are *length equivalent* on a language $L$ if

$$|f(w)| = |g(w)|, \text{ for all } w \in L.$$

**Lemma 4 ([11]).** *Two morphisms are length equivalent on $L$ iff they are length equivalent on any basis of $L$.*

*Proof.* Let $\Sigma = \{a_1, \ldots, a_n\}$ be an alphabet and $h, g : \Sigma^* \to \Delta^*$ be two morphisms that are length equivalent on a basis $B$ of $L$. Denote $r_i = |h(a_i)|$ and $s_i = |g(a_i)|$. Then for each $v$ in $B$ $\Phi(v)(r_1, \ldots, r_n)^T = |h(v)| = |g(v)| = \Phi(v)(s_1, \ldots, s_n)^T$ where $T$ is a transposition operator. Take $u \in L$. Since $B$ is a basis of $L$ there are real numbers $\alpha_v$ such that $\Phi(u) = \sum_{v \in B} \alpha_v \Phi(v)$. Then

$$|h(u)| = \Phi(u)(r_1, \ldots, r_n)^T = \sum_{v \in B} \alpha_v \Phi(v)(r_1, \ldots, r_n)^T =$$

$$\sum_{v \in B} \alpha_v \Phi(v)(s_1, \ldots, s_n)^T = \Phi(u)(s_1, \ldots, s_n)^T = |g(u)|.$$

The construction is based on two lemmas. We omit their proofs. What is important is that they are originally proved for free monoids. The first one is true also for free groups [26]. The second one probably is also true for free groups [26] but its proof is much more complicated than the one for free monoids.

**Lemma 5 ([11]).** *Let $x, y, z, \bar{x}, \bar{y}$ and $\bar{z}$ be words such that*

$$\begin{cases} xy = \bar{x}\bar{y}, \ xz = \bar{x}\bar{z}, \ yz = \bar{y}\bar{z} \\ yx = \bar{y}\bar{x}, \ zx = \bar{z}\bar{x}, \ zy = \bar{z}\bar{y} \end{cases}.$$

*Then either $x = \bar{x}$ and $y = \bar{y}$ and $z = \bar{z}$ or $x, y, z, \bar{x}, \bar{y}, \bar{z}$ are powers of the same word.*

**Lemma 6 ([11]).** *Let $x, y, z, \bar{x}, \bar{y}$ and $\bar{z}$ be words such that $x\bar{x} \neq 1$, $y\bar{y} \neq 1$ and $z\bar{z} \neq 1$ and*

$$\begin{cases} xyz = \bar{x}\bar{y}\bar{z}, \ xzy = \bar{x}\bar{z}\bar{y} \\ yzx = \bar{y}\bar{z}\bar{x}, \ yxz = \bar{y}\bar{x}\bar{z} \\ zyx = \bar{z}\bar{y}\bar{x}, \ zxy = \bar{z}\bar{x}\bar{y} \end{cases}.$$

*Then either $x = \bar{x}$ and $y = \bar{y}$ and $z = \bar{z}$ or $x, y, z, \bar{x}, \bar{y}, \bar{z}$ are powers of the same word.*

Note that above two lemmas say something about properties of two small fixed systems of word equations.

Let $L$ be a commutative language over $\Sigma$. Define a language $L_{a,b}$ for two different symbols $a, b$ of $\Sigma$ in the following way.

- Case A: there is no word $z \in L$ which contains both letters $a$ and $b$. Then we define $L_{a,b} = \emptyset$.

- Case B: there is a word $z \in L$ such that $|z|_a \geq 2$ and $|z|_b \geq 1$. Then we take any word $x$ such that $a^2bx \in L$ and we define

$$L_{a,b} = \{ab(ax), ba(ax), a(ax)b, b(ax)a, (ax)ab, (ax)ba\}$$

- Case C: there is a word $z \in L$ such that $|z|_b \geq 2$ and $|z|_a \geq 1$. Then we take a word $x$ such that $ab^2x \in L$ and define

$$L_{a,b} = \{ab(bx), ba(bx), a(bx)b, b(bx)a, (bx)ab, (bx)ba\}.$$

- Case D: there is no word in $z \in L$ such that $|z|_a \geq 2$ and $|z|_b \geq 1$ or $|z|_a \geq 1$ and $|z|_b \geq 2$ but there is a word in $L$ which contains both letters $a$ and $b$. Then we take a word $x$ such that $abx \in L$ and define

$$L_{a,b} = \{abx, bax, axb, bxa, xab, xba\}.$$

Cases B and C are not disjoint. If we can apply both we apply arbitrary.

Let $B$ be a basis of $L$. Then we define

$$T_L = B \cup \bigcup_{a,b \in \Sigma, a \neq b} L_{a,b}$$

Clearly, the size of $T_L$ is bounded by $6 \binom{n}{2} + n = 3n^2 - 2n$. We omit the proof of the following lemma.

**Lemma 7 ([11]).** $T_L$ is a test set for $L$.

As an easy consequence we get.

**Theorem 7 ([11]).** *Each commutative language over n-letter alphabet contains a test set of size $O(n^2)$.*

## 4.2   Lower Bound

We define the family of commutative languages $L_n$ over $3n$-letter alphabet $\{a_1, \ldots, a_n, b_1, \ldots, b_n, c_1, \ldots, c_n\}$ as

$$L_n = \{a_i b_j c_j : i = 1..n, j = 1..n\}$$

We are not able to prove that it is a test set language. Probably it is not. We will prove however that no of its subset $Y$ of size $< n^2$ can be a test set of it. Indeed take $Y \subseteq L_n$ of size $< n^2$. Then there are $i'$, $j'$ such that no permutation of $a_{i'} b_{j'} c_{j'}$ is in $Y$. Define two morphisms $f$, $g$ in the following way

$$f(a_i) = \begin{cases} a & \text{if } i \neq i' \\ b & \text{if } i = i' \end{cases}, \quad f(b_i) = \begin{cases} a & \text{if } i \neq j' \\ a^2 & \text{if } i = j' \end{cases}, \quad f(c_i) = a, \text{for all } i,$$

$$g(a_i) = \begin{cases} a & \text{if } i \neq i' \\ b & \text{if } i = i' \end{cases}, \quad g(b_i) = a, \text{for all } i, \quad g(c_i) = \begin{cases} a & \text{if } i \neq j' \\ a^2 & \text{if } i = j' \end{cases}.$$

Then, if $i \neq i'$ and $j \neq j'$, then for any permutation $x$ of $a_i b_j c_j$, we have $f(x) = a^3 = g(x)$. If $j \neq j'$, then, for any permutation $x$ of $a_{i'} b_j c_j$, we have $f(x) = g(x) \in \{baa, aba, aab\}$. If $i \neq i'$, then, for any permutation $x$ of $a_i b_{j'} c_{j'}$, we have $f(x) = a^4 = g(x)$. Hence, the morphisms are equivalent on $Y$, but they are not equivalent on $L_n$ since

$$f(b_{j'} a_{i'} c_{j'}) = a^2 ba \neq aba^2 = g(b_{j'} a_{i'} c_{j'}).$$

**Theorem 8** ([11]). *For each $n$, there is a commutative language over $3n$-letter alphabet containing $6n^2$ words and such that the smallest test set for it is of size $n^2$.*

## 5   Test Sets for Regular Languages

### 5.1   Upper Bound

We start by proving a simple fact which is the base of our further consideration in case of regular languages.

**Lemma 8.** *If two morphisms $f$, $g$ are equivalent on the set $\{w_1 w_2, w_1 u_2, u_1 w_2\}$, then they are equivalent on $u_1 u_2$.*

*Proof.* We will prove it in each group (in particular free).

$$f(u_1 u_2) = f((u_1 w_2)(w_1 w_2)^{-1}(w_1 u_2)) = f(u_1 w_2) f(w_1 w_2)^{-1} f(w_1 u_2) =$$

$$g(u_1 w_2) g(w_1 w_2)^{-1} g(w_1 u_2) = g((u_1 w_2)(w_1 w_2)^{-1}(w_1 u_2)) = g(u_1 u_2).$$

Note that Lemma 8 can be formulated using the construction of Lemma 1 in the following way.

**Lemma 9.** *If*

$$w_1 w_2 = w_1' w_2', \quad w_1 u_2 = w_1' u_2', \quad u_1 w_2 = u_1' w_2'$$

*then*

$$u_1 u_2 = u_1' u_2'$$

In other words three equations over 8 variables imply the fourth one.

The rest of our construction has nothing to do with semigroup theory but rather with graph theory.

Let $A = (\Sigma, Q, q_0, F, \delta)$ be a nondeterministic automaton accepting $L$ where $\Sigma$ is an alphabet, $Q$ is a set of states, $q_0 \in Q$ is the initial state, $F \subseteq Q$ is the set of accepting states, and $\delta \subseteq Q \times \Sigma \cup \{1\} \times Q$ is the transition relation. We assume that $A$ is trim i.e. each state of $A$ is accessible from the initial state and from each state we can reach one of the final states. We start by replacing $A$ by equivalent automaton $A'$ having one accepting state $q_F$ by adding for each state $q \in F$ transition $(q, 1, q_F)$.

Each automaton $A$ can be represented by (oriented) graph $G_A$ whose edges are labeled by elements of $\Sigma \cup \{1\}$.

Let $T_1$ be a (oriented) spanning tree of $G_{A'}$ rooted at $q_0$ whose edges lead from parents to children. Let $T_2$ be a (oriented) spanning tree of $G_{A'}$ rooted at $q_F$ whose edges lead from children to parents. We associate with each accepting path

$$\pi = (u_0, u_1), (u_1, u_2), \ldots, (u_{l-1}, u_l), (u_l, u_{l+1})$$

of $G_{A'}$ (clearly $u_0 = q_0$ and $u_{l+1} = q_F$) a subseqence of $k \geq 0$ edges

$$\lambda = (u_{l_1}, u_{l_1+1}), (u_{l_2}, u_{l_2+1}), \ldots, (u_{l_k}, u_{l_k+1})$$

in the following $(u_{l_1}, u_{l_1+1})$ is the first edge of $\pi$ which does not belong to $T_1$. If such an edge does not exist, then $k = 0$ and $\lambda$ is the empty subsequence. Now, if the edges following $(u_{l_1}, u_{l_1+1})$ in $\pi$ are edges of $T_2$, then $k = 1$ otherwise $(u_{l_2}, u_{l_2+1})$ is the second edge in $\pi$ which does not belong to $T_1$. Again, if the edges following $(u_{l_2}, u_{l_2+1})$ are in $T_2$, then $k = 2$ otherwise $(u_{l_3}, u_{l_3+1})$ is the third edge in $\pi$ which does not belong to $T_1$ and so on.

Note that any sequence of edges $(u_1, v_1), (u_2, v_2), \ldots, (u_k, v_k)$ is associated to at most one path $\pi$. Indeed, the path $\pi$ has to start in $q_0$ and go in tree $T_1$ to the node $u_1$ (there is only one such path) then it goes via $(u_1, v_1)$. Then it has to go inside $T_1$ from $v_1$ to $u_2$ and via $(u_2, v_2)$ and so on. The last part of $\pi$ has to go inside $T_2$ from $v_k$ to $q_F$.

Denote by $F_k$ the set of words which are accepted by a path in $G_{A'}$ associated to a sequence of at most $k$ edges.

**Lemma 10 ([16]).** $F_1$ *is a test set for* $L$.

*Proof.* Since for each $w \in L$ there is an accepting path $\pi$ such that $w = w(\pi)$, for each $w \in L$ there is $k \geq 0$ such that $w \in F_k$. Hence, it is enough to prove that for each $k \geq 1$, $F_k$ is a test set for $F_{k+1}$. Suppose that two morphisms $f$ and $g$ are equivalent on $F_k$. Take a word $w \in F_{k+1}$. We will prove that $f$ and $g$ are equivalent on $w$. The word $w$ is accepted by a path $\pi$ associated to at most $k+1$ edges. If $\pi$ is associated to at most $k$ edges, then there is nothing to prove. Assume then that $\pi$ is associated to exactly $k+1$ edges $(u_1, v_1), \ldots, (u_{k+1}, v_{k+1})$. We divide it into two subpaths $\pi_1$ from $q_0$ to $v_1$ and $\pi_2$ from $v_1$ to $q_F$, see Fig. 1. Then there is a path $\pi_1'$ in $T_1$ from $q_0$ to $v_1$ and a path $\pi_2'$ in $T_2$ from $v_1$ to $q_F$. Now we have four accepting paths: $\pi = \pi_1 \pi_2$, $\pi_1 \pi_2'$, $\pi_1' \pi_2$ and $\pi_1' \pi_2'$. Path $\pi_1' \pi_2$ is associated to $k$ edges, path $\pi_1 \pi_2'$ to one edge and path $\pi_1' \pi_2'$ to at most one edge. Hence,

$$w(\pi_1' \pi_2), w(\pi_1 \pi_2'), w(\pi_1' \pi_2') \in F_k.$$

Denote $w_1 = w(\pi_1')$, $w_2 = w(\pi_2')$, $u_1 = w(\pi_1)$ and $u_2 = w(\pi_2)$. The morphisms $f$ and $g$ are equivalent on words $w_1 w_2, w_1 u_2, u_1 w_1$ since they are in $F_k$. By Lemma 8 they are equivalent on $u_1 u_2 = w(\pi_1 \pi_2) = w(\pi) = w$. This completes the proof.

Now we are ready to prove.

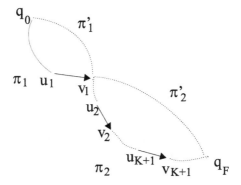

**Fig. 1.** Paths $\pi = \pi_1\pi_2$, $\pi_1\pi_2'$, $\pi_1'\pi_2$, $\pi_1'\pi_2'$.

**Theorem 9.** *Let $m$ be the number of edges in $G_A$, $n$ the number of vertices of $G_A$ and $f$ be the number of accepting vertices of $G_A$. Then, there is a test set for $L$ possesing at most $m - n + f + 1$ words.*

*Proof.* It is enough to prove that $F_1$ is of size at most $m - n + f + 1$. The number of paths associated to 0 edges is of course 1. The number of paths associated to exactly one edge is the number of edges of $G_{A'}$ which does not belong to $T_1$. The number of edges of $G_{A'}$ is exactly $m + f$ and the number of edges of $T_1$ is exactly $n$ so the number of edges associated with at most one edge is at most $m + f - n + 1$. Hence the size of $F_1$ is at most $m + f - n + 1$.

### 5.2   Lower Bound

We will prove that the bound in Theorem 9 is tight if $f \leq n-1$ and $m \geq n-1$. The first inequality is a natural one: $f$ is the size of a subset of accepting states so $f \leq n$. Moreover, if $f = n$ then $q_0$ is an accepting state which means that the empty word is in the language. We can remove the empty word from considerations since each two morphisms are equivalent on it. The second inequality is also natural if $m < n - 1$ then the automaton is not trim - it contains parts which can be removed.

Let $\Sigma_t = \{a_1, a_2, \dots, a_t\}$. Consider the language

$$L_t = \{a_1, a_1 a_2, \dots, a_1 a_2 \dots a_t\}$$

containing $t$ words. The language $L_t$ is accepted by the automaton (note that it is minimal automaton accepting $L_t$) in Fig. 2 containing $n = t + 1$ states

$$q_0 \xrightarrow{\ a_1\ } \xrightarrow{\ a_2\ } \cdots \xrightarrow{\ a_t\ }$$

**Fig. 2.** The minimal automaton accepting the language $L_t$.

including $f = t$ accepting states and $m = t$ edges so $m - n + f + 1 = t = |L_t|$. On the other hand $L_t$ is a test set language. Indeed, morphisms

$$f(a_i) = \begin{cases} 1 \text{ if } i \neq k \\ a \text{ if } i = k \end{cases}, \quad g(a_i) = \begin{cases} 1 \text{ if } i \neq k + 1 \\ a \text{ if } i = k + 1 \end{cases}$$

are equivalent on $L_t - \{a_1 \ldots a_k\}$ but they are not equivalent on $a_1 \ldots a_k$.

Since $L_t$ is a test set language each subset of it is a test set language, too. If such a language contains the word $a_1 \ldots a_t$, then the minimal automaton for it can be obtained from the automaton for $L_t$ by changing its set of accepting states. The size of such a language is also $m - n + f + 1$ but now there is no connection between $f$ and $m$ or $n$ except $f \leq n - 1$.

Denote by $L_{t,f}$ any language of size $f$ being a subset of $L_t$ containing the word $a_1 \ldots a_t$. Let $\Sigma'_k = \{b_1, \ldots, b_k\}$ be a set of new symbols. Then the language $L_{t,f} \cup \Sigma'_k$ is a test set language. Moreover the minimal automaton for $L_{t,f} \cup \Sigma'_k$ consists of $m = t + k$ transitions, $n = t + 1$ states including $f \leq t$ final states. Moreover the language consists of $m - n + f + 1 = f + k$ words so the bound in Theorem 9 is tight.

**Theorem 10.** *Let $m \geq n - 1$ and $f \leq n - 1$. Then there is a language accepted by an automaton with $m$ transitions, $n$ states including $f$ accepting states such that the smallest test set for it is of size $m - n + f + 1$.*

## 6   Test Sets for Context-Free Languages

### 6.1   Upper Bound

Let $G$ be a context-free grammar generating a language $L$. In the first step we construct a grammar $G'$ in Chomsky normal form which generates $L$ and does not contain useless nonterminals. The size of the grammar $G'$ measured as the total length of right hand sides of all productions is linear with respect to the size of grammar $G$. Next, to each nonterminal symbol $A$ of $G'$ we associate a word $w_A$ derivable from $A$ in $G'$. Now, we replace each production $A \to BC$ where $B$ and $C$ are nonterminals by three productions $A \to w_B C$, $A \to B w_C$ and $A \to w_B w_C$. In this way we obtain a linear context-free grammar $lin(G')$. We omit the proof of the following lemma. Its proof uses induction on the length of a derivation of a word in $G'$ and uses Lemma 8.

**Lemma 11 ([19]).** *The language generated by $lin(G')$ is a test set for $L$.*

Each linear context-free grammar $linG$ is associated to a graph $graph(linG)$ in the following way. The vertices of $graph(linG)$ correspond to nonterminals of $linG$. There is also one special vertex $v$ which does not correspond to a nonterminal. It is called *the sink vertex*. Edges of $graph(linG)$ are labeled by pairs of terminal words and correspond to productions of $linG$. There is an edge leading from nonterminal $A$ to nonterminal $B$ which is labeled $(u, v)$ iff there is a production $A \to uBv$ in $linG$. Similarly, there is an edge leading from nonterminal $A$ to the sink $v$ which is labeled $(u, 1)$ iff there is a production $A \to u$ in $linG$.

A *source vertex* of $graph(linG)$ is the start symbol of $linG$. Each path

$$\pi = (A_0, A_1), (A_1, A_2), \ldots, (A_l, A_{l+1})$$

in $graph(linG)$ ending in a nonterminal corresponds in a natural way to a derivation in $linG$

$$A_0 \to u_0 A_1 v_0 \to u_0 u_1 A_2 v_1 v_0 \to^* u_0 \ldots u_l A_{l+1} v_l \ldots v_0$$

where $(u_i, v_i)$ is the label of edge $(A_i, A_{i+1})$ Similarly, each path

$$\pi = (A_0, A_1), (A_1, A_2), \ldots, (A_l, A_{l+1})$$

in $graph(linG)$ ending in the sink vertex corresponds in a natural way to a derivation of a terminal word in $linG$

$$A_0 \to u_0 A_1 v_0 \to u_0 u_1 A_2 v_1 v_0 \to^* u_0 \ldots u_l v_l \ldots v_0$$

where $(u_i, v_i)$ is the label of edge $(A_i, A_{i+1})$. A path which leads from the source vertex to the sink vertex is called *accepting path*. Denote by $w(\pi)$, for accepting path $\pi$, the terminal word derived by a derivation corresponding to $\pi$. Clearly, the language generated by $linG$ is the set

$$\{w(\pi) : \pi \text{ is an accepting path in } graph(linG)\}.$$

Define a language $L_4$ over

$$\{a_1, b_1, a_2, b_2, \bar{a}_2, \bar{b}_2, a_3, b_3, \bar{a}_3, \bar{b}_3, a_4, b_4, \bar{a}_4, \bar{b}_4\}$$

which is generated by the following linear context-free grammar

$$A_4 \to a_4 A_3 \bar{a}_4 | b_4 A_3 \bar{b}_4$$

$$A_3 \to a_3 A_2 \bar{a}_3 | b_3 A_2 \bar{b}_3$$

$$A_2 \to a_2 A_1 \bar{a}_2 | b_2 A_1 \bar{b}_2$$

$$A_1 \to a_1 | b_1$$

The language $L_4$ consists of 16 words. Denote $T_4 = L_4 - \{b_4 b_3 b_2 b_1 \bar{b}_2 \bar{b}_3 \bar{b}_4\}$.

We omit the proof of next lemma. Original proof for free monoids occupies 3.5 pages A4 format [19]. The proof for free groups occupies 0.5 page A4 format [24, 25]. So in contrary to commutative languages the proof of the crucial lemma is shorter in free groups than in free monoids.

**Lemma 12 ([19,24,25]).** *$T_4$ is a test set for $L_4$.*

Note that Lemma 12 formulated in terms of systems of word equations as in Lemma 1 says that 15 equations imply sixteen equation.

For each nonterminal $A$ construct a spanning tree $T_A$ of a subgraph of graph $graph(lin(G'))$ consisting of vertices reachable from $A$. The tree $T_A$ is rooted at $A$ and all edges in it lead from parents to children. Similarly as in case of regular languages to each accepting path

$$\pi = (A_0, A_1), (A_1, A_2), \ldots, (A_{l-1}, A_l), (A_l, A_{l+1})$$

we associate a subsequence of edges

$$\lambda = (A_{l_1}, A_{l_1+1}), (A_{l_2}, A_{l_2+1}), \ldots, (A_{l_k}, A_{l_k+1})$$

in the following way. The edge $(A_{l_1}, A_{l_1+1})$ is the first edge of $\pi$ which does not belong to the tree $T_{A_0}$. If such an edge does not exist then $k = 0$ and $\lambda$ is the empty sequence. The edge $(A_{l_2}, A_{l_2+1})$ is the first edge of $\pi$ which follows $(A_{l_1}, A_{l_1+1})$ and does not belong to the tree $T_{A_{l_1+1}}$. If such an edge does not exist then $k = 1$. The edge $(A_{l_3}, A_{l_3+1})$ is the first edge of $\pi$ which follows $(A_{l_2}, A_{l_2+1})$ and does not belong to the tree $T_{A_{l_2+1}}$ and so on.

Clearly, each sequence of edges $(A_1, B_1), (A_2, B_2), \ldots, (A_k, B_k)$ is associated to at most one path. Indeed, the path has to start in the source vertex $A_0$ and go to the vertex $A_1$ inside the tree $T_{A_0}$ then it has to traverse $(A_1, B_1)$ and go inside $T_{B_1}$ from $B_1$ to $A_2$, traverse $(A_2, B_2)$ and so on. Finally it has to go inside $T_{B_k}$ from $B_k$ to the sink vertex.

Denote by $F_k$ the set of words $w(\pi)$ where $\pi$ is an accepting path associated to a sequence of at most $k$ edges.

**Lemma 13.** $F_6$ is a test set for $L$.

*Proof.* Since for each $w \in L$ there is an accepting path $\pi$ such that $w = w(\pi)$, for each $w \in L$ there is $k \geq 0$ such that $w \in F_k$. Hence, it is enough to prove that for each $k \geq 1$, $F_k$ is a test set for $F_{k+1}$. Suppose that two morphisms $f$ and $g$ are equivalent on $F_k$. Take a word $w \in F_{k+1}$. We will prove that $f$ and $g$ are equivalent on $w$. The word $w$ is accepted by a path $\pi$ associated to at most $k+1$ edges. If $\pi$ is associated to at most $k$ edges, then there is nothing to prove. Assume then that $\pi$ is associated to exactly $k+1$ edges $(u_1, v_1), \ldots, (u_{k+1}, v_{k+1})$. We divide it into four subpaths, see Fig. 3: $\pi_1$ from the source vertex to $u_2$, $\pi_2$ from $v_2$ to $u_4$, $\pi_3$ from $v_4$ to $u_6$ and $\pi_4$ from $v_6$ to the sink node. Then there is a path $\pi_1'$ in $T_{A_0}$ from $A_0$ to $u_2$, a path $\pi_2'$ in $T_{v_2}$ from $v_2$ to $u_4$, a path $\pi_3'$ in $T_{v_4}$ from $v_4$ to $u_6$ and a path $\pi_4'$ from $v_6$ to the sink vertex. Now we have 16 accepting paths of the form

$$(\pi_1 \text{ or } \pi_1'), (u_2, v_2), (\pi_2 \text{ or } \pi_2'), (u_4, v_4), (\pi_3 \text{ or } \pi_3'), (u_6, v_6), (\pi_4 \text{ or } \pi_4')$$

All of 16 paths except $\pi$ is associated to at most $k$ edges so that $w(\pi') \in F_k$ for $\pi' \neq \pi$. Hence, morphisms $f$ and $g$ are equivalent on these 15 words. By Lemma 12 they are also equivalent on $w(\pi) = w$. This completes the proof.

As an easy consequence we have.

**Theorem 11.** *Let $m$ be a size of a context-free grammar generating a language $L$. Then $L$ posseses a test set consisting of $O(m^6)$ words.*

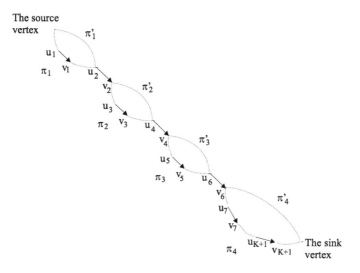

**Fig. 3.** 16 accepting paths.

## 6.2   Lower Bound

Define first a language $L_3$ over $\{a_1, b_1, a_2, b_2, \bar{a}_2, \bar{b}_2, a_3, b_3, \bar{a}_3, \bar{b}_3\}$ which is generated by the following linear context-free grammar

$$A_3 \to a_3 A_2 \bar{a}_3 | b_3 A_2 \bar{b}_3$$

$$A_2 \to a_2 A_1 \bar{a}_2 | b_2 A_1 \bar{b}_2$$

$$A_1 \to a_1 | b_1$$

**Lemma 14 ([19]).** $L_3$ *is a test set language.*

*Proof.* Because of the symmetry it is enough to prove that $L_3 - \{b_3 b_2 b_1 \bar{b}_2 \bar{b}_3\}$ is not a test set for $L_3$. Indeed, morphisms

$$g(a_3) = 1, \ g(\bar{a}_3) = p, g(b_3) = p, \ g(\bar{b}_3) = 1$$

$$g(a_2) = 1, \ g(\bar{a}_2) = q, g(b_2) = q, \ g(\bar{b}_2) = 1$$

$$g(a_1) = 1, \ g(b_1) = qp,$$

$$h(a_3) = q, \ h(\bar{a}_3) = 1, h(b_3) = 1, \ h(\bar{b}_3) = q$$

$$h(a_2) = p, \ h(\bar{a}_2) = 1, h(b_2) = 1, \ h(\bar{b}_2) = p$$

$$h(a_1) = 1, \ h(b_1) = qp,$$

are equivalent on $L_3 - \{b_3 b_2 b_1 \bar{b}_2 \bar{b}_3\}$ but they are not equivalent on $L_3$.

Note that the above lemma formulated in terms of systems of equations as in Lemma 1 says that some system of 8 equations is independent.

Now we define a language $L'_m$ over $\{a_{1,i} : i = 1..m\} \cup \{a_{2,i}, \bar{a}_{2,i} : i = 1..m\} \cup \{a_{3,i}, \bar{a}_{3,i}, : i = 1..m\}$ generated by the following grammar

$$A_3 \to a_{3,1} A_2 \bar{a}_{3,1} | \ldots | a_{3,m} A_2 \bar{a}_{3,m}$$

$$A_2 \to a_{2,1} A_1 \bar{a}_{2,1} | \ldots | a_{2,m} A_1 \bar{a}_{2,m}$$

$$A_1 \to a_{1,1} | \ldots | a_{1,m}$$

**Lemma 15 ([19]).** $L'_m$ *is a test set language.*

*Proof.* Define morphism $g'$, $h'$ in the following way

$$g'(a_{3,i}) = \begin{cases} g(a_3) & \text{if } i \neq i' \\ g(b_3) & \text{if } i = i' \end{cases} , \quad g'(\bar{a}_{3,i}) = \begin{cases} g(\bar{a}_3) & \text{if } i \neq i' \\ g(\bar{b}_3) & \text{if } i = i' \end{cases}$$

$$g'(a_{2,i}) = \begin{cases} g(a_2) & \text{if } i \neq j' \\ g(b_2) & \text{if } i = j' \end{cases} , \quad g'(\bar{a}_{2,i}) = \begin{cases} g(\bar{a}_2) & \text{if } i \neq j' \\ g(\bar{b}_2) & \text{if } i = j' \end{cases}$$

$$g'(a_{1,i}) = \begin{cases} g(a_1) & \text{if } i \neq k' \\ g(b_1) & \text{if } i = k' \end{cases} ,$$

$$h'(a_{3,i}) = \begin{cases} h(a_3) & \text{if } i \neq i' \\ h(b_3) & \text{if } i = i' \end{cases} , \quad h'(\bar{a}_{3,i}) = \begin{cases} h(\bar{a}_3) & \text{if } i \neq i' \\ h(\bar{b}_3) & \text{if } i = i' \end{cases}$$

$$h'(a_{2,i}) = \begin{cases} h(a_2) & \text{if } i \neq j' \\ h(b_2) & \text{if } i = j' \end{cases} , \quad h'(\bar{a}_{2,i}) = \begin{cases} h(\bar{a}_2) & \text{if } i \neq j' \\ h(\bar{b}_2) & \text{if } i = j' \end{cases}$$

$$h'(a_{1,i}) = \begin{cases} h(a_1) & \text{if } i \neq k' \\ h(b_1) & \text{if } i = k' \end{cases} ,$$

where $g$ and $h$ are the morphisms from the proof of the previous lemma. They are equivalent on $L'_m - \{a_{3,i'} a_{2,j'} a_{1,k'} \bar{a}_{2,j'} \bar{a}_{3,i'}\}$ but they are not equivalent on $L'_m$.

As an easy consequence we have

**Theorem 12 ([19]).** *For each $m \geq 1$, there is a language generated by a linear context-free grammar with $3m$ productions such that the smallest test set for it contains $m^3$ words.*

# 7    Conclusions

We would like to point out that the constructions of small test sets for the family of all languages, regular languages and context-free languages are the same in free monoids and free groups. The construction of small test sets for commutative languages is probably the same, too [26]. Moreover, the construction of languages which lead to lower bounds are also the same for the families of commutative languages, regular languages and context-free languages. Moreover, the author does not know an example of a subset of a language which is a test set for the language in a free monoid and not a test set in a free group. This would suggest that the the notions of test sets in free groups and free monoids coincide. But what about the family of all languages where there is a difference between free monoids and free groups: the existing lower bound for free monoids is polynomial and for free groups unbounded? The reason for this difference is that in case of free groups we may define the notion of a commutator which has very strong properties. The language which can be derived from this construction are very complicated and very difficult to handle without using the properties of commutator. This may suggest that the lower boud for free monoids in case of the family of all languages is unbounded as in case of free groups but the proof of it is difficult.

The author would like to point out another interesting thing: constructions of small test sets for commutative languages, regular langauges and context-free langauges are based on properties of small systems of word equations. In case of commutative languages such a system of six equations, in case of regular languages 4 equations, and in case of context-free languages 16 equations. Similarly, lower bounds are based on properties of small sets of equations.

# References

1.  K.I. Appel, F.M. Djorup, On the equation $z_1^n z_2^n \ldots z_k^n = y^n$ in a free semigroup, *Trans. Amer. Math. Soc.* **134**(1968), 461–471.
2.  M.H. Albert, J. Lawrence, A proof of Ehrenfeucht's Conjecture, *Theoret. Comput. Sci.* **41**(1985) 121–123.
3.  M.H. Albert, J. Lawrence, The descending chain condition on solution sets for systems of equations in groups, *Proc. Edinburg Math. Soc.*29(1985), 69–73.
4.  G. Baumslag, A. Myasnikov, V. Romankov, Two theorems about equationally noetherian groups, *J. Algebra* **194**, 654–664, 1997.
5.  Choffrut, J. Karhumäki, Combinatorics of words, chapter in G. Rozenberg, A. Salomaa (Eds.), *Handbook of formal languages*, Vol. 1, 329–438, Springer-Verlag, 1997.
6.  K. Culik II, J. Karhumäki, Systems of equations over free monoids and Ehrenfeucht's Conjecture, *Discrete. Math.* **43**(1983), 139–153.
7.  V. Guba, The equivalence of infinite systems of equations in free groups and semigroups to finite subsystems, *Matematiczeskije Zametki* **40**(3), September 1986 (in Russian).
8.  I. Hakala, *On word equations and the morphism equivalence problem for loop languages*, Ph.D. Thesis, University of Oulu, Finland, 1997.

9. I. Hakala, J. Kortelainen, On the system of word equations $x_1^i x_2^i \ldots x_m^i = y_1^i y_2^i \ldots y_n^i$ ($i=1, 2,\ldots$) in a free monoid, *Acta Inform.* **34**(1997), 217–230.

10. I. Hakala, J. Kortelainen, On the system of word equations $x_0 u_1^i x_1 u_2^i x_2 u_3^i x_3 = y_0 v_1^i y_1 v_2^i y_2 v_3^i y_3$ ($i=0, 1, 2,\ldots$) in a free monoid, *Theoret. Comput. Sci.* **225**(1999), 149–161.

11. I. Hakala, J. Kortelainen, Polynomial size test sets for commutative languages, *Rairo Informatique Theorique et Applications* **31**(1997), 191–304.

12. T. Harju, J. Karhumäki, M. Petrich, Compactness of equations on completely regular semigroups, *in* J. Mycielski, G. Rozenberg, A. Salomaa (Eds.), *Structures in Logic and Computer Science*, LNCS 1261, 268–280, Springer-Verlag, 1997.

13. T. Harju, J. Karhumäki, W. Plandowski, Compactness of systems of equations in semigroups, *Intern. J. Algebra Comput.* **7**, 457–470, 1997.

14. M. A. Harrison, *Introduction to Formal Language Theory*, Addison-Wesley, Reading, MA, 1978.

15. S. Holub, Local and global cyclicity in free semigroups, *Theoret. Comput. Sci.* **262**(2001), 25–36.

16. S. Jarominek, J. Karhumäki, W. Rytter, Efficient constructions of test sets for regular and context-free languages, *Theoret. Comput. Sci.* **116**(1993), 305–316.

17. M.I. Kargapolov, Ju.I. Merzlakov, *Basics of group theory*, Nauka, 1982 (in Russian).

18. J. Karhumäki, W. Plandowski, On the Size of Independent Systems of Equations in Semigroups *Theoret. Comput. Sci.* **168**(1): 105–119 (1996).

19. J. Karhumäki, W. Plandowski, W. Rytter, Polynomial size test sets for context-free languages, *J. Comput. Syst. Sci* **50**(1995), 11–19.

20. J. Kortelainen, On the system of word equations $x_0 u_1^i x_1 \ldots u_m x_m = y_0 v_1^i y_1 \ldots v_n^i y_n$ ($i = 0, 1, 2, \ldots$) in a free monoid, *Journal of Automata, Languages, and Combinatorics* **3**(1998), 43–58.

21. J. Lawrence, The nonexistence of finite test set for set-equivalence of finite substitutions, *Bull. EATCS* **28**, 34–37, 1986.

22. M. Lothaire, *Combinatorics on words*, Vol. 17 of *Encyclopedia of Mathematics and its Apllications*, Addison-Wesley, 1983. Reprinted in the *Cambridge Mathematical Library*, Cambridge University Press, 1997.

23. M. Lothaire II, *Algebraic Combinatorics on words*, Vol. 90 of *Encyclopedia of Mathematics and its Applications*, Cambridge University Press, 2002.

24. W. Plandowski, Testing equivalence of morphisms on context-free languages, *in* Lect. Notes in Comput. Sci. 855, 460–470, 1994.

25. W. Plandowski, The complexity of the morphism equivalence problem for context-free languages, Ph.D. Thesis, Warsaw University, 1995.

26. W. Plandowski, personal communication.

27. E.T. Poulsen, *The ehrenfeucht conjecture: an algebra framework for its proof* Aarhus Universitet, Matematisk Institut, 1985.

28. A. Salomaa, The ehrenfeucht conjecture: a proof for language theorist, manuscript.

29. J. Stalings, Finiteness properties of matrix representations, manuscript.

30. P. Turakainen, The equivalence of deterministic gsm replications on $Q$-rational languages is decidable, *Math. Systems Theory* **20**(1987), 273–282.

31. P. Turakainen, The equivalence of dgsm replications on $Q$-rational languages is decidable, *Proc. ICALP'88*, LNCS 317, 654–666, Springer-Verlag, 1988.

# Complexity Theory Made Easy
## The Formal Language Approach to the Definition of Complexity Classes

Heribert Vollmer

Theoretische Informatik, Universität Hannover
Appelstraße 4, D-30167 Hannover
`vollmer@informatik.uni-hannover.de`

**Abstract.** In recent years generalized acceptance criteria for different nondeterministic computation models have been examined. Instead of the common definition where an input word is said to be accepted if in the corresponding computation tree an accepting path exists, more general conditions on this tree are used. We survey some recent results from this context, paying particular attention to nondeterministic finite automata as well as nondeterministic polynomial-time Turing machines.

## 1   Introduction

Let $M$ be a nondeterministic finite automaton and $w$ be an input of $M$. By definition, $w$ is accepted by $M$ if and only if there is at least one possible computation path of $M$ which accepts $w$. Let us look at the tree $T_M(w)$ of all computations that automaton $M$ on input $w$ can possibly perform. A node $v$ in this tree is labeled by a configuration $C$ of $M$ at a certain point during its computation on input $w$, where such a configuration is given by the state of $M$ and the portion of the input which is still unscanned. The children of $v$ in the computation tree are associated with the successor configurations of $C$, i.e., if the transition function of $M$ has several entries for this particular $C$, then each of these will lead to a successor configuration and a child of $v$ in the computation tree. The leaves in the tree are associated to those configurations that $M$ reaches when all input symbols are consumed.

Now the acceptance criterion of nondeterministic automata can be rephrased as follows: An input word $w$ is accepted by $M$ if and only if in the computation tree of $M$ on $x$ there is at least one leaf labeled with an accepting state.

Using the concept of computation trees, modified acceptance criteria for NFAs may be studied, e.g. the following: Say that a word is accepted by $M$ if and only if the number of accepting leaves in the computation tree is divisible by a fixed prime number $p$. Can non-regular languages be recognized in this way? The acceptance here is given by a more complicated condition on the cardinality of the set of accepting paths in the computation tree. (For the definition of the class REG we just require that this cardinality is non-zero.) However, one may even go beyond such cardinality conditions: If we attach certain symbols to the leaves in $T_M(w)$, e.g., the symbol 1 to an accepting leaf and 0 to a non-accepting

Z. Ésik and Z. Fülöp (Eds.): DLT 2003, LNCS 2710, pp. 95–110, 2003.

leaf, then the computation tree of $M$ on input $w$ defines a word, which we get by concatenating the symbols attached to the leaves, read from left to right (in a natural order of the paths to be made precise below). We call this string the *leaf word* of $M$ on $w$. Observe that the length of the leaf word can be exponential in the length of $w$. Generally, an acceptance criterion is nothing other than the set of those leaf words that make $M$ accept its input; that is, such a criterion is defined by a so called *leaf language* $L$ over the alphabet of the leaf symbols. By definition a word is accepted by $M$ if and only if the leaf word of $M$ on input $w$ is in $L$. In the example above we used as leaf language the set $L$ of all binary words with a number of 1's divisible by $p$. As a generalization of the above question, one might now ask what languages can be accepted using arbitrary regular leaf languages.

The concept of leaf languages can be examined for essentially every nondeterministic computation model. Historically the power of *nondeterministic Turing machines* whose acceptance is given by a leaf language was first studied [BCS92, Ver93,HLS+93,JMT96]. This lead to an "amusing and instructive way of looking at [...] diverse complexity classes" [Pap94a, p. 504] that are of current focal interest in computational complexity theory.

As an example, let us look at the class NP. By definition, a language $A \in$ NP is given by a nondeterministic polynomial-time machine (NPTM) $M$ such that for all inputs $x$, we have that $x$ belongs to $A$ if and only if in the computation tree that $M$ produces when working on $x$, we find at least one accepting path. A language $A \in$ US is given by an NPTM $M$ such that for all inputs $x$, we have that $x$ belongs to $A$ if and only if in the computation tree of $M$ on $x$, there is exactly one accepting path. A language $A \in \text{Mod}_p$P is given by an NPTM $M$ such that for all inputs $x$, we have that $x$ belongs to $A$ if and only if in the computation tree of $M$ on $x$, the number of accepting paths is not divisible by $p$. A language $A \in$ PP is given by an NPTM $M$ such that for all inputs $x$, we have that $x$ belongs to $A$ if and only if in the computation tree of $M$ on $x$, we have more accepting than rejecting paths. (For background on these and other classes mentioned in the present paper without definition, we refer the reader to [Joh90].) All these acceptance definitions can easily be formulated in terms of leaf languages: The classes NP, US and $\text{Mod}_p$P can be defined by regular leaf languages: NP via $(0 + 1)^*10^*$, US via $0^*10^*$, and $\text{Mod}_p$P via $\left\{ w \mid |w|_1 \equiv 0 \pmod{p} \right\}$. The characterization given for PP only gives a context-free leaf language: $\{ w \mid |w|_1 > |w|_0 \}$). Also for PSPACE, the class of languages accepted by Turing machines that operate in polynomial space, a context-free leaf languages can easily be found (recalling that PSPACE corresponds to polynomial time on alternating Turing machines [CKS81] we see that the language of all true Boolean expressions involving the constants 0 and 1 and the connectives $\land$ and $\lor$ will do). We will address the question if the latter two classes can be defined by regular leaf languages below.

This concept was developed by Papadimitriou and Sipser around 1979 when teaching a complexity class at MIT [Pap94b]. It was later rediscovered and published independently in [BCS92,Ver93] and has since then been used actively in the study of complexity classes mostly in between $\text{NC}^1$ and PSPACE, leading to remarkable characterizations of complexity classes as well as (absolute and relativized) separations, cf. the surveys [Vol99b,Vol00]. Much of the literature on leaf languages can be found following the links on the leaf languages homepage [LLH]. In this paper we present—after formally introducing the above concepts in Sect. 1—some of the original motivation behind the examination of leaf languages for NPTMs and a few early results in Sect.2, and then turn to the discussion some more recent developments: Sect. 3 relates leaf languages to generic oracle separations of complexity classes and discusses consequences for the P-NP-problem. The definability of function classes using the leaf language concept is exemplified in Sect. 4 for the class of functions computable in polynomial space. In Sect. 5 we will finally come back to leaf languages for nondeterministic finite automata.

## 2    Leaf Languages for Turing Machines

As described in the introduction, in the leaf language approach to the characterization of complexity classes, the acceptance of a word given as input to a nondeterministic machine depends only on the values printed at the leaves of the computation tree. To be more precise, let $M$ be a nondeterministic Turing machine, halting on each path printing a symbol from an alphabet $\Sigma$, with some order on the nondeterministic choices. Then, leafstring$^M(x)$ is the concatenation of the symbols printed at the leaves of the computation tree of $M$ on input $x$ (according to the order of $M$'s paths induced by the order of $M$'s choices).

Call a computation tree of a machine $M$ *balanced* if, informally, it is a full binary tree from which a right part is cut off. Formally we require that $M$ branches at most binary, all of its computation paths have the same length (i.e., same number of nondeterministic choices), and moreover, if we identify every path with the string over $\{0,1\}$ describing the sequence of the nondeterministic choices on this path, then there is some ("maximal", "rightmost") path $z$ such that all paths $y$ with $|y| = |z|$ and $y \preceq z$ (in lexicographic ordering) exist, but no path $y$ with $y \succ z$ exists. (The role and importance of using balanced computation trees in the upcoming development is studied in [HVW96], see also [Vol00].)

Given now a pair of languages $A, R \subseteq \Sigma^*$ such that $A \cap R = \emptyset$, this defines a complexity class BLeaf$^P(A, R)$ as follows: A language $L$ belongs to BLeaf$^P(A, R)$ if there is a polynomial-time NTM $M$ whose computation tree is always balanced, such that for all $x$, $x \in L \implies$ leafstring$^M(x) \in A$ and $x \notin L \implies$ leafstring$^M(x) \in R$. In the case that $A = \overline{R}$ we also simply write BLeaf$^P(A)$ for BLeaf$^P(A, R)$. The classes which can be defined by a pair $(A, \overline{A})$ are *syntactic classes* in the terminology of Papadimitriou [Pap94a], while those which cannot are *semantic classes*.

This computation model was introduced, as already mentioned above, by Papadimitriou and Sipser around 1979, and published for the first time by Bovet, Crescenzi, and Silvestri, and independently by Vereshchagin [BCS92,Ver93]. The name "leaf language" appears first in [HLS$^+$93] and [Pap94a, pp. 504f].

Let $\mathcal{C}$ be a class of languages. The class $\text{BLeaf}^P(\mathcal{C})$ consists of the union over all $B \in \mathcal{C}$ of the classes $\text{BLeaf}^P(B)$. In a sequence of papers ([HLS+93,JMT96, CMTV98] and more) the complexity of the classes $\text{BLeaf}^P(\mathcal{C})$ was studied as a function of the complexity of $\mathcal{C}$. It was obtained that generally an exponential jump in complexity can be observed, e.g., $\text{BLeaf}^P(\text{LOGSPACE}) = \text{PSPACE}$, $\text{BLeaf}^P(P) = \text{EXPTIME}$, etc.

The examples given in Sect. 1 show that the classes NP, US and $\text{Mod}_p\text{P}$ can be defined by regular leaf languages: $\text{NP} = \text{BLeaf}^P((0 + 1)^*10^*)$, $\text{US} = \text{BLeaf}^P(0^*10^*)$, and $\text{Mod}_p\text{P} = \text{BLeaf}^P(\{ w \mid |w|_1 \equiv 0 \pmod{p} \})$. For the classes PP and PSPACE we pointed out context-free leaf languages in Sect. 1. The question if these latter two classes are also definable via a regular language was raised in [HLS+93].

**Theorem 1 [HLS+93].** *Let $B$ be a regular language whose syntactic monoid is non-solvable. Then $\text{BLeaf}^P(B) = \text{PSPACE}$.*

*Proof sketch.* The left to right inclusion is clear, since a PSPACE machine can traverse a whole computation tree. For the converse direction, recall the characterization of PSPACE by alternating machines. Hence we can think of languages in PSPACE as being defined by machines whose computation tree is a tree with leaves 0 and 1 and inner nodes associated with $\wedge$ or $\vee$. An input is accepted if and only if this tree evaluates to 1. Barrington [Bar89] noticed that the commutator $(g, h)$ of two elements taken from a non-solvable group $G$ has the character of a logical conjunction: $(g, h)$ is equal to the identity iff at least one of $g, h$ is the identity. Here the identity corresponds to truth value 0 and every non-identity element corresponds to 1. Building on this he proved that an $\wedge$-$\vee$-expression $\phi$ can be transformed to a sequence of elements over $G$ which multiplies out to the identity if and only if $\phi$ evaluates to 1. In our case, a computation tree of a polynomial-time machine yields an exponentially long sequence of group elements. This sequence can be produced as a leaf string of a polynomial-time machine. Now since the syntactic monoid of $B$ has a non-solvable subgroup, the result follows.    □

A consequence of this theorem is that every PSPACE-computation can be cut into an exponential sequence of computations, each seeing only the input and obtaining a constant amount of information from the previous computation. The remarkable point is that each computation in this sequence is polynomial-time bounded (in fact, the low level circuit class $\text{AC}^0$ is already sufficient). We say that PSPACE is *P-serializable* (or *$\text{AC}^0$-serializable*) [CF91,HLS+93].

Let PH be the union of all classes of the polynomial-time hierarchy, i.e., $\text{PH} = \text{NP} \cup \text{NP}^{\text{NP}} \cup \text{NP}^{\text{NP}^{\text{NP}}} \cup \cdots$. Let MOD-PH be the union of all classes of the oracle hierarchy constructed similarly using as building blocks not only NP but also all classes $\text{Mod}_p\text{P}$ for $p \in \mathbb{N}$. While $\text{BLeaf}^P(\text{REG}) = \text{PSPACE}$ and in fact, for this class one single (non-solvable) regular language is sufficient, [HLS+93] proved that taking arbitrary regular leaf languages with a solvable monoid, the union of the classes of the MOD-PH hierarchy is obtained ($\text{BLeaf}^P(\text{SOLVABLE}) = \text{MOD-PH}$), and that taking arbitrary regular leaf languages with an aperiodic monoid, the union of the classes of the polynomial-time hierarchy is obtained ($\mathrm{BLeaf}^{\mathrm{P}}(\mathrm{APERIODIC}) = \mathrm{PH}$).

Let us mention that many more complexity classes can be defined via regular leaf languages, for instance higher levels of the polynomial hierarchy (for example, $\Sigma_2^{\mathrm{p}}$ can be defined by $(0 + 1)^*11(010)^+11(0 + 1)^*$, intuitively: there is a pair of occurrences of the substring 11 such that in between we have only 010's; an aperiodic language) and all classes of the Boolean hierarchy over NP (for example, NP $\wedge$ coNP, the class of languages that can be obtained as an intersection of some NP-set with some coNP-set, can be defined by the language of all words in which the substring 010 appears at least once, but the substring 0110 never appears). On the other hand (and coming back to the question of regular characterizability of PP), we conclude from the above that if PP can be characterized by a regular leaf language (say $L$), then either PP = PSPACE (if $L$ is non-solvable) or PP $\subseteq$ MOD-PH (if $L$ is solvable); both possibilites are considered unlikely by most complexity theorists.

## 3    The P-NP-Problem

In 1975 Baker, Gill, and Solovay constructed an oracle $B$ that separates P from NP: $\mathrm{P}^B \neq \mathrm{NP}^B$ [BGS75]. We say that P $\neq$ NP holds *relative to the oracle B*. On the other hand, it is not too hard to see that there is an oracle $A$ relative to which P and NP coincide (it suffices to pick as $A$ any PSPACE-complete language). Hence, the existence of oracles $A$ and $B$ shows that the P-NP-question is difficult to solve in the sense that non-relativizing proof techniques will be necessary. Conflicting relativizations like this give a sort of *independence result*: Relativizing arguments will neither be able to prove nor to disprove P $\neq$ NP. This is a remarkable contrast to the situation we find in recursion theory where usually proofs relativize and obtained inclusions thus hold relative to any oracle [Rog67].

While the results presented in the previous section (asking about definability of specific complexity classes) were very much in the spirit of Papadimitriou and Siper who used leaf languages to define a "panorama of complexity classes" [Pap94a], the first published papers that use leaf languages had another goal, namely to develop a tool to obtain independence results as just mentioned. While oracles that make classes equal are usually no problem, the construction of relativized separations can be very hard; and exactly here leaf languages can be of help, in particular the so-called *plt-reductions* among leaf languages. Say that a function $f\colon \Sigma^* \to \Sigma^*$ is *polylog-time bit-computable* if there exist two polynomial time oracle transducers $R\colon \Sigma^* \times \mathbb{N} \to \Sigma$ and $l\colon \Sigma^* \to \mathbb{N}$ such that, for any $x \in \Sigma^*$, $f(x) = R^x(|x|, 1)R^x(|x|, 2)\cdots R^x(|x|, l^x(|x|))$. Machines $R$ and $l$ access the input $x$ of function $f$ as an oracle, i.e., writing a number $i$ in binary on the oracle query tape will result in the $i$th bit of $x$ as oracle answer. Note that the run-time of $R$ and $l$ is polynomial in the length of their input $|x|$ and thus polynomial in the logarithm of the length of $x$; hence every bit of $f(x)$ can be computed in time polylogarithmic in the length of $x$, which explains the name "polylog-time bit-computable".

Now we define $\leq_m^{\text{plt}}$-*reductions* as follows: Let $(A, R)$ and $(A', R')$ be two pairs of languages. $(A, R)$ is *polylog-time reducible* to $(A', R')$ (in symbols: $(A, R) \leq_m^{\text{plt}}$ $(A', R')$), if there exists a polylog-time bit-computable function $f$, such that $f(A) \subseteq A'$ and $f(R) \subseteq R'$. We note that for the case $R = \overline{A}$ and $R' = \overline{A'}$ this is equivalent to $x \in A \Leftrightarrow f(x) \in A'$.

The importance of plt-reductions stems from that fact that these relate to oracle separations of leaf language definable classes, as proved in [BCS92, Ver93]. First, if in the definition of $\text{BLeaf}^{\text{P}}(A, R)$ above, $M$ is a nondeterministic polynomial-time oracle Turing machine (NPOTM) with access to oracle $O$, we denote the obtained class by $\text{BLeaf}^{\text{P}}(A, R)^O$. The main result of [BCS92, Ver93] can now be stated formally as:

**Theorem 2 [BCS92,Ver93].** *Let $(A, R)$ and $(A', R')$ be pairs of leaf languages. Then $(A, R) \not\leq_m^{\text{plt}} (A', R')$ if and only if for some oracle $O$ we have $\text{BLeaf}^{\text{P}}(A, R)^O \not\subseteq \text{BLeaf}^{\text{P}}(A', R')^O$.*

Many relativization results have been established in complexity theory, and many separations with the help of this theorem. Certainly in case of conflicting oracles one would like to know which of the two is nearer to the "actual" unrelativized world, i.e., the world without an oracle. In this vein one has looked for notions of "typical" oracles with the hope that separations for these also hold, if no oracle is present. Particular oracles are mostly constructed to make a certain statement true; thus they are of an "intentional nature" [Kur83]. *Random oracles* on the other hand, where membership of words in the oracle is essentially determined by independent fair coin tosses, are not of this kind but structureless by nature. Bennett and Gill [BG81], introducing random oracles, established the *random oracle hypothesis*, stating that every separation relative to a random oracle also holds in the unrelativized case. Unfortunately, the random oracle hypothesis is false [CCG+94]; in fact, it even fails very badly, namely in the case that both occurring classes are defined with the same machine model [VW97]. Specifically, PSPACE and IP (the class of languages that have "interactive protocols") can both be relativizably defined in the leaf language model, and within this model a separating oracle exists but, as shown in the nineties, IP = PSPACE [LFKN90,Sha92].

Another type of oracles that have attracted a lot of attention are *generic oracles*, introduced in computational complexity theory by Blum and Impagliazzo [BI87] (but going back to much elder developments in mathematical logic). Generic oracles are "typical" in the sense, that they have *all* the properties that can be enforced by a stage construction. Going back to recursion theory, stage constructions have been the main technique to obtain oracles with certain desired properties (see e.g. [Rog67]). To obtain their above mentioned oracle $B$, Baker, Gill, and Solovay also relied on this method; thus, a generic oracle separates P and NP, but since it has additional properties enforced by other stage constructions, it even makes the polynomial hierarchy infinite. When looking at the characteristic sequence of generic oracles, "anything even remotely possible will happen, and happen infinitely often" [BI87, p. 120]. For example, generic oracles will have infinitely often intervals of consecutive zeroes, whose length cannot be bounded recursively. In this sense, generic oracles are arbitrary, but not random.

Though a corresponding *generic oracle hypothesis* also does not hold [Fos93], there are a number of very good motivations to study generic oracles. We only mention two of them: First, generic relativizations (i.e., relativizations via generic oracles) do make statements about the real world. Blum and Impagliazzo showed that any language acceptable by a (time- or space-bounded) Turing machine with a generic oracle can be accepted without an oracle in essentially the same resources. This is a "half-sided" positive relativization result, since it implies that if resource-bounded classes coincide relative to a generic oracle, then they coincide absolutely. Even more important for us here is the following theorem, relating generic oracles immediately to a main motivation for a lot of current research in complexity theory, namely the P-NP-problem.

**Theorem 3 [HH90,BI87].** *If there is a generic oracle $G$ such that $\mathrm{P}^G \neq \mathrm{UP}^G$ then $\mathrm{P} \neq \mathrm{NP}$.*

Second, generic oracles proved to be very useful when looking for simultaneous collapses and separations; a number of such results can be found in [BI87]. Corollary 5 below should be seen in this context.

The existence of generic oracles separating complexity classes was related to relations among the corresponding leaf languages in [GKV03]. In order to present their result, we have to get a bit more precise and formal. We will use the notion of *generic oracles* as used by Blum and Impagliazzo in [BI87]. (It should be remarked that many different notions of genericity have been studied in computational complexity theory; for an overview the reader might consult [FFKL93,AS96]. We briefly repeat the central definitions from [BI87]. An *oracle* $O$ is a set of natural numbers. We will identify such sets with their characteristic functions, i.e., an oracle is nothing else than a total function $O \colon \mathbb{N} \to \{0,1\}$. We will also consider *finite oracles*, partial functions $v \colon \mathbb{N} \to \{0,1\}$ whose domain, is a finite prefix of the natural numbers, i.e., a set $\{0,1,\ldots,n\}$ for some $n \in \mathbb{N}$. Finite oracles will be identified with the binary string $v(0)v(1)\ldots v(n)$.

We say that a finite oracle $w$ *extends* a finite oracle $v$, if the domain of $v$ is a subset of the domain of $w$ and the functions $v$ and $w$ agree on the domain of $v$. A (total) oracle $O$ *extends* a finite oracle $v$ if the functions $O$ and $v$ agree on the domain of $v$. We also say that $v$ is a finite prefix of $O$.

As mentioned, a generic oracle is intuitively an oracle that in a sense has all properties that can be enforced by a stage construction. During the stages of these constructions, usually finite oracles are extended. In order to be able to complete the next stage, at every stage we must have still enough possibilities for such an extension. Formally we need a property called "denseness": A set $D$ of finite oracles is *dense*, if every finite oracle $v$ has an extension to a finite oracle $w \in D$.

Dense sets $D$ should be thought of as those finite prefixes of oracles that fulfill (or, *meet*) the condition aimed at during a specific stage of a usual oracle construction. The result of a stage construction as a whole has to meet a countable number $D_1, D_2, \ldots$ of such conditions. In the separation of P from NP by

Baker, Gill, and Solovay, the sets $D_i$ consist of those finite prefixes of oracles, for which a *test language* $L(O)$ (depending on oracle $O$) cannot be decided by the $i$-th deterministic polynomial time oracle Turing machine. Hence, if $O$ meets all $D_i$, the resulting $L(O)$ cannot be in $P^O$. From the way $L(O)$ is defined, namely $L(O)$ consists of those words $0^n$ for which there is some string of length $n$ in $O$, it is very easy to see that for every $O$, $L(O) \in NP^O$, and thus the desired relativized separation was obtained.

Generic oracles now are oracles that meet all conditions describable in a certain language. Here, we consider all conditions that result from sets in the arithmetical hierarchy (see, e.g., [Rog67]). Let $\mathcal{C} = \{D_1, D_2, \dots\}$ be a countable collection of dense sets of finite oracles. An oracle is $\mathcal{C}$-*generic* if, for each $i$, $O$ has a finite prefix $w_i \in D_i$. A set $D$ of finite oracles (or a set of finite words) is *arithmetic*, if membership in $D$ can be expressed as a finite-length first order formula with recursive predicates. Let $\mathcal{A}$ be the collection of dense, arithmetic sets of finite oracles (in other words, the class of all dense sets from the arithmetical hierarchy). An oracle is *generic* if it is $\mathcal{A}$-generic.

The following analogue of Theorem 2 has been obtained:

**Theorem 4 [GKV03].** *Let $A$ and $B$ be arithmetic sets. Then $A \not\leq_m^{plt} B$ if and only if for some generic oracle $G$, $\mathrm{BLeaf}^P(A)^G \not\subseteq \mathrm{BLeaf}^P(B)^G$, if and only if for all generic oracles $G$, $\mathrm{BLeaf}^P(A)^G \not\subseteq \mathrm{BLeaf}^P(B)^G$.*

In other words, to show that two syntactic complexity classes can be separated with a generic oracle it suffices to show that the corresponding leaf languages are not plt-reducible to one-another.

*Proof sketch.* The implication from right to left (of the first "iff") immediately follows from Theorem 2. For the left to right implication a proof using standard arguments is as follows: For any oracle $O$, we define (following Baker, Gill, and Solovay) a *test language* $L(O)$ that is easily seen to be in $\mathrm{BLeaf}^P(B)^O$. The desired oracle $G$ is constructed by a stage construction such that in stage $2m+1$ the $m$-th dense arithmetic set is met, and in stage $2m$ it is ensured that the $m$-th $\mathrm{BLeaf}^P(A)$-NPOTM does not accept $L(G)$. (At this stage of the proof we have to demand that we are dealing with classes $\mathrm{BLeaf}^P(A, \overline{A})$ and $\mathrm{BLeaf}^P(B, \overline{B})$ since otherwise, if we restrict the sets $\overline{A}$ or $\overline{B}$ to some subsets, it may happen that the sets of oracles meeting the conditions of stage $2m$ are no longer dense. We come back to this point in a moment.) Thus the odd stages will ensure that $G$ is generic while the even stages will ensure that $\mathrm{BLeaf}^P(A)^G \not\subseteq \mathrm{BLeaf}^P(B)^G$. We remark that a similar construction was used by Foster in [Fos93], where a generic oracle separating IP from PSPACE was constructed. In that paper, the odd stages ensure that the resulting oracle is generic while the even stages ensure IP $\neq$ PSPACE relative to the constructed oracle.

The second "iff" follows from a standard result in recursion theory [BI87] stating that for many interesting properties $\Pi$, either all generic oracles have property $\Pi$ or none has; for details we refer the reader to [GKV03].     □

Combining Theorems 2 and 4, we obtain a consequence concerning simultaneous oracle separations.

**Corollary 5 [GKV03].** *Any oracle separation between arithmetic syntactic complexity classes (i.e., classes* $\mathrm{BLeaf}^{\mathrm{P}}(A)$ *for some arithmetic leaf language A) holds relative to all generic oracles.*

Applications of the just given theorem concerning (besides simultaneous separations) type-2 complexity and further topics, can be found in [GKV03]. The in our opinion main contribution of Theorem 4, however, is a partial answer to the question of how much generic oracles can be of help in a solution toward the P-NP-question. Because it is known that P and UP can be separated by *some* oracle, it is intriguing to conclude that, by Corollary 5, there exists a generic separating oracle; thus by Theorem 3 it follows that P $\neq$ NP. However, for this argument to be valid it is necessary that UP can be defined (relativizably) by a leaf language. Thus, what we do obtain is a result like "if there is an arithmetic language $A$ such that for all oracles $O$, $\mathrm{BLeaf}^{\mathrm{P}}(A)^O = \mathrm{UP}^O$, then P $\neq$ NP." Unfortunately, it is known that the prerequisite of this statement does not hold: There is no $A$ such that relativizably, UP $= \mathrm{BLeaf}^{\mathrm{P}}(A)$ [BCS92]. To characterize UP relativizably, we need a leaf language pair $(A, R)$ with $R \neq \overline{A}$, and in this case, Theorem 4 does not apply.

The main obstacle in the proof of Theorem 4, when dealing with pairs $(A, R)$, $R \subsetneq \overline{A}$, is the denseness condition for the sets $C_{2m}$, which can only be proven for $R = \overline{A}$. But maybe there are suitable restrictions to general pairs $(A, R)$ with a very sparse difference $\overline{A} \setminus R$ for which the $C_{2m}$ are still dense, and maybe even some pair defining UP can be found that fulfills these restrictions. The main open issue in this context thus is to find sufficient conditions for leaf language pairs such that Theorem 4 still holds. If we now could prove that UP has a characterization via such a suitable pair of leaf languages, then P $\neq$ NP.

## 4   Leaf Functions

In [GV03] the class FPSPACE of all *functions* computable in polynomial space was examined, in particular in search for a characterization of FPSPACE similar to the leaf language characterization of the (language) class PSPACE (Theorem 1). Is there a "simple" function $\mathfrak{F}$, such that for every $h \in$ FPSPACE there is a polynomial-time nondeterministic Turing machine $M$ with the property that for every input word $x$, the value $h(x)$ is the value of $\mathfrak{F}$ applied to the leaf string of $M$ on $x$.

A bit more precise, given any function $F$ that evaluates leaf strings (we call $F$ a *leaf function*), define the class $\mathrm{FBLeaf}^{\mathrm{P}}(F)$ as the class of all functions $h$ for which there is a NPTM $M$ such that $h(x)$ equals the value of $F$ applied to the leaf string of $M(x)$. A more formal definition is given in [KSV00], together with an extensive list of leaf function characterizations of many counting, optimization, and other function classes. That paper also proves an oracle separation result for leaf functions, analogous to Theorem 2 for leaf languages, leading to previously not known oracle separations.

The question now is if there is a "simple" function $\mathfrak{F}$ such that FPSPACE $= \mathrm{FBLeaf}^{\mathrm{P}}(\mathfrak{F})$. $\mathfrak{F}$ should be "multiplication like" in the sense that each path contributes with some finite information that is "multiplied" to the result obtained from the path to the left, and the "product" obtained as a result of this multiplication will then be propagated to the path to the right.

From the leaf language characterization of the (language) class PSPACE, it is actually not too hard to conclude that there is a finite automaton that, given a leaf string as input, can compute the output of the corresponding FPSPACE-function, as was pointed out by Klaus Wagner.

**Proposition 6.** *There is a finite automaton $M$ with output such that for the function $f_M$ computed by $M$, we have:* $\text{FBLeaf}^{\text{P}}(f_M) = \text{FPSPACE}$.

*Proof sketch.* Let $f \in \text{FPSPACE}$. Then, the set of all pairs $(x, i)$ such that the $i$-th bit of $f(x)$ is on is in PSPACE. Express this PSPACE language using NPTM $M$ and some regular leaf language $B$ with non-solvable syntactic monoid. Now define a NPTM $M'$ operating as follows: On input $x$, $M'$ first branches for all values of $i$ in a certain exponential range, and then simulates $M$ on input $(x, i)$. Also, we have to make sure that the blocks for different values of $i$ are separated by a certain symbol # in the leaf string. Consider the finite automaton $M''$ reading the leaf string of $M'$ and operating as follows: While reading leaf symbols within a block for some $i$, it simulates a finite automaton $M'''$ for $B$. When $M''$ encounters a block marker #, it outputs 1 iff $M'''$ is in a final state, and 0 otherwise. In the next block, $M''$ resumes the simulation of $M'''$ in its initial state. Thus, $M''$ outputs a binary value when it reads a # and produces no output for other leaves. The outputs at a block marker, however, are exactly the bits of the value $f(x)$.    □

This result gives a very simple leaf function for FPSPACE that is much of a formal language theoretic nature. However, it does not address the nice algebraic properties that the class FPSPACE shares. A paper by Richard Ladner [Lad89] shows that the class FPSPACE coincides with the counting class #APTIME. Say that a *proof tree* of an alternating Turing machine $M$ is a minimal edge-induced subtree of the computation tree of $M$ that proves that $M$ accepts its input, in the same vein as an accepting path of a nondeterministic machine proves that the machine accepts its input. Formally, a proof tree of $M$ on input $x$ is a subtree $T'$ of the computation tree $T_{M(x)}$ of $M$ on $x$ whose root is the root of $T_{M(x)}$, whose leaves are accepting leaves of $T_{M(x)}$ and whose nodes can all be obtained respecting the following conditions:

- Every existential node in $T'$ has exactly one successor node in $T'$.
- Every universal node in $T'$ has all its successors in $T_{M(x)}$ also in $T'$.

Now, a function $f$ belongs to #APTIME if there is an alternating polynomial-time machine $M$ such that, for every $x$, $f(x)$ equals the number of proof trees of $M$ working on input $x$.

Counting proof trees of alternating machines corresponds in a nice way to evaluating arithmetic circuits over the natural numbers; hence Ladner's result indirectly yields an algebraic characterization of FPSPACE. The following theorem gives a leaf function characterization of FPSPACE that uses the spirit of Ladner's result.

**Theorem 7 [GV03].** *Let $\mathfrak{F}$ be the function that, given a sequence of $3 \times 3$ matrices with entries from $\{-1, 0, 1\}$, multiplies out this sequence and outputs as result the upper left entry in the product. Then, $\mathrm{FBLeaf}^P(\mathfrak{F}) = \mathrm{FPSPACE}$.*

*Proof sketch.* The proof of this result relies on a stronger characterization of FPSPACE in terms of counting functions than the one given in Ladner's paper [Lad89]. We first move to the slightly more general setting of polynomial space Turing machines that compute integer valued functions, i.e., we allow negative values, and show that this class coincides with the class of functions that can be obtained as differences of two functions counting proof trees of alternating machines. These difference functions can be computed by corresponding arithmetic circuits over the integers [Jia92]. Arithmetic circuits can be evaluated by straight-line programs using only 3 registers [BOC92]. Finally, the computation of these programs can be mimicked by multiplication of $3 \times 3$ matrices with entries from $\{-1, 0, 1\}$ [CMTV98]. This shows $\mathrm{FPSPACE} \subseteq \mathrm{FBLeaf}^P(\mathfrak{F})$; we remark that the way of going from counting functions via arithmetic circuits and straight-line programs to a matrix product is more or less the same as the way from $\mathrm{GapNC}^1$ to polynomial matrix products from [CMTV98], see also [Vol99a, Chapter 5.3].

To prove the inclusion $\mathrm{FBLeaf}^P(\mathfrak{F}) \subseteq \mathrm{FPSPACE}$, note that an $\mathrm{FBLeaf}^P(\mathfrak{F})$ computation can be seen as a polynomial-depth exponential-size arithmetic circuit over the integers that making use of $\mathrm{NC}^1$ circuits for addition and multiplication [Vol99a] can be transformed into a Boolean circuits of roughly the same size. This latter circuit evaluated by a polynomial-space Turing machine in depth-first manner. $\qquad\square$

Along the way, one thus obtains characterizations of FPSPACE using all the different computation models appearing in the proof; e.g., FPSPACE-functions are exactly those that can be computed by exponential size straight-line programs that use only 3 registers.

As pointed out in Sect. 2, regular leaf languages are particularly interesting since they allow characterizations of many important classes between P and PSPACE. We think it is worthwhile to identify a corresponding class of "easy" leaf functions that allows characterizations of classes including FPSPACE, #P, GapP, the classes of the min-max-hierarchy, and others.

## 5    Leaf Languages for Finite Automata

In this section we come back to the topic of leaf languages for finite nondeterministic automata. Since we want to talk about computation trees with output symbols in the leaves, we adjust the usual definition of an NFA as follows:

A *finite leaf automaton* $M$ is given by an input alphabet $\Sigma$, a finite set of states $Q$, a transition function $\delta \colon Q \times \Sigma \to Q^+$, an initial state $s \in Q$, a leaf alphabet $\Gamma$, and a function $v \colon Q \to \Gamma$ that associates a state $q$ with its *value* $v(q)$. In contrast to the definition of nondeterministic finite automata where we have that $\delta(q, a)$ is a *set* of states, we here additionally fix an ordering on the possible successor states by arranging them in a string from $Q^+$.

The computation tree $T_M(w)$ of $M$ on input $w$ is a labeled directed rooted tree defined as follows: The root of $T_M(w)$ is labeled $(s, w)$. If $i$ is a node in $T_M(w)$ labeled by $(q, x)$, where $x \neq \lambda$, $x = ay$ for $a \in \Sigma$, $y \in \Sigma^*$, and if $\delta(q, a) = q_1 q_2 \cdots q_k$, then $i$ has $k$ children in $T_M(w)$ which are labeled by $(q_1, y), (q_2, y), \ldots, (q_k, y)$ in this order.

If we look at the tree $T_M(w)$ and attach the symbol $v(q)$ to a leaf in this tree with label $(q, \varepsilon)$, then leafstring$^M(w)$ is the string of symbols attached to the leaves, read from left to right in the order induced by $\delta$.

Now let $A \subseteq \Gamma^*$. The language Leaf$^M(A) =_{\text{def}} \{ w \in \Sigma^* \mid \text{leafstring}^M(w) \in A \}$ is the language accepted by $M$ with leaf language $A$. The class Leaf$^{\text{FA}}(A)$ consists of all languages $B \subseteq \Sigma^*$, for which there is a leaf automaton $M$ with input alphabet $\Sigma$ and leaf alphabet $\Gamma$ such that $B = \text{Leaf}^M(A)$. If $\mathcal{C}$ is a class of languages then Leaf$^{\text{FA}}(\mathcal{C}) =_{\text{def}} \bigcup_{A \in \mathcal{C}} \text{Leaf}^{\text{FA}}(A)$.

In [PV01], Leaf$^{\text{FA}}(\mathcal{C})$ was identified for many classes $\mathcal{C}$, either complexity classes (given by resource bounds, e.g., time, space) or formal language classes (e.g., the classes of the Chomsky hierarchy). Sometimes the class Leaf$^{\text{FA}}(\mathcal{C})$ is as powerful as the class BLeaf$^P(\mathcal{C})$, sometimes a large gap between the two can be observed. The following theorem, that should be compared with Theorem 1, gives an example for the latter situation.

**Theorem 8 [GR75,PV01].** Leaf$^{\text{FA}}(\text{REG}) = \text{REG}$.

*Proof sketch.* We sketch the proof from [PV01] and remark that the result is essentially already present in a paper by Ginsburg and Rozenberg [GR75], studying Lindenmayer systems with regular control sets.

For the right to left inclusion, we note that for every language $A$, $A \in \text{Leaf}^{\text{FA}}(A)$, using the leaf automaton that copies its input to the leaves of its computation tree.

The left to right inclusion relies on a divide-and-conquer strategy, similar to one used to prove Savitch's Theorem (cf., e.g., [BDG95]). Given a language $B \in \text{Leaf}^{\text{FA}}(A)$, where $A$ is a regular language accepted by DFA $N$, we define an alternating finite automaton $M$ recognizing $B$ as follows: $M$ verifies that $N$, given the leaf string as input, is transfered from its initial to a final state. For that, $M$ guesses an intermediate state of this computation and then checks both halves of the computation recursively. The result follows, since alternating finite automata accept exactly the regular sets, see [Yu97].    □

One possible application of this result concerns the descriptive power of *monadic second-order logic with generalized quantifiers*. In the present paper we cannot give an adequate introductory treatment of topic of logical definability of language classes and general quantifiers, but we refer the reader to [EF95], in particular Chapter 10.

Consider a language $L$ over the alphabet $\{0, 1\}$. Let $X$ be a unary second-order variables, in other words: a set variable. There are $2^n$ different instances (assignments) of $X$, given a universe with $n$ elements (that is: an input word of length $n$. We assume an ordering of those instances, denoted by $X^1 < X^2 < \cdots < X^{2^n}$, given by the lexicographic ordering of the characteristic strings of the instances seen as simple sets. The *monadic second-order Lindström quantifier* $Q_L$ binding $X$ then is defined as follows. Let $\varphi(X)$ be a formula with free occurrence of $X$. Then $Q_L X \varphi(X)$ *holds on a string* $w = w_1 \cdots w_n$ ($w \models Q_L X \varphi(X)$) iff the word of length $2^n$ whose $i$th letter, $1 \le i \le 2^n$, is 1 if $w \models \varphi(X^i)$, and 0 otherwise, belongs to $L$. This definition generalizes perfectly to the case of languages $L$ over non-unary alphabets as well as quantifiers binding $k > 1$ variables (the string "tested" for membership in $L$ will then have length $2^{nk}$), see [GV03].

As examples, if $L$ is chosen to be the language $L_\exists =_{\text{def}} 0^*1(0+1)^*$ or the language $L_\forall =_{\text{def}} 1^*$, we obtain the usual second-order existential and universal quantifiers. If $L$ is a monoid word-problem (equivalently: a regular language), then we say that $Q_L$ is a monadic second-order monoidal quantifier; if $L$ is a groupoid word-problem (equivalently: a context-free language) then $Q_L$ is called monadic second-order groupoidal quantifier.

Denote by SOM the class of languages that can be defined in monadic second-order logic using only existential and universal quantifiers.

**Theorem 9 [Büc62,Tra61].** *The class* SOM *equals the class* REG *of regular languages.*

It is known that every regular language can even be defined using one existential quantifier binding only one set-variable, applied to a formula without any further second-order quantifiers [Tho82].

The following correspondence between the power of monadic second-order Lindström quantifiers and leaf automata was proven in [GV01]:

**Lemma 10.** *Let* $L \subseteq \Sigma^*$ *be a language with a neutral letter, i.e., there is a letter* $e \in \Sigma$ *such that, for all* $u, v \in \Sigma^*$, *we have* $uv \in L \iff uev \in L$. *Then* $\text{Leaf}^{\text{FA}}(L)$ *consists exactly of those languages definable by a formula that starts with a monadic second-order* $Q_L$ *quantifier and besides this has no occurrences of further second-order quantifiers.*

Let $\text{SOM}(Q_{\text{Mon}})$ be the class of all languages definable in monadic second-order logic with arbitrarily nested existential, universal and monoidal quantifiers. A consequence of Lemma 10 and Theorem 8 is a result concerning the expressive power of $\text{SOM}(Q_{\text{Mon}})$.

**Theorem 11 [GV03].** $\text{SOM}(Q_{\text{Mon}})$ *equals the class* REG *of regular languages.*

This is an extension of Theorem 9 to monadic second-order logics with monoidal quantifiers. It shows that these quantifiers do not extend the expressive power. Furthermore making use of the mentioned result by Thomas from [Tho82], it follows that a natural hierarchy (expressed in the possible number, position, and types of Lindström and usual quantifiers) collapses to the $\Sigma_1$ level.

Finally, we only briefly mention two other results about expressive power of related logics: Monadic second-order logic with groupoidal ("context-free") quantifiers can express a strict superclass CFL: It was shown that they can at least define every language in LOGCFL [GV01]. (LOGCFL is the class of languages logspace-reducible to context-free languages; it is known that these are exactly those languages accepted by auxiliary pushdown-automata operating simultaneously in logarithmic space and polynomial time [Coo71,Sud78]). The corresponding logic defines exactly those languages in Leaf$^{\mathrm{FA}}$(CFL); if this class coincides with a more familiar language class is open.

If we drop the restriction that our Lindström quantifiers are monadic but allow generalized quantifiers binding variables for arbitrary predicates (of arity possibly greater than 1), we obtain a logic whose expressive power relates to classes defined by leaf languages for nondeterministic Turing-machines in a way very similar to Lemma 10 above [BV98].

**Acknowledgement.** For many wonderful and productive discussions that lead to the results reported in this paper I am grateful to Matthias Galota, Sven Kosub, and Timo Peichl, my co-authors of the publications [PV01,GV01,GKV03, GV03] surveyed here.

# References

[AS96]   K. Ambos-Spies. Resource-bounded genericity. In S. B. Cooper, T. A. Sla-man, and S. S. Wainer, editors, *Computability, Enumerability, Unsolvabil-ity. Directions in Recursion Theory*, volume 224 of *London Mathematical Society Lecture Notes*, pages 1–59. Cambride University Press, 1996.

[Bar89]   D. A. Mix Barrington. Bounded-width polynomial size branching pro-grams recognize exactly those languages in NC$^1$. *Journal of Computer and System Sciences*, 38:150–164, 1989.

[BCS92]   D. P. Bovet, P. Crescenzi, and R. Silvestri. A uniform approach to define complexity classes. *Theoretical Computer Science*, 104:263–283, 1992.

[BDG95]   J. L. Balcázar, J. Díaz, and J. Gabarró. *Structural Complexity I*. Texts in Theoretical Computer Science. Springer Verlag, Berlin Heidelberg, 2nd edition, 1995.

[BG81]   C. Bennett and J. Gill. Relative to a random oracle P$^A$ ≠ NP$^A$ ≠ coNP$^A$ with probability 1. *SIAM Journal on Computing*, 10:96–113, 1981.

[BGS75]   T. Baker, J. Gill, and R. Solovay. Relativizations of the P=NP problem. *SIAM Journal on Computing*, 4:431–442, 1975.

[BI87]   M. Blum and R. Impagliazzo. Generic oracles and oracle classes. In *Proceedings 28th IEEE Symposium on Foundations of Computer Science*, pages 118–126. IEEE Computer Society Press, 1987.

[BOC92]   M. Ben-Or and R. Cleve. Computing algebraic formulas using a constant number of registers. *SIAM Journal on Computing*, 21:54–58, 1992.

[Büc62]   J. R. Büchi. On a decision method in restricted second-order arithmetic. In *Proceedings Logic, Methodology and Philosophy of Sciences 1960*, Stan-ford, CA, 1962. Stanford University Press.

[BV98]   H.-J. Burtschick and H. Vollmer. Lindström quantifiers and leaf language definability. *International Journal of Foundations of Computer Science*, 9:277–294, 1998.

[CCG$^+$94]   R. Chang, B. Chor, O. Goldreich, J. Hartmanis, J. Hastad, D. Ranjan, and P. Rohatgi. The random oracle hypothesis is false. *Journal of Computer and System Sciences*, 49:24–39, 1994.

[CF91]      J.-Y. Cai and M. Furst. PSPACE survives constant-width bottlenecks. *International Journal of Foundations of Computer Science*, 2:67–76, 1991.

[CKS81]     A. K. Chandra, D. Kozen, and L. J. Stockmeyer. Alternation. *Journal of the Association for Computing Machinery*, 28:114–133, 1981.

[CMTV98]    H. Caussinus, P. McKenzie, D. Thérien, and H. Vollmer. Nondeterministic $NC^1$ computation. *Journal of Computer and System Sciences*, 57:200–212, 1998.

[Coo71]     S. A. Cook. Characterizations of pushdown machines in terms of time-bounded computers. *Journal of the Association for Computing Machinery*, 18:4–18, 1971.

[EF95]      H.-D. Ebbinghaus and J. Flum. *Finite Model Theory*. Perspectives in Mathematical Logic. Springer Verlag, Berlin Heidelberg, 1995.

[FFKL93]    S. Fenner, L. Fortnow, S. Kurtz, and L. Li. An oracle builder's toolkit. In *Proceedings 8th Structure in Complexity Theory*, pages 120–131. IEEE Computer Society Press, 1993.

[Fos93]     J. A. Foster. The generic oracle hypothesis is false. *Information Processing Letters*, 45:59–62, 1993.

[GKV03]     M. Galota, S. Kosub, and H. Vollmer. Generic separations and leaf languages. *Mathematical Logic Quarterly*, 2003. To appear.

[GR75]      S. Ginsburg and G. Rozenberg. TOL schemes and control sets. *Information and Computation*, 27:109–125, 1975.

[GV01]      M. Galota and H. Vollmer. A generalization of the Büchi-Elgot-Trakhtenbrot-theorem. In *Computer Science Logic*, Lecture Notes in Computer Science, pages 355–368, Berlin Heidelberg, 2001. Springer Verlag.

[GV03]      M. Galota and H. Vollmer. Functions computable in polynomial space. Technical Report 03-18, Electronic Colloqium in Computational Complexity, 2003.

[HH90]      J. Hartmanis and L. Hemachandra. Robust machines accept easy sets. *Theoretical Computer Science*, 74(2):217–226, 1990.

[HLS⁺ 93]   U. Hertrampf, C. Lautemann, T. Schwentick, H. Vollmer, and K. W. Wagner. On the power of polynomial time bit-reductions. In *Proceedings 8th Structure in Complexity Theory*, pages 200–207, 1993.

[HVW96]     U. Hertrampf, H. Vollmer, and K. W. Wagner. On balanced vs. unbalanced computation trees. *Mathematical Systems Theory*, 29:411–421, 1996.

[Jia92]     J. Jiao. Some questions concerning circuit counting classes and other low-level complexity classes. Master's essay, 1992.

[JMT96]     B. Jenner, P. McKenzie, and D. Thérien. Logspace and logtime leaf languages. *Information and Computation*, 129:21–33, 1996.

[Joh90]     D. S. Johnson. A catalog of complexity classes. In J. van Leeuwen, editor, *Handbook of Theoretical Computer Science*, volume A, pages 67–161. Elsevier, 1990.

[KSV00]     S. Kosub, H. Schmitz, and H. Vollmer. Uniform characterizations of complexity classes of functions. *International Journal of Foundations of Computer Science*, 11(4):525–551, 2000.

[Kur83]     S. Kurtz. On the random oracle hypothesis. *Information and Control*, 57:40–47, 1983.

[Lad89]     R. Ladner. Polynomial space counting problems. *SIAM Journal on Computing*, 18(6):1087–1097, 1989.

[LFKN90]    C. Lund, L. Fortnow, H. Karloff, and N. Nisan. Algebraic methods for interactive proof systems. In *Proceedings 31st Symposium on Foundations of Computer Science*, pages 2–10. IEEE Computer Society Press, 1990.

110     H. Vollmer

[LLH]      Leaf Languages Homepage. http://www-thi.informatik.uni-hannover.de/
           forschung/leafl.php.
[Pap94a]   C. H. Papadimitriou. *Computational Complexity*. Addison-Wesley, Read-
           ing, MA, 1994.
[Pap94b]   C. H. Papadimitriou. Personal communication, 1994.
[PV01]     T. Peichl and H. Vollmer. Finite automata with generalized acceptance
           criteria. *Discrete Mathematics and Theoretical Computer Science*, 4:179–
           192, 2001.
[Rog67]    H. Rogers Jr. *Theory of Recursive Functions and Effective Computability*.
           McGraw-Hill, New York, 1967.
[Sha92]    A. Shamir. IP = PSPACE. *Journal of the Association for Computing
           Machinery*, 39:869–877, 1992.
[Sud78]    I. H. Sudborough. On the tape complexity of deterministic context-free
           languages. *Journal of the Association for Computing Machinery*, 25:405–
           414, 1978.
[Tho82]    W. Thomas. Classifying regular events in symbolic logic. *Journal of
           Computer and Systems Sciences*, 25:360–376, 1982.
[Tra61]    B. A. Trakhtenbrot. Finite automata and logic of monadic predicates.
           *Doklady Akademii Nauk SSSR*, 140:326–329, 1961. In Russian.
[Ver93]    N. K. Vereshchagin. Relativizable and non-relativizable theorems in the
           polynomial theory of algorithms. *Izvestija Rossijskoj Akademii Nauk*,
           57:51–90, 1993. In Russian.
[Vol99a]   H. Vollmer. *Introduction to Circuit Complexity – A Uniform Approach*.
           Texts in Theoretical Computer Science. Springer Verlag, Berlin Heidel-
           berg, 1999.
[Vol99b]   H. Vollmer. Uniform characterizations of complexity classes. *Complexity
           Theory Column 23, ACM-SIGACT News*, 30(1):17–27, 1999.
[Vol00]    H. Vollmer. A generalized quantifier concept in computational complexity
           theory. In J. Väänänen, editor, *Generalized Quantifiers and Computation*,
           volume 1754 of *Lecture Notes in Computer Science*, pages 99–123. Springer
           Verlag, Berlin Heidelberg, 2000. An extended version appeared in J. Vogel
           and K. W. Wagner, editors, *Komplexität, Graphen und Automaten*, Gerd
           Wechsung zum 60. Geburtstag, Jena, 1999.
[VW97]     H. Vollmer and K. W. Wagner. Measure one results in computational com-
           plexity theory. In D.-Z. Du and K.-I. Ko, editors, *Advances in Algorithms,
           Languages, and Complexity*. Kluwer Academic Publishers, 1997.
[Yu97]     S. Yu. Regular languages. In R. Rozenberg and A. Salomaa, editors, *Hand-
           book of Formal Languages*, volume I, chapter 2, pages 41–110. Springer
           Verlag, Berlin Heidelberg, 1997.

# Synchronizing Monotonic Automata[*]

Dimitry S. Ananichev and Mikhail V. Volkov

Department of Mathematics and Mechanics
Ural State University, 620083 Ekaterinburg, RUSSIA
{Dimitry.Ananichev, Mikhail.Volkov}@usu.ru

**Abstract.** We show that if the state set $Q$ of a synchronizing automaton $\mathcal{A} = \langle Q, \Sigma, \delta \rangle$ admits a linear order such that for each letter $a \in \Sigma$ the transformation $\delta(\_, a)$ of $Q$ preserves this order, then $\mathcal{A}$ possesses a reset word of length $|Q| - 1$. We also consider two natural generalizations of the notion of a reset word and provide for them results of a similar flavour.

## 1 Motivation and Overview

Let $\mathcal{A} = \langle Q, \Sigma, \delta \rangle$ be a DFA (deterministic finite automaton), where $Q$ denotes the state set, $\Sigma$ stands for the input alphabet, and $\delta : Q \times \Sigma \to Q$ is the transition function defining an action of the letters in $\Sigma$ on $Q$. The action extends in a unique way to an action $Q \times \Sigma^* \to Q$ of the free monoid $\Sigma^*$ over $\Sigma$; the latter action is still denoted by $\delta$. The automaton $\mathcal{A}$ is called *synchronizing* if there exists a word $w \in \Sigma^*$ whose action resets $\mathcal{A}$, that is, leaves the automaton in one particular state no matter which state in $Q$ it started at: $\delta(q, w) = \delta(q', w)$ for all $q, q' \in Q$. Any word $w$ with this property is said to be a *reset* word for the automaton. It is rather natural to ask how long such a word may be. Černý conjectured in 1964 (see [1]) that for any synchronizing automaton with $n$ states there exists a reset word of length $(n-1)^2$. By now this simply looking conjecture is arguably the most longstanding open problem in the theory of finite automata. It is however confirmed for several special types of automata. Clearly, the present paper is not an appropriate place for listing and analyzing all related results so we refer to the recent survey [6] and mention here only three typical examples involving restrictions of rather different sorts.

In Kari's elegant paper [5] the restriction has been imposed on the underlying digraphs of automata in question, namely, Černý's conjecture has been verified for automata with Eulerian digraphs. In contrast, Dubuc [2] has proved the conjecture under the assumption that there is a letter which acts on the state set $Q$ as a cyclic permutation of order $|Q|$. A condition of yet another type has

---

[*] The authors acknowledge support from the Education Ministry of Russian Federation, grants E02-1.0-143 and 04.01.059, from the Russian Foundation of Basic Research, grant 01-01-00258, and from the INTAS through the Network project 99-1224 'Combinatorial and Geometric Theory of Groups and Semigroups and its Applications to Computer Science'.

Z. Ésik and Z. Fülöp (Eds.): DLT 2003, LNCS 2710, pp. 111–121, 2003.

been used by Eppstein [3] who has confirmed Černý's conjecture for automata whose states can be arranged in some cyclic order which is preserved by the action of each letter in $\Sigma$. Eppstein has called those automata *monotonic*; we will refer to them as to *oriented* automata since we prefer to save the term 'monotonic' for a somewhat stronger notion which is in fact the object of the present paper.

We call a DFA $\mathcal{A} = \langle Q, \Sigma, \delta \rangle$ *monotonic* if its state set $Q$ admits a linear order $\leq$ such that for each letter $a \in \Sigma$ the transformation $\delta(\_, a)$ of $Q$ preserves $\leq$ in the sense that $\delta(q, a) \leq \delta(q', a)$ whenever $q \leq q'$. It is clear that monotonic automata form a (proper) subclass of the class of oriented automata, and therefore, by Eppstein's result any synchronizing monotonic automaton with $n$ states possesses a reset word of length $(n-1)^2$. We will radically improve this upper bound by showing that such an automaton can be in fact reset by a word of length $n-1$. It is easy to see that the latter bound is already exact. (Observe that for general oriented automata the bound $(n-1)^2$ is exact: for each $n \geq 3$ Černý has constructed in [1] an $n$-state synchronizing automaton whose shortest reset word is of length $(n-1)^2$, and one can easily check that all these automata are oriented.)

In fact, we will prove a much stronger result in the flavour of Pin's generalization [7,8] of Černý's conjecture. Given a DFA $\mathcal{A} = \langle Q, \Sigma, \delta \rangle$, we define the *rank* of a word $w \in \Sigma^*$ as the cardinality of the image of the transformation $\delta(\_, w)$ of the set $Q$. (Thus, in this terminology reset words are precisely words with rank 1.) In 1978 Pin conjectured that for every $k$, if an $n$-state automaton admits a word of rank at most $k$, then it has also a word with rank at most $k$ and of length $(n-k)^2$. Pin [7,8] has proved the conjecture for $n-k = 1, 2, 3$ but Kari [5] has found a remarkable counter example in the case $n-k = 4$. It is not yet clear if the conjecture holds true for some restricted classes of automata such as, say, the class of oriented automata. For monotonic automata, however, the situation is completely clarified by the following

**Theorem.** *Let $\mathcal{A}$ be a monotonic DFA with $n$ states and let $k$ be an integer satisfying $1 \leq k \leq n$. If there is a word of rank at most $k$ with respect to $\mathcal{A}$, then some word of length at most $n - k$ also has rank at most $k$ with respect to $\mathcal{A}$.*

The proof (which, being elementary in its essence, is not easy) is presented in Section 2. In Section 3 we discuss a related problem arising when one replaces the above notion of the rank by a similar notion of the interval rank. Given a monotonic DFA $\mathcal{A} = \langle Q, \Sigma, \delta \rangle$, we define the *interval rank* of a word $w \in \Sigma^*$ as the cardinality of the least interval of the chain $\langle Q, \leq \rangle$ containing the image of the transformation $\delta(\_, w)$. Thus, when looking for a word of low interval rank, we aim at compressing the state set of an automaton into a certain small interval; in other words, if we have several copies of the automaton, each being in a distinct initial state, then applying such a word we can make the behaviour of all the copies be 'almost the same'.

It is to be expected that compressing to small intervals would require more effort than compressing to just small subsets whose elements can be scattered over the state set in arbitrary way. We provide a series of examples showing that in general no linear function of the size $n$ of the state set can serve as an upper bound for the length of a word of interval rank 2 (Propositions 1 and 2). This strongly contrasts with our main theorem. On the other hand, for any $k$ with $2 \leq k \leq n$ we give a rough quadratic upper bound for the length of a word of interval rank $k$ (Proposition 3).

## 2    Proof of the Main Theorem

Of course, without any loss we may assume that the state set $Q$ of our monotonic automaton $\mathcal{A} = \langle Q, \Sigma, \delta \rangle$ is the set $\{1, 2, \ldots, n\}$ of the first $n$ positive integers and that the linear order $\leq$ on $Q$ is the usual order $1 < 2 < \cdots < n$. For $x, y \in Q$ with $x \leq y$ we denote by $[x, y]$ the interval $\{x, x + 1, x + 2, \ldots, y\}$. Then for any non-empty subset $X \subseteq Q$ we have $X \subseteq [\min(X), \max(X)]$ where $\max(X)$ and $\min(X)$ stand respectively for the maximal and the minimal elements of $X$. Given a word $w \in \Sigma^*$ and non-empty subset $X \subseteq Q$, we write $X.w$ for the set $\{\delta(x, w) \mid x \in X\}$. Also observe that since the composition of order preserving transformations is order preserving, all transformations $\delta(\_, w)$ where $w \in \Sigma^*$ are order preserving. We say that a subset $X \subseteq Q$ is *invariant* with respect to a transformation $\varphi$ of the set $Q$ if $X\varphi \subseteq X$.

**Lemma 1.** *Let $X$ be a non-empty subset of $Q$ such that $\max(X.w) \leq \max(X)$ for some $w \in \Sigma^*$. Then for each $p \in [\max(X.w), \max(X)]$ there exists a word $\mathcal{D}(X, w, p) \in \Sigma^*$ of length at most $\max(X) - p$ such that $\max(X.\mathcal{D}(X, w, p)) \leq p$.*

*Proof.* If $p = \max(X)$, then the empty word satisfies all the properties to be fulfilled by the word $\mathcal{D}(X, w, p)$. Therefore for the rest of the proof we may assume that $p < \max(X)$ and, therefore, $\max(X.w) < \max(X)$. Take an arbitrary $q$ in the interval $[\max(X.w) + 1, \max(X)]$. We want to show that there is a letter $\alpha(q) \in \Sigma$ such that $\delta(q, \alpha(q)) < q$. Arguing by contradiction, suppose that for some $q \in [\max(X.w) + 1, \max(X)]$ we have $\delta(q, a) \geq q$ for all letters $a \in \Sigma$. Since all transformations $\delta(\_, a)$ are order preserving, this would mean that the interval $Y = [q, n]$ is invariant with respect to all these transformations whence it is also invariant with respect to all transformations $\delta(\_, w)$ with $w \in \Sigma^*$. But $\max(X) \in Y$ while $\delta(\max(X), w) = \max(X.w) \notin Y$, a contradiction.

Now we construct a sequence of words as follows: let $u_1 = \alpha(\max(X))$ and, as long as $\delta(\max(X), u_i) > p$, let $u_{i+1} = u_i \alpha(\delta(\max(X), u_i))$. Observe that by the construction the length of the word $u_i$ equals $i$ and the last word $u_s$ in the sequence must satisfy $\delta(\max(X), u_s) \leq p$. Besides that we have $s \leq \max(X) - p$ because by the construction

$$\max(X) > \delta(\max(X), u_1) > \delta(\max(X), u_2) > \ldots$$

$$\cdots > \delta(\max(X), u_{s-1}) > p.$$

Thus, the word $u_s$ can be chosen to play the role of $\mathcal{D}(X, w, p)$ from the formulation of the lemma.

By symmetry, we also have the following dual statement:

**Lemma 2.** *Let $X$ be a non-empty subset of $Q$ such that $\max(X.w) \geq \max(X)$ for some $w \in \Sigma^*$. Then for each $p \in [\min(X), \min(X.w)]$ there exists a word $\mathcal{U}(X, w, p) \in \Sigma^*$ of length at most $p - \min(X)$ such that $\min(X.\mathcal{U}(X, w, p)) \geq p$.*

Now we can begin with the proof of the main theorem. We induct on $n$ with the induction base $n = 1$ being obvious. Thus, suppose that $n > 1$ and consider the set $X = \{\min(Q.w) \mid w \in \Sigma^*, |Q.w| \leq k\}$. (This set is not empty because by the condition of the theorem there exists a word of rank $\leq k$ with respect to $\mathcal{A}$.) Let $m = \max(X)$ and let $v \in \Sigma^*$ be such that $\min(Q.v) = m$ and $|Q.v| \leq k$.

Consider the interval $Y = [1, m]$. It is invariant with respect to all transformations $\delta(\_, w)$, $w \in \Sigma^*$. Indeed, arguing by contradiction, suppose that there are $q \in Y$ and $w \in \Sigma^*$ such that $\delta(q, w) > m$. Since the transformation $\delta(\_, w)$ is order preserving, $\min(Q.vw) = \delta(m, w) \geq \delta(q, w) > m$. At the same time, $|Q.vw| \leq |Q.v| \leq k$ whence $\min(Q.vw)$ belongs to the set $X$. This contradicts the choice of $m$.

Now consider the set $Z = \{q \in Q \mid \delta(q, w) \leq m$ for some $w \in \Sigma^*\}$. Observe that $Z$ is an interval and that $Y \subseteq Z$ since for $q \in Y$ the empty word can serve as $w$ with $\delta(q, w) \leq m$. Therefore $\max(Z) \geq m$. Fix a word $u \in \Sigma^*$ such that $\delta(\max(Z), u) \leq m$. Then $\delta(q, u) \leq m$ for each $q \in Z$ as the transformation $\delta(\_, u)$ is order preserving.

Finally, consider the interval $T = [\max(Z) + 1, n] = Q \setminus Z$. It is invariant with respect to all transformations $\delta(\_, w)$, $w \in \Sigma^*$. Indeed, arguing by contradiction, suppose that there exist $q \in T$ and $w \in \Sigma^*$ such that $\delta(q, w) \leq \max(Z)$. This means that $\delta(q, wu) \leq m$ whence $q \in Z$, in a contradiction to the choice of $q$.

The following picture should help the reader to keep track of the relative location of the intervals introduced so far. We have also depicted the actions of the words $u$ and $v$ introduced above on the states $\max(Z)$ and 1 respectively.

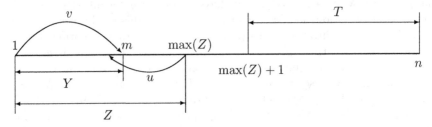

Now consider the state $p \in Q$ defined as follows:

$$p = \begin{cases} k - |T| & \text{if } |T| + m \leq k, \\ m & \text{if } |T| + m > k. \end{cases} \tag{1}$$

Observe that $m \leq p \leq n - |T| = \max(Z)$. Therefore we can apply Lemma 1 to the set $Z$, the state $p$ and the word $u \in \Sigma^*$. Let $w_1 = \mathcal{D}(Z, u, p)$; then the length of $w_1$ is at most $\max(Z) - p$ and $\max(Z.w_1) \le p$. Therefore $Z.w_1 = (Q \setminus T).w_1 \subseteq [1, p]$. Since the interval $T$ is invariant with respect to $\delta(\_, w_1)$, we conclude that $Q.w_1 \subseteq [1, p] \cup T$. From (1) we see that in the case when $|T| + m \le k$ the length of word $w_1$ does not exceed $\max(Z) - k + |T| = n - k$ and $|Q.w_1| \le q + |T| = k$. We have thus found a word of length at most $n - k$ and rank at most $k$ with respect to $\mathcal{A}$. This means that for the rest of the proof we may assume that $|T| + m > k$ and $p = m$. In particular, the length of $w_1$ is at most $\max(Z) - m$ and $Q.w_1 \subseteq Y \cup T$.

Consider the following state $r \in Q$:

$$r = \begin{cases} m + 1 + |T| - k & \text{if } |T| + 1 \le k, \\ m & \text{if } |T| + 1 > k. \end{cases} \tag{2}$$

Clearly, $1 \le r \le m$, and we can apply Lemma 2 to the set $Q$, the state $r$ and the word $v$. Let $w_2 = \mathcal{U}(Q, v, r)$. The length of $w$ is at most $r - \min(Q) = r - 1$ and $\delta(1, w_2) \ge r$. Since the interval $Y = [1, m]$ is invariant with respect to $\delta(\_, w_2)$, we conclude that $Y.w_2 \subseteq [r, m]$. From (2) we see that in the case when $|T| + 1 \le k$ the length of the word $w_1 w_2$ does not exceed $\max(Z) - m + r - 1 = \max(Z) + |T| - k = n - k$ and $|Q.w_1 w_2| \le m - r + 1 + |T| = k$. Again we have found a word of length at most $n - k$ and rank at most $k$. Thus, from now on we assume that $|T| + 1 > k$ and $r = m$. This means that $Q.w_1 w_2 \subseteq \{m\} \cup T$ and the length of the word $w_1 w_2$ is at most $(\max(Z) - m) + (m - 1) = \max(Z) - 1$.

Consider now the automaton $\mathcal{A}_T = \langle T, \Sigma, \delta_T \rangle$ where $\delta_T$ is $\delta$ restricted to the set $T \times \Sigma$. We have observed that the set $T$ is invariant with respect to all transformations $\delta(\_, w)$, $w \in \Sigma^*$, whence $\mathcal{A}_T$ is a DFA which obviously is monotonic. We claim that there is a word of rank at most $k - 1$ with respect to $\mathcal{A}_T$. Indeed, suppose that $|T.w| \ge k$ for each word $w \in \Sigma^*$. Since $T \cap Y = \emptyset$ and both $T$ and $Y$ are invariant, we obtain $T.w \cap Y.w = \emptyset$ for every $w \in \Sigma^*$. Therefore

$$|Q.w| \ge |Y.w| + |T.w| \ge 1 + k > k.$$

This contradicts to the condition that there exists a word of rank at most $k$ with respect to the automaton $\mathcal{A}$.

We see that we are in a position to apply the induction assumption to the automaton $\mathcal{A}_T$. Hence there exists a word $w_3 \in \Sigma^*$ of length at most

$$|T| - (k - 1) = n - \max(Z) - k + 1$$

such that $|T.w_3| \le k - 1$. Then the word $w_1 w_2 w_3$ has the length at most $(\max(Z) - 1) + (n - \max(Z) - k + 1) = n - k$ and $Q.w_1 w_2 w_3 \subseteq \{\delta(m, w_3)\} \cup T.w_3$ whence $|Q.w_1 w_2 w_3| \le 1 + |T.w_3| = k$.

For the sake of completeness we mention that it is pretty easy to find examples showing that the upper bound $n - k$ for the length of a word of rank $\le k$ with respect to a monotonic automaton is tight. Given $n$ and $k$ with $1 \le k \le n$, one can consider, for instance, the automaton on the set $\{1, 2, \ldots, n\}$ with the input alphabet $\{a\}$ and the transition function

$$\delta(i, a) = \begin{cases} i - 1 & \text{if } i > k, \\ i & \text{if } i \leq k. \end{cases}$$

Clearly, the word $a^{n-k}$ is the shortest word of rank $\leq k$ with respect to this automaton.

## 3    Compressing to Intervals

We start with presenting a series of examples of monotonic automata $\mathcal{A}_\ell$, where $\ell = 2, 3, \ldots$, that cannot be efficiently compressed to a 2-element interval. The state set $Q_\ell$ of the automaton $\mathcal{A}_\ell$ consists of $3\ell$ elements and can be conveniently identified with the chain

$$1 - 2\ell < 2 - 2\ell < \cdots < -1 < 0 < 1 < \cdots < \ell. \tag{3}$$

The input alphabet $\Sigma_\ell$ of $\mathcal{A}_\ell$ contains three groups of letters. The first group consists of $\ell$ 'non-increasing' letters $D_1, \ldots, D_\ell$ whose action on the set $Q_\ell$ is defined as follows:

$$\delta(j, D_i) = \begin{cases} i - 1 & \text{if } j = i, \\ 1 - 2i & \text{if } 1 - 2i < j < 0, \\ j & \text{in all other cases.} \end{cases} \tag{4}$$

The second group consists of $\ell - 1$ 'non-decreasing' letters $U_1, \ldots, U_{\ell-1}$ that act on the state set by the rule:

$$\delta(j, U_i) = \begin{cases} 1 - 2i & \text{if } j = -2i, \\ i & \text{if } 0 \leq j < i, \\ j & \text{in all other cases.} \end{cases} \tag{5}$$

Finally, we need a 'special' letter $S$ whose action is described by

$$\delta(j, S) = \begin{cases} \ell & \text{if } j > 0, \\ j + 1 & \text{if } j = 1 - 2i \text{ where } 2 \leq i \leq \ell, \\ j & \text{in all other cases.} \end{cases} \tag{6}$$

The following picture shows the action of $\Sigma_\ell$ on $Q_\ell$ for $\ell = 3$.

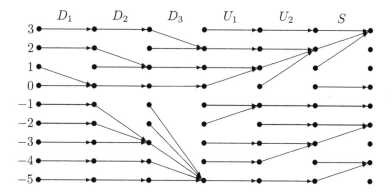

It is easy to verify that the actions (4)–(6) preserve the ordering (3) of the set $Q_\ell$, and therefore, $\mathcal{A}_\ell = \langle Q_\ell, \Sigma_\ell, \delta \rangle$ is a monotonic DFA. The intervals $[1 - 2\ell, -1]$ and $[0, \ell]$ are invariant with respect to the action of $\Sigma_\ell$. Therefore, for any word $w \in \Sigma_\ell^*$, the set $Q_\ell.w$ contains at least two states: a negative state and a non-negative one whence the rank of $w$ is at least 2. It is easy to find rather short words of rank 2 with respect to $\mathcal{A}_\ell$: one can verify that, for instance, the word $D_\ell U_1 S$ of length 3 has rank 2. The interval rank of this word is however the biggest possible as the image of the transformation $\delta(\_, D_\ell U_1 S)$ consists of the two extremes of the chain (3). Still we have the following

**Proposition 1** *There exists a word over $\Sigma_\ell$ whose interval rank with respect to the automaton $\mathcal{A}_\ell$ is equal to 2.*

*Proof.* For each $m$ such that $1 \leq m \leq \ell$ consider the interval $I_m = [1 - 2m, m]$. By the definition of $D_1$ we have

$$I_1.D_1 = \{-1, 0\} \tag{7}$$

Now consider the word

$$w_m = D_m D_{m-1} \cdots D_1 S U_{m-1}.$$

It is straightforward to verify that

$$I_m.w_m = [1 - 2m, m].D_m D_{m-1} \cdots D_1 S U_{m-1} =$$
$$= \{1 - 2m, 0\}.S U_{m-1} = \{2 - 2m, 0\}.U_{m-1} = \{3 - 2m, m - 1\}.$$

We see that $I_m.w_m \subseteq I_{m-1}$ for each $m = 1, \ldots, \ell$. Hence

$$I_m.w_m w_{m-1} \cdots w_2 \subseteq I_1. \tag{8}$$

Now for each $m$ such that $2 \leq m \leq \ell$ consider the word

$$W_m = w_m w_{m-1} \cdots w_2 D_1.$$

From (8) and (7) we get

$$I_m.W_m \subseteq I_1.D_1 = \{-1,0\}$$

for each $m = 2,\ldots,\ell$ and, in particular,

$$Q.W_\ell \subseteq \{-1,0\}.$$

Thus the interval rank of the word $W_\ell$ is at most 2. As we observed above, the rank of any word with respect to the automaton $\mathcal{A}_\ell$ is at least 2 whence the interval rank of $W_\ell$ is precisely 2.

Since the length of the word $w_m$ is $m + 2$, it is easy to calculate that the length of the word

$$W_\ell = w_\ell w_{\ell-1}\cdots w_2 D_1$$

is equal to $(\ell + 6)(\ell - 1)/2 + 1$. This is in fact the best possible result as our next proposition shows.

**Proposition 2** *The length of any word $w \in \Sigma_\ell^*$ whose interval rank with respect to the automaton $\mathcal{A}_\ell$ is 2 is at least $(\ell + 6)(\ell - 1)/2 + 1$.*

*Proof.* From the discussion preceding the formulation of Proposition 1 it should be clear that the only interval of size 2 to which the set $Q_\ell$ can be compressed is the interval $[-1,0]$. Let $v = a_1 a_2 \cdots a_m$, $a_i \in \Sigma_\ell$, be a word such that $Q.v = [-1,0]$. Since both the intervals $[1 - 2\ell, -1]$ and $[0,\ell]$ are invariant with respect to the transformation $\delta(\_,v)$, we must have $[1 - 2\ell, -1].v = \{-1\}$ and $[0,\ell].v = \{0\}$.

Let $v_i = a_1 \cdots a_i$ be the prefix of length $i$ of the word $v$, $i = 1,\ldots,m$. We denote the set $[1 - 2\ell, -1].v_i$ by $I_i$ and the set $[0,\ell].v_i$ by $J_i$.

For each $k \in [1, \ell - 1]$, let $\sigma(k)$ be the least number such that the sets $I_i$ for all $i \geq \sigma(k)$ are contained in the interval $[1 - 2k, -1]$ (This number exists because $I_m = [1 - 2\ell, -1].v = \{-1\} \subseteq [1 - 2k, -1]$ for each $k \in [1, \ell - 1]$.) Observe that if $\sigma(k) = s$ then the letter $a_s$ in the word $v$ coincides with the letter $U_k$. Indeed, by the choice of $s$ we must have

$$I_{s-1} \not\subseteq [1 - 2k, -1] \quad \text{and} \quad I_{s-1}.a_s = I_s \subseteq [1 - 2k, -1].$$

The first condition shows that $\min(I_{s-1}) < 1 - 2k$ whence $\min(I_{s-1}) \leq -2k$ while the second one implies that $\delta(\min(I_{s-1}), a_s) = \min(I_s) \geq 1 - 2k$. Since the transformation $\delta(\_, a_s)$ is order preserving, we have that $\delta(-2k, a_s) \geq 1 - 2k$. The only letter in $\Sigma_\ell$ satisfying this property is $U_k$.

Similarly, for each $k \in [1, \ell - 1]$, let $\tau(k)$ be the least number such that the sets $I_i$ for all $i \geq \tau(k)$ are contained in the interval $[-2k, -1]$. As this interval contains $[1 - 2k, -1]$, we clearly have

$$\tau(k) \leq \sigma(k). \tag{9}$$

Observe that if $\tau(k) = t$ then the letter $a_t$ in the word $v$ coincides with the letter $S$. (It shows, in particular, that the equality in (9) is impossible whence (9) is in fact strict.) Indeed, by the choice of $t$ we should have

$$I_{t-1} \nsubseteq [-2k, -1] \quad \text{and} \quad I_{t-1}.a_t = I_t \subseteq [-2k, -1].$$

The former condition means that $\min(I_{t-1}) < -2k$ and so $\min(I_{t-1}) \leq -1 - 2k$ while from the latter one we see that $\delta(\min(I_{t-1}), a_t) = \min(I_t) \geq -2k$. As above, employing the fact that the action of the letter $a_t \in \Sigma_\ell$ preserves the ordering (3), we conclude that $\delta(-1 - 2k, a_t) \geq -2k$. Checking through the rules (4)–(6), we see that $S$ is the only letter verifying this requirement.

Further, if $\tau(k) = t$ then none of the letters $D_{k+1}, \ldots, D_\ell$ may occur in the suffix $a_t \cdots a_m$ of the word $v$ because each of these letters sends the whole negative interval $[1 - 2\ell, -1]$ out of the interval $[-2k, -1]$.

For the next step in clarifying the possible structure of the word $v$, we apply a similar argument to the sets $J_i$. Namely, for each $k \in [1, \ell - 1]$ we let $\pi(k)$ be the least number such that the sets $J_i$ for all $i \geq \pi(k)$ are contained in the interval $[0, k]$. (This number exists because $J_m = [0, \ell].v = \{0\} \subseteq [0, k]$ for each $k \in [1, \ell - 1]$.) We want to show that if $\pi(k) = p$ then the letter $a_p$ necessarily coincides with $D_{k+1}$. Indeed, the choice of $p$ ensures that

$$J_{p-1} \nsubseteq [0, k] \quad \text{and} \quad J_{p-1}.a_p = J_p \subseteq [0, k].$$

From the first condition we conclude that $\max(J_{p-1}) > k$, or in other words, $\max(J_{p-1}) \geq k + 1$ while the second condition means that $\delta(\max(J_{p-1}), a_p) = \max(J_p) \leq k$. Hence $\delta(k + 1, a_p) \leq k$ and $D_{k+1}$ is the only letter that has this property. In particular, this means that $\delta(\max(J_{p-1}), a_p) = k$ whence $k \in J_p$.

Further, if $\pi(k) = p$ then none of the letters $U_{k+1}, \ldots, U_{\ell-1}$ occur in the suffix $a_p \cdots a_m$ of the word $w$ because each of these letter sends the whole non-negative interval $[0, \ell]$ beyond the interval $[0, k]$. This means that

$$\sigma(k + 1) < \pi(k). \tag{10}$$

Now we note that if $\tau(k) = t$ then $J_{t-1} = \{0\}$. Indeed, recall that we have shown that $a_t = S$ and by (6) $S$ sends each $j > 0$ to the state $\ell$. Thus, if $J_{t-1} \neq \{0\}$ then $\ell \in J_{t-1}.a_t = J_{t-1}.S$. However, all letters in $\Sigma_\ell$ except $D_\ell$ fix $\ell$ and the letter $D_\ell$ does not occur in the word $v$ after $a_t$.

Recall that we have proved that if $\pi(k) = k$ then $k \in J_p$. No letter in $\Sigma_\ell$ can move a positive state down by more than 1, hence we need at least $k$ letters in order to transform the set $J_p$ into the set $J_{t-1} = \{0\}$. We thus have obtain the inequality $(t - 1) - p \geq k$ or

$$\tau(k) - \pi(k) \geq k + 1. \tag{11}$$

The inequalities (9) (which we recall is in fact strict), (10) and (11) imply that

$$\pi(k) - \pi(k + 1) \geq \sigma(k + 1) + 1 - \pi(k + 1)$$
$$\geq \tau(k + 1) + 2 - \pi(k + 1) \geq k + 4. \tag{12}$$

The equality $[0, \ell].v_m = [0, \ell].v = \{0\}$ implies that

$$m \geq \pi(0) = \big(\pi(0) - \pi(1)\big) + \big(\pi(1) - \pi(2)\big) + \dots$$
$$\dots + \big(\pi(\ell - 2) - \pi(\ell - 1)\big) + \pi(\ell - 1). \quad (13)$$

From (13) and (12) we conclude that

$$m \geq \sum_{k=0}^{\ell-2}(k + 4) + 1 = (\ell + 6)(\ell - 1)/2 + 1$$

as required.

Recall that the number of states of the automaton $\mathcal{A}_\ell$ is equal to $3\ell$. Thus, Propositions 1 and 2 show that for any $n \geq 6$ and divisible by 3 there exists a monotonic DFA with $n$ states for which the shortest word of interval rank 2 is of length $\approx n^2/18$. On the other hand, we have a quadratic upper bound for the lengths of words of interval rank $k \geq 2$ in monotonic automata:

**Proposition 3** *Let $\mathcal{A} = \langle Q, \Sigma, \delta \rangle$ be a monotonic DFA with $n$ states and let $k$ be an integer satisfying $2 \leq k \leq n$. If there is a word of interval rank at most $k$ with respect to $\mathcal{A}$, then some word of length at most $(n - k)(n - k + 1)/2$ also has interval rank at most $k$ with respect to $\mathcal{A}$.*

*Proof.* Consider the automaton $\mathcal{I}$ whose states are the intervals of the chain $\langle Q, \leq \rangle$. The automaton $\mathcal{I}$ has the same input alphabet $\Sigma$ and the transition function $\delta'$ defined by the rule: for each $I$ being an interval of $\langle Q, \leq \rangle$ and for each letter $a \in \Sigma$

$$\delta'(I, a) = [\min(I.a), \max(I.a)].$$

It is easy to see that the existence of a word of interval rank at most $k$ with respect to $\mathcal{A}$ implies that there is a path in $\mathcal{I}$ from $Q$ to an interval of size at most $k$. Conversely, if we read the consecutive labels of a minimum length path from $Q$ to an interval of size at most $k$ in the automaton $\mathcal{I}$ then we get a word of minimum length with interval rank at most $k$ with respect to $\mathcal{A}$. Thus, it remains to estimate the length of such a path. Clearly, a minimum length path from $Q$ to an interval of size at most $k$ goes only through intervals of size $k + 1, \dots, n - 1$ between its extreme points and visit each of these intermediate intervals at most once. Therefore the length of such a path exceeds the number of intervals of size $k + 1, \dots, n - 1$ at most by one, and this gives the desired upper bound $(n - k)(n - k + 1)/2$.

Clearly, the gap between the lower bound provided by Propositions 1 and 2 and the upper bound of Proposition 3 is quite big. We think that the lower bound gives a more realistic approximation than the upper bound which has been deduced from a rather straightforward approach.

**Acknowledgments.** Several useful comments of anonymous referees are gratefully acknowledged. In particular, one of the referees has discovered a flaw in the initial version of the proof of our main theorem.

# References

1. Černý, J.: Poznámka k homogénnym eksperimentom s konecnými avtomatami. Mat.-Fyz. Cas. Slovensk. Akad. Vied. **14** (1964) 208–216 [in Slovak].
2. Dubuc, L.: Sur les automates circulaires et la conjecture de Černý. RAIRO Inform. Theor. Appl. **32** (1998) 21–34 [in French].
3. Eppstein, D.: Reset sequences for monotonic automata. SIAM J. Comput. **19** (1990) 500–510.
4. Kari, J.: A counter example to a conjecture concerning synchronizing words in finite automata. EATCS Bull. **73** (2001) 146.
5. Kari, J.: Synchronizing finite automata on Eulerian digraphs. Math. Foundations Comput. Sci.; 26th Internat. Symp., Marianske Lazne, 2001. Lect. Notes Comput. Sci. **2136** (2001) 432–438.
6. Mateescu, A., Salomaa, A.: Many-valued truth functions, Černý's conjecture and road coloring. EATCS Bull. **68** (1999) 134–150.
7. Pin, J.-E.: Le Problème de la Synchronisation. Contribution à l'Étude de la Conjecture de Černý. Thèse de 3éme cycle. Paris, 1978 [in French].
8. Pin, J.-E.: Sur les mots synchronisants dans un automate fini. Elektronische Informationverarbeitung und Kybernetik **14** (1978) 283–289 [in French].

# Covering Problems from a Formal Language Point of View

Marcella Anselmo[1*] and Maria Madonia[2**]

[1] Dip. di Informatica ed Applicazioni, Università di Salerno
I-84081 Baronissi (SA) ITALY
anselmo@dia.unisa.it
[2] Dip. di Matematica ed Informatica, Università di Catania
Viale Doria 6/A, Catania ITALY
madonia@dmi.unict.it

**Abstract.** We consider the formal language of all words that are 'covered' by words in a given language. This language is said cov-free when any word has at most one minimal covering over it. We study the notion of cov-freeness in relation with its counterpart in classical monoids and in monoids of zig-zag factorizations. In particular cov-freeness is characterized by the here introduced notion of cov-stability. Some more properties are obtained using this characterization. We also show that the series counting the minimal coverings of a word over a regular language is rational.

## 1 Introduction

Recently several problems related to covers were examined, specially motivated by molecular biology, but also by data compression and computer-assisted music analysis. A wide literature exists; see for example [3,5,9,11,12,17,18] and all other papers there cited, namely in [18]. Rougly speaking, a covering of a word $w$ over a language $X$ is a way to cover $w$ by concatenations and overlaps of elements of $X$. The above cited literature affords the problems from an algorithmic point of view: given a word $x$ find an 'optimal' set of words of a certain type that 'covers' (or 'approximately covers') $x$, where 'optimal', 'covers', 'approximately covers' can have different meaning.

We think that a formal and deep study of this notion from a formal language theoretical point of view is needed. The paper [15] formalizes the intuitive concept of covering, introducing the notions of covering submonoid, covering code and investigating some related algebraic properties. In [15], the set of all words that are covered by words in a given set $X$, is denoted $X^{cov}$. Following this terminology, each above mentioned problem can be restated. For example,

* Partially supported by "Progetto Cofinanziato MIUR: Linguaggi formali e automi: teoria ed applicazioni"
** Partially supported by "Progetto Cofinanziato MIUR: Ragionamento su aggregati e numeri a supporto della programmazione e relative verifiche"

Z. Ésik and Z. Fülöp (Eds.): DLT 2003, LNCS 2710, pp. 122–133, 2003.

a word $x$ is quasi-periodic as in [3,5] if $x \in \{w\}^{cov}$ for some word $w$; in the same conditions, $w$ is referred to as a cover of $x$ in [17]; on the other hand in [11], $w$ covers $x$ if $x$ is factor of a word in $\{w\}^{cov}$. In [12] an algorithm is presented for computing a minimum set of $k$-covers for a given word, where a set of $k$-covers of word $x \in A^*$ is $X \subseteq A^k$ such that $x \in X^{cov}$. In [18], $p$ is an approximate period of $x$ if $x \in X^{cov}$, and the distance between $p$ and each word in $X$ is bounded by a constant. Also the shortest superstring problem [8], the one of finding the shortest string having each of the strings in a given set $X$ as factors, can be restated in terms of $X^{cov}$. If $X = \{x_1, x_2, \cdots, x_n\}$, then the problem is to find the shortest element in $X^{cov} \cap A^* x_1 A^* \cap \cdots \cap A^* x_n A^*$. Also remark that an approximate solution for the shortest superstring problem makes use of some cycle covers of $X$.

Moreover we show in this paper that language $X^{cov}$ with $X$ finite, can be very naturally described by the help of splicing systems as defined in [9] (see Section 2). We recall that splicing systems are a generative formalism introduced by Head to describe the set of DNA molecules that may potentially arise from an original set of DNA molecules under enzymatic activities [9].

We emphasize that above mentioned papers deal with efficient algorithms for solving problems related to the notion of covering. We want here afford a formal study of some properties of the set $X^{cov}$ of words covered by words in $X$, from a formal language theoretical point of view. Remark that in [15], coverings are presented as a generalization of z-decompositions (or zig-zag decompositions); indeed a covering can be viewed as a z-decomposition with steps to the right in $X$ and steps to the left in $A^*$. We emphasize that nevertheless, the most of the new definitions and results here presented are not mere generalizations of analogous results on z-decompositions. When considering coverings instead of z-decompositions, new problems arise (see for instance Lemma 1) and different results are obtained (see for instance Examples 7, 8 and Proposition 5).

In this paper we consider the formal definition of covering, as given in [15]. A covering over $X$ is a sequence of steps to the right on $X$ and to the left on $A^*$, alternatively. Particular meaning have minimal coverings, that is those coverings in which every step is necessary: eliminating a step we no longer have a covering of the whole word. And particular meaning have those languages for which any word has at most one minimal covering over them: we call them covering codes. Remark that these covering codes are different from covering codes as defined in [6] and that covering codes can be defined in terms of coding pairs of [14]. A covering submonoid generated by a covering code is said to be cov-free.

In this paper we study covering submonoids and in particular their cov-freeness, in analogy to the study of freeness in classical monoids. In the literature, the notion of freeness is related to the one of stability, in order to characterize it without considering the generating system, but by a global notion on the monoid. Indeed in the theory of monoids, a submonoid is free iff it is stable and in the theory of zig-zag submonoids a z-submonoid is z-free iff it is z-stable. Here (Section 3) we introduce the notion of cov-stability with the same motivations and in analogy with the definitions of classical stability [4] and z-stability [13]. We show that a covering submonoid is cov-stable iff it is cov-free.

This equivalence result allows us to prove many other properties of covering submonoids showing analogies and differences with their classical counterpart (Section 4). Recall that any covering submonoid is also a z-submonoid [15] and any z-submonoid is also a submonoid and that if $M$ is z-free then $M$ is free [16]. In particular, we show that if a covering submonoid $M$ is cov-free then it is free, very pure and also $X^*$ is very pure, where $X$ is the minimal covering generating system of $M$. On the other hand the intersection of two cov-free covering submonoids is not necessarily cov-free. We also compare the three properties of cov-freeness, z-freeness and (classical) freeness for covering submonoids. Despite both cov-freeness and z-freeness imply freeness, we give examples pointing out that cov-freeness and z-freeness are not related each other. Once again, coverings are not a mere generalization of z-decompositions.

Another problem we deal with in this paper (Section 5) is counting minimal coverings a word has over a given language. The main result is that the series associating to a word the number of its minimal coverings over a given language $X$ is always rational when $X$ is regular. Observe that this is not the case for z-factorizations. The series counting the different z-factorizations of a word is not always rational, even in the case of a regular language [1]. This result in particular allows us to decide whether a language is a covering code.

## 2    Preliminaries

Let us recall the following classical definitions [4].

Let $A$ be a finite alphabet, and $(A^*, ., \lambda)$ be the free monoid generated by $A$. As usual, $|u|$ denotes the length of $u$, $x \leq y$ means $x$ prefix of $y$; $x < y$ means $x$ proper prefix of $y$ ($x \neq y$); $xy^{-1} = v$ if $vy = x$. The set of proper prefixes (suffixes, resp.) of a language $X$ is denoted $Pref(X)$ ($Suff(X)$, resp.). It is well known that any submonoid $M$ of $A^*$ admits an unique minimal generating system, which from now on, we denote by $G(M)$. In particular, $G(M) = (M - \lambda) \setminus (M - \lambda)^2$. Language $X \subseteq A^*$ is a code if every word $w \in A^*$ has at most one factorization on $X$. Let $M$ be a submonoid of $A^*$, $M$ is stable if for all $u, v, w \in A^*$, $uv, vw, w, u \in M \Rightarrow v \in M$. Recall that a submonoid of $A^*$ is free iff it is stable.

Now, let us recall some definitions concerning the zig-zag factorizations (or z-factorizations) that can be found in [14] and [7]. Let $\overline{A}$ be a disjoint alphabet in bijection with $A$ and $w = a_1 \ldots a_n \in (A \cup \overline{A})^*$. The inverse of $w$ (denoted by $\overline{w}$) is the word $\overline{a}_n \ldots \overline{a}_1$ and for every $X \subseteq (A \cup \overline{A})^*$, we denote by $\overline{X}$ the set $\{\overline{x} : x \in X\}$. We denote by $red_A(w)$ (or simply $red(w)$) the canonical representative of the class of $w$ in the free group generated by $A$. A z-decomposition of $w$ over $(X, Y) \subseteq A^* \times A^*$ is a $n$-uple $(w_1, \ldots, w_n) \in (X \cup \overline{Y})^n$ such that $red_A(w_1 \ldots w_n) = w$ and for any $1 \leq i \leq n$, $red_A(w_1 \ldots w_i)$ is a prefix of $w$. A z-factorization of $w$ over $(X, Y)$ is a z-decomposition $(w_1, \ldots, w_n)$ of $w$ such that: for any $1 \leq i < j \leq n$, $red_A(w_1 \ldots w_i) \neq red_A(w_1 \ldots w_j)$. We denote by $(X, Y)^\dagger$ the set of words of $A^*$ having a z-decomposition over $(X, Y)$. Language $X$ is a z-code when any word has one z-factorization over $(X, X)$ at most. $M$ is z-free when its minimal z-generating system $ZG(M)$ is a z-code.

The following definitions and results about coverings are in [15].

A *covering* of $w$ over $X$ is a $n$-uple $\delta = (w_1, \cdots, w_n)$ where: $n$ is odd; for any odd $i$, $w_i \in X$; for any even $i$, $w_i \in \bar{A}^*$, $red(w_1 \cdots w_n) = w$ and for any $i$, $1 \leq i \leq n$, $red(w_1 \cdots w_i)$ is a prefix of $w$. The length of $\delta$ is $n$ and is denoted $\| \delta \|$. A covering is trivial when $\| \delta \| = 1$. A covering can be thus regarded to as a z-decomposition either over $(X, A^*)$ or over $(X, A)$. In $\delta = (w_1, \cdots, w_n)$, $w_i \in X$ ($w_i \in \bar{A}^*$, resp.) is referred to as a *step to the right, (left, resp.)*. If $w = a_1 \cdots a_m$, we say that $a_j$, $1 \leq j \leq m$, is *covered* by some step to the rigth (left, resp.) $w_i$ if $red(w_1 \cdots w_{i-1}) \leq a_1 \cdots a_{j-1}$, and $a_1 \cdots a_j \leq red(w_1 \cdots w_i)$ (if $a_1 \cdots a_j \leq red(w_1 \cdots w_{i-1})$ and $red(w_1 \cdots w_i) \leq a_1 \cdots a_{j-1}$, resp.) (by convention $a_1 \cdots a_0 = red(w_1 \cdots w_0) = \lambda$). The set of all words having a covering over $X$ is denoted $X^{cov}$. A minimal covering of $w \in A^*$ over $X$ is a covering that cannot be simplified: it means that it is not possible, by eliminating some steps, to obtain, still, a way to cover $w$. More precisely, a covering $\delta = (w_1, \cdots, w_n)$ of $w$ over $X$ is *minimal* if for any odd $i$ we have $w_i = x_1 x_2 x_3$ with $x_2 \neq \lambda$ and: $|red(w_1 \cdots w_j)| \leq |red(w_1 \cdots w_{i-1} x_1)|$ for any $j < i$; $|red(w_1 \cdots w_k)| \geq |red(w_1 \cdots w_{i-1} x_1 x_2)|$ for any $k > i$.

*Remark 1.* Note that minimal coverings correspond to z-factorizations of $w$ over $(X, A)$ [15]. This allows us in particular to claim that in a minimal covering of $w = a_1 \cdots a_n$, each $a_i$, for $1 \leq i \leq n$ is covered by three steps at most.

*Example 1.* Let $X = \{ababb, bab, ba, bba\}$ and let $w = ababbab$.
Then $\zeta = (ababb, \overline{bb}, bba, \overline{ba}, bab)$ and $\delta = (ababb, \overline{b}, bab)$ are coverings of $w$ over $X$ (see Fig. 1). Moreover $\zeta$ is not a minimal covering of $w$ over $X$, but $\delta$ is a minimal covering of $w$ over $X$.

$M \subseteq A^*$ is a *covering submonoid* of $A^*$ if $M^{cov} = M$. Note that any covering submonoid of $A^*$ is a submonoid of $A^*$. The *minimal covering generating system* of $M$ is $X \subseteq A^*$ such that $X^{cov} = M$ and, for any $Y \subseteq A^*$ such that $Y^{cov} = M$, $X \subseteq Y$ holds. Any covering submonoid $M$ of $A^*$ admits a minimal covering generating system and such a system is unique; we denote it by $cov\text{-}G(M)$. Moreover we have that $cov\text{-}G(M) = G(M) - O(M)$, where $O(M) = \{w \in G(M) : w$ has a non trivial covering over $G(M)\}$. $X$ is a *covering code* iff any $w \in A^*$ has at most one minimal covering over $X$. A covering submonoid is *cov-free* iff its minimal covering generating system is a covering code. Remark that these covering codes are different from covering codes as defined in [6]. Further, following the terminology of [14], $X$ is a covering code iff $(X, A)$ is a coding pair.

**Fig. 1.** Coverings of Example 1

*Example 2.* Let $X = \{aabab, abb\}$. It is easy to see that $X$ is a covering code. Indeed, we can only overlap the word $aabab$ and the word $abb$ on the factor $ab$ to obtain the word $aababb \in X^{cov}$ and all the possible concatenations of this word with itself or with other words of $X$ are without ambiguity.

We want to close this section showing how a splicing system can be used to generate a language associated to $X^{cov}$, when $X$ is finite. In Proposition 1 below, splicing system $S$ defines $\natural X^{cov}\$$, unless for some finite set $OV(X)$. Remark that when $X$ is finite, then $OV(X)$ is finite, but $X^{cov}$ is never finite.

A splicing system $S = (A, I, B, C)$, as defined by Head [9], consists of a finite alphabet $A$, a finite set $I$ of initial strings in $A^*$, and finite sets $B$ and $C$ of triples $(c, x, d)$ with $c, x, d \in A^*$. The language $L(S)$ generated by $S$ consists of the strings in $I$ and all strings that can be obtained by adjoining to $L(S)$ $ucxfq$ and $pexdv$ whenever $ucxdv$ and $pexfq$ are in $L(S)$ and $(c, x, d)$ and $(e, x, f)$ both belong to $B$ or $C$.

**Proposition 1.** *Let $X \subseteq A^*$ be finite, $\natural, \$ \notin A$ and $OV(X) = Pref(X) \cap Suff(X) \cup \{\lambda\}$. Then there exists a splicing system $S$ such that $L(S) = \natural X^{cov}\$ \cup \natural OV(X)\$$.*

*Proof.* Let $S = (A, \{\natural x\$|\ x \in X\}, \{(\lambda, v, \$), (\natural, v, \lambda)|\ v \in OV(X)\}, \emptyset)$. Any word in $\natural X^{cov}\$$ is generated by $S$, referring to its minimal covering and considering Remark 1. $\square$

# 3    Cov-freeness and Cov-stability

Classically the notion of freeness is related to the one of stability, that characterizes freeness without considering the generating system, but by a global notion on the submonoid. In fact in the theory of monoids, a submonoid is free iff it is stable and in the theory of zig-zag submonoids a z-submonoid is z-free iff it is z-stable. Here we introduce the notion of cov-stability with the same motivations and in analogy with the definitions of classical stability [4] and z-stability given in [13]. We show that a covering submonoid is cov-stable iff it is cov-free.

We emphasize that when coverings are dealt with, new problems arise in stating the definition of stability, that therefore will be a little more involved, since it needs the distinctions between different cases. The main difference with analogous definitions comes from the new situation pointed out in Lemma 1 and Example 3: in a minimal word with two different minimal coverings, the last (first, resp.) step in a covering is not necessarily different from the last (first, resp.) step in the other covering. This new situation implies the distinction of two cases, following that last steps are different or not (Fig. 2 illustrates these cases). Furthermore, some examples are given after the definition, in order to clarify where a definition of cov-stability simpler than the one here below, would not correctly work.

Let us now investigate some properties of minimal coverings. Remark 2 analyzes covering arising from the union of two minimal coverings.

*Remark 2.* Suppose $X \subseteq A^*$, $u, v, w \in A^*$, $\delta_1 = (u_1, u_2, \cdots, u_n)$ a mini-mal covering of $uv$ over $X$ and $\delta_2 = (w_1, w_2, \cdots, w_m)$ a minimal covering of $vw$ over $X$. The union of coverings $\delta_1$ and $\delta_2$ gives raise to the covering $\delta_3 = (u_1, u_2, \cdots, u_n, \bar{v}, w_1, w_2, \cdots, w_m)$ of $uvw$. Covering $\delta_3$ is not necessarily a minimal one. From $\delta_3$, we can obtain a minimal covering of $uvw$ by elimi-nating some steps that are not necessary for covering the word. This process can be done in at least two ways; informally: either by eliminating some of the last steps of $\delta_1$, or some of the first steps of $\delta_2$. As an example, let $u = ab$, $v = c$, $w = de$, $\delta_1 = (ab, \bar{b}, bc)$, $\delta_2 = (cd, \bar{d}, de)$. We have $\delta_4 = (ab, \lambda, cd, \bar{d}, de)$, $\delta_5 = (ab, \bar{b}, bc, \lambda, de)$ are minimal coverings of $uvw$ obtained from the union of $\delta_1$ and $\delta_2$ and, moreover, $\delta_4 \neq \delta_5$.

The following lemma and related example highlight one of the fundamental difference between coverings, z-decompositions and factorizations. This differ-ence will be crucial in defining the notion of cov-stability.

**Lemma 1.** *Let $X \subseteq A^*$. Let us suppose that $X$ is not a covering code and let $w$ be a minimal word with two different minimal coverings $\delta_1 = (w_1, w_2, \cdots, w_n)$ and $\delta_2 = (z_1, z_2, \cdots, z_m)$. Then $|red(w_1 \cdots w_i)| = |red(z_1 \cdots z_j)|$ implies $(i = j = 1)$ or $(i = n$ and $j = m)$ or $(i = n - 1$ and $j = m - 1)$.*

*Proof.* Suppose there exists $i, j$ s.t. $|red(w_1 \cdots w_i)| = |red(z_1 \cdots z_j)|$ and we have not $i = j = 1$ or $(i = n$ and $j = m)$ or $(i = n - 1, j = m - 1)$. Let $i, j$ minimum with such properties. One can show that if $i, j$ are both odd then $red(w_1 \cdots w_i)$ has two minimal different coverings; if $i, j$ are both even then $red(w_{i+1} \cdots w_n)$ has two minimal different coverings; if $i$ is odd and $j$ is even then $red(w_1 \cdots w_i z_{j+1})$ has two minimal different coverings. In each case we contradict the minimality of $w$. $\square$

*Example 3.* Let $X = \{abc, bcde, cdef, efg\}$ and $abcdefg$ be a minimal word with two different minimal coverings over $X$: $\delta_1 = (w_1, \cdots, w_5) = (abc, \overline{bc}, bcde, \bar{e}, efg)$ and $\delta_2 = (z_1, \cdots, z_5) = (abc, \bar{c}, cdef, \overline{ef}, efg)$. As in Lemma 1, we have that $|red(w_1 \cdots w_i)| = |red(z_1 \cdots z_j)|$ for $i = j = 1$; $i = j = 4$; and $i = j = 5$: the last (first, resp.) step in $\delta_1$ is equal to the last (first, resp.) step in $\delta_2$ but $\delta_1 \neq \delta_2$ .

**Definition 1.** *Let $M \subseteq A^*$ be a covering submonoid of $A^*$.*
    *Then $M$ is* cov-stable *if $\forall u, v, w \in A^*$ such that $w, vw \in M$ and $uvx, uy \in M$ for some $\lambda \leq x < w$, $\lambda \leq y < vw$, we have:*

1. *$vz \in M$ for some $1 \leq z < w$ if $v \neq \lambda$; moreover $vx \in M$ if $|y| < |v|$*
2. *$t \in M$ for some $t$, proper suffix of $ux$, $|t| \geq |x| - |y|$ if $v = \lambda$, $u \neq \lambda$ and $|x| > |y|$.*

*Remark 3.* If a covering submonoid is cov-stable, then it is stable. In fact, in Definition 1, the case when $v \neq \lambda$ and $x = y = \lambda$ is the stability for the ordinary submonoids.

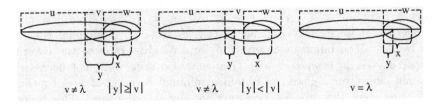

**Fig. 2.** Cases arising in Definition 1

*Example 4.* Let $X = \{abcd, defg, bcde, cdef\}$ and $M = X^{cov}$. One can easily show that $X$ is a covering code and thus $M$ is cov-free. According to Definition 1, $M$ is also cov-stable. Consider $u = ab$, $v = c$, $w = defg$, $x = de$, $y = cd$. Words $u, v, w, x, y$ satisfy the hypothesis of cov-stability for $M$ with $v \neq \lambda$, and indeed $vz \in M$ for $z = def$. Observe that if we defined that $vx \in M$ each time $v \neq \lambda$, then $M$ would not be cov-stable, against $M$ is cov-free.

*Example 5.* Let $X = \{abc, bcd, cde\}$ and $M = X^{cov}$. One can easily show that $X$ is a covering code and thus $M$ is cov-free. According to Definition 1, $M$ is also cov-stable. Consider $u = ab$, $v = \lambda$, $w = cde$, $x = cd$, $y = c$. Words $u, v, w, x, y$ satisfy the hypothesis of cov-stability for $M$ with $v = \lambda$, and indeed $t \in M$ for $t = bcd$. Observe that no $z$, $1 \leq z < w$ exists such that $vz \in M$. Hence the necessity of distinguishing the case when $v = \lambda$ from the case when $v \neq \lambda$.

**Theorem 1.** *Let $M \subseteq A^*$ be a covering submonoid of $A^*$. Then $M$ is cov-stable iff it is cov-free.*

*Proof.* (*Sketch*) (Recall Remark 2 about the union of minimal coverings.)

Suppose $M$ is cov-stable and not cov-free. Hence $X = cov\text{-}G(M)$ is not a covering code. Let $m$ be a minimal word with two different minimal coverings $\delta_1 = (w_1, w_2, \cdots, w_n)$ and $\delta_2 = (z_1, z_2, \cdots, z_l)$ over $X$. Define $w = w_n$, $v = z_l w_n^{-1}$, $u = red(z_1 \cdots z_{l-1})$, $x = \bar{w}_{n-1}$, $y = \bar{z}_{l-1}$. We have $w, vw, uvx, uy \in M$, $x < w$ and $y < vw$.

First consider the case $w_n \neq z_l$ and suppose $|w_n| < |z_l|$. Since $M$ is cov-stable, $\exists z < w$ s.t. $vz \in M$. From the union of a minimal covering of $vz$ with the covering $(w_n)$ of $w$ we can obtain a minimal covering of $vw = z_l$ of length greater than 1, against $z_l \in cov\text{-}G(M)$.

Consider now the case $w_m = z_l$. We have $v = \lambda$, $u \neq \lambda$ and $x \neq y$. We can obtain two different minimal coverings for $ux$: $\delta_3$ containing some first steps of $\delta_1$; $\delta_4$ obtained from the union of $\delta_1$ with a minimal covering of $t$. If $\delta_3 \neq \delta_4$ then we contradict the minimality of $m$. If $\delta_3 = \delta_4$, defining $v' = red(z_3 \cdots z_l)w^{-1}$, two cases arise following that $|v'x| < |t|$ or $|v'x| > |t|$. In both cases, applying Lemma 1 and Remark 2, one can find a word contradicting the minimality of $m$.

Suppose now $M$ is cov-free. Then $X = cov\text{-}G(M)$ is a covering code. Let $u, v, w, x, y$ such as in the hypothesis of cov-stability. Let $\delta_x = (x_1, \cdots, x_l)$, $\delta_w = (w_1, \cdots, w_n)$, $\delta_y = (y_1, \cdots, y_p)$, and $\delta_v = (v_1, \cdots, v_q)$ be minimal coverings over $X$ of $uvx$, $w$, $uy$, and $vw$, respectively. From the union of $\delta_x$ with $\delta_w$ and the union of $\delta_y$ with $\delta_v$ we can obtain two minimal coverings $\delta_1$, $\delta_2$ of $uvw$.

Suppose $v \neq \lambda$. Observe that $\delta_1$ has last step $w_n$ and $\delta_2$ has last step $v_q$. If there exists odd integer $k$, $1 \leq k < q$ s.t. $v \leq red(v_1 \cdots v_k) < w$ then define $z = v^{-1}red(v_1 \cdots v_k)$ and the goal is achieved. Otherwise $red(v_1 \cdots v_{q-2}) < v$ and $|v_q| > |w|$. On the other hand $|w_n| \leq |w|$. Therefore $\delta_1 \neq \delta_2$, since they differ at least in the last step, against $M$ is cov-free.

Suppose now that also $|y| < |v|$ and let $v = yy'$ with $y' \neq \lambda$. Let $i$ be the maximum odd integer $1 \leq i \leq l$ s.t. $|red(x_i \cdots x_l)| \geq |y'x|$ and $j$ the minimum odd integer $1 \leq j \leq l$ s.t. $|red(v_1 \cdots v_j)| \leq |v|$. One can show that $|red(x_i \cdots x_l)|$, $|red(v_1 \cdots v_j)| \leq |vx|$ and hence, we can obtain a covering of $vx$ over $X$, from the union of $(v_1, \cdots, v_j)$ and $(x_i, \cdots, x_n)$.

Finally suppose that $v = \lambda$, $u \neq \lambda$ and $|x| > |y|$. If there does not exist $t \in M$ proper suffix of $ux$ s.t. $|t| \geq |x| - |y|$ then $|x_1| > |uy|$. On the other hand $|y_1| \leq |uy|$. In particular $x_1 \neq y_1$. Since $u \neq \lambda$, the first step of $\delta_1$ is $x_1$ and the first step of $\delta_2$ is $y_1$. Moreover $\delta_1 = \delta_2$, because $X$ is a covering code. This is a contradiction since $x_1 \neq y_1$.                                    □

The proof of the above theorem is a little involved since many cases and sub-cases are to be handled. We think that this difficulty is inherent in the definition of cov-stability as given in Section 3.

*Remark 4.* Following [14], $X$ is a covering code iff $(X, A)$ is coding pair and then it is decidable whether a regular language is a covering code. Hence from Theorem 1 we also have that it is decidable whether a regular covering submonoid is cov-stable. Also note that a different proof that it is decidable whether a regular language is a covering code can be obtained as in Remark 5, and that a simpler proof is given in [15] for the finite case.

## 4    Some Properties of Covering Submonoids

We give here some properties of covering submonoids regarding their cov-freeness. All the proofs in this section make use of Theorem 1. Moreover, Proposition 3 and Corollary 1 could also be proved without using Theorem 1, but, in this case, the proofs are rather long and involved.

Recall that any covering submonoid is also a z-submonoid [15] and any z-submonoid is also a submonoid [16]. The class of covering submonoids is closed under intersection [15], but, despite the class of free submonoids, this closure does not hold for cov-free covering submonoids as we shall show in the following proposition.

**Proposition 2.** *The intersection of two cov-free covering submonoids of $A^*$ is not necessarily cov-free.*

*Proof.* We show the result by an example. Let $X_1 = \{abcde, bc, cdef\}$, $X_2 = \{abcde, bcd, cdef\}$ and let us set $M_1 = X_1{}^{cov}$ and $M_2 = X_2{}^{cov}$. $X_1$ and $X_2$

are covering codes, so $M_1$ and $M_2$ are cov-free covering submonoids. But $N = M_1 \cap M_2$ is not cov-free since it is not cov-stable (Theorem 1). In fact, let us set $u = a$, $v = b$, $w = cdef$, $x = cde$, $y = bcde$: we have that $w, vw \in N$, $uvx, uy \in N$ and $v \neq \lambda$. However, there is no $z$ such that $1 \leq z < w$ and $vz \in M$.    □

Now we relate cov-freeness to the property of being very pure. We recall that a submonoid $M$ of $A^*$ is very pure if for all $s, t \in A^*$, $st, ts \in M \Rightarrow s, t \in M$ and that if a submonoid is very pure then it is free [4]. Let $M$ be a covering submonoid and let $X = cov\text{-}G(M)$. We have $M = X^{cov}$ and $X^* \subseteq X^{cov}$ where this inclusion can be strict. Now, suppose that $M$ is cov-free: in Propositions 3 and 4 we show that, in this case, $X^{cov}$ and $X^*$ are both very pure. Note that when $X^{cov} = X^*$, Proposition 4 is a special case of Proposition 3. Moreover, these conditions are not sufficient for the cov-freeness of $M$ as shown in Example 6.

**Proposition 3.** *Let $M$ be a covering submonoid of $A^*$. If $M$ is cov-free, then $M$ is very pure.*

*Proof.* Let us set $X = cov\text{-}G(M)$. Suppose $st, ts \in X^{cov}$. It follows that $sts, tst \in X^{cov}$. Now from Theorem 1, we have that $X^{cov}$ is cov-stable and then, from Remark 3, it is stable. Therefore using condition of stability we have: for $u = st$, $v = s$, $w = ts$ then $s \in X^{cov}$ and for $u = ts$, $v = t$, $w = st$ then $t \in X^{cov}$.    □

**Corollary 1.** *Let $M \subseteq A^*$ be a covering submonoid. If $M$ is cov-free then it is free.*

*Proof.* Since $M$ is cov-free, by Proposition 3, we have that $M$ is very pure and therefore free.    □

**Proposition 4.** *Let $M$ be a covering submonoid of $A^*$ and $X = cov\text{-}G(M)$. If $M$ is cov-free, then $X^*$ is very pure.*

*Proof.* Since $M$ is cov-free, we have that $X$ is a covering code. Let us suppose that $X^*$ is not very pure; then there exist two words $s, t \in A^*$ such that $st, ts \in X^*$ and $t \notin X^*$. Let $st = s_1 s_2 \ldots s_n$ and $ts = t_1 t_2 \ldots t_m$, with $s_1, \ldots, s_n, t_1, \ldots, t_m \in X$. But $X^* \subseteq X^{cov}$ and therefore $st, ts \in X^{cov}$. From Proposition 3 we have that $X^{cov}$ is very pure and therefore $t \in X^{cov}$: let $\delta = (w_1, w_2, \ldots, w_k)$ be a minimal covering of $t$ over $X$. Now, let us consider the word $x = tst \in X^{cov}$. This word admits two different minimal coverings $\xi_1$ and $\xi_2$. In fact we can set $\xi_1 = (w_1, \ldots, w_k, \lambda, s_1, \lambda, \ldots, s_n)$ and $\xi_2 = (t_1, \lambda, \ldots, t_m, \lambda, w_1, \ldots, w_k)$. We have surely $\xi_1 \neq \xi_2$ and this is against $X$ covering code.    □

*Example 6.* The converse of Propositions 3 and 4 is not true. Consider $X = \{abcd, cdef, def\}$ and $M = X^{cov}$. Note that $X^* \subset X^{cov}$ since we have $abcdef \in X^{cov}$ and $abcdef \notin X^*$. It is possible to show that $X^{cov}$ and $X^*$ are very pure. Moreover we have $X = cov\text{-}G(M)$, but $X$ is not a covering code (the word $abcdef$ has two different minimal coverings over $X$) and, then, $M$ is not cov-free.

We want now to compare the three properties of cov-freeness, z-freeness and (classical) freeness for covering submonoids. From Corollary 1 we have that if $M$ is a cov-free covering submonoid then $M$ is free and in [16] it was shown that if $M$ is a z-free z-submonoid then $M$ is free. On the other hand the cov-freeness and the z-freeness are not related each other, once again showing that coverings are not mere generalizations of z-decompositions. In the following examples, $M$ is a covering submonoid that is cov-free but not z-free and $M'$ is a covering submonoid that is z-free but not cov-free. Moreover since freeness and stability are equivalent notions in classical submonoids, in z-submonoids and also in covering submonoids, same results hold for cov-stability, z-stability and (classical) stability.

*Example 7.* Let $X = \{ab, bc, cd\}$ and $M = X^{cov}$; from results in [15], we have that $cov\text{-}G(M) = X$. But $M$ is also a z-submonoid; let $Z = ZG(M)$. We have $Z = \{ab, bc, cd, abc, bcd\}$ and $Z$ is not a z-code (for example the word $w = abcd$ has two different z-factorizations over $Z$). Hence $M$ is cov-free but not z-free. Moreover $M$ is cov-stable but not z-stable.

*Example 8.* Let $T = \{abcd, abc, bcd, cd\}$ and $M' = (T, T)^{\uparrow}$. We have that $T$ is a z-code and $M'$ is a covering submonoid. Moreover following [15], $cov\text{-}G(M) = X$ with $X = \{abc, bcd, cd\}$. But $X$ is not a covering code since $w = abcd$ has two different minimal coverings. Hence $M'$ is z-free but not cov-free. Moreover $M'$ is z-stable but not cov-stable.

## 5   Counting Minimal Coverings

Minimal coverings of a word are those coverings in which every step is useful; that is eliminating a step we no longer have a covering. Whenever every possible word has at most one minimal covering over some language $X$, we say that $X$ is a covering code. If $X$ is not a covering code (and this property is decidable; see Remark 5), some word has more than one minimal covering. In such case, it can be interesting to count the number of different minimal coverings a word has i.e. to study the (formal power) series counting minimal coverings. In this section the main result shows that the series associating to a word the number of its minimal coverings over $X$ is always rational when $X$ is regular. Observe that this is not the case for z-factorizations: the series counting the different z-factorizations of a word is not always rational, even in the case of a regular language [1]. The main result thus shows a situation in which coverings and z-factorizations behave differently.

The reader is referred to [4,10] for following notions, namely for what concerns one-way finite automata (1FA), two-way finite automata (2FA), and their equivalence using crossing sequences. We recall that the behaviour of a 1FA is the series that counts the number of different successful paths in the automaton labelled by a word. A series is rational when it has a linear representation; in particular the behaviour of a 1FA is a rational series. On the other hand, the behaviour of a 2FA is not always defined in the set $N$ of natural numbers: a word can admit an infinite number of successful paths labelled by it. If this is the case, one can consider only a finite number of paths of a restricted type (see for example [2]).

Let $X \subseteq A^*$ and denote $cov_X : A^* \to N$ the series such that for $w \in A^*$, $cov_X(w)$ is the number of different minimal coverings of $w$ over $X$.

**Proposition 5.** *Let $X \subseteq A^*$. If $X$ is a regular language then $cov_X$ is a rational series.*

*Proof.* If $X$ is regular then there is a deterministic 1FA $\mathcal{A} = (Q, q_0, \delta, F)$ recognizing $X$. Let us construct the following 2FA: $\mathcal{B} = (Q \cup \{1\}, 1, \delta', 1)$ where $1 \notin Q$ is the initial and final state, and $\delta'$ is a partial function defined for any $a \in A, q \in Q$, by: $\delta'(1, \bar{a}) = 1$; $\delta'(1, a) = \delta(q_0, a)$ if $\delta(q_0, a) \notin F$; $\delta'(1, a) = 1$ if $\delta(q_0, a) \in F$; $\delta'(q, a) = \delta(q, a)$ if $\delta(q, a) \notin F$; $\delta'(q, a) = 1$ if $\delta(q, a) \in F$.

We have that successful paths in $\mathcal{B}$ correspond to (generic) coverings. Since we are interested in counting only minimal coverings, consider the set $CS_3$ of crossing sequences of $\mathcal{B}$ of length 3 at most and without a repetition of state 1. The main observation is now that any minimal covering of a word corresponds to a path in $\mathcal{B}$ whose crossing sequences all belong to $CS_3$ (see Remark 1). Therefore the following 1FA recognizes the series $cov_X$: $\mathcal{C} = (CS_3, (1), \delta'', (1))$ where $\delta''(cs, a) = cs'$ iff $cs$ matches $cs'$ on $a$. $\qquad\square$

*Remark 5.* From the proof of Proposition 5 we have in particular that it is decidable whether a regular language is a covering code, since it is decidable whether a 1FA (namely $\mathcal{C}$) is ambiguous or not. This proof is different from the one that can be obtained following results in [14] (see Remark 4).

Further the proof of Proposition 5 shows that if $X$ is a regular language then $X^{cov}$ is regular too, since the automaton $\mathcal{B}$ recognizes $X^{cov}$ (this result is given in a more general framework in [14]).

*Example 9.* Let $X = \{ab, ba\}$ and $\mathcal{A} = (\{1, 2, 3, 4\}, 1, \delta, \{4\})$ where $\delta(1, a) = 2$, $\delta(1, b) = 3$, $\delta(2, b) = \delta(3, a) = 4$, a determistic 1FA recognizing $X$. The trim

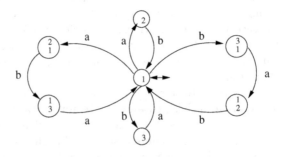

**Fig. 3.** Automaton of Example 9

part of automaton $\mathcal{C}$ constructed as in the proof of Proposition 5 is given in Fig. 3. The behaviour of $\mathcal{C}$ is $cov_X$ and the language recognized by $\mathcal{C}$ is $X^{cov}$. $\mathcal{C}$ is ambiguous and hence $X$ is not a covering code. As an example, the word $ababab$ has two minimal coverings and two corresponding successful paths in $\mathcal{C}$.

# References

1. M. Anselmo: The zig-zag power-series: a two-way version of the star operator. Theor. Comp. Sc. **79** n° 1 (1991) 3–24
2. M. Anselmo: Two-way Automata with Multiplicity. Procs. ICALP 90, LNCS **443** Springer-Verlag (1990) 88–102
3. A. Apostolico and A. Ehrenfeucht: Efficient Detection of quasiperiodicities in strings. Theor. Comp. Sc. **119** (1993) 247–265.
4. J. Berstel and D. Perrin: Theory of codes. Academic Press (1985).
5. G. S. Brodal and C. N. S. Pedersen: Finding Maximal Quasiperiodities in Strings. Procs. 11 Annual Symp. on Comb. Pattern Matching LNCS **1848** (2000) 347–411
6. G. Cohen, I. Honkala, S. Litsyn and A. Lobstein: Covering Codes. Elsevier, North-Holland Mathematical Library **54** (1997)
7. Do Long Van, B. Le Saëc and I. Litovsky: Stability for the zigzag submonoids. Theor. Comp. Sc. **108** (1993) 237–249.
8. D. Gusfield: Algorithms on Strings, Trees, and Sequences: Computer Science and Computational Biology. Cambridge Univ. Press (1997)
9. T. Head: Formal language theory and DNA: an analysis of the generative capacity of specific recombinant behaviors. Bull. Math. Biol. **49** (1987) 737–759.
10. J. E. Hopcroft and J. D. Ullman: Introduction to Automata Theory, Languages, and Computation. Addison-Wesley Reading, MA, (1995).
11. C. S. Iliopoulos, D. W. G. Moore and K. Park: Covering a string. Proc. 4th Symp. Combinatorial Pattern Matching LNCS **684** (1993) 54–62.
12. C. S. Iliopoulos and W. F. Smyth: On-line Algorithms for $k$-Covering. Procs. ninth AWOCA (1998) 97–106.
13. B. Le Saec, I. Litovsky and B. Patrou: A more efficient notion of zigzag stability. RAIRO Informatique Théorique **30** n.3 (1996) 181–194.
14. I. Litovsky and B. Patrou: On a binary zigzag operation. Proc. 3rd Int. Conf. Developments in Languages Theory, Thessaloniki, July 1997, Bozapalidis Ed., Aristotle University of Thessaloniki, 273–289.
15. M. Madonia, S. Salemi and T. Sportelli: Covering submonoids and covering codes. Journal of Automata, Languages and Combinatorics **4** n.4 (1999) 333–350.
16. M. Madonia, S. Salemi and T. Sportelli: On z-submonoids and z-codes. RAIRO Informatique théorique **25** n.4 (1991) 305–322.
17. D. Moore and W. F. Smyth: An optimal algorithm to compute all the covers of a string. Inf. Proc. Letters **50** (5) (1994) 239–246 and further corrections in Inf. Proc. Letters **50** (2) (1995) 101–103.
18. J. S. Sim, C. S. Iliopoulos, K. Park and W. F. Smyth: Approximate Periods of Strings. Theor. Comp. Sc. **262** (2001) 557–568.

# Regular Languages Generated by Reflexive Finite Splicing Systems*

Paola Bonizzoni[1], Clelia De Felice[2], Giancarlo Mauri[1], and Rosalba Zizza[2]

[1] Dipartimento di Informatica Sistemistica e Comunicazione
Università degli Studi di Milano - Bicocca
Via Bicocca degli Arcimboldi 8, 20126 Milano - Italy
{bonizzoni, mauri}@disco.unimib.it
[2] Dipartimento di Informatica ed Applicazioni,
Università di Salerno, 84081 Baronissi (SA), Italy
{defelice, zizza}@unisa.it

**Abstract.** *Splicing systems* are a generative device inspired by a cut and paste phenomenon on DNA molecules, introduced by Head in 1987 and subsequently defined with slight variations also by Paun and Pixton respectively [8,13,17]. We will face the problem of characterizing the class of regular languages generated by finite splicing systems. We will solve this problem for the special class of the *reflexive finite splicing systems* introduced in [9,10]. As a byproduct, we give a characterization of the regular languages generated by *finite Head splicing systems*. As in already known results, the notion of *constant*, given by Schützenberger in [19], intervenes.

## 1 Introduction

In this paper we will face the problem of characterizing the class of regular languages generated by finite linear splicing systems. *Splicing systems* are a generative device introduced by Head in 1987 as a formal model of certain cut and paste biochemical transformation processes of an initial collection of DNA strands under the simultaneous influence of enzymes [8]. This topic can be considered a part of the more general area of Molecular Computing [15].

A splicing system (or $H$-system) is a triple $H = (A, I, R)$, where $A$ is a finite alphabet, $I \subseteq A^*$ is the initial language and $R$ is the set of rules, $R \subseteq (A')^*$, $A \subseteq A'$ (see Section 2.2 for the definitions). The formal language generated by the splicing system is the smallest language containing $I$ and closed under the splicing operation, which makes the rule intervene. Different variants of the original definition of the splicing operation given by Head have been proposed briefly; in particular two of them have been introduced by Paun and Pixton respectively [13,14,15,18].

* Partially supported by MIUR Project *"Linguaggi Formali e Automi: teoria ed applicazioni"*, by the contribution of EU Commission under The Fifth Framework Programme (*project MolCoNet* IST-2001-32008) and by 60% Project *"Linguaggi Formali e Modelli di Calcolo"* (University of Salerno).

Z. Ésik and Z. Fülöp (Eds.): DLT 2003, LNCS 2710, pp. 134–145, 2003.
© Springer-Verlag Berlin Heidelberg 2003

The computational power of (iterated linear) splicing systems has been investigated thoroughly and it mainly depends on which level of the Chomsky hierarchy $I, R$ belong to. Precisely, let $F_1, F_2$ be two families of languages in the Chomsky hierarchy. Following [10,15], we set $H(F_1, F_2) = \{L(H) \mid H = (A, I, R)$ with $I \in F_1, R \in F_2\}$. In [10] it is proved that either $H(F_1, F_2)$ is a specific class of languages in the Chomsky hierarchy or it is strictly intermediate between two of them. More precisely, according to some hypotheses on $F_1, F_2$, splicing systems can reach the same power of the Turing machines [10,13,14]. On the other hand, let $FIN$ (resp. $REG$) be the class of finite (resp. regular) languages. In [18] the author proved that $H(REG, FIN) = REG$, while the class $H(FIN, FIN)$, as shown in [10,15,18], turns out to be strictly intermediate between $FIN$ and $REG$. As an example, $(aa)^*b$ is a regular language which belongs to $H(FIN, FIN)$ [4,9], whereas $(aa)^* \notin H(FIN, FIN)$ [6,10]. Now two problems arise and Problem 2 is the special question we will focus on in this paper.

*Problem 1.* Given $L \in REG$, can we decide whether $L \in H(FIN, FIN)$?


*Problem 2.* Characterize regular languages which belong to $H(FIN, FIN)$.

The paper [3] is a complete survey on the state of the art of the results concerning the two above-mentioned problems. In particular, the search for a characterization of the regular languages in $H(FIN, FIN)$ is a largely investigated but difficult open problem. The difficulty is also due to the fact that, as observed in [5], the computational power of finite splicing systems (i.e., splicing systems $H = (A, I, R)$ with $I, R \in FIN$) increases when we substitute Head's systems with Paun's systems and Paun's systems with Pixton's systems.

The main result of this paper is the solution of Problem 2 for the special class of the *reflexive finite splicing systems* introduced in [9,10]. As a byproduct, we give a characterization of the regular languages generated by *finite Head splicing systems*. To be more precise, let us first consider the more largely used Paun's definition of splicing. When the binary relation induced by the set of rules is reflexive, we have a reflexive Paun splicing system. One of the results in this paper characterizes finite Head splicing systems as reflexive Paun splicing systems which, in addition, satisfy a transitive hypothesis, i.e., we will state the following result:

**Main result 1.** A regular language $L$ is generated by a finite Head splicing system if and only if $L$ is generated by a finite Paun reflexive splicing system $H = (A, I, R_B \cup R_C)$ which is $R_B$-transitive and $R_C$-transitive (Theorem 2).

So, finite Head systems are all reflexive and this explains why reflexivity is a quite natural property of splicing systems.

The characterization of reflexive splicing languages we give here, extends the description of languages generated by a special class of reflexive Paun splicing systems (with rules with one-sided contexts, see Section 3) obtained in [9]. It is worthy of note that all these results are obtained by using only the classical notion of *constant* introduced by Schützenberger [19]. Indeed, we prove that, when we take into account Paun's and Pixton's definitions, $L$ is a reflexive splicing language if and only if there exist a finite set $\mathcal{M}$ of constants for $L$ such that $L$ is of the following form: $L = Y \cup \bigcup_{m \in \mathcal{M}} L(m) \cup \bigcup_{(\alpha,\beta) \in J} L_{(\alpha,\beta)}$, where $L(m)$ is a set of words having $m$ as a factor (*constant languages*), $Y$ is a finite set of words such that no $m$ is a factor of a word of $Y$. The (finite) set $J$ and the structure of $L_{(\alpha,\beta)}$ depend on the splicing operation we choose. $L_{(\alpha,\beta)}$ is a language obtained by extending the splicing operation to two constant languages $L(m)$ and $L(m')$. As a consequence, we have three definitions for $L_{(\alpha,\beta)}$. Depending on which definition of splicing we take into account, $L_{(\alpha,\beta)}$ will be termed a $X$-split language, whereas $L$ will be a $X$-con-split language, with $X \in \{H, PA, PI\}$ (see Definitions 7 and 13). In conclusion, we will state the following result:

**Main result 2.** A regular language is generated by a finite Paun (resp. Pixton) reflexive splicing system if and only if $L$ is a $PA$-con-split (resp. $PI$-con-split) language (Theorems 1, 3).

and, as a consequence, we give a structural property of the regular languages generated by finite Head's systems (Corollary 1). In the proofs of these results we exhibit the splicing systems that generate reflexive languages. However, Problem 1 remains open even for reflexive languages. As a final observation, we strongly guess the existence of finite Paun (or Pixton) splicing languages which are not reflexive.

This paper is organized as follows. Basics on words and linear splicing are gathered in Section 2, together with a decidability property for a regular language used in the proofs of our results. Definitions and preliminary properties of reflexive Paun splicing languages are collected in Section 3. The description of our class of splicing languages and the characterization of reflexive Paun splicing languages is presented in Section 4. Regular languages generated by finite Head splicing systems are presented in Section 5 and the characterization of reflexive Pixton splicing languages is given in Section 6. An extended version of this paper contains all the missing proofs of the results presented in this abstract [2].

# 2    Preliminaries

## 2.1    Words

Let $A^*$ be the free monoid over a finite alphabet $A$ and let $A^+ = A^* \setminus 1$, where 1 is the empty word. In the following $\mathcal{A} = (Q, A, \delta, q_0, F)$ will be a finite state automaton, where $Q$ is a finite set of states, $q_0 \in Q$ is the initial state, $F \subseteq Q$ is the set of final states, $\delta$ is the transition function and $L(\mathcal{A})$ is the language recognized by $\mathcal{A}$ [1,11,16]. A finite state automaton $\mathcal{A}$ is *deterministic* if, for each $q \in Q$, $a \in A$, there is at most one state $q' \in Q$ so that $\delta(q, a) = q'$. Furthermore, $\mathcal{A}$ is *trim* if each state is accessible and coaccessible, i.e., for each state $q \in Q$ there are $x, y \in A^*$ such that $\delta(q_0, x) = q$ and $\delta(q, y) \in F$. Given a regular language $L \subseteq A^*$ it is well known that there is a *minimal* finite state automaton $\mathcal{A}$ recognizing it. This automaton is unique up to a possible renaming of the states, is deterministic, trim and has the minimal number of states. As usual, in the transition diagram of a finite state automaton $\mathcal{A}$, each final state will be indicated by a double circle and the initial state will be indicated by an arrow without a label going into it. If it is not differently supposed, $L \subseteq A^*$ will always be a regular language.

Let us recall the definition of a constant, already introduced by Schützenberger in [19]. Let $\mathcal{A}$ be the minimal finite state automaton recognizing a regular language $L$ and let $w \in A^*$. We will set $Q_w(\mathcal{A}) = \{q \in Q \mid \delta(q, w) \text{ is defined }\}$, simply indicated $Q_w$ when the context makes the meaning evident. The *left* and *right* contexts of a word $w \in A^*$ are therefore defined as follows: $C_{\mathcal{L}}(w, L) = \{z \in A^* \mid \exists q \in Q_w : \delta(q_0, z) = q\}$, $C_{\mathcal{R},q}(w, L) = \{y \in A^* \mid \delta(q, wy) \in F\}$, $C_{\mathcal{R}}(w, L) = \bigcup_{q \in Q_w} C_{\mathcal{R},q}(w, L)$.

Notice that these definitions are slightly different from the ones given in [1]. Furthermore, we denote $C(w, L) = \{(x, y) \in A^* \times A^* \mid xwy \in L\}$ the set of *contexts* of $w$ with respect to $L$. We recall that two words $w, w'$ are equivalent with respect to the *syntactic congruence* of $L$ if they have the same set of contexts with respect to $L$, i.e., $w \equiv_L w' \Leftrightarrow [\forall x, y \in A^*, xwy \in L \Leftrightarrow xw'y \in L] \Leftrightarrow C(w, L) = C(w', L)$ [12,16].

A word $w \in A^*$ is a *constant* for a regular language $L$ if $A^* w A^* \cap L \neq \emptyset$ and $C(w, L) = C_{\mathcal{L}}(w, L) \times C_{\mathcal{R}}(w, L)$ [19]. A characterization of constants, which is more or less folklore, is stated below.

**Proposition 1.** *Let $L \subseteq A^*$ be a regular language and let $\mathcal{A}$ be the minimal finite state automaton recognizing $L$. A word $w \in A^*$ is a constant for $L$ if and only if $Q_w \neq \emptyset$ and there exists $q_w \in Q$ such that for all $q \in Q_w$ we have $\delta(q, w) = q_w$.*

If it is not differently supposed, for a regular language $L \subseteq A^*$, we will always refer to the minimal finite state automaton $\mathcal{A}$ recognizing $L$. Thus, the definitions of $C_{\mathcal{L}}(w, L)$ and $C_{\mathcal{R}}(w, L)$ can be given as follows: $C_{\mathcal{L}}(w, L) = \{z \in A^* \mid \exists y \in A^* : zwy \in L\}$, $C_{\mathcal{R}}(w, L) = \{z \in A^* \mid \exists y \in A^* : ywz \in L\}$. Obviously, $Q_1 = Q$ and $\delta(q, 1) = q$, for each $q \in Q$. Thus, in virtue of Proposition 1, if $w$ is a constant for $L$ we have $w \neq 1$ unless $Q$ has only one element.

## 2.2  Linear Splicing

As we have already said, there are three definitions of linear splicing operation, given by Head, Paun and Pixton respectively [10,15]. The difference among them depends on the biological phenomena that they want to model, an aspect which will not be discussed here. Paun's definition is given below, whereas Head's (resp. Pixton's) definition is given in Section 5 (resp. 6).

**Paun's definition [13].** A *Paun splicing system* is a triple $S_{PA} = (A, I, R)$, where $I \subset A^*$ is a set of strings, called *initial language*, $R$ is a set of *rules* $r = u_1 \# u_2 \$ u_3 \# u_4$, with $u_i \in A^*, i = 1, 2, 3, 4$ and $\#, \$ \notin A$. Given two words

$x = x_1 u_1 u_2 x_2, y = y_1 u_3 u_4 y_2, x_1, x_2, y_1, y_2 \in A^*$ and the rule $r = u_1 \# u_2 \$ u_3 \# u_4$, the splicing operation produces $w' = x_1 u_1 u_4 y_2$ and $w'' = y_1 u_3 u_2 x_2$, denoted $(x, y) \vdash_r (w', w'')$. We also say that $u_1 u_2, u_3 u_4$ are *sites* of splicing and we denote $SITES(R)$ the set of sites of the rules in $R$.

Our study will consider having an unlimited number of copies of each word in the set, so that a pair of strings $(x, y)$ can generate more than one pair of words with the use of different rules. Let $L \subseteq A^*$. We denote $\sigma'(L) = \{w', w'' \in A^* \mid (x, y) \vdash_r (w', w''), x, y \in L, r \in R\}$. The (iterated) splicing operation is defined as follows: $\sigma^0(L) = L, \sigma^{i+1}(L) = \sigma^i(L) \cup \sigma'(\sigma^i(L)), i \geq 0, \sigma^*(L) = \bigcup_{i \geq 0} \sigma^i(L)$.

**Definition 1 (Paun splicing language).** *Given a splicing system* $S_{PA} = (A, I, R)$, *the language* $L(S_{PA}) = \sigma^*(I)$ *is the language generated by* $S_{PA}$. *A language* $L$ *is* $S_{PA}$ *generated (or is a* Paun splicing language*) if a splicing system* $S_{PA}$ *exists such that* $L = L(S_{PA})$.

A splicing system is *finite* when $I, R$ are finite sets.

## 2.3   Languages Closed with Respect to a Set of Rules

In the next part of this paper we suppose that each rule $r$, in a given splicing system $S_{PA}$, is *useful*, i.e., there exist $x, y, w', w'' \in L(S_{PA})$ such that $(x, y) \vdash_r (w', w'')$. A notion which will be used in the sequel is that of *languages closed with respect to a rule*. This definition was formally given in [4] and in [7] and is reported below.

**Definition 2.** *[4,7] A language* $L \subset A^*$ *is* closed with respect to *a rule* $r$ *if and only if for each* $x, y \in L$, *if* $(x, y) \vdash_r (w', w'')$ *then* $w', w'' \in L$.

Indeed, in order to prove that a language $L$ is generated by a splicing system, in particular we must find a language $I \subseteq L$, such that, starting from $I$, the application of the splicing rules generates words in $L$. Problem 3 is a natural question which follows.

*Problem 3.* Let $L$ be a regular language and let $R$ be a set of rules in a finite Paun splicing system $S_{PA} = (A, I, R)$. Is $L$ closed with respect to $R$?

In [4], the authors provided a characterization of languages closed with respect to a rule via automata. This characterization, reported in Lemma 1 below, gives a decision algorithm for Problem 3.

**Lemma 1.** *[4] Let* $S_{PA} = (A, I, R)$ *be a finite Paun splicing system, let* $L \subseteq A^*$ *be a regular language and let* $\mathcal{A}$ *be the minimal finite state automaton recognizing* $L$. *Then* $L = L(\mathcal{A})$ *is closed with respect to a rule* $r = u_1 \# u_2 \$ u_3 \# u_4 \in R$ *if and only if for each pair* $(p, q) \in Q_{u_1 u_2} \times Q_{u_3 u_4}$, *we have*
    (1) $C_{\mathcal{R}, p}(u_1 u_2, L) \subseteq C_{\mathcal{R}, q}(u_3 u_2, L)$,
    (2) $C_{\mathcal{R}, q}(u_3 u_4, L) \subseteq C_{\mathcal{R}, p}(u_1 u_4, L)$.

Lemma 2 shows a relationship between languages which are closed with respect to a rule and splicing languages. This relationship will be used to characterize reflexive splicing languages.

**Lemma 2.** *Let* $S_{PA} = (A, I, R)$ *be a splicing system and let* $L \subseteq A^*$. *If* $I \subseteq L$ *and* $L$ *is closed with respect to each rule in* $R$, *then* $L(S_{PA}) \subseteq L$.

## 3 Reflexive Paun Splicing Languages

In this section we define the class of splicing languages we deal with when we restrict ourselves to Paun's definition of splicing. Let $S_{PA} = (A, I, R)$ be a finite Paun splicing system, let $R' \subseteq R$, let $SITES(R')$ be the set of sites of the rules in $R'$. We denote $Rel(R')$ the binary relation over $\{u_1 \# u_2 \mid u_1 u_2 \in SITES(R')\}$ induced by $SITES(R')$, that is $u_1 \# u_2 \ Rel(R') \ u_3 \# u_4$ if and only if $u_1 \# u_2 \$ u_3 \# u_4 \in R'$.

**Definition 3 (reflexive Paun splicing system).** *[9] A finite splicing system* $S_{PA} = (A, I, R)$ *is a* reflexive Paun splicing system *if and only if the relation* $Rel(R)$ *is reflexive, i.e., if* $u_1 \# u_2 \$ u_3 \# u_4 \in R$, *then* $u_1 \# u_2 \$ u_1 \# u_2$ *and* $u_3 \# u_4 \$ u_3 \# u_4 \in R$.

A language $L$ is called *PA-reflexive* if there exists a finite reflexive Paun splicing system $S_{PA}$ such that $L = L(S_{PA})$. PA-reflexive languages have been introduced for the first time in [9,10] and Lemma 3 shows how they can be easily characterized in terms of constants.

**Lemma 3.** *A regular language* $L \subseteq A^*$ *is PA-reflexive if and only if there exists a finite splicing system* $S_{PA} = (A, I, R)$ *so that each site of the rules in* $R$ *is a constant for* $L$ *and* $L = L(S_{PA})$.

A special class of reflexive Paun splicing languages, the *constant languages* have been considered in [9]. Constant languages are the simplest regular Paun reflexive languages and we recall their definition below.

**Definition 4 (constant language).** *Let* $L$ *be a regular language and let* $m \in A^*$ *be a constant for* $L$. *The* constant language *associated with* $m$ *is the language* $L(m) = \{y \in L \mid y = y_1' m y_2', y_1', y_2' \in A^*\}$.

We also need Proposition 2 concerning languages which are finite union of constant languages.

**Proposition 2 (union of constant languages).** *[9] Let* $L \subseteq A^*$ *be a regular language and let* $\mathcal{M} \subseteq A^*$ *be a finite set of constants for* $L$. *Then* $L' = \cup_{m \in \mathcal{M}} L(m)$ *is a PA-reflexive language. Furthermore,* $L$ *can be generated by a Paun splicing system in which each rule has either the form* $u \# 1 \$ v \# 1$ *or* $1 \# u \$ 1 \# v$ *(one-sided contexts), where* $u, v$ *are constants for* $L$.

## 4   Main Result

In this section we illustrate our first main result, i.e., a complete characterization of the class of reflexive Paun splicing languages (Theorem 1). Given two constants $m, m'$ for $L$ and the two sets of the factorizations of these constants into two words, we begin with the definition of a language obtained by splicing the two constant languages $L(m), L(m')$ (Definition 6). By means of this operation, in Definition 7 we introduce the class $\mathcal{D}_{PA-con-split}$ which we prove to be the class of the reflexive Paun splicing languages. Roughly, a language $L$ is in $\mathcal{D}_{PA-con-split}$ if and only if $L$ can be obtained by the application of the above-mentioned operation to a finite number of pairs of constant languages and starting with the union of a finite set and a finite union of constant languages. Example 1 shows that this operation must be necessarily introduced.

**Definition 5 (split of a constant).** *Let $L$ be a regular language and let $m$ be a constant for $L$. A split of the constant $m$ is a pair $(x_1, x_2)$ of words in $A^*$ such that $x_1 x_2 = m$.*

We denote by $F(m) = \{(x_1, x_2) \mid x_1 x_2 = m\}$ the set of the splits of the constant $m$. Given two constants $m, m'$ for $L$ and the sets $F(m), F(m')$ of the splits for these constants, we can define a language obtained "by splicing" the two constant languages $L(m), L(m')$.

**Definition 6 ($PA$-split language).** *Let $L$ be a regular language and let $m$ and $m'$ be two constants for $L$. Given $\alpha \in F(m)$ and $\beta \in F(m')$, such that $\alpha = (\alpha_1, \alpha_2)$ and $\beta = (\beta_1, \beta_2)$, the $PA$-split language generated by $\alpha$ and $\beta$ is the language:*

$$L_{(\alpha,\beta)} = C_{\mathcal{L}}(m, L)\ \alpha_1 \beta_2\ C_{\mathcal{R}}(m', L) \cup C_{\mathcal{L}}(m', L)\ \beta_1 \alpha_2\ C_{\mathcal{R}}(m, L) .$$

The above definition is necessary to complete the characterization of Paun reflexive languages. Indeed, in Theorem 1, we will show that $L$ is a $PA$-reflexive language if and only if $L \in \mathcal{D}_{PA-con-split}$, where $\mathcal{D}_{PA-con-split}$ is defined below. We recall that given two words $w, x \in A^*$, $w$ is a *factor* of $x$ if there exist $y, z \in A^*$ so that $x = ywz$.

**Definition 7 ($PA$-con-split language).** *Let $L$ be a regular language and let $\mathcal{M}$ be a finite set of constants for $L$. Let $Y$ be a finite subset of $L$ such that $m$ is not a factor of a word in $Y$, for each $m \in \mathcal{M}$. Let $J \subseteq \{(\alpha, \beta) \mid \alpha \in F(m), \beta \in F(m'), m, m' \in \mathcal{M}\}$. $L$ is a $PA$-con-split language (associated with $Y, \mathcal{M}, J$) if and only if*

$$L = Y \cup \bigcup_{m \in \mathcal{M}} L(m) \cup \bigcup_{(\alpha,\beta) \in J} L_{(\alpha,\beta)} .$$

$\mathcal{D}_{PA-con-split}$ *is the class of $PA$-con-split languages.*

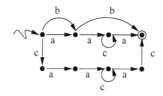

**Fig. 1.** A $PA$-reflexive splicing language which is not a finite union of constant languages

*Example 1.* In Figure 1 we report the transition diagram of a finite state automaton recognizing the language $L = ab \cup bb \cup aac^*a \cup bac^*a \cup caac^*ac$. Let us check that $L$ is a $PA$-con-split language which is not a finite union of constant languages. Observe that $ba$, $caa$, $bb$ and $ab$ are constants for $L$ and so $bac^*a$, $caac^*ac$, $bb$ and $ab$ are constant languages. On the contrary, it is not too difficult to see that $aac^*a$ is not a finite union of constant languages. However, we have that $aac^*a \cup bb = L_{(\alpha,\beta)}$ where $m = ba$, $m' = ab$, $L(m) = bac^*a$, $L(m') = ab$, $\alpha = (b,a)$, $\beta = (a,b)$. Thus, $L$ is a $PA$-con-split language. Finally, by using the characterization given in [9], we stress that $L$ is not a finite union of constant languages since it is not too difficult to see that, for any finite set $F$ of constants for $L$, we have that $aac^*a \subseteq L \setminus \cup_{f \in F} A^* f A^*$.

We end this section with one of the main results in this paper.

**Theorem 1.** *A regular language $L \subseteq A^*$ is a $PA$-reflexive language if and only if $L$ belongs to $\mathcal{D}_{PA-con-split}$.*

*Remark 1.* Notice that $L_{(\alpha,\beta)}$ can be considered as the result of an extension of the splicing operation to languages $L(m) = C_{\mathcal{L}}(m, L) \alpha_1\alpha_2 C_{\mathcal{R}}(m, L)$ and $L(m') = C_{\mathcal{L}}(m', L) \beta_1\beta_2 C_{\mathcal{R}}(m', L)$, $m = \alpha_1\alpha_2 \in \mathcal{M}$, $m' = \beta_1\beta_2 \in \mathcal{M}$. Indeed, for proving that a $PA$-con-split language $L$ is $PA$-reflexive, we construct a splicing system $S_{PA} = (A, I, R)$ in which, among the rules in $R$, we find rules with the form $\alpha_1 \# \alpha_2 \$ \beta_1 \# \beta_2$, for each $\alpha = (\alpha_1, \alpha_2) \in F(m)$, $\beta = (\beta_1, \beta_2) \in F(m')$. Conversely, for showing that a $PA$-reflexive language $L = L(S_{PA})$ is a $PA$-con-split language, we define $J$ as above starting with the rules in $S_{PA}$.

*Example 2.* Let $L$ be the language considered in Example 1. Following the proof of Theorem 1, we have that $L = L(S_{PA})$, where $S_{PA} = (A, I, R)$, $I = \{ab, bb, caaac, caacac, aaa, aaca, baa, baca\}$ and $R = \{caa\#1\$caac\#1, ba\#1\$ bac \#1, b\#a\$a\#b\}$.

As far as we know, it is not known whether we can decide if a language $L$ belongs to $\mathcal{D}_{PA-con-split}$, i.e., whether $L$ is a $PA$-reflexive language. Observe that this problem generalizes the analogous question, proposed in [9], concerning languages which are union of a finite set and a finite union of constant languages.

*Remark 2.* It seems that reflexive Head (resp. Pixton) splicing systems have not been considered in the literature. On the other hand for each Head (resp. Paun) splicing system $S_H$ (resp. $S_{PA}$) there exists a canonical transformation $\phi$ such that $\phi(S_H)$ is a Paun (resp. $\phi(S_{PA})$ is a Pixton) splicing system $S'_{PA}$ (resp. $S'_{PI}$) and $L(S_H) = L(S'_{PA})$ (resp. $L(S_{PA}) = L(S'_{PI})$) [15]. In the next two sections, we give the definition of a reflexive Head (resp. Pixton) splicing system which allows us to preserve these canonical transformations.

## 5    A Description of Finite Head Splicing Languages

In this section we prove the other main results of the paper (Theorem 2, Corollary 1). Corollary 1 is a consequence of Theorem 2 which characterizes the structure of finite Head splicing systems inside the class of finite Paun splicing systems. Indeed, Theorem 2 shows the equivalence between finite Head splicing systems and a subclass of reflexive Paun splicing systems satisfying a transitive hypothesis. We also state that Head splicing languages are a proper subset of $PA$-reflexive languages (see Example 1 and [5]). Let us first recall the original definition of splicing operation given by Head.

**Head's definition [8].** A *Head splicing system* is a 4-uple $S_H = (A, I, B, C)$, where $I \subset A^*$ is a finite set of strings, called *initial language*, $B$ and $C$ are finite sets of triples $(\alpha', \mu, \beta')$, called *patterns*, with $\alpha', \beta', \mu \in A^*$ and $\mu$ called the *crossing* of the triple. Given two words $u\alpha'\mu\beta'v$, $p\alpha''\mu\beta''q \in A^*$ and two patterns $p_1 = (\alpha', \mu, \beta')$ and $p_2 = (\alpha'', \mu, \beta'')$ that have the same crossing and are both in $B$ or both in $C$, the splicing operation produces $w' = u\alpha'\mu\beta''q$ and $w'' = p\alpha''\mu\beta'v$, denoted $(x, y) \vdash_{p_1, p_2} (w', w'')$. We also say that $\alpha'\mu\beta'$, $\alpha''\mu\beta''$ are *sites* of splicing.

Clearly, a Head splicing system is *finite* when $I, B, C$ are finite sets. Furthermore, we set $\sigma'(L) = \{w', w'' \in A^* \mid (x, y) \vdash_{p_1, p_2} (w', w''), \ x, y \in L, p_1, p_2 \text{ both in } B \text{ or both in } C\}$ and, as in Paun's systems, given $S_H = (A, I, B, C)$, the language $L(S_H) = \sigma^*(I)$ is the language generated by $S_H$. A language $L$ is $S_H$ *generated* (or $L$ is a *Head splicing language*) if a splicing system $S_H$ exists such that $L = L(S_H)$.

*Remark 3.* Looking for a definition of a reflexive Head splicing system which takes into account Remark 2, we find that each Head splicing system $S_H$ can be implicitly supposed to be reflexive. This property is satisfied since for each pattern $p = (\alpha', \mu, \beta')$ we can always apply the splicing operation to two copies of a word having $\alpha'\mu\beta'$ as a factor [10]. Furthermore, we can easily prove that each site of a pattern (in $B$ or $C$) is a constant for $L = L(S_H)$ (see [2]).

In order to prove the main results presented in this section, we consider again Paun's systems and we define a new relation $Rel_\mu(R')$.

**Definition 8** ($Rel_\mu(R')$ **relation**). *Let $S_{PA} = (A, I, R)$ be a finite reflexive Paun splicing system and let $\mu \in A^*$. For $R' \subseteq R$, we set $Rel_\mu(R')$ the binary* *relation over* $\{u_1\#u_2 \mid u_1u_2 \in SITES(R')\}$ *defined as follows:*

$$u_1\#u_2 \ Rel_\mu(R') \ u_3\#u_4 \ \textit{if and only if } u_1 = u'_1\mu, u_3 = u'_3\mu \ .$$

**Definition 9 ($Rel_\mu$-full).** *Let $S_{PA} = (A, I, R)$ be a finite reflexive Paun splicing system and let $\mu \in A^*$. The set $R' \subseteq R$ is $Rel_\mu$-full if and only if for each $u_1\#u_2, u_3\#u_4 \in \{x_1\#x_2 \mid x_1x_2 \in SITES(R')\}$, we have $u_1\#u_2 \ Rel_\mu(R') \ u_3\#u_4$ and $u_1\#u_2 \ Rel(R') \ u_3\#u_4$.*

We denote $R'_\mu = \{u_1\#u_2\$u_3\#u_4 \in R' \mid u_1\#u_2 \ Rel_\mu(R') \ u_3\#u_4\}$. Roughly, $R'_\mu$ contains rules in $R'$ such that for all $u_1u_2, u_3u_4 \in SITES(R'_\mu)$, $u_1$, $u_3$ satisfy the condition on the existence of the common suffix $\mu$. Furthermore, the set $R'_\mu$ is $Rel_\mu$-full if for all $u_1u_2, u_3u_4 \in SITES(R'_\mu)$, there exists a rule $u_1\#u_2\$u_3\#u_4$ in $R'_\mu$.

**Definition 10 ($R'$-transitive).** *Let $S_{PA} = (A, I, R)$ be a finite reflexive Paun splicing system and let $R' \subseteq R$. $S_{PA}$ is $R'$-transitive if there exist $m \geq 1$, $\mu_1, \ldots, \mu_m \in A^*$ such that $\bigcup_{i=1}^m R'_{\mu_i} = R'$ and each $R'_{\mu_i}$ is $Rel_{\mu_i}$-full, for $1 \leq i \leq m$.*

**Theorem 2.** *1) Each finite Head splicing system $S_H = (A, I, B, C)$ is equivalent to a finite Paun splicing system $S_{PA} = (A, I, R_B \cup R_C)$ which is reflexive and is $R_B$-transitive and $R_C$-transitive.*
  *2) Each finite Paun splicing system $S_{PA} = (A, I, R_B \cup R_C)$ which is reflexive, $R_B$-transitive and $R_C$-transitive is equivalent to a finite Head splicing system $S_H = (A, I, B, C)$.*

Thanks to Theorems 1 and 2, we can give a structural property of regular languages generated by finite Head splicing systems. Indeed, the required $R_B$-transitivity and $R_C$-transitivity for the set $R = R_B \cup R_C$ of the rules in a Paun system imply an additional hypothesis on the set $J$ introduced in Definition 7. Precisely, an *H-split language* is a *PA-split language*. An *H-con-split language* is a *PA-con-split language* such that $\mathcal{M} = \mathcal{M}' \cup \mathcal{M}''$, $\mathcal{M}', \mathcal{M}'' \subseteq \mathcal{M}$ and $J = J_{\mathcal{M}'} \cup J_{\mathcal{M}''}$, where $J_{\mathcal{M}'} = \{(\alpha, \beta) \mid \alpha = (\alpha_1, \alpha_2) \in F(m), \beta = (\beta_1, \beta_2) \in F(m')$, $m = x\mu y \in \mathcal{M}'$, $\alpha_1 = x\mu$, $m' = x'\mu y' \in \mathcal{M}'$, $\beta_1 = x'\mu\}$ and $J_{\mathcal{M}''} = \{(\alpha, \beta) \mid \alpha = (\alpha_1, \alpha_2) \in F(m), \beta = (\beta_1, \beta_2) \in F(m')$, $m = x\mu y \in \mathcal{M}''$, $\alpha_1 = x\mu$, $m' = x'\mu y' \in \mathcal{M}''$, $\beta_1 = x'\mu\}$. Roughly, in Head's case the pairs $((\alpha_1, \alpha_2), (\beta_1, \beta_2))$ in $J_{\mathcal{M}'}$ (or in $J_{\mathcal{M}''}$) must satisfy the additional condition on the existence of a common suffix for the first components $\alpha_1, \beta_1$ of the factorizations.

**Corollary 1.** *Let $L \subseteq A^*$ be a finite Head splicing language. Then $L$ is an H-con-split language (generated by a finite Paun splicing system $S_{PA} = (A, I, R_B \cup R_C)$ which is reflexive, $R_B$-transitive and $R_C$-transitive).*
  *Let $L \subseteq A^*$ be an H-con-split language generated by a finite Paun splicing system $S_{PA} = (A, I, R_B \cup R_C)$ which is reflexive, $R_B$-transitive and $R_C$-transitive. Then $L$ is a finite Head splicing language.*

# 6   Reflexive Pixton Splicing Languages

In order to end the characterization of reflexive splicing languages, we briefly present their structure when we take into account Pixton's definition of the splicing operation.

**Pixton's definition [18].** A *Pixton splicing system* is a triple $S_{PI} = (A, I, R)$, where $I \subset A^*$ is a set of strings, called *initial language*, $R$ is a finite collection of rules $r = (\alpha', \alpha''; \beta')$, $\alpha', \alpha'', \beta' \in A^*$. Given two words $x = \epsilon \alpha' \eta$, $y = \epsilon' \alpha'' \eta'$ and the rule $r = (\alpha', \alpha''; \beta')$, the splicing operation produces $w = \epsilon \beta' \eta'$, denoted $(x, y) \vdash_r w$. We also say that $\alpha', \alpha''$ are *sites* of splicing.

A Pixton splicing system is *finite* if $I, R$ are finite sets and a language $L$ is $S_{PI}$ *generated* (or $L$ is a *Pixton splicing language*) if a splicing system $S_{PI}$ exists such that $L = L(S_{PI})$.

**Definition 11 (reflexive Pixton splicing system).** *A finite splicing system* $S_{PI} = (A, I, R)$ *is a PI-reflexive splicing system if and only if for each rule* $(\alpha', \alpha''; \beta') \in R$ *we also have* $(\alpha', \alpha'; \alpha')$, $(\alpha'', \alpha''; \alpha'') \in R$.

A language $L$ is called *PI-reflexive* if there exists a finite Pixton reflexive splicing system $S_{PI}$ such that $L = L(S_{PI})$. As in Paun's case we can prove that a Pixton splicing language $L$ is *PI*-reflexive if and only if there exists a finite splicing system $S_{PI} = (A, I, R)$ so that each site of the rules in $R$ is a constant for $L$ and $L = L(S_{PI})$ [2].

The definitions and results which follow are the extension to Pixton's systems of the analogous definitions and results given in Section 4.

**Definition 12 (PI-split language).** *Let* $L \subseteq A^*$ *be a regular language, let* $\mathcal{M}$ *be a finite set of constants for* $L$, *let* $J \subseteq \{(m, m') \mid m, m' \in \mathcal{M}\}$. *Let* $\gamma : J \to A^*$ *be a mapping. The PI-split language generated by* $m$ *and* $m'$ *is the language*

$$L_{(m, m')} = C_{\mathcal{L}}(m, L) \, \gamma(m, m') \, C_{\mathcal{R}}(m', L) .$$

**Definition 13 (PI-con-split language).** *Let* $L \subseteq A^*$ *be a regular language, let* $\mathcal{M} \subseteq A^*$ *be a finite subset of constants for* $L$. *Let* $J \subseteq \{(m, m') \mid m, m' \in \mathcal{M}\}$ *and let* $\gamma : J \to A^*$ *be a mapping. Let* $Y$ *be a finite subset of* $L$ *such that no* $m \in \mathcal{M}$ *is a factor of a word in* $Y$.
*$L$ is a PI-con-split language (associated with* $Y, \mathcal{M}, \gamma$) *if and only if*

$$L = Y \cup \bigcup_{m \in \mathcal{M}} L(m) \cup \bigcup_{(m, m') \in J} L_{(m, m')} .$$

$\mathcal{D}_{PI-con-split}$ *is the class of the PI-con-split languages.*

**Theorem 3.** *A regular language* $L \subseteq A^*$ *is a PI-reflexive language if and only if* $L$ *belongs to* $\mathcal{D}_{PI-con-split}$.

# References

1. Berstel, J., Perrin, D.: Theory of codes. Academic Press, New York (1985)
2. Bonizzoni, P., De Felice, C., Mauri, G., Zizza, R.: The structure of reflexive regular splicing languages via Schützenberger constants. *manuscript* (2003)
3. Bonizzoni, P., De Felice, C., Mauri, G., Zizza, R.: Decision Problems on Linear and Circular Splicing. In: Ito, M., Toyama, M. (eds.): DLT 2002. Lecture Notes in Computer Science, Springer-Verlag, New York (2003)
4. Bonizzoni, P., De Felice, C., Mauri, G., Zizza, R.: On the power of linear and circular splicing. *submitted* (2002)
5. Bonizzoni, P., Ferretti, C., Mauri, G., Zizza, R.: Separating some splicing models. Information Processing Letters **79:6** (2001) 255–259
6. Gatterdam, R.W.: Algorithms for splicing systems. SIAM Journal of Computing **21:3** (1992) 507–520
7. Goode, E., Head, T., Pixton, D. *private communication* (2002)
8. Head, T.: Formal Language Theory and DNA. An analysis of the generative capacity of specific recombinant behaviours. Bull. Math. Biol. **49** (1987) 737–759
9. Head, T.: Splicing languages generated with one sided context. In: Paun, Gh. (ed.): Computing with Bio-molecules. Theory and Experiments, Springer-Verlag Singapore (1998)
10. Head, T., Paun, Gh., Pixton, D.: Language theory and molecular genetics. Generative mechanisms suggested by DNA recombination. In: Rozenberg, G., Salomaa, A. (eds.): Handbook of Formal Languages, Vol. 2. Springer-Verlag (1996) 295–360
11. Hopcroft, J.E., Motwani, R., Ullman, J.D.: Introduction to Automata Theory, Languages, and Computation. 2nd edn. Addison-Wesley, Reading, Mass. (2001)
12. McNaughton, R., Papert, S.: Counter-Free Automata. MIT Press, Cambridge, Mass. (1971)
13. Paun, Gh.: On the splicing operation. Discrete Applied Mathematics **70** (1996) 57–79
14. Paun, Gh., Rozenberg, G., Salomaa, A.: Computing by splicing. Theoretical Computer Science **168:2** (1996) 321–336
15. Paun, Gh., Rozenberg, G., Salomaa, A.: DNA computing, New Computing Paradigms. Springer-Verlag (1998)
16. Perrin, D.: Finite Automata. In: van Leeuwen, J. (ed.): Handbook of Theoretical Computer Science, Vol. B. Elsevier (1990) 1–57
17. Pixton, D.: Linear and Circular Splicing Systems. In: Proc. of 1st Int. Symp. on Int. in Neural and Biological Systems (1996) 181–188
18. Pixton, D.: Regularity of splicing languages. Discrete Applied Mathematics **69** (1996) 101–124
19. Schützenberger, M.-P.: Sur certaines opérations de fermeture dans le langages rationnels. Symposia Mathematica **15** (1975) 245–253

# The Myhill-Nerode Theorem for Recognizable Tree Series

Björn Borchardt[*]

Dresden University of Technology
Faculty of Computer Science
D-01062 Dresden
borchard@tcs.inf.tu-dresden.de

**Abstract.** In this paper we prove a Myhill-Nerode theorem for recognizable tree series over commutative semifields and thereby present a minimization of bottom-up finite state weighted tree automata over a commutative semifield, where minimal means with respect to the number of states among all equivalent, deterministic devices.

## 1 Introduction

The concept of finite state weighted tree automata is a generalization of (i) finite state (unweighted) tree automata and (ii) finite state weighted (string) automata in the sense that trees are accepted with a weight, which is taken from a semiring. Let us briefly recall the two latter underlying models.

Trees can be traversed in two directions: from the leaves towards the root, which is modeled by bottom-up (also: frontier-to-root) finite state tree automata [22,17,5], or from the root towards the leaves, which is modeled by top-down (also: root-to-frontier) finite state tree automata [17,19]. The class of recognizable tree languages is the class of tree languages accepted by these devices. For more details on finite state tree automata we refer the reader to [12,13]. An algebraic characterization of recognizable tree languages is provided by the Myhill-Nerode theorem, which, as stated in [13] (Chapter II, Theorem 7.1, also cf. [10,14]), says that for every tree language $L \subseteq T_\Sigma$ the following statements are equivalent: (i) $L$ is recognizable (by a bottom-up finite state tree automaton), (ii) the term algebra $\mathcal{F}_\Sigma(X)$ has a congruence of finite index saturating $L$ (i.e., $L$ is the union of some equivalence classes), (iii) the index of the Nerode congruence $\equiv_L$ is finite (where for every $s, t \in T_\Sigma$, $s \equiv_L t$ iff for all trees $C \in T_\Sigma(\{x\})$ with exactly one occurrence of the variable $x$, the equivalence $C[x \leftarrow s] \in L \Leftrightarrow C[x \leftarrow t] \in L$ holds).

As the second underlying model, finite state weighted (string) automata [20] (also: string-to-weight transducers) arise from unweighted (string) automata by associating a weight with each of its transitions; the weight is taken from a semiring. For more details on this concept the reader may consult [16,2,18,6,7]. The

[*] Research was financially supported by the German Research Council under grant (DFG, GRK 433/2).

Z. Ésik and Z. Fülöp (Eds.): DLT 2003, LNCS 2710, pp. 146–158, 2003.
© Springer-Verlag Berlin Heidelberg 2003

second mentioned paper also investigates the minimization in terms of syntactic algebras and proves that a power series is recognizable iff its syntactic algebra has finite rank (cf. [2], Chapter II, Theorem 1.1). Besides others [18] provides a minimization of deterministic weighted automata over the tropical semiring, where minimization means with respect to the number of states among all equivalent, deterministic devices. This construction is based on a generalization of the Myhill-Nerode congruence and of the Myhill-Nerode construction for weighted automata (cf. [18], Theorem 14).

As mentioned above, a finite state weighted tree automaton $M = (Q, \Sigma, Q_d, A, \mu)$ accepts trees with a weight, which is taken from a semiring $(A, \oplus, \odot, 0, 1)$. Similar to unweighted tree automata we distinguish between bottom-up (for short: bu-w-fta) and top-down devices. In this paper we will only consider the model bu-w-fta. Every transition has a weight, which is implemented in the tree representation $\mu = (\mu_k : \Sigma^{(k)} \to A^{Q^k \times Q} \mid k \geq 0)$. $M$ can be entered by reading an arbitrary symbol $\alpha$ of rank 0 in every state $q$ with the weight $\mu_0(\alpha)_{(),q}$. $M$ can be left in a final state $q \in Q_d$. Now we let $M$ run on a tree $t \in T_\Sigma$. We obtain the weight of a run on $t$ by multiplying the weights of the applied transitions. The weight of $t$ is the sum of the weights of all runs accepting $t$ and the tree series accepted by $M$ is a mapping $S_M \in A\langle\!\langle T_\Sigma \rangle\!\rangle$, where $A\langle\!\langle T_\Sigma \rangle\!\rangle = \{S \mid S : T_\Sigma \to A\}$ is the class of formal tree series over $\Sigma$ and $A$. We formalize this procedure by a $\Sigma$-algebra, where we perform the summation arising from the nondeterminism at each step (compare [11] for a proof of the equivalence of automata-theoretic semantics, in which the summation is only performed on successful runs, and the initial algebra semantics, in which the summation is done at every transition step, in the more general context of tree series transducers). We note that there exist other weighted tree automata models, e.g. the concept of recognizable tree series of [1] (which if the underlying semiring is commutative, is equivalent to the concept of bu-w-fta), $R$-cost automata of [21], and $A'$-automata of [15] (cf. [3] for a discussion on the relationship between the latter two models and our model of weighted tree automata). As a survey paper on formal tree series we refer the reader to [9].

In the present paper we consider bu-w-fta as an extension of classical bottom-up finite state tree automata and generalize the Myhill-Nerode congruence and the Myhill-Nerode theorem as stated above. More precisely, we show that for every tree series $S$ over a commutative semifield the following statements are equivalent (cf. Theorem 2):

(i)   $S \in A\langle\!\langle T_\Sigma \rangle\!\rangle$ is recognizable by a deterministic bu-w-fta.
(ii)  There exists a congruence relation $\equiv$ of the term algebra $(T_\Sigma, \Sigma)$ of finite index satisfying Properties (MN1) and (MN2).
(iii) The Myhill-Nerode congruence $\equiv_S$ is of finite index.

We note that, by Theorem 6.2 of [3] if the underlying semiring is a locally finite semifield, then $S$ is recognizable by a nondeterministic bu-w-fta iff $S$ is recognizable by a deterministic bu-w-fta. Hence Theorem 2 gives algebraic characterizations for tree series which are accepted by (a) deterministic bu-w-fta over a commutative semifield and (b) nondeterministic bu-w-fta over a locally finite, commutative semifield. Roughly speaking, (MN1) and (MN2) are two technical conditions which, applied in the Boolean semiring and thereby considering unweighted tree automata, collapse to the classical property that "≡ saturates the tree language $supp(S)$". Hence Theorem 2 is a canonical generalization of the Myhill-Nerode theorem for tree languages. Similar to unweighted tree automata we derive from the Myhill-Nerode theorem a minimization for (a) deterministic bu-w-fta over a commutative semifield and (b) nondeterministic bu-w-fta over a locally finite, commutative semifield, where minimization means minimal with respect to the number of states among all deterministic, equivalent devices.

[4] considers weighted tree automata as a generalization of weighted (string) automata and proves the following characterization for a tree series $S$ over a field (cf. [4], corollary on page 353): $S$ is recognizable (in the sense of [1]) iff the syntactic algebra $\mathcal{A}_S$ is of finite dimension. Although the above stated theorem of Bozapalidis and Theorem 2 of the present paper seem to be very similar, none of these results implies the other one: we show that the tree series $(S, t) = size(t) \in Real\langle\langle T_\Sigma \rangle\rangle$ is recognizable by a nondeterministic bu-w-fta (cf. Example 1), but not by a deterministic bu-w-fta (cf. Example 3). Hence the syntactic algebra $\mathcal{A}_S$ is finite dimensional (by the corollary on page 353 of [4]), but the Myhill-Nerode congruence $\equiv_S$ is of infinite index (by Theorem 2 of this paper). We note that the characterization of [4] holds for tree series over a field, while our result only assumes a commutative semifield.

## 2     Preliminaries

### 2.1     Notions and Trees

We denote the set of all nonnegative integers by $\mathbb{N} = \{0, 1, \dots\}$. For a set $S$, the cardinality and the power set of $S$ are denoted by $card(H)$ and $\mathfrak{P}(S)$, respectively. If $\Sigma$ is a nonempty finite set and $rk : \Sigma \to \mathbb{N}$, then the tuple $(\Sigma, rk)$ is called *ranked alphabet*. In order to be short in notation we will write $\Sigma$ rather than $(\Sigma, rk)$. For every $k \geq 0$ we define the set $\Sigma^{(k)} = \{\sigma \in \Sigma \mid rk(\sigma) = k\}$ of all symbols of $\Sigma$, which have rank $k$. An element $\sigma \in \Sigma^{(k)}$ is also written as $\sigma^{(k)}$.

For the rest of this paper let $\Sigma$ be a ranked alphabet. Moreover, let $X$ be a set (of variables) disjoint with $\Sigma$. The set of *trees* over $\Sigma$ indexed by $X$, denoted by $T_\Sigma(X)$, is the smallest set $T_\Sigma(X) \subseteq (\Sigma \cup X \cup \{(,),,\})^*$ such that (i) $X \cup \Sigma^{(0)} \subseteq T_\Sigma(X)$ and (ii) if $k \geq 1$, $\sigma \in \Sigma^{(k)}$, and $s_1, \dots, s_k \in T_\Sigma(X)$, then $\sigma(s_1, \dots, s_k) \in T_\Sigma(X)$. The set $T_\Sigma(\emptyset)$ is denoted by $T_\Sigma$. Clearly, $T_\Sigma(X) = T_{\Sigma \cup X}$. The *size* of a tree $t \in T_\Sigma(X)$ is given by the mapping $size : T_\Sigma(X) \to \mathbb{N}$, where $size(\sigma(s_1, \dots, s_k)) = \sum_{i=1}^{k} size(s_i) + 1$ for every $k \geq 0$, $\sigma \in \Sigma^{(k)}$, and $s_1, \dots, s_k \in T_\Sigma$. Now let $x_1, x_2, \dots$ be variables and $X_m = \{x_1, \dots, x_m\}$ for every $m \geq 0$. Let $t \in T_\Sigma(X_m)$ for some $m \geq 0$. The *substitution of* $x_1, \dots, x_m$ *by* $s_1, \dots, s_m \in T_\Sigma(X_m)$ *in* $t$, denoted by $t[s_1, \dots, s_m]$, is the tree, which we obtain from $t$ by replacing every occurrence of $x_j$ in $t$ by $s_j$ for every $1 \leq j \leq m$. Moreover, a tree $t \in T_\Sigma(X_m)$ is called a *$\Sigma$-m-context* if every variable $x \in X_m$ occurs precisely once in $t$. We denote the class of all *$\Sigma$-m-contexts* by $C_{\Sigma,m}(X_m)$.

## 2.2  Algebraic Notions

In this section we introduce semirings, semifields, term algebras, matrices, and equivalence and congruence relations (also cf. [16]).

Let $A$ be a nonempty set, $0, 1 \in A$, and $\oplus$ and $\odot$ binary, associative operations on $A$. The tuple $(A, \oplus, \odot, 0, 1)$ is called *semiring* if (SR1) $\oplus$ is commutative, (SR2) 0 is the neural element of $\oplus$, (SR3) 1 is the neural element of $\odot$, (SR4) $\odot$ is distributive over $\oplus$, and (SR5) $0 \odot a = a \odot 0 = 0$ for every $a \in A$. Whenever $\oplus$, $\odot$, 0, and 1 are clear from the context, then we denote the semiring $(A, \oplus, \odot, 0, 1)$ by $A$. We call a semiring *commutative* if $\odot$ is commutative. $A$ is *zero divisor free* if for every $a, b \in A$, $a \odot b = 0$ implies $a = 0$ or $b = 0$.

Let $^{-1}$ be a unary operation on $A \setminus \{0\}$. The tuple $(A, \oplus, \odot, ^{-1}, 0, 1)$ is called *semifield* if (SF1) $(A, \oplus, \odot, 0, 1)$ is a semiring and, (SF2) for every $a \in A \setminus \{0\}$, $a^{-1}$ is the multiplicative inverse of $a$. Whenever it does not lead to confusion, we will write $A$ rather than $(A, \oplus, \odot, ^{-1}, 0, 1)$. Let $A = (A, \oplus, \odot, ^{-1}, 0, 1)$ be a semifield. $A$ is called *commutative* (*zero divisor free*) if $(A, \oplus, \odot, 0, 1)$ is a commutative (zero divisor free, respectively) semiring. $A$ is called *locally finite* if, for every finite $B \subseteq A$, there exists a semifield $(\tilde{B}, \oplus, \odot, ^{-1}, 0, 1)$ such that $B \subseteq \tilde{B}$ and $\tilde{B}$ is finite (in other words: the closure of $B$ under $\{\oplus, \odot, ^{-1}\}$ is finite). Typical (commutative) semifields are

- the *Boolean semifield* $\mathbb{B} = (\{0, 1\}, \vee, \wedge, ^{-1}, 0, 1)$, where the operations are conjunction and disjunction, and $1^{-1} = 1$,
- the *tropical semifield* $Trop = (\mathbb{R} \cup \{+\infty\}, min, +, ^{-1}, +\infty, 0)$ with minimum and addition, where $min(a, +\infty) = a$ and $a + (+\infty) = (+\infty) + a = +\infty$ for every $a \in \mathbb{R} \cup \{+\infty\}$, and the inverse is defined for every $a \in \mathbb{R}$ to be $-a$,
- the *real numbers Real* $= (\mathbb{R}, +, \cdot, ^{-1}, 0, 1)$ with the usual operations,

By standard arguments one proves the following lemma.

**Lemma 1.** *Every semifield is zero divisor free.*

Now consider an arbitrary ranked alphabet $\Sigma$ and let $k \geq 0$. The *term algebra (with respect to $\Sigma$)* is the algebra $(T_\Sigma, \Sigma)$, where every $\sigma \in \Sigma^{(k)}$ is identified with the mapping $\sigma : T_\Sigma{}^k \to T_\Sigma$ (*top concatenation with $\sigma$*), which is defined for every $t_1, \ldots, t_k \in T_\Sigma$ to be the tree $\sigma(t_1, \ldots, t_k)$.

We now recall some definitions concerning matrices. An $I \times J$-*matrix* is a mapping $M : I \times J \to A$, where $A, I$, and $J$ are arbitrary sets. As usual, $M(i, j)$ is also denoted by $M_{i,j}$. The set of all $I \times J$-matrices over $A$ is denoted by $A^{I \times J}$. If either $I = \{i\}$ or $J = \{j\}$ are singletons, then we write $A^J$ and $M_j$ ($A^I$ and $M_i$, respectively) rather than $A^{I \times J}$ and $M_{i,j}$.

Let us now recall some definitions concerning relations. Let $S$ be a set. A *binary relation $\rho$ (on $S$)* is a subset $\rho \subseteq S \times S$. If $(s, t) \in \rho$, then we also write $s \, \rho \, t$. We call a *reflexive* ($(a, a) \in \rho$ for every $a \in S$), *symmetric* ($(a, b) \in \rho$ implies $(b, a) \in \rho$), and *transitive* ($(a, b), (b, c) \in \rho$ implies $(a, c) \in \rho$) relation $\rho$ an *equivalence relation*.

Let $\rho$ be an equivalence relation and $a \in S$. We call the set $[a]_\rho = \{b \in S \mid (a, b) \in \rho\}$ the *equivalence class of $a$ (with respect to $\rho$)*. The set of all equivalence classes is denoted by $S/_\rho = \{[a]_\rho \mid a \in S\}$ and the *index of $\rho$* is defined to be the cardinality of $S/_\rho$. A set $S' \subseteq S$ is *saturated or respected by $\rho$* if $S$ is the union of some equivalence classes of $\rho$. Moreover, a *representative mapping (with respect to $\rho$)* is a mapping $\varphi : S/_\rho \to S$ such that $\varphi([a]_\rho) \in [a]_\rho$ for every $[a]_\rho \in S/_\rho$. We denote the *set of all representative mappings with respect to $\rho$* by $RepMap(\rho)$. Clearly, $RepMap(\rho) \neq \emptyset$.

Now consider a ranked alphabet $\Sigma$ and an equivalence relation $\rho \subseteq T_\Sigma \times T_\Sigma$. $\rho$ is a *congruence relation (of the term algebra $(T_\Sigma, \Sigma)$)* if for every $k \geq 0$, $s \in \Sigma^{(k)}$, and $s_1, \dots, s_k, t_1, \dots, t_k \in T_\Sigma$, $s_1 \rho t_1, \dots, s_k \rho t_k$ implies $\sigma(s_1, \dots, s_k) \rho \sigma(t_1, \dots, t_k)$.

## 2.3   Formal Tree Series

Let $A$ be a semiring and $\Sigma$ be a ranked alphabet. A *(formal) tree series (over $\Sigma$ and $A$)* is a mapping $S : T_\Sigma \to A$. The image $S(t) \in A$ of $t \in T_\Sigma$ is called *coefficient* of $t$ and we write $(S, t)$ rather than $S(t)$. The tree series $S$ can be written as the sum $\sum_{t \in T}(S, t)\, t$ (cf. [8], Chapter VI, 3.1). The *support of $S$* is defined to be the set $supp(S) = \{t \in T_\Sigma \mid (S, t) \neq 0\}$.

# 3   Bottom-Up Finite State Weighted Tree Automata

We now briefly recall the model of bottom-up finite state weighted tree automata. For more details we refer the reader to [3].

**Definition 1 ([3], Definition 3.1).** *Let $Q$ be a finite set, $\Sigma$ a ranked alphabet, and $A$ a semiring. A* bottom-up tree representation *(over $Q$, $\Sigma$, and $A$) is a family $\mu = (\mu_k \mid k \geq 0)$ of mappings $\mu_k : \Sigma^{(k)} \longrightarrow A^{Q^k \times Q}$. $\mu$ is called* deterministic *if for every $k \geq 0$, $\sigma \in \Sigma^{(k)}$, and $(q_1, \dots, q_k) \in Q^k$ there is at most one $q \in Q$ such that $\mu_k(\sigma)_{(q_1, \dots, q_k), q} \neq 0$.*

Every tree representation $\mu$ over a finite set $Q$ induces a family of mappings $(\overline{\mu_k(\sigma)} : A^Q \times \cdots \times A^Q \to A^Q \mid k \geq 0, \sigma \in \Sigma^{(k)})$ in the following way: for every $q \in Q$ and $V_1, \dots, V_K \in A^Q$,

$$\overline{\mu_k(\sigma)}(V_1, \dots, V_k)_q = \sum_{(q_1, \dots, q_k) \in Q^k} (V_1)_{q_1} \odot \cdots \odot (V_k)_{q_k} \odot \mu_k(\sigma)_{(q_1, \dots, q_k), q}.$$

Clearly, $(A^Q, (\overline{\mu_k(\sigma)} \mid k \geq 0, \sigma \in \Sigma^{(k)}))$ is a $\Sigma$-algebra. Its initial homomorphism $h_\mu : T_\Sigma \to A^Q$ is given for every $k \geq 0$, $\sigma \in \Sigma^{(k)}$, and $t_1, \dots, t_k \in T_\Sigma$ by $h_\mu(\sigma(t_1, \dots t_k)) = \overline{\mu_k(\sigma)}(h_\mu(t_1), \dots, h_\mu(t_k))$. We call $h_\mu(t)$ the *characteristic vector of $t$ (with respect to $\mu$)*. Let us now define bottom-up finite state weighted tree automata.

**Definition 2 ([3], Definition 3.3).** *Let $Q$ be a finite set, $\Sigma$ a ranked alphabet, $Q_d \subseteq Q$, $A$ a semiring, and $\mu$ a bottom-up tree representation. The tuple $M = (Q, \Sigma, Q_d, A, \mu)$ is called* bottom-up finite state weighted tree automaton *(for* short bu-w-fta). A bu-w-fta is called deterministic *if its tree representation is deterministic. The tree series $S_M$ accepted by $M$ is defined for every $t \in T_\Sigma$ by* $(S_M, t) = \sum_{q \in Q_d} h_\mu(t)_q$. *We denote by $A^{n,bu}\langle\langle T_\Sigma \rangle\rangle$ and $A^{d,bu}\langle\langle T_\Sigma \rangle\rangle$ the classes of all tree series, which are accepted by a bu-w-fta and a deterministic bu-w-fta, respectively.*

*Example 1.* Consider the bu-w-fta $M = (Q, \Sigma, Q_d, A, \mu)$ given by $Q = \{p, q\}$, $\Sigma = \{\alpha^{(0)}, \sigma^{(2)}\}$, $Q_d = \{p\}$, $A = Real$, and

$$\mu_0(a) = \begin{matrix} p & q \\ (1 & 1) \end{matrix}, \qquad \mu_2(\sigma) = \begin{matrix} p & q \\ \begin{pmatrix} 0 & 0 \\ 1 & 0 \\ 1 & 0 \\ 1 & 1 \end{pmatrix} & \begin{matrix} (p,p) \\ (p,q) \\ (q,p) \\ (q,q) \end{matrix} \end{matrix}.$$

One can easily show that $h_\mu(t)_p = size(t)$ and $h_\mu(t)_q = 1$ for every $t \in T_\Sigma$. Thus every input tree is accepted by $M$ with its size: $(S_M, t) = size(t)$.    ◇

Later on, for a given $t \in T_\Sigma$, we work with the set of all those states $q \in Q$ such that there is a "non zero derivation" of $t$ in $M$ ending up in $q$. Formally, let $\hat{\mu} : T_\Sigma \to \mathfrak{P}(Q)$ be the mapping, which is defined for every $t \in T_\Sigma$ by $\hat{\mu}(t) = \{q \in Q \mid h_\mu(t)_q \neq 0\}$. We conclude this section by presenting three properties of the mapping $\hat{\mu}$.

**Lemma 2 ([3], Lemma 3.6).** *Let $M = (Q, \Sigma, Q_d, A, \mu)$ be a deterministic bu-w-fta. Then $card(\hat{\mu}(t)) \leq 1$ for every tree $t \in T_\Sigma$. If $\hat{\mu}(t) = \{q\}$ for some $q \in Q$, then we write $\hat{\mu}(t) = q$.*

**Lemma 3 ([3], Lemma 3.7).** *Let $M = (Q, \Sigma, Q_d, A, \mu)$ be a bu-w-fta. Also, let $k \geq 0$, $\sigma \in \Sigma^{(k)}$, and $t_1, \ldots, t_k \in T_\Sigma$. If there exists an $1 \leq i \leq k$ such that $\hat{\mu}(t_i) = \emptyset$, then $\hat{\mu}(\sigma(t_1, \ldots, t_k)) = \emptyset$.*

By a repeated application of Lemma 3 we obtain the following corollary:

**Corollary 1.** *Let $M = (Q, \Sigma, Q_d, A, \mu)$ be a bu-w-fta. Moreover, let $m \geq 0$, $s_1, \ldots, s_m \in T_\Sigma$, and $t \in T_\Sigma(X_m)$. If $\hat{\mu}(s_j) = \emptyset$ for some $1 \leq j \leq m$, then $\hat{\mu}(t[s_1, \ldots, s_m]) = \emptyset$.*

## 4  Substitution and Weights

In this section we investigate the behavior of the characteristic vectors if we substitute variables by trees. In particular, we consider a finite set $X_m$ of variables, a tree $t \in T_\Sigma(X_m)$, and trees $s_1, \ldots, s_m \in T_\Sigma$ and show a relationship between the characteristic vectors of $t$, $s_1, \ldots, s_m$, and $t[s_1, \ldots, s_m]$. Intuitively a run on $t[s_1, \ldots, s_m]$ can be decomposed into runs on each of the copies of $s_j$, which are substituted in $t[s_1, \ldots, s_m]$, and a run on $t$. For our purpose it suffices to prove the relationship under the assumption that, for every $1 \leq j \leq m$, the non zero runs on the copies of $s_j$ end up in the same state $q_j$. This additional requirement we can ensure by assuming that $M$ is a deterministic bu-w-fta. Then the weight of a run on $t[s_1, \ldots, s_m]$ is the product of the weight of the run on $t[q_1, \ldots, q_m] \in T_\Sigma(Q)$ and the powers (according to the number of occurrences of $x_j$) of the weights of the runs on $s_1, \ldots, s_m$. In order to let an automaton run on $t[q_1, \ldots, q_m]$ we extend the given bu-w-fta.

**Definition 3.** Let $M = (Q, \Sigma, Q_d, A, \mu)$ be a bu-w-fta. The extension of $M$ (with respect to $Q$) is the bu-w-fta $M_Q = (Q, \Sigma \cup \{q^{(0)} \mid q \in Q\}, Q_d, A, \mu_Q)$, where for every $k \geq 0$, $\sigma \in \Sigma^{(k)}$, and $p, q, q_1, \ldots, q_k \in Q$,

$$(\mu_Q)_0(p)_q = \begin{cases} 1 & \text{if } p = q, \\ 0 & \text{otherwise,} \end{cases}$$

$$(\mu_Q)_k(\sigma)_{(q_1, \ldots, q_k), q} = \mu_k(\sigma)_{(q_1, \ldots, q_k), q}.$$

Let us now prove the relationship between the characteristic vectors of $s_1, \ldots, s_m \in T_\Sigma$, $t[s_1, \ldots, s_m] \in T_\Sigma$, and $t[q_1, \ldots, q_m] \in T_\Sigma(Q)$ with respect to a deterministic bottom-up tree representation.

**Theorem 1.** Let $M = (Q, \Sigma, Q_d, A, \mu)$ be a deterministic bu-w-fta over a commutative semiring, $M_Q$ the exention of $M$ with respect to $Q$, and $t \in T_\Sigma(X_m)$ for some $m \geq 0$. Moreover, let $\varepsilon_j$ denote the number of occurences of $x_j$ in $t$ for every $1 \leq j \leq m$. Then, for every $s_1, \ldots, s_m \in T_\Sigma$ and $q \in Q$,

$$h_\mu(t[s_1, \ldots, s_m])_q = \begin{cases} h_{\mu_Q}(t[\hat{\mu}(s_1), \ldots, \hat{\mu}(s_m)])_q \odot \prod_{j=1}^{m} h_\mu(s_j)_{\hat{\mu}(s_j)}{}^{\varepsilon_j} \\ \qquad \text{if } \hat{\mu}(s_j) \neq \emptyset \text{ for every } 1 \leq j \leq m, \\ 0 \qquad \text{otherwise.} \end{cases}$$

We note that $h_\mu(t[s_1, \ldots, s_m])$ can also be expressed by the result of the o-substitution of tree series (cf. [11], Definition 3.2): the monomial $d\, t[s_1, \ldots, s_m]$ with $d = h_\mu(t[s_1, \ldots, s_m])_q$ is equal to

$$c\, t \xleftarrow{\quad o \quad} (c_1\, s_1, \ldots, c_m\, s_m)$$

where $c = h_{\mu_Q}(t[\hat{\mu}(s_1), \ldots, \hat{\mu}(s_m)])$ and $c_j = h_\mu(s_j)_{\hat{\mu}(s_j)}$ for every $1 \leq j \leq m$.

**Corollary 2.** Let $M = (Q, \Sigma, Q_d, A, \mu)$ be a deterministic bu-w-fta over a zero divisor free, commutative semiring and $t \in T_\Sigma(X_m)$ for some $m \geq 0$. Also, let $s_1, s_1', \ldots, s_m, s_m' \in T_\Sigma$ such that $\hat{\mu}(s_1) = \hat{\mu}(s_1'), \ldots, \hat{\mu}(s_m) = \hat{\mu}(s_m')$. Then $\hat{\mu}(t[s_1, \ldots, s_m]) = \hat{\mu}(t[s_1', \ldots, s_m'])$.

## 5 The Myhill-Nerode Theorem for Tree Series over Commutative Semifields

In this section we generalize the Myhill-Nerode theorem from tree languages to tree series. For every tree series $S \in A\langle\langle T_\Sigma \rangle\rangle$ over a commutative semifield

$A$ we introduce a congruence relation $\equiv_S$ of the term algebra $(T_\Sigma, \Sigma)$, which extends the Myhill-Nerode congruence for tree languages in a natural way. $\equiv_S$ can be used to characterize tree series, which are accepted by deterministic bu-w-fta over a commutative semifield. We show that the following statements are equivalent: (i) $S \in A^{d,bu}$, (ii) there exists a congruence relation of the underlying term algebra which has finite index and satisfies Properties (MN1) and (MN2) (which we define on page 155), and (iii) $\equiv_S$ is of finite index. Similar to the proof of the classical Myhill-Nerode theorem we show the implications (i) $\Rightarrow$ (iii) $\Rightarrow$ (ii) $\Rightarrow$ (i). From the above equivalences we obtain a characterization for tree series $S \in A^{n,bu}$ over a locally finite, commutative semifield by applying the equality $A^{d,bu}\langle\!\langle T_\Sigma\rangle\!\rangle = A^{n,bu}\langle\!\langle T_\Sigma\rangle\!\rangle$ (cf. Theorem 6.2 of [3]). Finally we present a minimization of (a) deterministic bu-w-fta over a commutative semifield and (b) nondeterministic bu-w-fta over a locally finite, commutative semifield, which is a straightforward application of the Myhill-Nerode theorem. We note that minimization means minimal with respect to the number of states among all equivalent, deterministic devices.

We also note that the proof of (ii) $\Rightarrow$ (i) (Myhill-Nerode construction) requires multiplicative inverses, which in general do not exist in a semiring. Hence, for the rest of this section let $A$ be a commutative semifield. Also, let $S \in A\langle\!\langle T_\Sigma\rangle\!\rangle$. Recall that $C_{\Sigma,1}(X_1)$ denotes the class of all $\Sigma$-1-contexts, i.e., the class of all trees over $\Sigma \cup X_1$ with precisely one occurrence of the variable $x_1$. We start by defining the binary relation $\equiv_S \subseteq T_\Sigma \times T_\Sigma$ by

$$\equiv_S = \{(s,t) \in T_\Sigma \times T_\Sigma \mid \exists a \in A\setminus\{0\} \,\forall C \in C_{\Sigma,1}(X_1) : (S, C[s]) = a \odot (S, C[t])\}.$$

Let us now investigate $\equiv_S$. In particular, we show that $\equiv_S$ is a congruence relation of the term algebra $(T_\Sigma, \Sigma)$, which is stated in Lemma 5.

**Lemma 4.** Let $k \geq 0$, $\sigma \in \Sigma^{(k)}$, $s_1,\ldots,s_k,t_1,\ldots,t_k \in T_\Sigma$, and $a_1,\ldots,a_k \in A\setminus\{0\}$ such that $s_i \equiv_S t_i$ and $(S, C[s_i]) = a_i \odot (S, C[t_i])$ for every $1 \leq i \leq k$ and $C \in C_{\Sigma,1}(X_1)$. Then $\big(S, C[\sigma(s_1,\ldots,s_k)]\big) = a_1\odot\cdots\odot a_k\odot\big(S, C[\sigma(t_1,\ldots,t_k)]\big)$.

Clearly, $\equiv_S$ is an equivalence relation. From Lemma 4 follows that $\equiv_S$ is also a congruence relation of the term algebra.

**Lemma 5.** *$\equiv_S$ is a congruence relation of the term algebra $(T_\Sigma, \Sigma)$.*

Let us now prove that, if $S$ is recognizable by a deterministic bu-w-fta $M$, then $\equiv_S$ is of finite index. This we finally show in Corollary 3. We start the proof of the aforementioned statement by showing a tight relationship between $\equiv_S$ and the state behaviour of $M$. Recall that the tree series, which is recognized by $M$, is denoted by $S_M$.

**Lemma 6.** *Let $M = (Q, \Sigma, Q_d, A, \mu)$ be a deterministic bu-w-fta such that $S_M = S$. Also, let $s,t \in T_\Sigma$. If $\hat{\mu}(s) = \hat{\mu}(t)$, then $s \equiv_S t$.*

Does the claim of Lemma 6 also hold for nondeterministic bu-w-fta? The answer is no, which follows from Example 2.

*Example 2.* In Example 1 we have presented a nondeterministic bu-w-fta $M$ over the commuatative semifield *Real* with $(S_M, t) = size(t)$ and $\hat{\mu}(t) = \{p, q\}$ for every $t \in T_\Sigma$. In particular, $\hat{\mu}(\alpha) = \{p, q\} = \hat{\mu}(\sigma(\alpha, \alpha))$. On the other hand, by applying the equality $size(C[s]) = size(s) + size(C) - 1$ for every $s \in T_\Sigma$ and $C \in C_{\Sigma,1}(X_1)$ which can be shown by induction on $size(C[t])$, one can easily prove that $s \equiv_{S_M} t$ iff $size(s) = size(t)$ for every $s, t \in T_\Sigma$. From $size(\alpha) = 1 \neq 3 = size(\sigma(\alpha, \alpha))$ follows that $\alpha \not\equiv_{S_M} \sigma(\alpha, \alpha)$ even though $\hat{\mu}(\alpha) = \hat{\mu}(\sigma(\alpha, \alpha))$. ◇

In order to obtain an upper bound for the index of $\equiv_S$ we introduce and investigate the set $L_S = \{s \in T_\Sigma \mid \forall C \in C_{\Sigma,1}(X_1) \ (S, C[s]) = 0)\}$.

**Observation 1.** *Either $L_S$ is the empty set or it is an equivalence class of $\equiv_S$.*

**Lemma 7.** *Let $M = (Q, \Sigma, Q_d, A, \mu)$ be a deterministic bu-w-fta such that $S_M = S$. Then $\hat{\mu}(t) \in Q$ for every $t \in T_\Sigma \setminus L_S$.*

**Corollary 3.** *For every deterministic bu-w-fta $M$ over $A$ such that $S_M = S$, $index(\equiv_S) \leq card(Q) + 1$. Moreover, if $L_S = \emptyset$ then $index(\equiv_S) \leq card(Q)$. In particular, if $S \in A\langle\langle T_\Sigma \rangle\rangle^{d,bu}$, then $\equiv_S$ is of finite index.*

Now let us prove that if $\equiv_S$ has finite index, then there exists a congruence relation of the term algebra $(T_\Sigma, \Sigma)$ of finite index satisfying Properties (MN1) and (MN2) (which we define on page 155). We prove this claim by showing that the particular relation $\equiv_S$ has these properties, which is stated in Lemmata 5 and 11. In order to define Property (MN1) and the extension of the Myhill-Nerode construction (cf. Definition 4) we would like to work with multiplicative inverses of $(S, t)$. But in general there exists a tree $t \in T_\Sigma$ such that $(S, t) = 0$ and then we can not take the inverse of $(S, t)$. Therefore we define a new tree series $S^\varphi \in A\langle\langle T_\Sigma \rangle\rangle$, where $\varphi$ is an arbitrary representative mapping of $\equiv_S$. This definition requires some preliminary considerations.

**Lemma 8.** *Let $\varphi \in RepMap(\equiv_S)$. There exists precisely one mapping $a_\varphi : T_\Sigma \to A \setminus \{0\}$, which satisfies the following two properties.*

(i) *For every $t \in T_\Sigma$ and $C \in C_{\Sigma,1}(X_1)$ the equation $(S, C[t]) = a_\varphi(t) \odot (S, C[\varphi([t]_{\equiv_S})])$ holds.*
(ii) *For every $t \in L_S$ we have $a_\varphi(t) = 1$.*

For the rest of this section let $\varphi \in RepMap(\equiv_S)$ and $a_\varphi : T_\Sigma \to A \setminus \{0\}$ the uniquely determined mapping, which satisfies Properties (i) and (ii) of Lemma 8. Now we define the tree series $S^\varphi \in A\langle\langle T_\Sigma \rangle\rangle$, which is given for every $t \in T_\Sigma$ by

$$(S^\varphi, t) = \begin{cases} (S, t) & \text{if } (S, t) \neq 0, \\ a_\varphi(t) & \text{otherwise.} \end{cases}$$

We note that $(S^\varphi, t) \neq 0$ for every $t \in T_\Sigma$, because $a_\varphi(t) \in A \setminus \{0\}$. Recall that in the Myhill-Nerode theorem for tree languages one of the characterizations of a recognizable tree language $L$ is that there exists a congruence relation which saturates $L$. Since we additionally have to handle the weights of the trees, this property has to be generalized appropriately. Therefore we introduce the following properties for equivalence relations $\equiv \subseteq T_\Sigma \times T_\Sigma$:

(MN1) For every $k \geq 0$, $\sigma \in \Sigma^{(k)}$, and $s_1, \ldots, s_k, t_1, \ldots, t_k \in T_\Sigma$ such that $\sigma(t_1, \ldots, t_k) \notin L_S$ and $s_i \equiv t_i$ for every $1 \leq i \leq k$,

$$(S^\varphi, s_k)^{-1} \odot \cdots \odot (S^\varphi, s_1)^{-1} \odot (S^\varphi, \sigma(s_1, \ldots, s_k))$$
$$= (S^\varphi, t_k)^{-1} \odot \cdots \odot (S^\varphi, t_1)^{-1} \odot (S^\varphi, \sigma(t_1, \ldots, t_k)).$$

(MN2) The relation $\equiv$ respects $supp(S)$, i.e., for every $s, t \in T_\Sigma$ with $s \equiv_S t$, $s \in supp(S)$ implies $t \in supp(S)$.

We note that if the underlying semifield is the Boolean semifield and thereby considering unweighted tree automata, then (MN1) and (MN2) are equivalent to the classical property "$\equiv$ respects $supp(S)$". Let us now prove that $\equiv_S$ satisfies (MN1) and (MN2). This requires the following two lemmata.

**Lemma 9.** *Let $s, t \in T_\Sigma$. If $s \equiv_S t$, then (i) $(S, s) = a_\varphi(s) \odot a_\varphi(t)^{-1} \odot (S, t)$ and (ii) $(S^\varphi, s) = a_\varphi(s) \odot a_\varphi(t)^{-1} \odot (S^\varphi, t)$.*

**Lemma 10.** *Let $s, t \in T_\Sigma$ such that $s = \sigma(s_1, \ldots, s_k)$ and $t = \sigma(t_1, \ldots, t_k) \notin L_S$ for some $k \geq 0$, $\sigma \in \Sigma^{(k)}$, and $s_1, \ldots, s_k, t_1, \ldots, t_k \in T_\Sigma$. If $s_i \equiv_S t_i$ for every $1 \leq i \leq k$, then $a_\varphi(s) \odot a_\varphi(t)^{-1} = a_\varphi(s_1) \odot a_\varphi(t_1)^{-1} \odot \cdots \odot a_\varphi(s_k) \odot a_\varphi(t_k)^{-1}$.*

**Lemma 11.** *$\equiv_S$ satisfies (MN1) and (MN2).*

Now assume that there is a congruence relation of the term algebra $(T_\Sigma, \Sigma)$ of finite index satisfying (MN1) and (MN2). In the following we present a deterministic bu-w-fta, which accepts $S$ (cf. Definition 4 and Lemma 13).

**Definition 4.** *Let $\equiv \subseteq T_\Sigma \times T_\Sigma$ be a congruence relation of the term algebra $(T_\Sigma, \Sigma)$ of finite index, which satisfies (MN1) and (MN2). The bu-w-fta associated with $\equiv$ is the bu-w-fta $M = (Q, \Sigma, Q_d, A, \mu)$, where $Q = (T_\Sigma/\equiv) \setminus \{L_S\}$, $Q_d = \{[s]_\equiv \in Q \mid \forall t \in [s]_\equiv\ t \in supp(S)\}$, and for every $k \geq 0$, $\sigma \in \Sigma^{(k)}$, and $[s_1]_\equiv, \ldots, [s_k]_\equiv, [s]_\equiv \in Q$,*

$$\mu_k(\sigma)_{([s_1]_\equiv, \ldots, [s_k]_\equiv), [s]_\equiv} = \begin{cases} (S^\varphi, s_k)^{-1} \odot \cdots \odot (S^\varphi, s_1)^{-1} \odot (S^\varphi, s) \\ \quad \textit{if } [s]_\equiv = [\sigma(s_1, \ldots, s_k)]_\equiv, \\ 0 \quad \textit{otherwise.} \end{cases}$$

We note that the tree representation $\mu$ of the bu-w-fta associated with $\equiv$ is independent from the choice of the representatives of the equivalence classes, because $\equiv$ satisfies (MN1). Hence $M$ is well defined.

**Lemma 12.** *Let* $\equiv \, \subseteq T_\Sigma \times T_\Sigma$ *as in Definition 4 and* $\mu$ *be the tree representation of the bu-w-fta associated with* $\equiv$. *Then, for every* $t \in T_\Sigma$,

$$(i) \ \hat{\mu}(t) = \begin{cases} \emptyset & \text{if } t \in L_S, \\ [t]_\equiv & \text{otherwise,} \end{cases}$$

$$(ii) \ \text{if } t \notin L_S, \text{ then } h_\mu(t)_{[t]_\equiv} = (S^\varphi, t).$$

**Lemma 13.** *Let* $\equiv \, \subseteq T_\Sigma \times T_\Sigma$ *be as in Definition 4. The bu-w-fta associated with* $\equiv$ *is a deterministic bu-w-fta recognizing* $S$.

Let us now sum up the results of this section. We thereby obtain a Myhill-Nerode theorem for recognizable tree series.

**Theorem 2.** *Let* $A$ *be a commutative semifield and* $S \in A\langle\langle T_\Sigma \rangle\rangle$. *The following statements are equivalent:*

(i)   $S \in A^{d,bu}\langle\langle T_\Sigma \rangle\rangle$.
(ii)  *There exists a congruence relation of the term algebra* $(T_\Sigma, \Sigma)$ *of finite index satisfying Properties (MN1) and (MN2).*
(iii) *The relation* $\equiv_S$ *is of finite index.*

*If* $A$ *is a locally finite, commutative semifield, then the following statement and (i), (ii), and (iii) are equivalent:*

(i)'  $S \in A^{n,bu}\langle\langle T_\Sigma \rangle\rangle$.

We note that the equivalence of Statements (i) and (i)' of Theorem 2 was shown in [3] (Theorem 6.2). Similar to classical tree automata we obtain a minimization for (a) deterministic bu-w-fta over a commutative semifield or (b) nondeterministic bu-w-fta over a locally finite, commutative semifield by applying the Myhill-Nerode construction with the Myhill-Nerode congruence. Recall that minimal means with respect to the number of states among all equivalent, deterministic devices.

**Theorem 3.** *If (a)* $A$ *is a commutative semifield and* $S \in A^{d,bu}\langle\langle T_\Sigma \rangle\rangle$ *or if (b)* $A$ *is a locally finite, commutative semifield and* $S \in A^{n,bu}\langle\langle T_\Sigma \rangle\rangle$, *then* $M_{\equiv_S}$ *recognizes* $S$. *Moreover,* $M_{\equiv_S}$ *is minimal with respect to the number of states among all equivalent, deterministic bu-w-fta.*

*Example 3.* In Example 1 we have presented a nondeterministic bu-w-fta $M$ with semantics $(S, t) = size(t)$ for every $t \in T_\Sigma$, where $S$ is a tree series over the commutative semifield *Real*. By applying Theorem 2 we show that there does not exist a deterministic bu-w-fta over *Real* accepting $S_M$. As mentioned in Example 2, $s \equiv_S t$ iff $size(s) = size(t)$ for every $s, t \in T_\Sigma$. Hence $\equiv_S$ is of infinite index and thus $S$ is not accepted by a deterministic bu-w-fta over *Real*.

On the other hand the size of a tree can be generated by a deterministic bu-w-fta over the tropical semifield *Trop*. Let $S' \in Trop\langle\langle T_\Sigma \rangle\rangle$ such that $(S', t) = size(t)$ for every $t \in T_\Sigma$. By applying the equality $size(C[s]) =$

$size(s) + size(C) - 1$ for every $s \in T_\Sigma$ and $C \in C_{\Sigma,1}(X_1)$ one can easily show that $s \equiv_{S'} t$ for every $s, t \in T_\Sigma$. Hence $index(\equiv_{S'}) = 1$ and thus, by Theorem 2, $S' \in Trop^{d,bu} \langle\!\langle T_\Sigma \rangle\!\rangle$. Let us now construct a minimal bu-w-fta over $Trop$, which recognizes $S'$. Theorem 3 states that the bu-w-fta $M_{\equiv_{S'}} = (Q', \Sigma, Q'_d, A, \mu)$ is minimal with respect to the number of states, where $Q' = Q'_d = \{[\alpha]_{\equiv_{S'}}\}$, $\mu'_0(\alpha) = (1)$, and $\mu'_2(\sigma) = (1)$.                                                 ◇

# References

[1] J. Berstel and C. Reutenauer. Recognizable formal power series on trees. *Theoretical Computer Science*, 18:115–148, 1982.

[2] J. Berstel and Ch. Reutenauer. *Rational Series and Their Languages*, volume 12 of *EATCS-Monographs*. Springer Verlag, 1988.

[3] B. Borchardt and H. Vogler. Determinization of Finite State Weighted Tree Automata. *J. Automata, Languages and Combinatorics, accepted*, 2003.

[4] S. Bozapalidis. Effective constuction of the syntactic algebra of a recognizable series on trees. *Acta Informatica*, 28:351–363, 1991.

[5] J. Doner. Tree acceptors and some of their applications. *J. Comput. System Sci.*, 4:406–451, 1970.

[6] M. Droste and P. Gastin. The Kleene-Schützenberger theorem for formal power series in partially commuting variables. *Information and Computation*, 153:47–80, 1999. extended abstract in: *24th ICALP*, LNCS vol. 1256, Springer, 1997, pp. 682–692.

[7] M. Droste and D. Kuske. Skew and infinitary formal power series. Technical Report 2002-38, Department of Mathematics and Computer Science, University of Leicester, 2002.

[8] S. Eilenberg. *Automata, Languages, and Machines, Vol.A*. Academic Press, 1974.

[9] Z. Esik and W. Kuich. Formal Tree Series. *J. Automata, Languages and Combinatorics, accepted*, 2003.

[10] Z. Fülöp and S. Vágvölgyi. Congruential tree languages are the same as recognizable tree languages. *Bulletin of EATCS*, 30:175–185, 1989.

[11] Z. Fülöp and H. Vogler. Tree series transformations that respect copying. *Theory of Computing Systems*, to appear, 2003.

[12] F. Gécseg and M. Steinby. *Tree Automata*. Akadémiai Kiadó, Budapest, 1984.

[13] F. Gécseg and M. Steinby. Tree languages. In G. Rozenberg and A. Salomaa, editors, *Handbook of Formal Languages*, volume 3, chapter 1, pages 1–68. Springer-Verlag, 1997.

[14] D. Kozen. On the Myhill-Nerode theorem for trees. *EATCS Bulletin*, 47:170–173, June 1992.

[15] W. Kuich. Formal power series over trees. In S. Bozapalidis, editor, *Proc. of the 3rd International Conference Developments in Language Theory*, pages 61–101. Aristotle University of Thessaloniki, 1998.

[16] W. Kuich and A. Salomaa. *Semirings, Automata, Languages*. EATCS Monographs on Theoretical Computer Science, Springer Verlag, 1986.

[17] M. Magidor and G. Moran. Finite automata over finite trees. Technical Report 30, Hebrew University, Jerusalem, 1969.

[18] M. Mohri. Finite-state transducers in language and speech processing. *Computational Linguistics*, 23:269–311 (1–42), 1997.

[19] M. O. Rabin. Decidability of second-order theories and automata in infinite trees. *Transactions of the Amer. Math. Soc.*, 141:1–35, 1969.

[20] M.P. Schützenberger. On the definition of a family of automata. *Information and Control*, 4:245–270, 1961.

[21] H. Seidl. Finite tree automata with cost functions. *Theoret. Comput. Sci.*, 126:113–142, 1994.

[22] J.W. Thatcher and J.B. Wright. Generalized finite automata theory with application to a decision problem of second-order logic. *Math. Systems Theory*, 2(1):57–81, 1968.

# Generating Series of the Trace Group

Anne Bouillard and Jean Mairesse

LIAFA, CNRS - Université Paris 7
Case 7014 - 2, place Jussieu - 75251 Paris Cedex 5 - France
{bouillard, mairesse}@liafa.jussieu.fr

**Abstract.** We prove an analog for trace groups of the Möbius inversion formula for trace monoids (Cartier-Foata, 1969). A by-product is to obtain an explicit and combinatorial formula for the growth series of a trace group. This is used to study the average height of traces.

## 1 Introduction

A trace group (monoid) is the quotient of a free group (monoid) by relations of commutation between some pairs of generators. Trace monoids are often used in computer science to model the occurrence of events in concurrent systems, see [7] and the references therein. Trace groups have been studied from several viewpoints (and under various names), see for instance [8,10,17]. An important motivation is that trace goups can 'approximate' braid groups [17].

A *decomposition* of an element $m$ of a monoid $M$ is a $n$-uple $(m_1, \ldots, m_n)$ satisfying $m_1 \cdots m_n = m$ with $m_i \in M \backslash \{1\}$ (where 1 is the unit of the monoid). The decomposition is *even* (*odd*) if $n$ is even (odd). The monoid $M$ has the *finite decomposition property* if any element has a finite number of decompositions. Assume $M$ has the property and set $(\mu_M | m)$ to be the number of even decompositions of $m$ minus the number of odd decompositions. View $\mu_M$ as a formal series of $\mathbb{Z}\langle\!\langle M \rangle\!\rangle$ and call it the *Möbius series of $M$*. A one line computation shows that $\mu_M$ is the formal inverse of the characteristic series:

$$\text{in } \mathbb{Z}\langle\!\langle M \rangle\!\rangle, \quad \Big( \sum_{m \in M} m \Big) \cdot \mu_M = \mu_M \cdot \Big( \sum_{m \in M} m \Big) = 1 \,. \tag{1}$$

This identity is called a *Möbius inversion formula*, see Cartier and Foata [3], Lallement [15], or Rota [16] in a different setting. The classical Möbius inversion principle in number theory (see [13], Chapter XVI) is a special instance of the identity. However, it is generally difficult, given a monoid with the finite decomposition property, to effectively compute the Möbius series.

It is easily checked that a trace monoid has the finite decomposition property. An element $m$ of the trace monoid is a *clique* if any letter appears at most once in $m$ and if all the letters of $m$ are commuting. Let $\mathfrak{C}$ denote the set of cliques. Let $|m|$ be the length of $m$, see (3). In [3, Theorem 2.4], it is proved that the Möbius series of a trace monoid $M$ is the polynomial:

$$\mu_M = \sum_{c \in \mathfrak{C}} (-1)^{|c|} c \,.$$

Z. Ésik and Z. Fülöp (Eds.): DLT 2003, LNCS 2710, pp. 159–170, 2003.

This is the starting point of the combinatorial study of the trace monoid [5,7].

Like any non-trivial group, a trace group does *not* have the finite decomposition property. Hence, one cannot follow the above path to get the formal inverse of the characteristic series. Nevertheless, we obtain in this paper an identity for the trace group which has the flavor of a Möbius inversion formula. Let $F$ be a trace group over the alphabet $\Sigma \cup \overline{\Sigma}$ ($\overline{a} \in \overline{\Sigma}$ is the inverse of $a \in \Sigma$). Let $M$ be the trace monoid over $\Sigma \cup \overline{\Sigma}$ whose monoid presentation is obtained from the one of $F$ by removing the relations $a\overline{a} = \overline{a}a = 1, a \in \Sigma$. Let $\phi$ be the canonical injection from $F$ to $M$. (More precisely, $\phi(t), t \in F$, is the projection on $M$ of any minimal length representative of $t$ over $(\Sigma \cup \overline{\Sigma})^*$.) We have (Theorem 4.1):

$$\text{in } \mathbb{Z}\langle\!\langle M \rangle\!\rangle, \quad \Big( \sum_{t \in \phi(F)} t \Big) \cdot \Big( \sum_{d \in \mathfrak{D}} (-1)^{|d|} d \Big) = \Big( \sum_{d \in \mathfrak{D}} (-1)^{|d|} d \Big) \cdot \Big( \sum_{t \in \phi(F)} t \Big) = 1 , \quad (2)$$

where $\mathfrak{D} = T(\mathfrak{C})$ and $T : M \to \mathcal{P}(M)$ is the map defined by $T(a) = a(\overline{a}a)^*(1 + \overline{a}), a \in \Sigma \cup \overline{\Sigma}$, and extended by morphism. As opposed to $\mathfrak{C}$, the set $\mathfrak{D}$ is infinite. However, $\mathfrak{D}$ is a rational subset of $M$. The identity (2) can be lifted to the free monoid under some conditions, extending a result of Diekert [5].

A nice consequence of the identity (2) is to get precise information about the growth of the trace group. The growth is probably the most popular way of quantifying infinite monoids or groups. Consider a monoid $M$ with a fixed and finite set of generators $S$ (when $M$ is a group, we assume that the generating set contains the inverse of each of its elements). The *length (with respect to $S$)* of an element $m$ in $M$ is by definition

$$|m| = \min\{k \mid m = m_1 \cdots m_k, m_i \in S\} . \quad (3)$$

The *growth series* of $M$ is the series $\mathrm{Gr}M \in \mathbb{Z}[\![x]\!]$ defined by

$$\mathrm{Gr}M = \sum_{m \in M} x^{|m|} = \sum_{n \in \mathbb{N}} \#\{m \in M \mid |m| = n\} x^n . \quad (4)$$

Set $a_n = \#\{m \in M \mid |m| = n\}$. The *growth rate* of $M$ is $\rho_M = \lim_n a_n^{1/n} \geqslant 1$ (the limit exists by sub-additivity of the sequence $(\log a_n)_n$). For $M$ and $F$ defined as in the paragraph preceding (2), we obtain as a by-product of (2) that

$$\mathrm{Gr}F(x) = \mathrm{Gr}M(x(1+x)^{-1}) . \quad (5)$$

This allows to transfer to the trace group several results initially proved for the trace monoid. In particular, $\mathrm{Gr}F$ is rational with a computable growth rate and we have $a_n \sim cn^k \rho_F^n$, where $c$ is a constant and where $k$ depends on the connected subgraphs of the dependence graph of $F$ (Proposition 4.1).

Finally, we use the identity (2) to study the height series in §5.

Let us comment on the article [17]. For proving only the rationality of the growth series, an easier way consists of considering lexicographic normal forms, see §3. In [17], Vershik and al. study particular trace groups: the generators $a_1, \ldots, a_k$ satisfy $a_i a_j = a_j a_i$ if $|i - j| > 1$. The authors compute Gr using lexicographic normal forms, and they derive a precise asymptotic analysis of the series. In particular, they prove that the limit value of the growth rate is 7 (resp. 4) when the number $k$ of generators of the trace group (resp. trace monoid) tends to infinity.

## 2   Preliminaries

Let $\Sigma$ be a non-empty finite set and denote by $\Sigma^*$ the free monoid over $\Sigma$. The empty word is denoted by 1. Consider a relation $R \subset \Sigma^* \times \Sigma^*$, and let $\sim_R$ be the least congruence on $\Sigma^*$ such that $u \sim_R v$ if $(u, v) \in R$. The quotient monoid $(\Sigma^*/\sim_R)$ is called the *monoid presented by $\Sigma$ and $R$* and is denoted $\langle \Sigma \mid u = v, (u, v) \in R \rangle$.

Let $\overline{\Sigma}$ be a copy of $\Sigma$, that is a set which is in bijection with $\Sigma$ and disjoint from $\Sigma$; to each letter $a \in \Sigma$ corresponds the letter $\bar{a}$ in $\overline{\Sigma}$. Set $\widetilde{\Sigma} = \Sigma \cup \overline{\Sigma}$. We extend the bar notation to $\widetilde{\Sigma}$, by setting $\bar{\bar{a}} = a$. Denote by $\mathbb{F}(\Sigma)$ the free group over $\Sigma$, with $\overline{\Sigma}$ as the set of inverses of the generators. A monoid presentation of the free group is

$$\mathbb{F}(\Sigma) = \langle \widetilde{\Sigma} \mid a\bar{a} = 1, \bar{a}a = 1, \forall a \in \Sigma \rangle. \tag{6}$$

Let $I \subset \Sigma \times \Sigma$ be an anti-reflexive and symmetric relation, called an *independence* (or *commutation*) relation. The *trace monoid* (or *free partially commutative monoid*) $\mathbb{M}(\Sigma, I)$ is defined by the monoid presentation

$$\mathbb{M}(\Sigma, I) = \langle \Sigma \mid ab = ba, \forall (a, b) \in I \rangle. \tag{7}$$

The elements of $\mathbb{M}(\Sigma, I)$ are called *traces*. Two words are representatives of the same trace if they can be obtained one from the other by repeatedly commuting independent adjacent letters. The graph $(\Sigma, I)$ is the *independence graph* of $\mathbb{M}(\Sigma, I)$. The *dependence graph* of $\mathbb{M}(\Sigma, I)$ is $(\Sigma, D)$ where $D = (\Sigma \times \Sigma) \setminus I$.

Let $\mathbb{F}(\Sigma, I)$ be defined by the monoid presentation

$$\mathbb{F}(\Sigma, I) = \langle \widetilde{\Sigma} \mid a\bar{a} = 1, \bar{a}a = 1, \forall a \in \Sigma; \quad ab = ba, \forall (a, b) \in I \rangle. \tag{8}$$

It is easily checked that $\mathbb{F}(\Sigma, I)$ is a group, called the *trace group* (or *free partially commutative group*). Notice that a *group presentation* of the trace group is $\mathbb{F}(\Sigma, I) = \langle \Sigma \mid ab = ba, \forall (a, b) \in I \rangle$. The elements of $\mathbb{F}(\Sigma, I)$ are also called traces.

Let $\iota : \widetilde{\Sigma} \longrightarrow \Sigma$ be defined by: $\forall a \in \Sigma, \iota(a) = \iota(\bar{a}) = a$. We associate with the graph $(\Sigma, I)$, the graph $(\widetilde{\Sigma}, \widetilde{I})$ where the relation $\widetilde{I} \subset \widetilde{\Sigma} \times \widetilde{\Sigma}$ is defined by $(u, v) \in \widetilde{I} \iff (\iota(u), \iota(v)) \in I$. The graph $(\widetilde{\Sigma}, \widetilde{I})$ is the *independence graph* of $\mathbb{F}(\Sigma, I)$. The *dependence graph* of $\mathbb{F}(\Sigma, I)$ is $(\widetilde{\Sigma}, \widetilde{D})$ where $\widetilde{D} = (\widetilde{\Sigma} \times \widetilde{\Sigma}) \setminus \widetilde{I}$.

The diagram below, where the applications are the canonical surjective morphisms, is commutative.

$$
\begin{array}{ccc}
\widetilde{\Sigma}^* & \xrightarrow{\;\pi_M\;} & \mathbb{M}(\widetilde{\Sigma}, \widetilde{I}) \\
\downarrow & \searrow{\scriptstyle \pi_F} & \downarrow{\scriptstyle \varphi} \\
\mathbb{F}(\Sigma) & \longrightarrow & \mathbb{F}(\Sigma, I)
\end{array}
$$

The notations $\pi_{\mathbb{F}}$ and $\pi_{\mathbb{M}}$ are shortened to $\pi$ when there is no possible confusion. In the sequel, we often simplify the notations by denoting a trace by any of its representatives, that is by identifying $w$ and $\pi(w)$. For instance, 1 denotes the empty trace.

When $I = \varnothing$, the corresponding trace monoid is the free monoid $\Sigma^*$ and the corresponding trace group is the free group $\mathbb{F}(\Sigma)$. When $I = \Sigma \times \Sigma \backslash \{(a,a), a \in \Sigma\}$, the corresponding trace monoid is the free commutative monoid $\mathbb{N}^{\Sigma}$ and the corresponding trace group is the free commutative group $\mathbb{Z}^{\Sigma}$.

Given a word $w \in \Sigma^*$ and a letter $a \in \Sigma$, denote by $|w|_a \in \mathbb{N}$ the number of occurrences of the letter in the word.

We now give several definitions which are stated in order to hold both in a trace monoid and in a trace group. For the trace monoid only, the same definitions take a simpler form, as the reader can convince himself.

The *length* of a trace $t$, denoted $|t|$, is the minimum of the lengths of its representatives: $|t| = \min\{|x|, x \in \pi^{-1}(t)\}$. This definition of the length is consistent with the one given in (3). Given a trace $t$ and a letter $a$, define $|t|_a = \min_{w \in \pi^{-1}(t)} |w|_a$, and define the *alphabet* of $t$ as $\text{alph}(t) = \{a \mid |t|_a > 0\}$.

A trace $t$ is a *clique* if

$$\forall a \in \text{alph}(t), |t|_a = 1 \quad \text{and} \quad \forall a, b \in \text{alph}(t), ab = ba .$$

We denote the set of cliques by $\mathfrak{C}$. Cliques are in one-to-one correspondence with the complete subgraphs (also called cliques in graph theory) of the independence graph.

An ordered pair $(u, v) \in \mathfrak{C} \backslash \{1\} \times \mathfrak{C} \backslash \{1\}$ is *Cartier-Foata (CF-) admissible* if

$$\forall b \in \text{alph}(v), \exists a \in \text{alph}(u), \ a = b \text{ or } ab \neq ba .$$

Observe that the above condition implies that $|uv| = |u| + |v|$.

Given a trace $t$, there is a uniquely defined sequence of non-empty cliques $(c_1, c_2, \ldots, c_m)$ such that $t = c_1 c_2 \cdots c_m$, and the ordered pair $(c_j, c_{j+1})$ is CF-admissible for all $j$ in $\{1, \ldots, m-1\}$. This sequence is called the *Cartier-Foata (CF) decomposition* of the trace. For a proof in the trace monoid, see [3, Chap. I], and for an illustration, see Example 2.1.

We visualize traces using Viennot's representation as heaps of pieces [18, 17]. In the trace monoid, a trace corresponds to a *heap*. In the trace group, a trace corresponds to a *colored heap* (letters of $\Sigma$ are associated with light gray pieces, letters of $\overline{\Sigma}$ with dark gray pieces, and, in the colored heap, consecutive pieces of type $a$ and $\bar{a}$ cancel each other). Observe that the existence of the heap representation implies that the word problem is solvable in polynomial time in a trace group or monoid.

*Example 2.1.* The *Basic Example* to be used as illustration throughout the paper consists of $\Sigma = \{a, b, c\}$ and $I = \{(a,b), (b,a)\}$. We have represented in Figure 1, the traces in $\mathbb{M}(\widetilde{\Sigma}, \widetilde{I})$ and $\mathbb{F}(\Sigma, I)$ corresponding to the word $u = aab\bar{c}c\bar{a}b$. We have $|\pi_{\mathbb{M}}(u)| = 7$ and $|\pi_{\mathbb{F}}(u)| = 3$. The CF decomposition of $\pi_{\mathbb{M}}(u)$ is $(a\bar{b}, a, \bar{c}, c, \bar{a}b)$ and the one of $\pi_{\mathbb{F}}(u)$ is $(a\bar{b}, b)$.

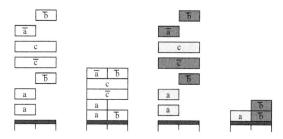

**Fig. 1.** A heap in $\mathbb{M}(\widetilde{\Sigma}, \widetilde{I})$ (left) and a colored heap in $\mathbb{F}(\Sigma, I)$ (right).

Let $M$ be either a trace monoid or a trace group. A function $T$ from $M$ to $\mathbb{Z}$ is called a *(formal power) series (over $M$)*. It is convenient to use the notation $T = \sum_{m \in M} (T|m)m$. The set of series is denoted by $\mathbb{Z}\langle\langle M \rangle\rangle$. When $M = \mathbb{N}^{\Sigma}$, the set of series is denoted by $\mathbb{Z}[[\Sigma]]$. As recalled in §1, the trace monoid $\mathbb{M}(\Sigma, I)$ has the finite decomposition property. It enables us to define the *Cauchy product* $TU$ of two series $T$ and $U$ of $\mathbb{Z}\langle\langle \mathbb{M}(\Sigma, I) \rangle\rangle$ by $(TU|m) = \sum_{tu=m}(T|t)(U|u)$. The set $\mathbb{Z}\langle\langle \mathbb{M}(\Sigma, I) \rangle\rangle$ equipped with the usual sum and the Cauchy product forms a ring. We denote the unit of this ring by 1. This ring is cancellative, in particular a right or left inverse is an inverse. A series $T \in \mathbb{Z}\langle\langle \mathbb{M}(\Sigma, I) \rangle\rangle$ is inversible if and only if $(T|1) = 1$ or $-1$. If $(T|1) = 1$, the inverse of $T$ is $(1-T)^* = \sum_{n \in \mathbb{N}} (1-T)^n$.

## 3   Rationality of the Growth Series

The results in this section are classical, at least for the trace monoid.

Let $M$ be a monoid with a finite set of generators $S$ and let $\pi$ be the canonical surjection from $S^*$ onto $M$. A set of *normal forms* of $M$ in $S^*$ is a subset $L$ of $S^*$ such that for each element $m$ of $M$, we have $\#(\pi^{-1}(m) \cap L) = 1$. Assume that the set of normal forms $L$ is a regular language of $S^*$ and consists of minimal length representatives, i.e. $\forall w \in L, |w| = |\pi(w)|$. Then the growth series of $M$ is rational. Indeed, project an unambiguous regular expression of $L$ into $\mathbb{Z}[x]$ by sending all the letters to $x$. The resulting rational series is $\mathrm{Gr} M$.

We now prove the rationality of the growth series for both the trace monoid and the trace group. It is convenient to denote by $S$ the set of generators, that is $S = \Sigma$ for the trace monoid and $S = \widetilde{\Sigma}$ for the trace group.

Consider a total order $\leqslant$ on $\Sigma$, and the induced total order $\leqslant$ on $\widetilde{\Sigma}$ given by $a \leqslant \bar{a} \leqslant b \leqslant \bar{b}$ if $a \leqslant b, a \neq b$. Denote by $\leqslant_{\mathrm{lex}}$ the corresponding lexicographic order of $S^*$. For a trace $t$, define $\mathrm{Lex}(t) \in S^*$ as follows

$$\mathrm{Lex}(t) \in \pi^{-1}(t), \ |\mathrm{Lex}(t)| = |t|, \ \forall u \in \pi^{-1}(t), [|u| = |t|] \implies [\mathrm{Lex}(t) \leqslant_{\mathrm{lex}} u] .$$

In words, $\mathrm{Lex}(t)$ is the representative of $t$ which is minimal for $\leqslant_{\mathrm{lex}}$ among all the representatives of minimal length. The set $\mathrm{Lex}(\mathbb{M}(\Sigma, I))$, resp. $\mathrm{Lex}(\mathbb{F}(\Sigma, I))$, is a set of normal forms of $M(\Sigma, I)$, resp $\mathbb{F}(\Sigma, I)$, called the set of *lexicographic normal forms*. This set is a regular language. Define $I_< = \{(a,b) \mid (a,b) \in I, a < b\}$. For $a \in \Sigma$, set $I(a) = \{c \mid (a,c) \in I\}$. Then we have

$$\mathrm{Lex}(M(\Sigma, I)) = \Sigma^* - \bigcup_{(a,b)\in I_<} \Sigma^* b I(a)^* a \Sigma^* \tag{9}$$

$$\mathrm{Lex}(\mathbb{F}(\Sigma, I)) = \widetilde{\Sigma}^* - [\ \bigcup_{(a,b)\in \widetilde{I}_<} \widetilde{\Sigma}^* b \widetilde{I}(a)^* a \widetilde{\Sigma}^* \cup \bigcup_{a\in\widetilde{\Sigma}} \widetilde{\Sigma}^* a \bar{a} \widetilde{\Sigma}^* \ ] \,.$$

The justification is simple; for the trace monoid it can be found in [5, Chapter 1.2]. We conclude that $\mathrm{GrM}(\Sigma, I)$ and $\mathrm{GrF}(\Sigma, I)$ are rational.

Given a trace $t$ with CF decomposition $(c_1, c_2, \ldots, c_k)$, define $\mathrm{CF}(t) \in S^*$ by $\mathrm{CF}(t) = \mathrm{Lex}(c_1)\mathrm{Lex}(c_2)\cdots\mathrm{Lex}(c_k)$. The set $\mathrm{CF}(M(\Sigma, I))$, resp. $\mathrm{CF}(\mathbb{F}(\Sigma, I))$, is a set of normal forms called the set of *Cartier-Foata normal forms*. This set is a regular language of $S^*$. Indeed, consider the deterministic automaton $\mathcal{CF}$ with states $\mathfrak{C}$, initial state 1, final states $\mathfrak{C}$, and with transitions

$$1 \xrightarrow{u} u \text{ if } u \in \mathfrak{C}\backslash\{1\}, \text{ and } u \xrightarrow{v} v \text{ if } (u,v) \text{ is CF-admissible} \,. \tag{10}$$

The automaton $\mathcal{CF}$ recognizes a language $L$ over $\mathfrak{C}^*$. Let $\mu$ be the morphism from $\mathfrak{C}^*$ to $S^*$ defined by $\mu(c) = \mathrm{Lex}(c)$ for $c \in \mathfrak{C}$. The set $\mu(L)$ is clearly the set of Cartier-Foata normal forms. This provides an alternative proof of the rationality of the growth series.

*Remark.* Using either the Cartier-Foata or the lexicographic normal forms, it is not difficult to prove that trace groups (monoids) are automatic groups (monoids) in the sense of [11,2].

Now let us restate some results from [3] already given in §1. Let $M(\Sigma, I)$ be a trace monoid. In $\mathbb{Z}\langle\!\langle M(\Sigma, I)\rangle\!\rangle$, the following identity holds

$$(\sum_{t\in M(\Sigma,I)} t) \cdot (\sum_{c\in\mathfrak{C}}(-1)^{|c|}c) = 1 \,. \tag{11}$$

To be consistent with §1, we call this identity a Möbius inversion formula for the trace monoid. Let $\mu_{M(\Sigma,I)}(x) \in \mathbb{Z}[\![x]\!]$ be defined by

$$\mu_{M(\Sigma,I)}(x) = \sum_{c\in\mathfrak{C}}(-1)^{|c|}x^{|c|} \,. \tag{12}$$

An easy consequence of (11) is the following identity in $\mathbb{Z}[\![x]\!]$:

$$\mathrm{GrM}(\Sigma, I) \cdot \mu_{M(\Sigma,I)}(x) = 1 \,. \tag{13}$$

In words, $\mu_{M(\Sigma,I)}(x)$ is the formal inverse of the growth series. Hence, we get a third proof of the rationality of $\mathrm{GrM}(\Sigma, I)$, but also a much stronger result due to the explicit and combinatorial expression for $\mu_{M(\Sigma,I)}(x)$.

The objective is to get analogs of (11) and (13) for trace groups.

## 4   Möbius Formula for the Trace Group

Given $u, v \in \widetilde{\Sigma}^*$ such that $\pi_{\mathbb{F}}(u) = \pi_{\mathbb{F}}(v)$ and $|u| = |v| = |\pi_{\mathbb{F}}(u)|$, then we have $\pi_{\mathbb{M}}(u) = \pi_{\mathbb{M}}(v)$. In other words, a trace of $\mathbb{F}(\Sigma, I)$ admits a unique representative of minimal length in $\mathbb{M}(\widetilde{\Sigma}, \widetilde{I})$. We denote by

$$\phi : \mathbb{F}(\Sigma, I) \longrightarrow \mathbb{M}(\widetilde{\Sigma}, \widetilde{I})$$

the corresponding map. Using the language of heaps, if $t$ is a colored heap then $\phi(t)$ is the corresponding (non-colored) heap obtained by forgetting the colors. For a formal proof of this point, see [9, Prop. 2.4.7].

Define the set of *alternate traces* as the subset $\mathfrak{D}$ of $\mathbb{M}(\widetilde{\Sigma}, \widetilde{I})$ given by the regular expression

$$\mathfrak{D} = \sum_{c \in \mathfrak{C}} [\ \prod_{x \in \text{alph}(c)} x(\bar{x}x)^*(1 + \bar{x})\ ], \tag{14}$$

where $\mathfrak{C}$ is the set of cliques of $\mathbb{M}(\widetilde{\Sigma}, \widetilde{I})$. Figure 2 shows some alternate traces for the Basic Example.

**Fig. 2.** Examples of elements of $\mathfrak{D}$.

**Theorem 4.1.** *In* $\mathbb{Z}\langle\langle \mathbb{M}(\widetilde{\Sigma}, \widetilde{I}) \rangle\rangle$, *we have the following identity:*

$$\Big( \sum_{t \in \phi(\mathbb{F}(\Sigma, I))} t \Big) \cdot \Big( \sum_{d \in \mathfrak{D}} (-1)^{|d|} d \Big) = 1. \tag{15}$$

*Proof.* We are going to use the same type of *bijective* argument as in [18, Remark 5.2]. For a trace $t \in \mathbb{M}(\widetilde{\Sigma}, \widetilde{I})$, define

$$\text{Top}(t) = \{a \in \text{alph}(t) \mid \exists u, t = ua\}. \tag{16}$$

The set $\text{Top}(t)$ contains all the pieces which are fully visible when the heap $t$ is viewed from above. Given $t \in \mathbb{M}(\widetilde{\Sigma}, \widetilde{I})$ and $h \in \mathfrak{D}$, define $M(t, h) \subset \Sigma$ by $M(t, h) = \iota(\text{Top}(h) \cup \{a \in \text{Top}(t) \mid \forall b \in \text{Top}(h), a\widetilde{I}b\})$. Observe that $M(t, h) = \varnothing$ if and only if $(t, h) = (1, 1)$. We now are going to define an application

$$\psi : \phi(\mathbb{F}(\Sigma, I)) \times \mathfrak{D} \longrightarrow \phi(\mathbb{F}(\Sigma, I)) \times \mathfrak{D}.$$

We have $\psi(1, 1) = (1, 1)$. For $(t, h) \in \phi(\mathbb{F}(\Sigma, I)) \times \mathfrak{D}$ with $(t, h) \neq (1, 1)$, we distinguish between four cases. Let $a$ be the minimal letter of $M(t, h)$ (we assume that $\widetilde{\Sigma}$ is equipped with a total order defined as in §3).

1. If we have $t = t_1 a_1$ with $a_1 \in \{a, \bar{a}\}$ and $a, \bar{a} \notin \text{alph}(h)$, then $\psi(t, h) = (t_1, a_1 h)$.
2. If we have $h = a_1 h_1$ with $a_1 \in \{a, \bar{a}\}$ and $a, \bar{a} \notin \text{Top}(t)$, then $\psi(t, h) = (t a_1, h_1)$.
3. If we have $t = t_1 a_1$ with $a_1 \in \{a, \bar{a}\}$ and $h = a_1 h_1$, then $\psi(t, h) = (t a_1, h_1)$.
4. If we have $t = t_1 a_1$ with $a_1 \in \{a, \bar{a}\}$ and $h = \bar{a}_1 h_1$, then $\psi(t, h) = (t_1, a_1 h)$.

The dynamics of $\psi$ for the Basic Example is illustrated in Figure 3. Here the order is $a \leqslant \bar{a} \leqslant b \leqslant \bar{b} \leqslant c \leqslant \bar{c}$.

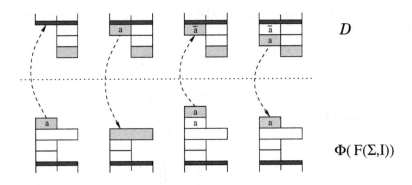

**Fig. 3.** The involution $\psi$. From left to right, the cases are 1,2,4, and 3.

Observe that the configurations of type $aa$ are allowed in $\phi(\mathbb{F}(\Sigma, I))$ but not in $\mathfrak{D}$, and the other way round the configurations of type $a\bar{a}$ are allowed in $\mathfrak{D}$ but not in $\phi(\mathbb{F}(\Sigma, I))$. Notice also that the minimal letters of $M(t, h)$ and $M\psi(t, h)$ are the same. It follows easily from these two points that $\psi^2 = \text{Id}$. It is also clear that the only fixed point of $\psi$ is $(1, 1)$. Furthermore, if $(t, h) \neq (1, 1)$ and $\psi(t, h) = (t', h')$, then we have $th = t'h'$ and $|h'| = |h| \pm 1$. So, we have

$$(-1)^{|h|} th = -(-1)^{|h'|} t'h' .$$

Set $S = \phi(\mathbb{F}(\Sigma, I)) \times \mathfrak{D}$. By decomposing the sum over the orbits of $\psi$, we get

$$\left( \sum_{t \in \phi(\mathbb{F}(\Sigma, I))} t \right) \cdot \left( \sum_{h \in \mathfrak{D}} (-1)^{|h|} h \right) = 1 + \sum_{(t, h) \in S \setminus \{(1, 1)\}} (-1)^{|h|} th = 1 . \qquad \square$$

**Corollary 4.1.** *In $\mathbb{Z}[\![x]\!]$, the following identity holds:*

$$\text{Gr}\mathbb{F}(\Sigma, I) \cdot \mu_{\text{M}(\widetilde{\Sigma}, \widetilde{I})}(x / (1 + x)) = 1 . \qquad (17)$$

Observe that $\mu_{\text{M}(\widetilde{\Sigma}, \widetilde{I})}(x) = \mu_{\text{M}(\Sigma, I)}(2x)$. Since the growth rate is the inverse of the modulus of a dominant singularity, we obtain immediately the following relations

$$\rho_{\mathbb{F}(\Sigma, I)} = \rho_{\text{M}(\widetilde{\Sigma}, \widetilde{I})} - 1 = 2\rho_{\text{M}(\Sigma, I)} - 1 .$$

A consequence is that the results originally proved for $\rho_{M(\Sigma,I)}$ in [12,14] can be directly transferred to $\rho_{\mathbb{F}(\Sigma,I)}$. For instance, $\rho_{\mathbb{F}(\Sigma,I)} = 1$ iff $\mathbb{F}(\Sigma,I) = \mathbb{Z}^\Sigma$. Furthermore, we get the following proposition.

**Proposition 4.1.** *Set* $a_n = \#\{t \in \mathbb{F}(\Sigma,I) \mid |t| = n\}$.

*1. If the dependence graph of* $\mathbb{F}(\Sigma,I)$ *is connected, then* $\mathrm{Gr}\mathbb{F}(\Sigma,I)$ *has a unique dominant singularity which is positive real and of order 1. Consequently* $a_n \sim c\rho^n_{\mathbb{F}(\Sigma,I)}$.

*2. If the dependence graph of* $\mathbb{F}(\Sigma,I)$ *is non-connected, denote by* $(\mathbb{F}_s)_s$ *the trace groups whose dependence graphs are the maximal connected subgraphs. The series* $\mathrm{Gr}\mathbb{F}(\Sigma,I)$ *has a unique dominant singularity equal to* $\rho_{\mathbb{F}(\Sigma,I)} = \min_s \rho_{\mathbb{F}_s}$, *and whose order is* $k = \#\{s, \rho_{\mathbb{F}_s} = \rho_{\mathbb{F}(\Sigma,I)}\}$. *Consequently* $a_n \sim cn^{k-1}\rho^n_{\mathbb{F}(\Sigma,I)}$.

*Example 4.1.* For the free group, we have $\mathrm{Gr}\mathbb{F}(\Sigma) = (1+x)/(1 - (2\#\Sigma - 1)x)$ and the growth rate is $\rho_{\mathbb{F}(\Sigma)} = (2\#\Sigma - 1)$. For the free commutative group, we have $\mathrm{Gr}\mathbb{Z}^\Sigma = (1+x)^{\#\Sigma}/(1-x)^{\#\Sigma}$ and $\rho_{\mathbb{Z}^\Sigma} = 1$.

Consider now the Basic Example. Applying Theorem 4.1, the formal inverse of the characteristic series $(\sum_{t \in \phi(\mathbb{F}(\Sigma,I))} t)$ is

$$1 - \sum_{x \in \widetilde{\Sigma}} x(\overline{x}x)^*(\overline{x} - 1) + \sum_{\iota(x)=a, \iota(y)=b} x(\overline{x}x)^*(\overline{x} - 1)y(\overline{y}y)^*(\overline{y} - 1).$$

Using (13), we have $\mathrm{Gr}\mathrm{M}(\widetilde{\Sigma}, \widetilde{I}) = (1 - 6x + 4x^2)^{-1}$. Applying Corollary 4.1, we get the growth series

$$\mathrm{Gr}\mathbb{F}(\Sigma, I) = (1+x)^2/(1 - 4x - x^2).$$

The Taylor expansion of the series around 0 is $\mathrm{Gr}\mathbb{F}(\Sigma,I) = 1 + 6x + 26x^2 + 110x^3 + \cdots + 635622x^9 + O(x^{10})$. For instance, there are 635622 traces of length 9. The growth rate of the trace group is $(\sqrt{5} + 2) = 4.236\ldots$

In the table below, we give the growth rate for the other trace groups such that $\#\Sigma \leq 4$. The value $\alpha$ in the table is the inverse of the smallest root of $1 - 5x - x^2 - 3x^3$. Numerically, we have $\alpha = 5.29\ldots$

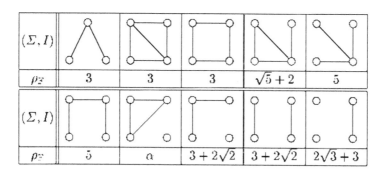

| $(\Sigma,I)$ | | | | | |
|---|---|---|---|---|---|
| $\rho_{\widetilde{z}}$ | 3 | 3 | 3 | $\sqrt{5}+2$ | 5 |
| $(\Sigma,I)$ | | | | | |
| $\rho_{\widetilde{z}}$ | 5 | $\alpha$ | $3+2\sqrt{2}$ | $3+2\sqrt{2}$ | $2\sqrt{3}+3$ |

*Lifting to the free monoid.* Set $M = \mathbb{F}(\Sigma, I)$ or $\mathbb{M}(\Sigma, I)$ and $S = \widetilde{\Sigma}$ or $\Sigma$ respectively. Recall that Lex was defined in §3 and define the application $\overline{\text{Lex}}$ : $M \to S^*$ as follows:

$$\overline{\text{Lex}}(t) \in \pi^{-1}(t), |\overline{\text{Lex}}(t)| = |t|, \forall u \in \pi^{-1}(t), [|u| = |\overline{\text{Lex}}(t)|] \Rightarrow [u \leqslant_{\text{lex}} \overline{\text{Lex}}(t)].$$

In words, $\overline{\text{Lex}}(t)$ is the maximal representative of minimal length of $t$. Recall that $\text{Lex}(M)$ is a regular language whose complement in $S^*$ is a two-sided ideal, see (9). We denote by $B$ the basis of this ideal. An *overlapping word* is a word $w \in S^*$ which can be written as $w = w_1 \cdots w_n$ with $w_1 \in S, w_i \in \text{Lex}(M), w_i w_{i+1} \in S^* B \backslash S^* B S^+$. The set of overlapping words is denoted by $\mathcal{O}(M)$. It is proved in [5, Theorem 4.4.2] that, in $\mathbb{Z}\langle\langle \Sigma^* \rangle\rangle$, we have

$$\left( \sum_{u \in \text{Lex}(M)} u \right) \cdot \left( \sum_{v \in \mathcal{O}(M)} (-1)^{|v|} v \right) = 1 . \tag{18}$$

In [5, Chapter 4.3] (see also [4]), more precise results are given for the trace monoid only: $\mathcal{O}(\mathbb{M}(\Sigma, I))$ is finite if and only if the relation $I_<$ is transitive, and in this case, $\mathcal{O}(\mathbb{M}(\Sigma, I)) = \overline{\text{Lex}}(\mathfrak{C})$. Then, the identity (18) can be viewed as a *lifting* to the free monoid of the Möbius inversion formula (11). Here we propose an analog of this last result for trace groups. Recall that the set $\mathfrak{D}$ of alternate traces was defined in (14).

**Proposition 4.2.** *If the relation $I_<$ is transitive, then $\mathcal{O}(\mathbb{F}(\Sigma, I)) = \overline{\text{Lex}}(\mathfrak{D})$.*

The proof is easy and left as an exercise. In particular, in $\mathbb{Z}\langle\langle \widetilde{\Sigma}^* \rangle\rangle$, we have

$$\left( \sum_{u \in \text{Lex}(\mathbb{F}(\Sigma, I))} u \right) \cdot \left( \sum_{v \in \overline{\text{Lex}}(\mathfrak{D})} (-1)^{|v|} v \right) = 1 . \tag{19}$$

This identity is a lifting to the free monoid of the identity (15). Observe that it is not always possible to find a total order $\leqslant$ such that $I_<$ is transitive.

*The monoid $\mathbb{M}\mathbb{I}(\widetilde{\Sigma}, \widetilde{I})$.* In between $\mathbb{M}(\widetilde{\Sigma}, \widetilde{I})$ and $\mathbb{F}(\Sigma, I)$ lies the *involutive trace monoid* $\mathbb{M}\mathbb{I}(\widetilde{\Sigma}, \widetilde{I}) = \langle \widetilde{\Sigma} \mid \bar{a}a = 1, \forall a \in \Sigma; ab = ba, \forall(a, b) \in \widetilde{I} \rangle$. Consider the trace monoid $\mathbb{M}(\widetilde{\Sigma}, \overline{I})$ where $\overline{I} = \widetilde{I} \cup \{(a, \bar{a}), a \in \widetilde{\Sigma}\}$. Consider the canonical injection $f : \mathbb{M}\mathbb{I}(\widetilde{\Sigma}, \widetilde{I}) \longrightarrow \mathbb{M}(\widetilde{\Sigma}, \widetilde{I})$ and the canonical surjection $g : \mathbb{M}(\widetilde{\Sigma}, \widetilde{I}) \longrightarrow \mathbb{M}(\widetilde{\Sigma}, \overline{I})$. Clearly $g \circ f$ is a bijection from $\mathbb{M}\mathbb{I}(\widetilde{\Sigma}, \widetilde{I})$ into $\mathbb{M}(\widetilde{\Sigma}, \overline{I})$. Hence, the generating series of $\mathbb{M}\mathbb{I}(\widetilde{\Sigma}, \widetilde{I})$ are obtained directly using $\mathbb{M}(\widetilde{\Sigma}, \overline{I})$.

## 5　The Height Generating Series

The set $\mathfrak{C}\backslash\{1\}$ is a set of generators for the trace monoid (group) since it contains $\Sigma$ ($\widetilde{\Sigma}$). To avoid confusions, we denote the length according to $\mathfrak{C}\backslash\{1\}$ (see (3)) by $h(.)$ and we call it the *height*. Consider a trace $t$ with CF decomposition $(c_1, c_2, \ldots, c_m)$, then the height of $t$ is $h(t) = m$. In the visualization of traces using heaps of pieces, the height corresponds precisely to the height of the heap. The motivation for studying the height is explained at the end of the section.

The *height series* is the growth series with respect to $\mathfrak{C} \setminus \{1\}$, i.e. the series of $\mathbb{Z}[\![x]\!]$ defined by $\sum_t x^{h(t)}$. This series is rational since it is recognized by the automaton $\mathcal{CF}$ (see (10)) modified by setting the labels to $x$. Consider a trace group $\mathbb{F}(\Sigma, I)$. Let $\mathbb{F}_n$, respectively $_n\mathbb{F}$, be the set of traces of length, respectively height, $n$. Assuming existence of the limits, define

$$\lambda_{\mathbb{F}(\Sigma,I)} = \lim_n \frac{\sum_{t \in \mathbb{F}_n} h(t)}{n \# \mathbb{F}_n}, \quad \gamma_{\mathbb{F}(\Sigma,I)} = \lim_n \frac{\sum_{t \in {}_n\mathbb{F}} |t|}{\#_n\mathbb{F}} \quad (20)$$

$$\lambda^*_{\mathbb{F}(\Sigma,I)} = \lim_n \frac{\sum_{w \in \widetilde{\Sigma}^n} h \circ \pi(w)}{n \# \widetilde{\Sigma}^n}. \quad (21)$$

Observe that $\lambda_{\mathbb{F}(\Sigma,I)}$ is the asymptotic average height with respect to the uniform distribution over $\mathbb{F}_n$. An analog interpretation holds for $\gamma_{\mathbb{F}(\Sigma,I)}$ and $\lambda^*_{\mathbb{F}(\Sigma,I)}$ with respect to uniform distributions over $_n\mathbb{F}$ and $\widetilde{\Sigma}^n$. Analogs of these quantities for the trace monoid have been studied by several authors, see [14] and the references therein.

**Proposition 5.1.** *The limits in (20) exist and the numbers $\lambda_{\mathbb{F}(\Sigma,I)}$ and $\gamma_{\mathbb{F}(\Sigma,I)}$ are algebraic.*

The proof is omitted here (it is given in [1]) since it is an adaptation of the argument given in [14, Proposition 5.1] for the trace monoid. The two starting ingredients of the proof are the Möbius formula of Corollary 4.1 and the rationality of the series $F = \sum_{t \in \mathbb{F}(\Sigma,I)} x^{|t|} y^{h(t)}$. Observe that $F$ is indeed rational since it is recognized by the automaton $\mathcal{CF}$ (see (10)) modified by setting the label of $u \to v$ to $x^{|v|} y$. The proof is constructive and provides an exact formula for $\lambda_{\mathbb{F}(\Sigma,I)}$ and $\gamma_{\mathbb{F}(\Sigma,I)}$:

$$\lambda_{\mathbb{F}(\Sigma,I)} = \frac{[G(x)(\rho - x)^{k+1}]_{|x=\rho}}{k\rho \cdot [F(x,1)(\rho - x)^k]_{|x=\rho}} \quad \gamma_{\mathbb{F}(\Sigma,I)} = \frac{[\tilde{G}(y)(\tilde{\rho} - y)^{\tilde{k}+1}]_{|y=\tilde{\rho}}}{\tilde{k}\tilde{\rho} \cdot [F(1,y)(\tilde{\rho} - y)^{\tilde{k}}]_{|y=\tilde{\rho}}}, \quad (22)$$

where $G = (\partial F/\partial y)(x,1)$, $\tilde{G} = (\partial F/\partial x)(1,y)$, where $\rho = \rho_{\mathbb{F}(\Sigma,I)}$ (resp. $\tilde{\rho}$) is the unique dominant singularity of $F(x,1) = \mathrm{Gr}\mathbb{F}(\Sigma, I)$ (resp. $F(1,y)$), and where $k$ (resp. $\tilde{k}$) is the order of $\rho$ (resp. $\tilde{\rho}$).

The existence of the limit in (21) follows directly from a subadditive argument. However $\lambda^*_{\mathbb{F}(\Sigma,I)}$ is in general not algebraic and its exact value can not be computed, except for trace groups which are direct products of free groups. In this last case, it is easily proved that $\lambda^*_{\mathbb{F}(\Sigma,I)} = \lambda^*_{\mathrm{M}(\Sigma,I)} - 1/\#\Sigma$ and that $\lambda^*_{\mathrm{M}(\Sigma,I)} = (\max_i \#\Sigma_i)/(\sum_i \#\Sigma_i)$ if $\mathrm{M}(\Sigma, I) = \prod_i \Sigma_i^*$. Otherwise, even for the Basic Example, the exact computation of $\lambda^*_{\mathbb{F}(\Sigma,I)}$ seems out of reach. In trace monoids, the situation is basically the same except that there are more cases for which $\lambda^*_{\mathrm{M}(\Sigma,I)}$ is exactly computable, see [14]. Let us mention as a curiosity that in all cases that we have checked numerically, we have $\lambda^*_{\mathrm{M}(\Sigma,I)} - \lambda^*_{\mathbb{F}(\Sigma,I)} \in [1/\#\Sigma - 2/100, 1/\#\Sigma]$.

*A Discrete Event System application.* It is natural to model the occurrence of events in a concurrent system using a trace monoid $\mathbb{M}(\Sigma, I)$ with $(a, b) \in I$ if the two events $a$ and $b$ can occur simultaneously (see [7]). Then, the two basic performance measures associated with a trace $t$ are its length $|t|$ (the 'sequential' execution time) and its height $h(t)$ (the 'parallel' execution time). The quantity $\lambda^*_{\mathbb{M}(\Sigma, I)}$ has been singled out by various authors as a good measure of the efficiency of the concurrent system in a random environment. To go further, one may want to quantify the robustness of the system with respect to breakdowns or cancellations. In [1], a model is proposed which leads to this unexpected interpretation: the resistance to cancellations can be parametrized by $\rho_{\mathbb{F}(\Sigma, I)}$.

# References

1. A. Bouillard. Rapport de DEA : Le groupe de traces. LIAFA research report 2002-13, Université Paris 7, 2002.
2. C. Campbell, E. Robertson, N. Ruškuc, and R. Thomas. Automatic semigroups. *Theoret. Comput. Sci.*, 250(1-2):365–391, 2001.
3. P. Cartier and D. Foata. *Problèmes combinatoires de commutation et réarrangements.* Number 85 in Lecture Notes in Mathematics. Springer, 1969.
4. C. Choffrut and M. Goldwurm. Determinants and Möbius functions in trace monoids. *Discrete Math.*, 194(1-3):239–247, 1999.
5. V. Diekert. *Combinatorics on traces.* Number 454 in LNCS. Springer Verlag, 1990.
6. V. Diekert and A. Muscholl. Solvability of equations in free partially commutative groups is decidable. In *ICALP'01*, n. 2076 in LNCS, pages 543–554. Springer, 2001.
7. V. Diekert and G. Rozenberg, editors. *The Book of Traces.* World Scientific, 1995.
8. C. Droms, B. Servatius, and H. Servatius. Groups assembled from free and direct products. *Discrete Math.*, 109:69–75, 1992.
9. C. Duboc. *Commutations dans les monoïdes libres : un cadre théorique pour l'étude du parallélisme.* PhD thesis, Univ. Rouen, 1986.
10. G. Duchamp and D. Krob. Partially commutative Magnus transformations. *Internat. J. Algebra Comput.*, 3:15–41, 1993.
11. D. Epstein, J. Cannon, D. Holt, S. Levy, M. Paterson, and W. Thurston. *Word processing in groups.* Jones and Bartlett, Boston, 1992.
12. M. Goldwurm and M. Santini. Clique polynomials have a unique root of smallest modulus. *Information Processing Letters*, 75(3):127–132, 2000.
13. G. Hardy and E. Wright. *An introduction to the theory of numbers.* Oxford at the Clarendon Press, 1979. 5-th edition.
14. D. Krob, J. Mairesse, and I. Michos. Computing the average parallelism in trace monoids. *Discrete Math.*, 2002. Accepted for publication. An abridged version appears in STACS'02, LNCS 2285, p. 477–488, Springer.
15. G. Lallement. *Semigroups and Combinatorial Applications.* Wiley, New-York, 1979.
16. G.-C. Rota. On the foundations of combinatorial theory. I. Theory of Möbius functions. *Z. Wahrscheinlichkeitstheor. Verw. Geb.*, 2:340–368, 1964.
17. A. Vershik, S. Nechaev, and R. Bikbov. Statistical properties of locally free groups with applications to braid groups and growth of random heaps. *Commun. Math. Phys.*, 212(2):469–501, 2000.
18. G.X. Viennot. Heaps of pieces, I: Basic definitions and combinatorial lemmas. In Labelle and Leroux, editors, *Combinatoire Énumérative*, number 1234 in Lect. Notes in Math., pages 321–350. Springer, 1986.

# Residual Finite Tree Automata

Julien Carme, Rémi Gilleron, Aurélien Lemay, Alain Terlutte, and
Marc Tommasi

Grappa – EA 3588 – Lille 3 University
http://www.grappa.univ-lille3.fr

**Abstract.** Tree automata based algorithms are essential in many fields
in computer science such as verification, specification, program analysis. They become also essential for databases with the development of
standards such as XML. In this paper, we define new classes of non deterministic tree automata, namely residual finite tree automata (RFTA).
In the bottom-up case, we obtain a new characterization of regular tree
languages. In the top-down case, we obtain a subclass of regular tree languages which contains the class of languages recognized by deterministic
top-down tree automata. RFTA also come with the property of existence
of canonical non deterministic tree automata.

## 1 Introduction

The study of tree automata has a long history in computer science; see the survey
of Thatcher [Tha73], and the texts of F. Gécseg and M. Steinby [GS84,GS96],
and of the TATA group [CDG+97]. With the advent of tree-based metalanguages
(SGML and XML) for document grammars, new developments on tree automata
formalisms and tree automata based algorithms have been done [MLM01,Nev02].
Also, because of the tree structure of documents, learning algorithms for tree
languages have been defined for the tasks of information extraction and information retrieval [Fer02,GK02,LPH00]. We are currently involved in a research
project dealing with information extraction systems from semi-structured data.
One objective is the definition of classes of tree automata satisfying two properties: there are efficient algorithms for membership and matching, and there are
efficient learning algorithms for the corresponding classes of tree languages.

In the present paper, we only consider finite ranked trees. There are bottom-
up (also known as frontier to root) tree automata and top-down (also known as
root to frontier) tree automata. The top-down version is particularly relevant for
some implementations because important properties such as membership[1] can
be solved without handling the whole input tree into memory. There are also
deterministic tree automata and non-deterministic tree automata. Determinism
is important to reach efficiency for membership and other decision properties.
It is known that non-deterministic top-down, non-deterministic bottom-up, and
deterministic bottom-up tree automata are equally expressive and define regular tree languages. But there is a tradeoff between efficiency and expressiveness because some regular (and even finite) tree languages are not recognized

---

[1] given a tree automaton $A$, decide whether an input tree is accepted by $A$.

Z. Ésik and Z. Fülöp (Eds.): DLT 2003, LNCS 2710, pp. 171–182, 2003.

by deterministic top-down tree automata. Moreover, the size of a deterministic bottom-up tree automaton can be exponentially larger than the size of a non-deterministic one recognizing the same tree language. This drawback can be dramatic when the purpose is to build tree automata. This is for instance the case in the problem of tree pattern matching and in machine learning problems like grammatical inference.

The process of learning finite state machines from data is referred as grammatical inference. The first theoretical foundations were given by Gold [Gol67] and first applications were designed in the field of pattern recognition. Grammatical inference mostly focused on learning string languages but recent works are concerned with learning tree languages [Sak90,Fer02,GK02]. In most works, the target tree language is represented by a deterministic bottom-up tree automaton. This is problematic because the time complexity of the learning algorithm depends on the size of the target automaton. Therefore, again it is crucial to define learning algorithms for non-deterministic tree automata. The reader should note that tree patterns [GK02] satisfy this property.

Therefore the aim of this article is to define non-deterministic tree automata corresponding to sufficiently expressive classes of tree languages and having nice properties from the algorithmic viewpoint and from the grammatical inference viewpoint. For this aim, we extend previous works from the string case [DLT02a] to the tree case and we define residual finite state automata (RFTA). The reader should note that learning algorithms for residual finite string automata have been defined [DLT01,DLT02b].

In Section 3, we study the bottom-up case. We define the residual language of a language $L$ w.r.t a ground term $t$ as the set of contexts $c$ such that $c[t]$ is a term in $L$. We define bottom-up residual tree automata as automata whose states correspond to residual languages. Bottom-up residual tree automata are non-deterministic and recognize regular tree languages. We prove that every regular tree language is recognized by a unique canonical bottom-up residual tree automaton, minimal according to the number of states. We give an example of regular tree languages for which the size of the deterministic bottom-up tree automata grows exponentially with respect to the size of the canonical bottom-up residual tree automata.

In Section 4, we study the top-down case. We define the residual language of a language $L$ w.r.t a context $c$ as the set of ground terms $t$ such that $c[t]$ is a term in $L$. We define top-down residual tree automata as automata whose states correspond to residual languages. Top-down residual tree automata are non-deterministic tree automata. Interestingly, the class of languages recognized by top-down residual tree automata is strictly included in the class of regular tree languages and strictly contains the class of languages recognized by deterministic top-down tree automata. We also prove that every tree language in this family is recognized by a unique canonical top-down residual tree automaton; this automaton is minimal according to the number of states.

The definition of residual finite state automata comes with new decision problems. All of them rely on properties of residual languages. It is proved that all residual languages of a given tree language $L$ can be built in both top-down and bottom-up cases. From these constructions we obtain positive answers to decision problems like 'decide whether an automaton is a (canonical) RFTA'. The exact complexity bounds are not given but we conjecture that are identical than in the string case.

The present work is connected with the paper by Nivat and Podelski [NP97]. They consider a monoid framework, whose elements are called pointed trees (contexts in our terminology, special trees in [Tho84]), to define tree automata. They define a Nerode congruence in the bottom-up case and in the top-down case. Their work leads to the generalization of the notion of deterministic to l-r-deterministic (context-deterministic in our terminology) for top-down tree automata. They have a minimization procedure for this class of automata. It should be noted that the class of languages recognized by context-deterministic tree automata (also called homogeneous tree languages) is strictly included in the class of languages recognized by residual top-down tree automata.

## 2   Preliminaries

We assume that the reader is familiar with basic knowledge about tree automata. We follow the notations defined in TATA [CDG+97].

A ranked alphabet is a couple $(\mathcal{F}, Arity)$ where $\mathcal{F}$ is a finite set and $Arity$ is a mapping from $\mathcal{F}$ into $\mathbb{N}$. The set of symbols of arity $p$ is denoted by $\mathcal{F}_p$. The set of *terms* over $\mathcal{F}$ is denoted by $\mathcal{T}(\mathcal{F})$. Let $\diamond$ be a special constant which is not in $\mathcal{F}$. The set of *contexts* (also known as pointed trees in [NP97] and special trees in [Tho84]), denoted by $\mathcal{C}(\mathcal{F})$, is the set of terms which contains exactly one occurrence of $\diamond$. The expression $c[\diamond]$ denotes a context, we only write $c$ when there is no ambiguity. We denote by $c[t]$ the term obtained from $c[\diamond]$ by replacing $\diamond$ by a term $t$.

A *bottom-up Finite Tree Automaton* ($\uparrow$-FTA) over $\mathcal{F}$ is a tuple $A = (Q, \mathcal{F}, Q_f, \Delta)$ where $Q$ is a finite set of states, $Q_f \subseteq Q$ is a set of final states, and $\Delta$ is a set of transition rules of the form $f(q_1, \ldots, q_n) \to q$ where $n \geq 0$, $f \in \mathcal{F}_n$, $q, q_1, \ldots, q_n \in Q$. In this paper, the size of an automaton refers to its size in number of states, so two automaton which have the same number of states but different number of rules are considered as having the same size. When $n = 0$ a rule is written $a \to q$, where $a$ is a constant. The *move relation* is written $\to_A$ and $\to_A^*$ is the reflexive and transitive closure of $\to_A$. A term $t$ reaches a state $q$ if and only if $t \to_A^* q$. A state $q$ *accepts* a context $c$ if and only if there exists a $q_f \in Q_f$ such that $c[q] \to_A^* q_f$. The automaton $A$ recognizes a term $t$ if and only if there exists a $q_f \in Q_f$ such that $t \to_A^* q_f$. The language recognized by $A$ is the set of all terms recognized by $A$, and is denoted by $L(A)$.

Two $\uparrow$-FTA are equivalent if they recognize the same tree language. A $\uparrow$-FTA $A = (Q, \mathcal{F}, Q_f, \Delta)$ is *trimmed* if and only if all its states can be reached by at least one term and accepts at least one context. A $\uparrow$-FTA is *deterministic* ($\uparrow$-DFTA) if and only if there are no two rules with the same left-hand side in its set of rules. A tree language is *regular* if and only if it is recognized by a bottom-up tree automaton. As any $\uparrow$-FTA can be changed into an equivalent trimmed $\uparrow$-DFTA, any regular tree language can be recognized by a trimmed $\uparrow$-DFTA.

Let $L$ be a tree language over a ranked alphabet $\mathcal{F}$ and $t$ a term. The bottom-up residual language of $L$ relative to a term $t$, denoted by $t^{-1}L$, is the set of all contexts in $\mathcal{C}(\mathcal{F})$ such that $c[t] \in L$: $t^{-1}L = \{c \in \mathcal{C}(\mathcal{F}) \mid c[t] \in L\}$.

Note that a bottom-up residual language is a set of contexts, and not a tree language. The Myhill-Nerode congruence for tree languages can be defined by two terms $t$ and $t'$ are equivalent if they define the same residual languages. From the Myhill-Nerode theorem for tree languages, we get the following result: a tree language is recognizable if and only if the number of residual languages is finite.

A *top-down finite tree automaton* ($\downarrow$-FTA) over $\mathcal{F}$ is a tuple $\mathcal{A} = (Q, \mathcal{F}, I, \Delta)$ where $Q$ is a set of states, $I \subseteq Q$ is a set of initial states, and $\Delta$ is a set of rewrite rules of the form $q(f) \to f(q_1, \dots, q_n)$ where $n \geq 0$, $f \in \mathcal{F}_n$, $q, q_1, \dots, q_n \in Q$. Again, if $n = 0$ the rule is written $q(a) \to a$. The *move relation* is written $\to_A$ and $\to_A^*$ is the reflexive and transitive closure of $\to_A$. A state $q$ accepts a term $t$ if and only if $q(t) \to_A^* t$. $\mathcal{A}$ recognizes a term $t$ if and only if at least one of its initial states accepts it. The language recognized by $A$ is the set of all ground terms recognized by $A$ and is denoted by $L(A)$. Any regular tree language can be recognized by a $\downarrow$-FTA. This means that $\downarrow$-FTA and $\uparrow$-FTA have the same expressive power. A $\downarrow$-FTA is *deterministic* ($\downarrow$-DFTA) if and only if its set of rules does not contain two rules with the same left-hand side. Unlike $\uparrow$-DFTA, $\downarrow$-DFTA are not able to recognize all regular tree languages.

Let $L$ be a tree language over a ranked alphabet $\mathcal{F}$, and $c$ a context of $\mathcal{C}(\mathcal{F})$. The top-down residual language of $L$ relative to $c$, denoted by $c^{-1}L$, is the set of ground terms $t$ such that $c[t] \in L$: $c^{-1}L = \{t \in \mathcal{T}(\mathcal{F}) \mid c[t] \in L\}$.

The definition of top-down residual languages comes with an equivalence relation on contexts. It is worth noting that it does not define a congruence over terms. Nonetheless, based on [NP97], it can be shown that a tree language $L$ is regular if and only if the number of top-down residual languages associated with $L$ is finite. In the proof, it is used that the number top-down residual languages is lower than the number of bottom-up residual languages.

The full proofs of theorems and propositions are given in [CGL+03].

## 3    Bottom-Up Residual Finite Tree Automata

In this section, we introduce a new class of bottom-up finite tree automata, called bottom-up residual finite tree automata ($\uparrow$-RFTA). This class of automata shares some interesting properties with both bottom-up deterministic and non-deterministic finite tree automata which both recognize the class of regular tree languages.

On the one hand, as $\uparrow$-DFTA, $\uparrow$-RFTA admits a unique canonical form, based on a correspondence between states and residual languages, whereas $\uparrow$-FTA does not. On the other hand, $\uparrow$-RFTA are non-deterministic and can be much smaller in their canonical form than their deterministic counter-parts.

### 3.1    Definition and Expressive Power of Bottom-Up Residual Finite Tree Automata

First, let us precise the nature of this correspondence, then let us give the formal definition of $\uparrow$-residual tree automata and describe their properties.

In order to establish the nature of this correspondence between states and residual languages, let us introduce the notion of state languages. The *state language* $C_q$ of a state $q$ is the set of contexts accepted by the state $q$:

$$C_q = \{c \in \mathcal{C}(\mathcal{F}) \mid \exists q_f \in Q_f, c[q] \to_A^* q_f\}.$$

As shown by the following example, state languages are generally not residual languages:

*Example 1.* Consider the tree language $L = \{f(a_1, b_1), f(a_1, b_2), f(a_2, b_2)\}$ over $\mathcal{F} = \{f(,), a_1, b_1, a_2, b_2\}$. This language $L$ is recognized by the tree automaton $A = (\{q_1, q_2, q_3, q_4, q_5\}, \mathcal{F}, \{q_5\}, \Delta)$ where $\Delta = \{a_1 \to q_1, b_1 \to q_2, b_2 \to q_3, a_2 \to q_4, a_1 \to q_4, f(q_1, q_2) \to q_5, f(q_4, q_3) \to q_5\}$. Residual languages of $L$ are $a_1^{-1}L = \{f(\diamond, b_1), f(\diamond, b_2)\}$, $b_1^{-1}L = \{f(a_1, \diamond)\}$, $b_2^{-1}L = \{f(a_1, \diamond), f(a_2, \diamond)\}$, $a_2^{-1}L = \{f(\diamond, b_2)\}$, $f(a_1, b_1)^{-1}L = \{\diamond\}$. The state language of $q_1$ is $\{f(\diamond, b_1)\}$, which is not a residual language. The tree $a_1$ reaches $q_1$, so each context accepted by $q_1$ is an element of the residual language $a_1^{-1}L$, which means that $C_{q_1} \subset a_1^{-1}L$. But the reverse inclusion is not true because $f(\diamond, b_2)$ is not an element of $C_{q_1}$. The reader should note that this situation is possible because $A$ is non-deterministic.

In fact, it can be proved (the proof is omitted) that residual languages are unions of state languages. For any $L$ recognized by a tree automaton $A$, we have

$$\forall t \in T(\mathcal{F}), t^{-1}L = \bigcup_{q \in Q,\ t \to_A^* q} C_q. \tag{1}$$

As a consequence, if $A$ is deterministic and trimmed, each residual language is a state language and conversely.

We can define a new class of non-deterministic automata stating that each state language must correspond to a residual tree language. We have seen that residual tree languages are related to the Myhill-Nerode congruence and we will show that minimization of tree automata can be extended in the definition of a canonical form for this class of non-deterministic tree automata.

**Definition 1.** *A bottom-up residual tree automaton ($\uparrow$-RFTA) is a $\uparrow$-FTA $A = (Q, \mathcal{F}, Q_f, \Delta)$ such that $\forall q \in Q$, $\exists t \in T(\mathcal{F})$, $C_q = t^{-1}L(A)$.*

According to the above definition and previous remarks, it can be shown that every trimmed $\uparrow$-DFTA is a $\uparrow$-RFTA. As a consequence, $\uparrow$-RFTA have the same expressive power than finite tree automata:

**Theorem 1.** *The class of tree languages recognized by $\uparrow$-RFTA is the class of regular tree languages.*

As an advantage of $\uparrow$-RFTA, the number of states of an $\uparrow$-RFTA can be much smaller than the number of states of any equivalent $\uparrow$-DFTA:

**Proposition 1.** *There exists a sequence $(L_n)$ of regular tree languages such that for each $L_n$, the size of the smallest $\uparrow$-DFTA which recognizes $L_n$ is an exponential function of $n$, and the size of the smallest $\uparrow$-RFTA which recognizes $L_n$ is a linear function of $n$.*

*Sketch of proof* We give an example of regular tree languages for which the size of the ↑-DFTA grows exponentially with respect to the size of the equivalent canonical ↑-RFTA. A path is a sequence of symbols from the root to a leaf of a tree. The length of a path is the number of symbols on the path, except the root. Let $\mathcal{F} = \{f(,), a\}$ and let us consider the tree language $L_n$ which contains exactly the trees with at least one path of length $n$. Let $A_n = (Q, \mathcal{F}, Q_f, \Delta)$ be a ↑-FTA defined by: $Q = \{q_*, q_0, \ldots, q_n\}, Q_f = \{q_0\}$ and

$$\Delta = \{a \to q_*, a \to q_n, f(q_*, q_*) \to q_*\} \cup$$
$$\bigcup_{k \in [1,\ldots,n], q \in Q \setminus \{q_0\}} \{f(q_k, q) \to q_{k-1}, f(q, q_k) \to q_{k-1}, f(q_k, q) \to q_*, f(q, q_k) \to q_*\}$$

Let $C_*$ be the set of contexts which contain at least one path of length $n$. Let $C_i$ be the set of contexts whose path from the root to $\diamond$ is of length $i$. Let $t_*$ be a term such that all its paths are of length greater than $n$. Note that the set of contexts $c$ such that $c[t_*]$ belongs to $L_n$ is exactly the set of contexts $C_*$. Let $t_0 \ldots t_n$ be terms such that for all $i \leq n$, $t_i$ contains exactly one path of length smaller than $n$, and the length of this path is $n - i$. Therefore, $t_i^{-1} L_n$ is the set of contexts $C_* \cup C_i$.

One can verify that $C_{q_*}$ is exactly $t_*^{-1} L_n = C_*$, and for all $i \leq n$, $C_{q_i}$ is exactly $t_i^{-1} L_n = C_* \cup C_i$. The reader should note that rules of the form $f(q_k, q) \to q_*$ and $f(q, q_k) \to q_*$ are not useful to recognize $L_n$ but they are required to obtain a ↑-RFTA (because $C_i$ is not a residual language of $L_n$). So $A_n$ is a ↑-RFTA and recognizes $L_n$. The size of $A_n$ is $n + 2$.

The construction of the smallest ↑-DFTA which recognizes $L(A_n)$ is left to the reader. But, it can easily be shown that the number of states is in $O(2^n)$ because states must store lengths of all paths smaller than $n$.     □

Unfortunately, the size of a ↑-RFTA can be exponentially larger than the size of an equivalent ↑-FTA.

## 3.2    The Canonical Form of Bottom-Up Residual Tree Automata

As ↑-DFTA, ↑-RFTA have the interesting property to admit a canonical form. In the case of ↑-DFTA, there is a one-to-one correspondence between residual languages and state languages. This is a consequence of the Myhill-Nerode theorem for trees.

A similar result holds for ↑-RFTA. In a canonical ↑-RFTA, the set of states is in one-to-one correspondence with a subset of residual languages called *prime residual languages*.

**Definition 2.** *Let $L$ be a tree language. A bottom-up residual language of $L$ is composite if and only if it is the union of the bottom-up residual languages that it strictly contains:*

$$t^{-1} L = \bigcup_{t'^{-1} L \subsetneq t^{-1} L} t'^{-1} L.$$

*A residual language is* prime *if and only if it is not composite.*

*Example 2.* Let us consider again the tree languages in the proof of Proposition 1. Let $Q_n$ be the set of states of $A_n$. All the $n+2$ states $q_*, q_0, \ldots, q_n$ of $Q_n$ have state languages which are prime residual languages. The subset construction applied on $A_n$ to build a $\uparrow$-DFTA $D_n$ leads to consider states which are subsets of $Q$. The state language of a state $\{q_{k_1} \ldots q_{k_n}\}$ is a composite residual language. It is the union of $t_{q_{k_1}}^{-1} L \ldots t_{q_{k_n}}^{-1} L$.

In canonical $\uparrow$-RFTAs, all state languages are prime residual languages.

**Theorem 2.** *Let $L$ be a regular tree language and let us consider the $\uparrow$-FTA $A_{can} = (Q, \mathcal{F}, Q_f, \Delta)$ defined by:*

- *$Q$ is in bijection with the set of all prime bottom-up residual languages of $L$. We denote by $t_q$ a ground term such that $q$ is associated with $t_q^{-1} L$ in this bijection*
- *$Q_f$ is the set of all elements $q$ of $Q$ such that $t_q^{-1} L$ contains the void context $\diamond$,*
- *$\Delta$ contains all the rules $f(q_1, \ldots, q_n) \rightarrow q$ such that $t_q^{-1} L \subseteq (f(t_{q_1}, \ldots, t_{q_n}))^{-1} L$ and all the rules $a \rightarrow q$ such that $a \in \mathcal{F}_0$ and $t_q^{-1} L \subseteq a^{-1} L$.*

*$A_{can}$ is a $\uparrow$-RFTA, it is the smallest $\uparrow$-RFTA in number of states which recognizes $L$, and it is unique up to a renaming of its states.*

*Sketch of proof* There are three things to prove in this theorem: the canonical $\uparrow$-RFTA $A_{can} = (Q, \mathcal{F}, Q_f, \Delta)$ of a regular tree language $L$ recognizes $L$, it is a $\uparrow$-RFTA, and there cannot be any strictly smaller $\uparrow$-RFTA which recognizes $L$. The three points are proved in this order.

We first have to prove the equality $L(A_{can}) = L$. It follows from the identity ($\circledast$) $\forall t,\ t^{-1} L = \bigcup_{q \in Q,\ t \rightarrow_{A_{can}}^* q} t_q^{-1} L$ which can be proved inductively on the height of $t$. Using this property, we have:

$$t \in L \Leftrightarrow \diamond \in t^{-1} L \underset{\circledast}{\Leftrightarrow} \diamond \in \bigcup_{q \in Q,\ t \rightarrow_{A_{can}}^* q} t_q^{-1} L \Leftrightarrow \exists q_f \in Q_f, t \rightarrow_{A_{can}}^* q_f \Leftrightarrow t \in L(A_{can})$$

The equality between $L$ and $L(A_{can})$ helps us to prove the characterization of $\uparrow$-RFTA: $t_q^{-1} L = C_q^{A_{can}}$ where $C_q^{A_{can}}$ is the state language of $q$ in $A_{can}$.

The last point can be proved in such a way. In a $\uparrow$-RFTA, any residual language is a union of state languages, and any state language is a residual language. So any prime residual language is a state language, so there is at least as much states in a $\uparrow$-RFTA as prime residual languages admitted by its corresponding tree language.

□

The canonical automaton is uniquely defined by the tree language under consideration, but there may be other automata which have the same number of states. The canonical $\uparrow$-RFTA is unique because it has the maximum number of rules. Even though all its states are associated to prime residual languages, the automaton considered in the proof of Proposition 1 is not the canonical one because some rules are missing: $\bigcup_{k=1}^n \{f(q_k, q_0) \rightarrow q_{k-1}, f(q_0, q_k) \rightarrow q_{k-1}\}$ and $\bigcup_{q \in Q} \{f(q, q_0) \rightarrow q_*, f(q, q_0) \rightarrow q_*\}$.

## 4     Top-Down Residual Finite Tree Automata

The definition of top-down residual finite tree automata ($\downarrow$-RFTA) is tightly correlated with the definition of $\uparrow$-RFTA. Similarly to $\uparrow$-RFTA, $\downarrow$-RFTA are defined as non-deterministic tree automata where each state language is a residual language. Any $\downarrow$-RFTA can be transformed in a canonical equivalent $\downarrow$-RFTA — minimal in the number of states and unique up to state renaming.

The main difference between the bottom-up and the top-down case is in the problem of the expressive power of tree automata. The three classes of bottom-up tree automata, $\uparrow$-DFTA, $\uparrow$-RFTA or $\uparrow$-FTA, have the same expressive power. In the top-down case, deterministic, residual and non-deterministic tree automata have different expressive power. This makes the canonical form of $\downarrow$-RFTA more interesting. Compared to the minimal form of $\downarrow$-DFTA, it can be smaller when both exist, and it exists for a wider class of tree languages.

Let us introduce $\downarrow$-RFTA through their similarity with $\uparrow$-RFTA, then study this specific problem of expressiveness.

### 4.1     Analogy with Bottom-Up Residual Tree Automata

Let us formally define state languages in the top-down case:

**Definition 3.** *Let $L$ be a regular tree language over a ranked alphabet $\mathcal{F}$, let $A$ be a top-down tree automaton which recognizes $L$, and let $q$ be a state of this automaton. The state language of $L$ relative to $q$, written $L_q$, is the set of terms which are accepted by $q$:*

$$L_q = \{t \in \mathcal{T}(\mathcal{F}) \mid q(t) \to_A^* t\}.$$

It follows from this definition some properties similar to those already studied in the previous section. Firstly, state languages are generally not residual languages. Secondly, residual languages are unions of state languages. Let us define $Q_c$:

$$Q_c = \{q \mid q \in Q, \exists q_i \in I, q_i(c[\diamond]) \to_A^* c[q(\diamond)]\}.$$

We have the following relation between state languages and residual languages.

**Lemma 1.** *Let $L$ be a tree language and let $A = (Q, \mathcal{F}, I, \Delta)$ be a top-down tree automaton which recognizes $L$. Then $\forall c \in \mathcal{C}(\mathcal{F}), \bigcup_{q \in Q_c} L_q = c^{-1}L$.*

These similarities lead us to this definition of top-down residual tree automata:

**Definition 4.** *A top-down Residual Finite Tree Automaton ($\downarrow$-RFTA) recognizing a tree language $L$ is a $\downarrow$-FTA $A = (Q, \mathcal{F}, I, \Delta)$ such that: $\forall q \in Q, \exists c \in \mathcal{C}(\mathcal{F}), L_q = c^{-1}L$.*

Languages defined in the proof of Proposition 1 are still interesting here to define examples of top-down residual tree automata:

*Example 3.* Let us consider again the family of tree languages $L_n$, and the family of corresponding ↑-RFTA $A_n$. For every $n$, let $A'_n$ be the ↓-RFTA defined by: $Q = \{q_*, q_0, \ldots, q_n\}, Q_i = \{q_0\}$ and $\Delta = \{q_*(a) \to a, q_n(a) \to a, q_*(f) \to f(q_*, q_*)\} \cup \bigcup_{k=1}^{n} \{q_{k-1}(f) \to f(q_k, q_*), q_{k-1}(f) \to f(q_*, q_k)\}$.

For every $k \leq n$, the state language of $q_k$ is equal to $L_{n-k}$. And, $L_{n-k}$ is the top-down residual language of $c_k$, where $c_k$ is a context whose height from the root to the special constant $\diamond$ is $k$ and $c_k$ does not contain any path whose length is smaller or equal to $n$. The state language of $q_*$ is $\mathcal{T}(\mathcal{F})$. And, $\mathcal{T}(\mathcal{F})$ is the top-down residual language of $L_n$ relative to $c_*$, where $c_*$ is a context who contains a path whose length is $n$. So $A'_n$ is a ↓-RFTA. Moreover, it is easy to verify that $A'_n$ recognizes $L_n$.

## 4.2 The Expressive Power of Top-Down Tree Automata

*Top-down deterministic automata and path-closed languages.* A tree language $L$ is *path-closed* if:

$$\forall c \in C(\mathcal{F}), c[f(t_1, t_2)] \in L \wedge c[f(t'_1, t'_2)] \in L \Rightarrow c[f(t_1, t'_2)] \in L.$$

The reader should note that the definition only considers binary symbols, the definition can easily be extended to $n$-ary symbols. The class of languages that ↓-DFTA can recognize is the class of path-closed languages [Vir81].

*Context-deterministic automata and homogeneous languages.* Podelski and Nivat in [NP97] have defined *l-r-deterministic* top-down tree automata. In the present paper, let us call them top-down *context-deterministic* tree automata.

**Definition 5.** *A top-down context-deterministic tree automaton (↓-CFTA) A is a ↓-FTA such that for every context $c \in C(\mathcal{F})$, $Q_c$ is either the empty set or a singleton set.*

An *homogeneous language* is a tree language $L$ satisfying:

$$\forall c \in C(\mathcal{F}), c[f(t_1, t_2)] \in L \wedge c[f(t_1, t'_2)] \in L \wedge c[f(t'_1, t_2)] \in L \Rightarrow c[f(t'_1, t'_2)] \in L.$$

Again, the definition can easily be extended from the binary case to $n$-ary symbols. They have shown that the class of languages recognized by ↓-CFTA is the class of homogeneous languages.

*The hierarchy.* A ↓-DFTA is a ↓-CFTA. For ↓-CFTA and ↓-RFTA, we have the following result:

**Lemma 2.** *Any trimmed ↓-CFTA is a ↓-RFTA.*

*Proof.* Let $A = (Q, \mathcal{F}, I, \Delta)$ be a trimmed ↓-CFTA recognizing a tree language $L$. As $A$ is trimmed, all states are reachable, so for every $q$, there exists a $c$ such that $q \in Q_c$. Then, by definition of a ↓-CFTA, for every $q$, there exists a $c$ such that $\{q\} = Q_c$. Using Lemma 1, we have:

$$\forall q \in Q, \exists c \in C(\mathcal{F}), L_q = c^{-1}L.$$

stating that $A$ is a ↓-RFTA.                                                □

Therefore, if we denote by $\mathcal{L}_\mathcal{C}$ the class of tree languages recognized by a class of automata $\mathcal{C}$, we obtain the following hierarchy:

$$\mathcal{L}_{\downarrow-DFTA} \subseteq \mathcal{L}_{\downarrow-CFTA} \subseteq \mathcal{L}_{\downarrow-RFTA} \subseteq \mathcal{L}_{\downarrow-FTA}$$

*The hierarchy is strict.*

- Let $L = \{f(a,b), f(b,a)\}$. $L_1$ is homogeneous but not path-closed. Therefore $L$ can be recognized by a $\downarrow$-CFTA, but can not be recognized by a $\downarrow$-DFTA.
- The tree languages $L_n$ in the proof of Proposition 1 are not recognized by $\downarrow$-CFTA. We can easily verify that $L_n$ is not homogeneous. Indeed, if $t$ is a term which has a path whose length is equal to $n-1$, and $t'$ a term which does not have any path whose length is smaller than $n$, $f(t,t)$, $f(t,t')$, $f(t',t)$ belong to $L_n$, but $f(t',t')$ does not. And, we have already shown that $L_n$ is recognized by a $\downarrow$-RFTA.
- Let $L' = \{f(a,b), f(a,c), f(b,a), f(b,c), f(c,a), f(c,b)\}$. $L'$ is a finite language, therefore it is a regular tree language which can be recognized by a $\downarrow$-FTA. $L'$ cannot be recognized by a $\downarrow$-RFTA. To prove that, let us consider $A'$ a $\downarrow$-FTA which recognizes $L'$. The top-down residual languages of $L'$ are $\{a,b\}$, $\{a,c\}$, $\{b,c\}$ and $L'$. As $A'$ recognizes $L'$, it recognizes $f(a,b)$. This implies the existence of three states $q_1$, $q_2$, $q_3$ and three rules $q_1(f) \rightarrow f(q_2,q_3)$, $q_2(a) \rightarrow a$, and $q_3(b) \rightarrow b$. If $A'$ was a $\downarrow$-RFTA, then $q_2$ would accept a residual language. As $q_2$ accepts $a$, it would accept either $\{a,b\}$ or $\{a,c\}$. Similarly, $q_3$ would accept either $\{a,b\}$ or $\{b,c\}$. In these conditions, and thanks to the rule $q_1(f) \rightarrow f(q_2,q_3)$, $A'$ would recognize $f(a,a)$, $f(b,b)$ or $f(c,c)$. So $A'$ cannot be a $\downarrow$-RFTA.

Therefore, we obtain the following result:

**Theorem 3.** $\mathcal{L}_{\downarrow-DFTA} \subsetneqq \mathcal{L}_{\downarrow-CFTA} \subsetneqq \mathcal{L}_{\downarrow-RFTA} \subsetneqq \mathcal{L}_{\downarrow-FTA}$

So top-down residual tree automata are strictly more expressive than context-deterministic tree automata. But as far as we know, there is no straightforward characterization of the tree languages recognized by $\downarrow$-RFTA.

## 4.3   The Canonical Form of Top-Down Residual Tree Automata

The problem of the canonical form of top-down tree automata is similar to the bottom-up case. Whereas there is no way to reduce a non-deterministic top-down tree automaton to a unique canonical form, a top-down residual tree automaton can take such a form. Its definition is similar to the definition of the canonical bottom-up tree automaton.

In the same way that we have defined composite bottom-up residual language, a top-down residual language of $L$ is *composite* if and only if it is the union of the top-down residual languages that it strictly contains and a residual language is *prime* if and only if it is not composite.

**Theorem 4.** *Let $L$ be a tree language in the class $\mathcal{L}_{\downarrow-RFTA}$. Let us consider the $\downarrow$-RFTA $A_{can} = (Q, \mathcal{F}, I, \Delta)$ defined by:*

- $Q$ is a set of state in bijection with the prime residual languages of $L$. For each of these residual languages, there exists a $c_q$ such that $q$ is associated with $c_q^{-1}L$ in this bijection.
- $I$ is the set of prime residuals which are subsets of $L$.
- $\Delta$ contains all the rules $q(a) \to a$ such that $a$ is a constant and $c_q[a] \in L$, and all the rules $q(f) \to f(q_1, \ldots, q_n)$ such that for all $t_1 \ldots t_n$ where $t_i \in c_{q_i}^{-1}L$, $c_q[f(t_1, \ldots, t_n)] \in L$.

$A_{can}$ is a $\downarrow$-RFTA, it is the smallest $\downarrow$-RFTA in number of states which recognizes $L$, and it is unique up to a renaming of its states.

*Sketch of proof*

The proof is mainly based on this lemma: $t \in c_q^{-1}L \Leftrightarrow t \in L_q^{A_{can}}$

where $L_q^{A_{can}}$ is the state language of $q$ in the automaton $A_{can}$.

This lemma is proved by induction on the height of $t$. This is not a straightforward induction. It involves the rules of a $\downarrow$-RFTA automaton $A'$ which recognizes $L$. Its existence is granted by the hypothesis of the theorem. Once this is proved, it can be easily deduced that $A_{can}$ recognizes $L$ and is a RFTA. As there is one state per prime residual in $A_{can}$, it is minimal in number of states.

$\square$

# 5   Decidability Issues

Some decision problems naturally arise with the definition of RFTA. Most of these problems are solved just noting that one can build all residual languages of a given regular language $L$ defined by a non-deterministic tree automaton. In the bottom-up case, the state languages of the minimal $\uparrow$-DFTA which recognizes $L$ are exactly the bottom-up residual languages of $L$, and this automaton can be built with the subset construction. In the top-down case, the subset construction does not necessarily gives us an automaton which recognizes exactly $L$, but there exists a way to construct the top-down residual languages of a tree language $L$ described in [CGL+03]. Therefore, knowing whether a tree automaton is a RFTA, whether a residual language is prime or composite, and whether a tree automaton is a canonical RFTA are decidable. These problems have not been deeply studied in terms of complexity, but they are at least as hard as the similar problems with strings, that is they are PSPACE-hard ([DLT02a]).

# 6   Conclusion

We have defined new classes of non-deterministic tree automata. In the bottom-up case, we get another characterization of regular tree languages. More interestingly, in the top-down case, we obtain a subclass of the regular tree languages. For both cases, we have a canonical form and the size of residual tree automata can be much smaller than equivalent (when exist) deterministic ones.

We are currently extending these results to the case of unranked trees because our application domain is concerned with html and xml documents. Also, we are designing learning algorithms for residual finite tree automata extending previous algorithms for residual finite string automata [DLT01,DLT02b].

182    J. Carme et al.

**Acknowledgements.** The authors wish to thank the anonymous reviewers for their critisms and suggestions. This research was partially supported by "TACT-TIC" région Nord-Pas-de-Calais — FEDER and the MOSTRARE IN-RIA project.

# References

[CDG+97]   H. Comon, M. Dauchet, R. Gilleron, F. Jacquemard, D. Lugiez, S. Tison, and M. Tommasi. Tree automata techniques and applications. Available on: http://www.grappa.univ-lille3.fr/tata, 1997.

[CGL+03]   J. Carme, R. Gilleron, A. Lemay, A. Terlutte, and M. Tommasi. Residual finite tree automata. Technical report, GRAPPA, 2003.

[DLT01]   F. Denis, A. Lemay, and A. Terlutte. Learning regular languages using rfsa. In *ALT 2001*, number 2225 in Lecture Notes in Artificial Intelligence. Springer Verlag, 2001.

[DLT02a]   F. Denis, A. Lemay, and A. Terlutte. Residual finite state automata. *Fundamenta Informaticae*, 51(4):339–368, 2002.

[DLT02b]   F. Denis, A. Lemay, and A. Terlutte. some language classes identifiable in the limit from positive data. In *ICGI 2002*, number 2484 in Lecture Notes in Artificial Intelligence, pages 63–76. Springer Verlag, 2002.

[Fer02]   Henning Fernau. Learning tree languages from text. In *Proc. 15th Annual Conference on Computational Learning Theory, COLT 2002*, pages 153–168, 2002.

[GK02]   Sally A. Goldman and Stephen S. Kwek. On learning unions of pattern languages and tree patterns in the mistake bound model. *Theorical Computer Science*, 288(2):237–254, 2002.

[Gol67]   E.M. Gold. Language identification in the limit. *Inform. Control*, 10:447–474, 1967.

[GS84]   F. Gécseg and M. Steinby. *Tree Automata*. Akademiai Kiado, 1984.

[GS96]   F. Gécseg and M. Steinby. Tree languages. In G. Rozenberg and A. Salomaa, editors, *Handbook of Formal Languages*, volume 3, pages 1–68. Springer Verlag, 1996.

[LPH00]   Ling Liu, Calton Pu, and Wei Han. XWRAP: An XML-enabled wrapper construction system for web information sources. In *ICDE*, pages 611–621, 2000.

[MLM01]   M. Murata, D. Lee, and M. Mani. "Taxonomy of XML Schema Languages using Formal Language Theory". In *Extreme Markup Languages*, Montreal, Canada, 2001.

[Nev02]   F. Neven. Automata, xml and logic. In *Proceedings of CSL*, pages 2–26, 2002.

[NP97]   M. Nivat and A. Podelski. Minimal ascending and descending tree automata. *SIAM Journal on Computing*, 26(1):39–58, February 1997.

[Sak90]   Yasubumi Sakakibara. learning context-free grammars from structural data in polynomial time. *Theorical Computer Science*, 76:223–242, 1990.

[Tha73]   J.W. Thatcher. Tree automata: an informal survey. In A.V. Aho, editor, *Currents in the theory of computing*, pages 143–178. Prentice Hall, 1973.

[Tho84]   Wolfgang Thomas. Logical aspects in the study of tree languages. In *Proceedings of the 9th International Colloquium on Trees in Algebra and Programming, CAAP '84*, pages 31–50, 1984.

[Vir81]   J. Viragh. Deterministic ascending tree automata. *Acta Cybernetica*, 5:33–42, 1981.

# From Glushkov WFAs to Rational Expressions

Pascal Caron[1] and Marianne Flouret[2]

[1] LIFAR, Université de Rouen, 76134 Mont-Saint-Aignan Cedex, France
Pascal.Caron@dir.univ-rouen.fr
[2] LIH, Université du Havre, 76058 Le Havre Cedex, France
Marianne.Flouret@univ-lehavre

**Abstract.** In this paper, we extend to the multiplicity case a characterization of Glushkov automata, and show the existence of a normal form for rational expressions. These results are used to obtain a rational expression of small size from a Glushkov WFA.

## 1 Introduction

Automata with multiplicities have many application areas. One of the most representative is doubtless linguistics. This application to speech processing can be found in Mohri's works [12], or Peireira *et al.* ones [13]. They play an important role too in image encoding, particularly in image compression. Let us cite Culik and Kari's works [9]. The automata theory is older than automata with multiplicities one. Some of automata theory results have been extended to multiplicity yet, sometimes with particular conditions. Schützenberger[14] laid its foundations giving the equivalence between rational and recognizable series, and started this long sequel. Among the most recents, we can cite the Lombardy and Sakarovitch result [10], which extend an algorithm by Antimirov on partial derivatives. Caron and Ziadi have provided a characterization of boolean Glushkov automata [7], allowing to obtain a short rational expression (in terms of symbols) from Glushkov automata. The Glushkov construction on automata with multiplicities has been extended by the authors [5]. This construction has importance, by the small size of the automata obtained. Indeed, the number of states is the number of occurrences of letters in the expression. If we reverse this procedure, we will get, from a Glushkov automaton with $n + 1$ states, a rational expression having $n$ occurences of letters. The conversion of an automaton into a rational expression is an important point of the automata theory. Concerning the multiplicities, the bloc decomposition, proved in [1], gives a large rational expression. In the boolean case, the result obtained with the Mac Naughton-Yamada's algorithm is a large expression too. This last method will be extended to the multiplicity case below.

In this paper, we extend the Glushkov characterization and, therefore, the conversion from a Glushkov automaton with multiplicities into a rational expression. Brüggemann-Klein defines a regular expression[1] in star normal form (SNF)

---

[1] We talk about *regular expression* in th boolean case and about *rational expression* in other cases.

Z. Ésik and Z. Fülöp (Eds.): DLT 2003, LNCS 2710, pp. 183–193, 2003.
© Springer-Verlag Berlin Heidelberg 2003

[3] as an expression for which all unions of *First* are disjoint. It means that, in the recursive calculus of the *First* and *Follow* functions, an element is added at most one time in these sets. Each edge of the underlying automaton is then generated at most one time. In the multiplicity case, the restriction to rational expressions in SNF has been presented in [6]. Concerning the non SNF case, as in the boolean case, the construction can generate edges superposing (which means that an element can be added several times in one of the Glushkov sets). This superposition corresponds to a cumulative calculus which can remove edges. Our goal, here, is to find again the trace of all edges which have been generated in the construction so as to find the rational expression of the beginning. We can observe that it exists various forms of rational expressions having the same Glushkov automaton. For the conversion, we then have to describe a normal form for the rational expressions with multiplicities.

Indeed, the expression obtained from a Glushkov automaton will be a normal form of the one used to compute this Glushkov automaton. The normal form is not unique up to a factorization of coefficients.

After having set our notations, we recall the Glushkov construction and its characterization. In the following section, we extend the McNaughton-Yamada algorithm. Sect. 4 deals with the properties linked to the non-SNF rational expression case. We end with an example illustrating, for this last case, the computation of a rational expression from a WFA.

## 2    Definitions

### 2.1    Classical Notions

Given a semiring $(\mathbb{K}, \oplus, \otimes, \overline{0}, \overline{1})$, a formal series [1] is a mapping $S$ from $\Sigma^*$ into $\mathbb{K}$. The image of a word $w$ by $S$ is denoted by $(S, w)$. The support of a series $S$ is the language $Supp(S) = \{w \in \Sigma^* | (S, w) \neq 0\}$. By extension, $Supp(E)$ is the support of the rational series defined by $E$.

A weighted finite automaton(WFA) is a 5-tuple $(\Sigma, Q, I, F, \delta)$ where $Q$ is the finite set of states, $\Sigma$ the set of symbols, $I$ the set of initial states, $F$ the set of final states and $\delta$ the set of transitions. The set $I$, $F$ and $\delta$ are rather viewed as mappings $I : Q \to \mathbb{K}$, $F : Q \to \mathbb{K}$, and $\delta : Q \times \Sigma \times Q \to \mathbb{K}$. We can always compute a WFA with a single initial state from such a construction. We denote $\mathcal{A}_{\mathbb{K}}$ the set of WFAs on $\mathbb{K}$.

*Rational expressions* are obtained from letters and coefficients by a finite number of combinations of rational laws ($+$, $\cdot$, $*$, and an external product $\times$).We denote $Rat_E$ the set of rational expressions. *Rational series* are formal series that can be described by *rational expressions*. Schutzenberger asserts that rational series are exactly recognized by WFAs [14].

### 2.2    From Rational Expressions to WFAs

We summarize here the extension to multiplicities of the Glushkov construction. The first step is to mark out each occurrence of the same symbol in a rational expression $E$. Therefore, each occurrence of letter will be indexed by its position in the expression. This set of index is denoted by $Pos(E)$. The resulting expression will be denoted $\overline{E}$, defined over the alphabet of indexed symbols $\overline{\Sigma}$, each one appearing at most once in $\overline{E}$. Glushkov defines four functions $First$, $Last$, $Follow$ and $Null$ on $E$ in order to compute a non necessarily deterministic automaton. These functions are extended by replacing, in the sets $First$, $Last$ and $Follow$, positions by couples associating a coefficient to each position. $First(E)$ represents the set of initial positions of words of $Supp(\overline{E})$ associated to their input cost, $Last(E)$ the set of final positions of words of $Supp(\overline{E})$ associated to their output cost. $Follow(E, i)$ is the set of positions appearing in words of $Supp(\overline{E})$ which immediately follows the position $i$ in the expression $\overline{E}$, associated to their transition costs. The $Null$ set represents the coefficient of the empty word. Each function should be computed recursively. The way to compute them is summarized in table 1. In order to understand it, we have to define how are made couples and what are the operations supported. A couple $(l, i)$ is made up with a coefficient $l$ and a position $i$. The characteristic function is defined by $\mathcal{I}_X(i)$. Let $k \in \mathbb{K}$, $X = \{(l_s, i_s) \mid l_s \in \mathbb{K}, i_s \in \mathbb{N}\}_{1 \leq s \leq p}$ and $Y = \{(l_t, i_t) \mid l_t \in \mathbb{K}, i_t \in \mathbb{N}\}_{1 \leq t \leq p'}$. The product can be written $k \cdot X = \{(k \otimes l_s, i_s)\}_{1 \leq s \leq p}$ if $k \neq 0$, $0 \cdot X = \emptyset$ and $X \cdot k = \{(l_s \otimes k, i_s)\}_{1 \leq s \leq p}$ if $k \neq 0$, $X \cdot 0 = \emptyset$. The union is defined by $X \cup Y = \{(l, i) \mid l \in \mathbb{K}, i \in \mathbb{N}\}$ with $l = l_s + l_t$ if $\mathcal{I}_X(i) \neq 0$ and $\mathcal{I}_Y(i) \neq 0$, $l = l_s$ and $i = i_s$ if $\mathcal{I}_Y(i) = 0$, and $l = l_t$ with $i = i_t$ if $\mathcal{I}_X(i) = 0$. These functions allow us to define the automaton $\overline{\mathcal{M}} = (\overline{\Sigma}, Q, \{0\}, F, \overline{\delta})$ where

1. $\overline{\Sigma}$ is the indexed alphabet,
2. $0$ is the single initial state with no incoming edge with $\overline{1}$ as input cost,
3. $Q = Pos(E) \cup \{0\}$
4. $F : Q \to \mathbb{K}$ such that $\begin{cases} 0 \to Null(E) \\ i \to Coeff_{Last(E)}(i) \end{cases}$
5. $\delta : Q \times \Sigma \times Q \to \mathbb{K}$

$$\delta(i, a, j) = \begin{cases} Coeff_{Follow(E,i)}(j) & \text{if } a_j \in \overline{\Sigma} \text{ and } i \neq 0 \\ Coeff_{First(E)}(j) & \text{if } a_j \in \overline{\Sigma} \text{ and } i = 0 \\ 0 & \text{elsewhere} \end{cases}$$

where the $Coeff_X$ function extracts the coefficient associated to a position in the set $X$ of couples of values and positions. Symetrically, the $P(X)$ function extracts the set of positions in the set $X$. The Glushkov automaton $\mathcal{M} = (\Sigma, Q, \{0\}, F, \delta)$ of $E$ is computed from $\overline{\mathcal{M}}$ by replacing the indexed letters on edges by the corresponding letters in the expression $E$. This application, proved in [5], is denoted by $\mathcal{M}_\mathbb{K}$ ($\mathcal{M}_\mathbb{B}$ for the boolean case).

## 2.3    Topology and Star Normal Form

We have to define the embedding $\widetilde{\mathcal{M}}$ (resp. $\widetilde{E}$) of a WFA $\mathcal{M}$ (resp. a rational expression $E$) in $\mathbb{B}$. Similarly, Buchsbaum *et al.* [4] define the topology of a WFA as "its underlying directed graph and labeling by input symbols, ignoring weights". The automaton $\widetilde{\mathcal{M}} = (\Sigma, Q, \widetilde{I}, \widetilde{F}, \widetilde{\delta})$ is then defined with

**Table 1.** Extended Glushkov functions

| E | Null(E) | First(E) | Last(E) | Follow(E,i) |
|---|---|---|---|---|
| $\emptyset$ | $0$ | $\emptyset$ | $\emptyset$ | $\emptyset$ |
| $k$ | $k$ | $\emptyset$ | $\emptyset$ | $\emptyset$ |
| $a_j$ | $\overline{0}$ | $\{(\overline{1},j)\}$ | $\{(\overline{1},j)\}$ | $\emptyset$ |
| $k\cdot F$ | $k\otimes Null(F)$ | $k\cdot First(F)$ | $Last(F)$ | $Follow(F,i)$ |
| $F\cdot k$ | $Null(F)\otimes k$ | $First(F)$ | $Last(F)\cdot k$ | $Follow(F,i)$ |
| $F+G$ | $Null(F) \oplus Null(G)$ | $First(F) \cup First(G)$ | $Last(F) \cup Last(G)$ | $\mathcal{I}_{Pos(F)}(i)\cdot Follow(F,i) \cup \mathcal{I}_{Pos(G)}(i)\cdot Follow(G,i)$ |
| $F\cdot G$ | $Null(F) \otimes Null(G)$ | $First(F) \cup Null(F)\cdot First(G)$ | $Last(F)\cdot Null(G) \cup Last(G)$ | $\mathcal{I}_{Pos(F)}(i)\cdot Follow(F,i) \cup \mathcal{I}_{Pos(G)}(i)\cdot Follow(G,i) \cup \text{Coeff}_{Last(F)}(i)\cdot First(G)$ |
| $F^+$ | $\overline{0}$ | $First(F)$ | $Last(F)$ | $Follow(F,i) \cup \text{Coeff}_{Last(F)}(i)\cdot First(F)$ |
| $F^*$ | $\overline{1}$ | $First(F)$ | $Last(F)$ | $Follow(F,i) \cup \text{Coeff}_{Last(F)}(i)\cdot First(F)$ |

$\widetilde{I}, \widetilde{F} \subset Q$ and $\widetilde{I} = \{q \in Q \mid I(q) \neq 0\}$, $\widetilde{F} = \{q \in Q \mid F(q) \neq 0\}$, and $\widetilde{\delta} = \{(p,a,q) \mid p,q \in Q,\ a \in \Sigma$ and $\delta((p,a,q)) \neq 0\}$. The regular expression $\widetilde{E}$ is obtained from $E$ without respect to the weights. Brüggemann-Klein defines expressions in *star normal form* (SNF) [3] as expressions for which all unions of *First* are disjoint. Formally, for all subexpression $H^*$ in $E$, we have $\forall x \in Last(H),\ Follow(H,x) \cap First(H) = \emptyset$. We extend this definition to multiplicities. A rational expression $E$ is in SNF if $\widetilde{E}$ is in SNF.

## 2.4   Graph Properties

In order to study the graphical structure of a Glushkov automaton, we have to suppress the label on edges. It is possible on Glushkov automata because they are homogeneous ($\forall \delta(p,a,q) \neq 0,\ \delta(p',a',q') \neq 0,\ q = q' \Rightarrow a = a'$). We can now define a $\mathbb{K}$-graph as a graph $G_{\mathbb{K}}(\mathcal{M}) = (X,U)$ obtained from an extended Glushkov automaton $\mathcal{M}$. Labels of $\Sigma$ are suppressed. The set of vertices is $X = Pos(E)\cup\{\Phi\}\cup\{0\}$ and $U = \{(p,k,q) \in X\times\mathbb{K}\times X \mid \exists a \in \Sigma, k = \delta(p,a,q),\ k \neq 0\}$ defines the edges. They are labeled with their corresponding coefficient, and new edges from each terminal state to $\Phi$, labelled with the corresponding output cost, are added. By extension, we define $G_{\mathbb{K}}(E)$ as the $\mathbb{K}$-graph obtained from the Glushkov automaton of $E$. In the boolean case, it is denoted $G_{\mathbb{B}}$ and is called the *Glushkov graph* in [7].

Let $G = (X,U)$ be a weighted graph. $\mathcal{O} = (X_{\mathcal{O}}, U_{\mathcal{O}}) \subseteq G$ is a *maximal orbit* of $G$ if and only if it is a strongly connected component (SCC) with at least one edge. The set of direct successors (resp. direct predecessors) of $x \in X$,

$\{y \in X \mid U(x,y) \neq 0\}$ (resp. $\{y \in X \mid U(y,x) \neq 0\}$), is denoted by $Q^+(x)$ (resp. $Q^-(x)$). For an orbit $\mathcal{O} \subset G$, $\mathcal{O}^+(x)$ denotes $Q^+(x) \cap (X \setminus X_{\mathcal{O}})$ and $\mathcal{O}^-(x)$ denotes the set $Q^-(x) \cap (X \setminus X_{\mathcal{O}})$. In other words, $\mathcal{O}^+(x)$ is the set of vertices which are directly reached from $x$ and which are not in $X_{\mathcal{O}}$. $In(\mathcal{O}) = \{x \in X_{\mathcal{O}} \mid \mathcal{O}^-(x) \neq \emptyset\}$ and $Out(\mathcal{O}) = \{x \in X_{\mathcal{O}} \mid \mathcal{O}^+(x) \neq \emptyset\}$ denote the *input* and the *output* of the orbit $\mathcal{O}$. By extension, we can define $In(G')$ and $Out(G')$ on a graph $G' = (X', U')$ where $G' \subseteq G$ by $In(G') = \{y \in X' \mid Q^-(y) \cap (X \setminus X') \neq \emptyset\}$, and $Out(G') = \{y \in X' \mid Q^+(y) \cap (X \setminus X') \neq \emptyset\}$.

A graph $G$ is $\mathbb{K}$-*reducible* if it has no orbit and if it can be reduced to one vertex by iterated applications of any of the three rules $R_1$, $R_2$, $R_3$ described below.

**Rule $R_1$:** If $x$ and $y$ are vertices such that $Q^-(y) = \{x\}$ and $Q^+(x) = \{y\}$, then delete $y$ and define $Q^+(x) := Q^+(y)$.

**Rule $R_2$:** If $x$ and $y$ are vertices such that $Q^-(x) = Q^-(y)$ and $Q^+(x) = Q^+(y)$, and $\exists k \in \mathbb{K} \mid \forall p \in Q^-(x), \forall q \in Q^+(x), \frac{U(p,x) \times U(x,q)}{U(p,y) \times U(y,q)} = k$, then delete $y$ and any edge connected to $y$.

**Rule $R_3$:** If $x$ is a vertex such that for all $y \in Q^-(x)$, $Q^+(x) \subset Q^+(y)$ and, $\exists k \in \mathbb{K} \mid \forall p \in Q^-(x), \forall q \in Q^+(x), (p,q) \neq (0,\Phi), \frac{U(p,q)}{U(p,x) \times U(x,q)} = k$, two cases have to be distinguished :

1. If $0 \notin Q^-(x)$, or $\Phi \notin Q^+(x)$, or $|X| = 3$, then delete edges in $Q^-(x) \times Q^+(x)$.
2. If $0 \in Q^-(x)$, $\Phi \in Q^+(x)$ and $|Q^-(x) \times Q^+(x)| \neq 1$ then delete edges in $Q^-(x) \times Q^+(x) \setminus \{(0,\Phi)\}$, and $U(0,\Phi) := U(0,\Phi) - k$.

An orbit $\mathcal{O}$ of a weighted graph $G$ is *transverse* if for all $x, y \in Out(\mathcal{O})$, $\mathcal{O}^+(x) = \mathcal{O}^+(y)$ and for all $x, y \in In(\mathcal{O})$, $\mathcal{O}^-(x) = \mathcal{O}^-(y)$. Let $G$ be a $\mathbb{K}$-graph, $\mathcal{O}$ a maximal orbit, $e$ a state of $In(\mathcal{O})$ and $s$ a state of $Out(\mathcal{O})$. Let $(p_1, \cdots, p_n)$ (resp. $(q_1, \cdots, q_m)$) be the ordered states of $\mathcal{O}^-(e)$ (resp. $\mathcal{O}^+(s)$). Let $(c_1, \cdots, c_n)$ (resp. $(c'_1, \cdots, c'_m)$) the set of coefficients to the state $e$ (resp. from the state $s$). $G$ is $\mathbb{K}$-*transverse* if

- $G$ is transverse.
- $\forall e_i \in In(\mathcal{O})$ (resp. $\forall s_j \in Out(\mathcal{O})$ ), it exists $k_i$ (resp. $k'_j$) such that $(c_1 k_i, \cdots, c_n k_i)$ (resp. $(k'_j c'_1, \cdots, k'_j c'_m)$) is the set of predecessors's (successors's) coefficients of $e_i$ (resp. $s_j$). $k_i$ is called the *input cost* of $e_i$ (resp. $k'_j$ the *output cost* of $s_j$).

## 3   Extension of McNaughton-Yamada Algorithm

There are various algorithms to compute a regular expression from an automaton [8]. In this section, we extend the McNaughton-Yamada's one [11]. As in the boolean case, this algorithm computes a large rational expression, as the method of blocks decomposition described in [1].

Let $\mathcal{M} = (\Sigma, Q, I, F, \delta)$ be an automaton of $n$ states, with at most a single edge between two states. If it is not the case, the multiple edges are gathered and the resulting one is labeled with the sum of each others. We then compute a graph from this automaton. Two states have to be added. The first one, denoted by 0, reaches all initial states. These edges are labeled with $\varepsilon$. The second one, denoted by $n + 1$, is reached from all final states by edges labeled $\varepsilon$. We remove all the $n$ primitive states in $n$ steps.

Let $\alpha_{ij}^0$ be the edge from the state $i$ to the state $j$ in the original automaton. Let $\alpha_{ij}^k$, $1 \leq k \leq n$, the label of the edge from $i$ to $j$ at the step $k$, obtained by replacing edges going from $i$ to $j$ through the state $k$ at the step $k - 1$. The algorithm is as follow, for $1 \leq i, j, \leq n$:

**For $k$ from 1 to $n$ do**
    Remove the state $k$
        replacing $\alpha_{ij}^{k-1}$ by
        $$\alpha_{ij}^k = \alpha_{ij}^{k-1} + \alpha_{ik}^{k-1}(\alpha_{kk}^{k-1})^*\alpha_{kj}^{k-1}$$
**endfor**

To extend this algorithm to multiplicities, we have to consider a bijection $\beta$ between a new alphabet and the labels of the edges composed of a multiplicity and a letter. We then compute the expression, from a WFA on which we have applied these substitutions, using the classical McNaughton-Yamada algorithm. We build a family of $\{(\varepsilon_i, \tau_i)_{1 \leq i \leq n}\}$, each couple corresponding to the multiplicity labels of input and output edges for each inner state.

The labels on edges $\alpha_{0i}^{n-1}$ are replaced by $\varepsilon_i \alpha_{0i}^{n-1}$ and, for output values, $\alpha_{in+1}^{n-1}$ becomes $\alpha_{in+1}^{n-1}\tau_i$. Then, $\alpha_{0\ n+1}^n$ is the rational expression denoting the same language as $\mathcal{M}$. The rational expression is then obtained by the reverse substitution on the result.

**Proposition 1.** *The extended McNaughton-Yamada algorithm computes a rational expression in star normal form.*

This extension implies the first point of the following lemma, the second one is used to prove some properties of Glushkov WFAs.

**Lemma 1.** *Let $E$ be a rational expression and $\mathcal{M}$ a $\mathbb{K}$-automaton. Let us define the mapping $MY : \mathcal{A}_{\mathbb{K}} \rightarrow Rat_E$ which associates for each WFA a rational expression using the McNaughton-Yamada algorithm. We have:*

1. $\widetilde{MY(\mathcal{M})} = MY(\widetilde{\mathcal{M}})$,
2. *If $E$ is in $SNF$, then $\widetilde{\mathcal{M}_{\mathbb{K}}(E)} = \mathcal{M}_{\mathbb{B}}(\widetilde{E})$.*

## 4    From Extended Glushkov Automata to Small Rational Expresssions

The authors have already shown some properties in the case of expressions in star normal form. We present in this section a generalization to the non SNF case which induces a new characterization. This characterization allows us to compute a rational expression with $n$ occurrences of letters from an automaton with $n+1$ states. In a first time, we define a *normal form* from rational expression with multiplicities and show that every expression can be turned into this normal form. Thanks to the existence of this normal form, we prove the uniqueness of the rational expression deduced from a Glushkov automaton.

## 4.1   Normal Form of Rational Expressions

Proposition 2 implies that a rational expression and its normal form have the same extended Glushkov automaton. An expression $E$ is a *star rational expression* if it exists $F$ such that $E = F^*$.

**Proposition 2.** *Let* $F, F_1, \dots, F_n$ *be non star rational expressions,* $\mathbb{K}$ *a field, and* $k_1, \cdots, k_n, k'_1, \cdots, k'_n \in \mathbb{K}$. *Then the two following propositions hold.*
  *It exists* $k, k' \in \mathbb{K}$ *and* $F'$ *a non star rational expression such that*

$$\underbrace{\left( k_n \left( \cdots \left( k_2 \left( k_1 F^* + k'_1 \right)^* + k'_2 \right)^* + \cdots \right)^* + k'_n \right)^*}_{(1)} = k F'^* + k',$$

*and* $\underbrace{(F_1^* F_2^* \dots F_n^* - 1)^*}_{(2)} = \frac{1}{2} + \frac{1}{2} \left( 2 - 2 \prod_{i=n}^{1} (1 - F_i) \right)^*.$

The rational expression we will obtain from a Glushkov automaton will be in *normal form*, i.e. without subexpression of the form (1) or (2).

## 4.2   Properties of Glushkov WFAs

To obtain a rational expression of small size, we need to extend some properties.

**Proposition 3.** *Let* $E$ *be a rational expression and* $G_E$ *its* $\mathbb{K}$-*graph. For all maximal orbit* $\mathcal{O}$ *of* $G_E$, *there exists a maximal star subexpression* $F$ *of* $E$ *such that* $\mathcal{O} = Pos(F)$, $In(\mathcal{O}) = P(First(F))$, $Out(\mathcal{O}) = P(Last(F))$.

**Lemma 2.** *For each subexpression* $F$ *of* $E$ *and every position* $x \in P(Last(F))$, *the set* $P(Follow(E, x)) \setminus Pos(F)$ *is the same one.*

**Proposition 4.** *Let* $G$ *be a* $\mathbb{K}$-*graph. Every maximal orbit of* $G$ *is* $\mathbb{K}$-*transverse.*

Let $\mathcal{O}$ be a maximal orbit of a weighted graph. The *∗-break* of $\mathcal{O}$ is the losing of the SCC property by applying a function modifying weights on edges. We now have to define such a function used to compute coefficients from the graph of an orbit, in order to obtain the rational expression. Let $G = (X, U)$ be a graph, $\mathcal{O}$ a $\mathbb{K}$-transverse maximal orbit and $n = Card(\mathcal{O})$. Let $\alpha \in \mathbb{K}$. Let $M = (M_{p,q})_{p,q \in \mathcal{O}} \in \mathbb{K}^{n \times n}$ be the matrix such that $M_{p,q}$ is the coefficient of the edge from the state $p$ to the state $q$. The *backward function* $BF$ is defined on orbits by :

$$BF_\alpha(\mathcal{O}) : \mathbb{K}^{n \times n} \longrightarrow \mathbb{K}^{n \times n}$$
$$M \qquad \mapsto M'$$

with $\forall p, q \in \mathcal{O}$, $M'_{p,q} = \begin{cases} M_{p,q} - k'k\alpha \text{ if } p \in Out(\mathcal{O}), q \in In(\mathcal{O}) \\ \qquad \text{with, } k' \text{ the output cost of } p \\ \qquad \text{and } k \text{ the input cost of } q \\ \\ M_{p,q} \qquad \text{elsewhere} \end{cases}$

$\mathcal{O}$ is *-stable if there exists $\alpha$ such that the function $BF_\alpha(\mathcal{O})$ induces an *-break of $\mathcal{O}$. $G$ is *-stable if each maximal orbit of $G$ is *-stable. If $G$ is a weighted graph, $G$ is strongly *-stable (resp. strongly $\mathbb{K}$-transverse) if (1) $G$ has no maximal orbit or (2) $G$ is *-stable (resp. $\mathbb{K}$-transverse) and if after applying a backward function to all orbits of $G$, it is strongly *-stable (resp. strongly $\mathbb{K}$-transverse).

**Proposition 5.** *Every maximal orbit of a $\mathbb{K}$-graph is *-stable.*

To prove this property, we need the following lemma, which means that, the only way to obtain an SCC is the existence of back edges. In fact, it means that we have the $BK$ property:

if $\exists n \geq 0, (\sigma_i)_{0 \leq i < n} \in \Sigma$, $(p_i)_{0 \leq i \leq n} \in Q, p_0 \in Out(\mathcal{O})$, $p_n \in In(\mathcal{O})$,

such that $\forall 0 \leq i \leq n-1$, $\delta(p_i, \sigma_i, p_{i+1}) \neq 0$, necessarily $n = 1$.

**Lemma 3.** *Let $G = (X, U)$ be a $\mathbb{K}$-graph. Let $\mathcal{O}$ be a maximal orbit. If edges from $Out(\mathcal{O}) \times In(\mathcal{O})$ are deleted then $\mathcal{O}$ loses the strong connectivity characteristic.*

The next proposition shows the uniqueness of the computed decomposition. It means that it exists only one rational expression corresponding to an extended Glushkov automaton.

**Proposition 6.** *Let $G$ be a $\mathbb{K}$-graph. For each maximal orbit $\mathcal{O}$ there exists only one backward function (BF) inducing an *-break.*

**Proposition 7.** *A $\mathbb{K}$-graph is strongly *-stable.*

Let $G$ be such a graph. Its *graph without orbit* is the graph obtained by the recursive *-break of each orbit $\mathcal{O}$.

**Theorem 1.** *Let $G = (X, U)$ be a weighted graph. $G$ is a $\mathbb{K}$- graph if and only if*

- *$G$ is strongly *-stable,*
- *the graph without orbit of $G$ is $\mathbb{K}$-reducible.*

## 5   Application

All the previous results are needed to build the underlying rational expression (in normal form) from a Glushkov automaton. Main steps are given in this section. Let $\mathcal{M}$ be the Glushkov automaton of Fig. 1.

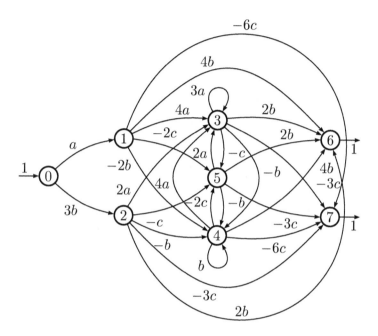

**Fig. 1.** A Glushkov $\mathbb{K}$-automaton

We present all steps for having the rational expression deduced from the automaton of figure 1. First of all, we have to compute the Glushkov $\mathbb{K}$-graph $G$ deduced from $\mathcal{M}$ by suppressing letters and by adding a new state $\Phi$ reached with their output cost from all the final states of $\mathcal{M}$.
Let $L = \{\mathcal{O}\}$ the set of all maximal orbits of $G$.

1- For each maximal orbit $\mathcal{O}$ of $L$, we disconnect this orbit. It is always possible (prop. 7), and there is only one way to do it (prop. 6).
 – We compute the matrix $s_i \times e_j, \forall s_i \in Out(\mathcal{O})$ and $\forall e_j \in In(\mathcal{O})$.
   Let $\mathcal{O} = \{3, 4, 5\}$. $M = \begin{pmatrix} 3 & -1 & -1 \\ 4 & 1 & -2 \\ 2 & -1 & 0 \end{pmatrix}$.
 – For each $e_i \in In(\mathcal{O})$, we compute the input coefficient $k_i$ such that $(c_1, \cdots, c_n)k_i$ is the set of predecessors coefficients and such that $(c_1, \cdots, c_n)$ is constant for each $e_i$.
   Then we have $(c_1, c_2) = (2, 1)$ and $(k) = (2, -1, -1)$ .

- For each $s_j \in Out(\mathcal{O})$, we compute the output coefficient $k'_j$ in the same way.
$(k') = (1, 2, 1)$, and $(c_6, c_7) = (2, -3)$.
- We can now compute the matrix $(k')(k)\alpha$ and apply the $BF_\alpha$ function on $\mathcal{O}$ with $\alpha = 1$.
$$(k')(k)\alpha = \begin{pmatrix} 2 & -1 & -1 \\ 4 & -2 & -2 \\ 2 & -1 & -1 \end{pmatrix}, \text{ and } BF_\alpha(\mathcal{O}) = \begin{pmatrix} 1 & 0 & 0 \\ 0 & 2 & 0 \\ 0 & 0 & 1 \end{pmatrix},$$
- We substract $c_i \times c_j$ to the coefficient of edges from $i \in \mathcal{O}^-$ to $j \in \mathcal{O}^+$. It leads to the graph of figure 2.

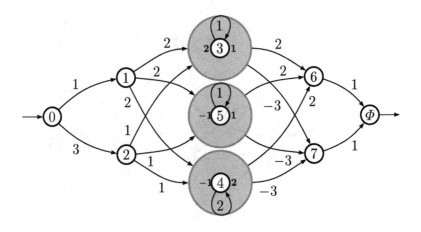

**Fig. 2.** The graph $G$ after the first step of transformation

We iterate the process for $\mathcal{O} = \{1\}$, $\mathcal{O} = \{2\}$ and $\mathcal{O} = \{3\}$.
**2 -** For each orbit $\mathcal{O}$ which is not reduced to a single state, apply the reduction rules $R_1$, $R_2$ and $R_3$.

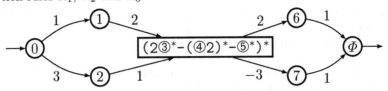

**3 -** Iterate the reduction rules for all the graph where each maximal orbit has been reduced to a single state.

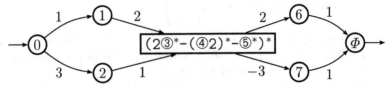

**4 -** Replace each state index by the corresponding letter.

$$(a2 + 3b)(2a^* - (b2)^* - c^*)^*(2b - 3c)$$

# References

1. J. Berstel and C. Reutenauer. *Rational series and their languages*. EATCS Monographs on Theoretical Computer Science. Springer-Verlag, Berlin, 1988.
2. R. Book, S. Even, S. Greibach, and G. Ott. Ambiguity in graphs and expressions. *IEEE Trans. Comput.*, c-20(2):149–153, 1971.
3. A. Brüggemann-Klein. Regular expressions into finite automata. *Theoret. Comput. Sci.*, 120(1):197–213, 1993.
4. A. Buchsbaum, R. Giancarlo, and J. Westbrook. On the determinization of weighted finite automata. *SIAM J. Comput.*, 30(5):1502–1531, 2000.
5. P. Caron and M. Flouret. Glushkov construction for series: the non commutative case. *Internat. J. Comput. Math.* To appear.
6. P. Caron and M. Flouret. Star normal form, rational expressions and glushkov WFAs properties. In *Seventh International Conference on Implementation and Application of Automata, CIAA'02*.
7. P. Caron and D. Ziadi. Characterization of Glushkov automata. *Theoret. Comput. Sci.*, 233(1–2):75–90, 2000.
8. S. Eilenberg. *Automata, languages and machines*, volume A. Academic Press, New York, 1974.
9. K. Culik II and J. Kari. Digital images and formal languages. In G. Rozenberg and A. Salomaa, editors, *Handbook of Formal Languages*, pages 599–616. svb, 1997.
10. S. Lombardy and J. Sakarovitch. Derivatives of regular expression with multiplicity. Technical Report 2001D001, ENST, Paris, 2001.
11. R.F. McNaughton and H. Yamada. Regular expressions and state graphs for automata. *IEEE Transactions on Electronic Computers*, 9:39–57, March 1960.
12. M. Mohri. Finite-state transducers in language and speech processing. *Computat. Ling.*, 23.2:269–311, 1997.
13. F. Pereira and M. Riley. Speech recognition by composition of weighted finite automata. In E. Roche and Y. Schabes, editors, *Finite state language processing*, pages 431–453, Cambridge, Massachusetts, 1997. M.I.T. Press.
14. M.P. Schützenberger. On the definition of a family of automata. *Inform. and Control*, 4:245–270, 1961.

# NFA Reduction Algorithms by Means of Regular Inequalities

Jean-Marc Champarnaud and Fabien Coulon

LIFAR, University of Rouen, France
{jmc,fcoulon}@dir.univ-rouen.fr

**Abstract.** We present different techniques for reducing the number of states and transitions in nondeterministic automata. These techniques are based on the two preorders over the set of states, related to the inclusion of left and right languages. Since their exact computation is $\mathcal{NP}$-hard, we focus on polynomial approximations which enable a reduction of the NFA all the same. Our main algorithm relies on a *first approximation*, which can be easily implemented by means of matrix products with an $\mathcal{O}(sn^4)$ time complexity, and optimized to an $\mathcal{O}(sn^3)$ time complexity, where $s$ is the average nondeterministic arity and $n$ is the number of states. This first algorithm appears to be more efficient than the known techniques based on equivalence relations as described by Lucian Ilie and Sheng Yu. Afterwards, we briefly describe some more accurate approximations and the exact (but exponential) calculation of these preorders by means of determinization.

## 1 Introduction

By NFA reduction algorithms, we mean algorithms which from a given NFA produce a smaller equivalent NFA w.r.t. the number of states. Actually, the main algorithms given in this paper also reduce the number of transitions, so that there is no ambiguity about which complexity measure is considered as being reduced. Among automata which recognize a given regular language $\mathcal{L}$, the problem of computing one or every minimal NFA w.r.t. the number of states has been shown to be $\mathcal{NP}$-hard in [11]. Indeed, known algorithms like in [12,13, 4] are quite not practicable. Our present aim is to provide reduction algorithms which have a polynomial complexity w.r.t. the initial number of states in the given NFA. Such algorithms may be useful, for instance, to prevent or moderate the blow up during the determinization of the NFA, or to speed up either its straightforward simulation, or its simulation based on a partial determinization [14,5].

Except for the method of Brzozowski [3], and some algorithms dedicated to acyclic automata like in Watson and Daciuk [6], the classical reduction algorithms for DFAs, e.g. in [7], are based on an equivalence relation over the set of states $Q$ which converges step by step toward the coarsest equivalence relation such that all states contained in one equivalence class have the same right language.

Z. Ésik and Z. Fülöp (Eds.): DLT 2003, LNCS 2710, pp. 194–205, 2003.

Whereas equivalence relations are fully appropriate to the DFAs, the related relations for NFAs seem to be preorders. Our new reduction methods are based on the two preorders $\overrightarrow{\subseteq}$ and $\overleftarrow{\subseteq}$ defined by $q\overrightarrow{\subseteq}p$ (resp. $q\overleftarrow{\subseteq}p$) if the right (resp. left) language of $q$ is included in the right (resp. left) language of $p$.

Valentin Antimirov has given a containment calculus for regular expressions, in the form of a term-rewriting system [2]. Once transposed into our context, this system gives us an inductive calculus of $\overrightarrow{\subseteq}$ and $\overleftarrow{\subseteq}$. See the definitions of $\overrightarrow{\subseteq}_n$ and $\overleftarrow{\subseteq}_n$ in Sect. 6.

Unfortunately, computing the exact relations $\overrightarrow{\subseteq}$ and $\overleftarrow{\subseteq}$ is known to be $\mathcal{NP}$-Hard, see e.g. [8], and indeed, the inductive construction of $\overleftarrow{\subseteq}_n$ and $\overrightarrow{\subseteq}_n$ in Sect. 6 has an exponential time and space complexity. So, the major part of our paper is devoted to a *first order* approximation of $\overleftarrow{\subseteq}$ and $\overrightarrow{\subseteq}$, which can be computed in time $\mathcal{O}(sn^3)$ where $s$ is the average number of transitions outgoing from one state and labeled by one letter, and $n$ is the number of states. Indeed, this first approximation is at least as efficient as the best results obtained with equivalence relations by Lucian Ilie and Sheng Yu in [9], in the sense that every equality detected by the equivalence relation is detected by our preorders through double inclusion. We provide a simple example where our approximations are strictly more efficient. Moreover, we give a detailed algorithm based on matrix products, which can be easily parallelized in the case of the $\mathcal{O}(sn^4)$ version [10]. Our algorithm can also be easily modified in order to compute the equivalence relation given by Lucian Ilie and Sheng Yu, so that we provide an $\mathcal{O}(sn^3)$ time complexity implementation of their algorithm.

After having presented a general framework for the merging operation over automata in Sect. 2, the Sect. 3 briefly describes the reduction technique based on equivalence relations. Section 4 introduces the preorders related with the inclusion of left and right languages and their application to detecting states which can be merged. Section 5 dwells on the first order approximation which can be computed in time $\mathcal{O}(sn^3)$. Section 6 mentions higher order approximations whose complexity increases as they get more accurate. Section 7 sums up the different reduction techniques deduced from the above-mentioned approximations, and finally, Sect. 8 discusses the problem of computing the exact preorders.

## 2   Definitions and Basic Properties

Let $X$ be a finite set, its cardinal is denoted $|X|$, and we let $\mathcal{P}(X)$ denote the collection of all its subsets.

An automaton is a quintuple $A = <Q, \Sigma, \delta, I, F>$ where $Q$ is a finite set of *states*, $\Sigma$ is the *alphabet*, $\delta : Q \times \Sigma \mapsto 2^Q$ is the *transition function*, $I$ (resp. $F$) is a subset of $Q$ whose elements are the *initial states* (resp. *final states*). The function $\delta$ is extended to $\mathcal{P}(Q) \times \Sigma^* \mapsto 2^Q$ by letting for all $E \subseteq Q$ and $a \in \Sigma$, $\delta(E,a) = \bigcup_{q \in E} \delta(q,a)$ and by the following recursive definition: we let $\delta(E,\varepsilon) = E$ and for all $w \in \Sigma^*$, $\delta(E,aw) = \delta(\delta(E,a),w)$. The language recognized by $A$, denoted $\mathcal{L}(A)$ is the set $\{w \in \Sigma^* \mid \delta(I,w) \cap F \neq \emptyset\}$. Two automata $A$ and $B$ are said to be *equivalent* if $\mathcal{L}(A) = \mathcal{L}(B)$.

We suppose that for all $q \in Q$ and $a \in \Sigma$, we have $\delta(q,a) \neq \emptyset$. This property is often referred to as the *completeness* of the automaton $A$. But in this paper, only complete automata are considered.

An automaton $A$ is *deterministic* if it has a unique initial state and for all $q \in Q$, $a \in \Sigma$, we have $|\delta(q,a)| = 1$. In the following, DFA stands for deterministic finite automaton, and NFA stands for nondeterministic finite automaton.

Let $A = <Q, \Sigma, \delta, I, F>$ be an NFA.

**Definition 1.** *The* left language *of a state* $q$, *denoted* $\overleftarrow{\mathcal{L}}^A(q)$ *is the set* $\{w \in \Sigma^* \mid q \in \delta(I,w)\}$.

*Symmetrically, the* right language *of a state* $q$, *denoted* $\overrightarrow{\mathcal{L}}^A(q)$ *is the set* $\{w \in \Sigma^* \mid \delta(q,w) \cap F \neq \emptyset\}$.

**Definition 2.** *The* reverse automaton *of* $A$, *denoted by* $\overline{A}$, *is the quintuple* $<Q, \Sigma, \overline{\delta}, F, I>$ *where* $q' \in \overline{\delta}(q,a)$ *is equivalent to* $q \in \delta(q',a)$ *for all* $q, q' \in Q$ *and* $a \in \Sigma$.

**Definition 3.** *Let* $q \in Q$. *The* fan out *of* $q$ *is denoted* $\overrightarrow{q}$ *and defined by* $\overrightarrow{q} = \{(a,q') \in \Sigma \times Q \mid q' \in \delta(q,a)\}$. *Symmetrically, the* fan in *of* $q$ *is denoted* $\overleftarrow{q}$ *and defined by* $\overleftarrow{q} = \{(q',a) \in Q \times \Sigma \mid q \in \delta(q',a)\}$.

*The elements of the fan in and fan out of* $q$ *are called* arrows.

Reduction algorithms which are considered in the following are algorithms which proceed by *merging* some states of the original NFA. Let us introduce a general framework for this kind of manipulation.

**Definition 4.** *Let* $q$ *and* $p$ *be two states of* $A$. *We let* $\mathrm{merge}(A,q,p)$ *stand for an automaton obtained from* $A$ *by merging* $q$ *and* $p$, *that is,* $p$ *is deleted and some arrows are added to* $\overleftarrow{q}$ *and* $\overrightarrow{q}$, *so that* $\overleftarrow{\mathcal{L}}^{\mathrm{merge}(A,q,p)}(q) = \overleftarrow{\mathcal{L}}^A(q) \cup \overleftarrow{\mathcal{L}}^A(p)$ *and* $\overrightarrow{\mathcal{L}}^{\mathrm{merge}(A,q,p)}(q) = \overrightarrow{\mathcal{L}}^A(q) \cup \overrightarrow{\mathcal{L}}^A(p)$. *We define a relation* $\sim$ *on* $Q$ *by letting* $q \sim p$ *if and only if* $\mathrm{merge}(A,q,p)$ *is equivalent to* $A$.

Whatever we know about $A$, we can always build $\mathrm{merge}(A,q,p)$ by adding $\overleftarrow{p}$ to $\overleftarrow{q}$ and adding $\overrightarrow{p}$ to $\overrightarrow{q}$.

**Proposition 1.** *Let* $q, p \in Q$, *we have* $q \sim p$ *if and only if* $\overleftarrow{\mathcal{L}}^A(q) \cdot \overrightarrow{\mathcal{L}}^A(p) \subseteq \mathcal{L}(A)$ *and* $\overleftarrow{\mathcal{L}}^A(p) \cdot \overrightarrow{\mathcal{L}}^A(q) \subseteq \mathcal{L}(A)$.

*Proof.* An automaton $B$ recognizes $\mathcal{L}(A)$ if and only if for all $u, v \in \Sigma^*$, we have

$$u \cdot v \in \mathcal{L}(A) \Longleftrightarrow \left[\, (\exists q' \in Q_B)\quad u \in \overleftarrow{\mathcal{L}}^B(q') \wedge v \in \overrightarrow{\mathcal{L}}^B(q')\,\right]$$

Hence the result.                                                                                        ∎

In particular, the relation $\sim$ is well defined: it is independent of the choice of $\mathrm{merge}(A,q,p)$. Unfortunately, it is trivially not an equivalence relation.

## 3   Equalities of Left and Right Languages

A first reduction algorithm is obtained by adapting the Moore's algorithm (which concerns the minimization of DFAs) to the case of NFAs. It has been described by Lucian Ilie and Sheng Yu in [9].

Define the equivalence relation $\equiv$ on $Q$ by letting $q \equiv p$ if and only if $\overrightarrow{\mathcal{L}}(q) = \overrightarrow{\mathcal{L}}(p)$. Indeed, $\equiv$ is the Nerode equivalence. Let $\equiv_0$ be another equivalence relation on $Q$ contained in $\equiv$, that is, $q \equiv_0 p$ implies $q \equiv p$. Then, for all $q, p \in Q$, we have $q \equiv_0 p \Longrightarrow q \sim p$, that is, $q$ and $p$ can be merged as soon as $q \equiv_0 p$.

Such a relation $\equiv_0$ can be computed as the coarsest equivalence relation on $Q$ such that $q \equiv_0 p$ implies the two following conditions:

1. $(q, p) \in F^2 \cup (Q \setminus F)^2$
2. $(\,\forall a \in \Sigma\,)(\,\forall q' \in \delta(q, a)\,)(\,\exists p' \in \delta(p, a)\,)$   $q' \equiv_0 p'$

Let $q, p, q', p' \in Q$, and suppose that $q \equiv_0 p$ and $q' \equiv_0 p'$. Then we still have $q' \equiv_0 p'$ in the automaton $\mathrm{merge}(A, q, p)$, so that the merging operations can be successively applied in any order and lead to a $k$-state automaton where $k$ is the number of equivalence classes for $\equiv_0$.

A dual relation can be obtained the same way by considering left languages instead of right languages.

## 4   Inequalities of Left and Right Languages

We shall see in this section that using preorders instead of equivalence relations gives better results for nondeterministic automata. Let us recall that a relation is a preorder if and only if it is both reflexive and transitive. An antisymmetric preorder is a partial order, and a symmetric preorder is an equivalence relation.

**Definition 5.** *Let $\overrightarrow{\sqsubseteq}$ and $\overleftarrow{\sqsubseteq}$ be two preorders on $Q$ defined by letting $q\overrightarrow{\sqsubseteq}p$ if and only if $\overrightarrow{\mathcal{L}}(q) \subseteq \overrightarrow{\mathcal{L}}(p)$, and $q\overleftarrow{\sqsubseteq}p$ if and only if $\overleftarrow{\mathcal{L}}(q) \subseteq \overleftarrow{\mathcal{L}}(p)$.*

**Definition 6.** *Let $R$ and $R'$ be two relations on $Q$. The relation $R$ is said to be smaller than $R'$ if for all $q, p \in Q$, $qRp$ implies $qR'p$.*

This partial order coincides with the inclusion if the relations are considered as subsets of $Q \times Q$.

**Proposition 2.** *Let $q$ and $p$ be two states of $A$. Let $\overrightarrow{\sqsubseteq}_0$ and $\overleftarrow{\sqsubseteq}_0$ be two relations on $Q$ respectively smaller than $\overrightarrow{\sqsubseteq}$ and $\overleftarrow{\sqsubseteq}$. We have $q \sim p$ as soon as one of the following conditions holds:*

1. *$q\overrightarrow{\sqsubseteq}_0p$ and $p\overrightarrow{\sqsubseteq}_0q$,*
2. *$q\overleftarrow{\sqsubseteq}_0p$ and $p\overleftarrow{\sqsubseteq}_0q$,*
3. *$q\overrightarrow{\sqsubseteq}_0p$ and $q\overleftarrow{\sqsubseteq}_0p$,*
4. *$p\overrightarrow{\sqsubseteq}_0q$ and $p\overleftarrow{\sqsubseteq}_0q$.*

*Proof.* Trivial from Proposition 1.    ∎

Such relations $\overset{\rightarrow}{\subseteq}_0$ and $\overset{\leftarrow}{\subseteq}_0$ are said to be *approximations* of the inequality relations $\overset{\rightarrow}{\subseteq}$ and $\overset{\leftarrow}{\subseteq}$.

When merging two states $q$ and $p$ which satisfy one of the conditions of Proposition 2, the relations $\overset{\rightarrow}{\subseteq}_0$ and $\overset{\leftarrow}{\subseteq}_0$ should be updated in order to remain smaller than $\overset{\rightarrow}{\subseteq}$ and $\overset{\leftarrow}{\subseteq}$ in the automaton obtained by merging. Let us enumerate the conditions of Proposition 2 and give the associated reduction operations:

1. In merge$(A, q, p)$, add $\overset{\leftarrow}{p}$ to $\overset{\leftarrow}{q}$, and let $\overset{\leftarrow}{\subseteq}_0 \leftarrow \overset{\leftarrow}{\subseteq}_0 \backslash \{(q, p') \mid p' \in Q,\ p \overset{\leftarrow}{\not\subseteq}_0 p'\}$,
2. In merge$(A, q, p)$, add $\overset{\rightarrow}{p}$ to $\overset{\rightarrow}{q}$, and let $\overset{\rightarrow}{\subseteq}_0 \leftarrow \overset{\rightarrow}{\subseteq}_0 \backslash \{(q, p') \mid p' \in Q,\ p \overset{\rightarrow}{\not\subseteq}_0 p'\}$,
3. Compute merge$(A, p, q)$,
4. Compute merge$(A, q, p)$.

## 5    First Order Appoximation of Inequalities

Let $\overset{\rightarrow}{\subseteq}_1$ be the greatest preorder on $Q$ which satisfies the two following axioms:

1. For all $q \in F$ and $p \in Q \backslash F$, we have $q \overset{\rightarrow}{\not\subseteq}_1 p$,
2. Let $q, p \in Q$ and $a \in \Sigma$,

$$q \overset{\rightarrow}{\subseteq}_1 p \Longrightarrow \left[ (\ \forall q' \in \delta(q, a)\ )(\ \exists p' \in \delta(p, a)\ )\quad q' \overset{\rightarrow}{\subseteq}_1 p' \right]$$

**Proposition 3.** *The relation* $\overset{\rightarrow}{\subseteq}_1$ *is smaller than* $\overset{\rightarrow}{\subseteq}$.

*Proof.* We shall prove by induction on the length of the word $w$ that for all $q, p \in Q,\ [q \overset{\rightarrow}{\subseteq}_1 p \wedge w \in \overset{\rightarrow}{\mathcal{L}}(q)] \Longrightarrow w \in \overset{\rightarrow}{\mathcal{L}}(p)$.

Let $q \overset{\rightarrow}{\subseteq}_1 p$ and $\varepsilon \in \overset{\rightarrow}{\mathcal{L}}(q)$. Since $q \in F$, the first axiom of $\overset{\rightarrow}{\subseteq}_1$ implies that $p \in F$ and $\varepsilon \in \overset{\rightarrow}{\mathcal{L}}(p)$.

Now, let $q \overset{\rightarrow}{\subseteq}_1 p$ and $aw \in \overset{\rightarrow}{\mathcal{L}}(q)$, where $a \in \Sigma$, $w \in \Sigma^*$ and $q, p \in Q$. There exists $q' \in \delta(q, a)$ such that $w \in \overset{\rightarrow}{\mathcal{L}}(q')$. The second axiom implies that there exists $p' \in \delta(p, a)$ such that $q' \overset{\rightarrow}{\subseteq}_1 p'$. By induction, we have $w \in \overset{\rightarrow}{\mathcal{L}}(p')$, hence $aw \in \overset{\rightarrow}{\mathcal{L}}(p)$.    ∎

Symmetrically, we have a preorder $\overset{\leftarrow}{\subseteq}_1$, smaller than $\overset{\leftarrow}{\subseteq}$, obtained by considering the reverse automaton.

Indeed, reducing $A$ by simply using the first order approximations $\overset{\rightarrow}{\subseteq}_1$ and $\overset{\leftarrow}{\subseteq}_1$ with the conditions 1 and 2 of Proposition 2 is at least as efficient as reducing $A$ by using the relation $\equiv_0$ of Sect. 3:

**Proposition 4.** *Let* $q, p \in Q$ *such that* $q \equiv_0 p$. *We have* $q \overset{\rightarrow}{\subseteq}_1 p$ *and* $p \overset{\rightarrow}{\subseteq}_1 q$.

*Proof.* Actually, the constraint "$R$ is a preorder" is weaker than "$R$ is an equivalence relation", and the axioms of $\overset{\rightarrow}{\subseteq}_1$ are weaker than the axioms of $\equiv_0$.    ∎

We can see on Fig. 1 an example where $\overrightarrow{\subseteq}_1$ and $\overleftarrow{\subseteq}_1$ are strictly more efficient than $\equiv_0$. We easily verify that $1 \overrightarrow{\subseteq}_1 2$ and $2 \overleftarrow{\subseteq}_1 1$, though we do not have $1 \equiv_0 2$. The sink state 7 is not drawn on the figure.

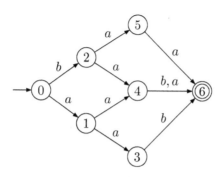

**Fig. 1.** States 1 and 2 have the same right language

## 5.1 Implementation of the First Order Approximation

Consider the semigroup $\mathcal{M}_2(n)$ of $n \times n$-matrices over the boolean semiring. Consider that the set $Q$ is equal to $\{1, 2, \dots, n\}$, so that each function from $Q$ to $Q$, as well as each binary relation $R$ on $Q$, is naturally associated with a matrix $X$ in $\mathcal{M}_2(n)$ by letting $X_{i,j} = 1 \iff jRi$, which is the same as $X_{i,j} = 1 \iff i \in R(j)$.

Let $X_0$ be the matrix associated with the relation $F \times (Q \setminus F)$. For all $a \in \Sigma$, let $\delta_a$ be the matrix associated with the application $q \mapsto \delta(q, a)$ and let ${}^t\delta_a$ be the transposition of $\delta_a$ in $\mathcal{M}_2(n)$. Indeed, ${}^t\delta_a$ is associated with $q \mapsto \overline{\delta}(q, a)$.

Let $\mathcal{M}_\mathbb{N}(n)$ be the semiring of $n \times n$-matrices over the semiring of natural integers. Let '.' be the matrix product in $\mathcal{M}_2(n)$ and 'o' be the matrix product in $\mathcal{M}_\mathbb{N}(n)$. Let $X$ be a matrix in $\mathcal{M}_\mathbb{N}(n)$ and $V$ be a vector of $\mathbb{N}^n$, we denote by $\overline{X}^V$ the matrix of $\mathcal{M}_2(n)$ defined by

$$\overline{X}^V_{i,j} = 1 \text{ if } X_{i,j} = V_j,$$
$$= 0 \text{ otherwise}$$

For all $a \in \Sigma$, let $V(a) \in \mathbb{N}^n$ be defined by

$$V(a)_q = |\delta(q, a)| \quad (\forall q \in Q)$$

Let $\Delta_a$, be defined for all $X \in \mathcal{M}_2(n)$ by $\Delta_a(X) = {}^t\delta_a.\overline{(X \circ \delta_a)}^{V(a)} + X$.

**Proposition 5.** *Let $X$ be the transposition of the matrix associated with the relation $\overrightarrow{\subseteq}_1$.*

*$X$ is the smallest matrix which is a fix point for all $\Delta_a$ ( $a \in \Sigma$ ) and is greater than ${}^tX_0$.*

*Proof.* Let $R$ be a relation on $Q$. For all $a \in \Sigma$, we let

$$A_a(R) \equiv \left[ (\, \forall q, p \in Q \,) \left[ (\, \exists q' \in \delta(q, a) \,)(\, \forall p' \in \delta(p, a) \,) \ q' R p' \right] \implies qRp \right]$$

Let $X$ be the transposition of the matrix associated with $R$. For all $q, p \in Q$, $(X \circ \delta_a)_{q',p}$ is the number of states $p'$ such that $p' \in \delta(p, a)$ and $q'Rp'$. If this number is equal to $|\delta(p, a)|$, that is, if $\left(\overline{X \circ \delta_a}^{V(a)}\right)_{q',p}$ is equal to 1, then we have $q'Rp'$ for all $p' \in \delta(p, a)$.

Hence, for all $q, p \in Q$, $\left({}^t\delta_a . \overline{X \circ \delta_a}^{V(a)}\right)_{q,p} = 1$ is equivalent to $(\, \exists q' \in \delta(q, a) \,)(\, \forall p' \in \delta(p, a) \,) \ q'Rp'$. So, $X$ is a fix point for $\Delta_a$ if and only if $A_a(R)$.

Now, let $X$ be the transposition of the matrix associated with $\overrightarrow{\mathbb{Z}}_1$. We know from the axioms of $\overrightarrow{\mathbb{C}}_1$ that $\overrightarrow{\mathbb{Z}}_1$ is the smallest relation on $Q$ which contains $F \times (Q \setminus F)$ and such that $A_a(\overrightarrow{\mathbb{Z}}_1)$ for all $a \in \Sigma$. Hence, $X$ is the smallest matrix which is a fix point for all $\Delta_a$ and contains ${}^tX_0$. ∎

Figure ?? illustrates the Proof. Let $q'Rp'$ for all states $p'$ in the figure. Then $(q, p)$ is added to $R$ because $(\exists q' \in \delta(q, a))(\forall p' \in \delta(p, a)) \ q'Rp'$.

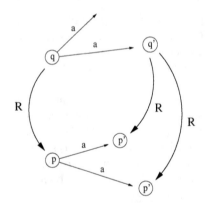

The following algorithm is then straightforward:

1. Let $X \leftarrow {}^tX_0$
2. Do
3.     For all $a \in \Sigma$,
4.         let $X \leftarrow \Delta_a(X)$
5. Until $X$ is a fix point for all $\Delta_a$ ($a \in \Sigma$).
${}^tX$ is now associated with $\overrightarrow{\mathbb{Z}}_1$.

Let us digress and get back on the first relation $\equiv_0$. Indeed, $\equiv_0$ can be computed by a slightly different algorithm:

- Line 1, replace $X \leftarrow {}^t X_0$ by $X \leftarrow Sym({}^t X_0)$ where $Sym$ stands for the symmetric closure,
- Line 4, replace $X \leftarrow \Delta_a(X)$ by $X \leftarrow Sym(\Delta_a(X))$,
- the algorithm ends with $X$ associated with $\neq_0$.

## 5.2  Complexity and Optimizations

Let $s$ be the average nondeterministic arity of $A$, that is,

$$s = \frac{1}{n|\Sigma|} \sum_{q \in Q, a \in \Sigma} |\delta(q, a)|$$

Hence, the average number of non-zero coefficients in the matrices $\delta_a$ and ${}^t\delta_a$ is $sn$.

A typical non optimized implementation of our algorithm leads to an $\mathcal{O}(sn^4)$ time complexity, either for computing $\overrightarrow{\subseteq}_1$ and $\overleftarrow{\subseteq}_1$ or $\equiv_0$. Indeed, the number of iterations is bounded by $n^2$. We prove in this section that the complexity can be reduced to $\mathcal{O}(sn^3)$. But indeed, a non optimized implementation of our algorithm might be rather as fast as the following optimized version, and might be more easily parallelized using standard parallelization techniques for matrix product [10].

Let us detail an optimized version of line 4. For all $a \in \Sigma$, let $M_a$ and $N_a$ be two $(n \times n)$-matrices which are initially empty.

4.1. **If $N_a$ is empty, then let $M_a \leftarrow \overline{X \circ \delta_a}^{V(a)}$ and $N_a \leftarrow \Delta_a(X)$**
      **Else**
4.2.      Let $L$ be the set of indices of rows which are
          different between $X$ and $N_a$.
4.3.      For all $l \in L$, compute the $l^{th}$ row of $\overline{X \circ \delta_a}^{V(a)}$
          and add it to the $l^{th}$ row of $M_a$.
4.4.      For all $l \in L$ and $i, j \in \{1, \ldots, n\}$ with $({}^t\delta_a)_{i,l} \neq 0$,
          add $({}^t\delta_a)_{i,l} \cdot (M_a)_{l,j}$ to $(N_a)_{i,j}$.
4.5.      For all $l \in L$, add the $l^{th}$ row of $X$ to
          the $l^{th}$ row of $N_a$.
      **Fi**
4.6. Let $X \leftarrow N_a$.

Notice that $X \leftarrow N_a$ does not require an $n^2$ time copy, since $X$ can be a reference, and notice that the calculation of $L$ can be done by updating the list of rows which have been modified during the $|\Sigma|$ last calls to Procedure 4.

We suppose that the matrix $\delta_a$ is stored as the list, row by row, of its non-empty coefficients, so that the matrix ${}^t\delta_a$ is stored as the list, column by column, of its non-empty coefficients. Consequently, Lines 4.3. and 4.4. have a $ksn$ time complexity where $k$ is the number of non-zero coefficients added to $X$ during the last iteration, or $k = n$ if this is the first iteration. The time complexity of one iteration of the algorithm is then $min(snk, sn^2)$. The maximal number of non-zero coefficients added to $X$ is $n^2$, hence we get an $\mathcal{O}(sn^3)$ time-complexity for this algorithm.

We can deduce a slightly different optimization for computing $\equiv_0$, which provides an $\mathcal{O}(sn^3)$ time complexity algorithm for the reduction method given by Lucian Ilie and Sheng Yu in [9].

## 6   Higher Order Approximations

For all finite sets $X$ and all $k \in \mathbb{N}$, we let $\mathcal{P}_k(X)$ stand for $\{P \subseteq X \mid |P| \leq k\}$.

For all $k \geq 1$, let $\overrightarrow{\subseteq}_k$ be the relation between $Q$ and $\mathcal{P}_k(Q)$ that is the greatest relation which satisfies the two following axioms:

1. $q \overrightarrow{\not\subseteq}_k P$ for all $q \in F$ and $P \in \mathcal{P}_k(Q \setminus F)$,
2. Let $q \in Q$, $P \in \mathcal{P}_k(Q)$ and $a \in \Sigma$,

$$q \overrightarrow{\subseteq}_k P \Longrightarrow \left[ (\; \forall q' \in \delta(q,a)\; )(\; \exists P' \in \mathcal{P}_k(\delta(P,a))\; )\quad q' \overrightarrow{\subseteq}_k P' \right]$$

Let $q, p \in Q$, we note $q \overrightarrow{\subseteq}_k p$ instead of $q \overrightarrow{\subseteq}_k \{p\}$, and we shall consider the trace of $\overrightarrow{\subseteq}_k$ on $Q^2$ as the relation $\{(q,p) \in Q^2 \mid q \overrightarrow{\subseteq}_k p\}$. For the case $k = 1$, we recover our first relation $\overrightarrow{\subseteq}_1$.

**Proposition 6.** *The trace of $\overrightarrow{\subseteq}_k$ on $Q^2$ is smaller than $\overrightarrow{\subseteq}$ for all $k \geq 1$.*

*Proof.* We shall prove by induction on the length of the word $w$ that for all $q \in Q$ and $P \in \mathcal{P}_k(Q)$, having $q \overrightarrow{\subseteq}_k P$ and $w \in \overrightarrow{\mathcal{L}}(q)$ implies $w \in \overrightarrow{\mathcal{L}}(p)$ for some $p \in P$.

Let $\varepsilon \in \overrightarrow{\mathcal{L}}(q)$. Since $q \in F$, the first axiom of $\overrightarrow{\subseteq}_k$ implies that there exists $p \in P \cap F$, hence $\varepsilon \in \overrightarrow{\mathcal{L}}(p)$.

Now, let $aw \in \overrightarrow{\mathcal{L}}(q)$. There exists $q' \in \delta(q,a)$ such that $w \in \overrightarrow{\mathcal{L}}(q')$. The second axiom implies that there exists $P' \subseteq \delta(P,a)$, with $|P'| \leq k$, such that $q' \overrightarrow{\subseteq}_k P'$. By induction, we have $w \in \overrightarrow{\mathcal{L}}(p')$ for some $p' \in P'$, hence, let $p$ in $P$ such that $p' \in \delta(p,a)$, we have $aw \in \overrightarrow{\mathcal{L}}(p)$. ∎

The trace of $\overrightarrow{\subseteq}_k$ on $Q^2$ is increasing w.r.t. $k$ and reaches $\overrightarrow{\subseteq}$. Indeed, the trace of $\overrightarrow{\subseteq}_n$ is equal to $\overrightarrow{\subseteq}$.

## 7   Reduction Methods

Let $\overleftarrow{\subseteq}_0$ and $\overrightarrow{\subseteq}_0$ be respectively two approximations of $\overleftarrow{\subseteq}$ and $\overrightarrow{\subseteq}$. One can simply apply Proposition 2. This method reduces the number of states as well as the number of transitions.

If one aims to reduce the number of states, without considering the number of transitions, the relations $\overleftarrow{\subseteq}_0$ and $\overrightarrow{\subseteq}_0$ can be grown first in the following way. Let $q$ and $p$ such that $q \overrightarrow{\subseteq}_0 p$, the fan in $\overleftarrow{p}$ can be added to the fan in $\overleftarrow{q}$. Then

$(p, q)$ can be added to $\overleftarrow{\sqsubseteq}_0$. Symmetrically, if $q \overleftarrow{\sqsubseteq} p$, then $\overrightarrow{p}$ can be added to $\overrightarrow{q}$, and then $(p, q)$ can be added to $\overrightarrow{\sqsubseteq}_0$.

If we have computed the approximations $\overleftarrow{\sqsubseteq}_k$ and $\overrightarrow{\sqsubseteq}_k$ for some $k \geq 2$, then we can consider merging one state $q$ with a set of states $P$:

**Definition 7.** *Let $q \in Q$ and $P \subseteq Q$, we define the automaton* $\mathrm{merge}(A, q, P)$ *obtained from $A$ by deleting $q$ and adding $\overrightarrow{q}$ to each $\overrightarrow{p}$ ($p \in Q$), and adding $\overleftarrow{q}$ to each $\overleftarrow{p}$ ($p \in P$).*

**Proposition 7.** *Let $q \in Q$ and $P \subseteq Q$. The automaton* $\mathrm{merge}(A, q, P)$ *is equivalent to $A$ as soon as one of the following conditions is verified:*

1. *$q \overrightarrow{\sqsubseteq}_k P$ and for all $p \in P$, $p \overrightarrow{\sqsubseteq}_k q$,*
2. *$q \overleftarrow{\sqsubseteq}_k P$ and for all $p \in P$, $p \overleftarrow{\sqsubseteq}_k q$.*

This kind of reduction was pointed out by Amilhastre, Janssen and Vilarem in [1] for the case of homogeneous languages, that is, languages which are contained in $\Sigma^h$ for some $h \in \mathbb{N}$. Let $k = n$ in Proposition 7 and consider that $\mathcal{L}(A)$ is an homogeneous language, a pair $(q, P)$ which satisfies condition 1 is called a *union reduction* by Amilhastre *et al.*

## 8   Computing the Exact Inequality Relations

Let $k \leq n$, computing the relations $\overrightarrow{\sqsubseteq}_k$ and $\overleftarrow{\sqsubseteq}_k$ has a polynomial complexity w.r.t. $n$, but the complexity grows exponentially w.r.t. $k$, which is unavoidable since computing $\overrightarrow{\sqsubseteq}$ and $\overleftarrow{\sqsubseteq}$ is known to be $\mathcal{N}P$-hard [8]. Nevertheless, $\overrightarrow{\sqsubseteq}$ and $\overleftarrow{\sqsubseteq}$ may be computed if we are able to determinize $A$ and $\overline{A}$ with a reasonable complexity. Of course, proceeding this way does not make sense if our aim is to prevent a combinatorial blow up during the determinization by first reducing $A$.

This method is a straightforward application of the notion of characteristic events, which can be found in [12] and is reformulated in a way closer to our context in [4].

Let $\mathcal{A} =< Q_\mathcal{A}, \Sigma, \delta_\mathcal{A}, I_\mathcal{A}, F_\mathcal{A} >$ be the minimal DFA of $\mathcal{L}$ obtained by subset determinization from $A$, so that states of $\mathcal{A}$ are subsets of $Q$. Let $\mathcal{B} =< Q_\mathcal{B}, \Sigma, \delta_\mathcal{B}, I_\mathcal{B}, F_\mathcal{B} >$ be the minimal DFA of $\overline{\mathcal{L}}$ obtained by subset determinization from $\overline{A}$, so that states of $\mathcal{B}$ are also subsets of $Q$.

For all $q \in Q$, let $\lambda(q) = \{P \in Q_\mathcal{A} \mid q \in P\}$ and $\rho(q) = \{P \in Q_\mathcal{B} \mid q \in P\}$.

**Proposition 8.** *Let $q, p \in Q$, we have $q \overrightarrow{\sqsubseteq} p$ if and only if $\rho(q) \subseteq \rho(p)$ and $q \overleftarrow{\sqsubseteq} p$ if and only if $\lambda(q) \subseteq \lambda(p)$.*

This property leads to an $\mathcal{O}(n^2 N)$-time complexity algorithm for computing $\overleftarrow{\sqsubseteq}$ and $\overrightarrow{\sqsubseteq}$, where $N = \max(|Q_\mathcal{A}|, |Q_\mathcal{B}|)$.

# 9   Conclusion

Our opinion is that the most interesting algorithm is the one based on the first order approximation, which has a reasonable worst case complexity, and should prove to be very fast in practice. To our knowledge, this algorithm is the most efficient one in this level of performance. Higher order approximations may be useful if one really cares about the succinctness of the NFA.

We tried to be the most exhaustive as we could in the domain of reduction by means of the preorders related to regular inequalities. But this does not exhaust the realm of reduction heuristics. In particular, we restrained us to algorithms which only proceed by merging states. Studies about the full minimization of NFAs [12,13,4] revealed that in order to reduce an NFA, we also have to split some states, in order to merge them differently. Heuristics in this domain may overflow the imagination, but we can hardly expect them to produce a fast algorithm.

# References

1. J. Amilhastre, P. Janssen, and M.C. Vilarem. FA minimization heuristics for a class of finite languages. In O. Boldt and H. Jürgensen, editors, *Automata Implementation, WIA'99, Lecture Notes in Computer Science*, volume 2214, pages 1–12. Springer, 2001.
2. V. Antimirov. Rewriting regular inequalities. *Lecture notes in computer science*, 965:116–125, 1995.
3. J.A. Brzozowski. Canonical regular expressions and minimal state graphs for definite events. *Mathematical Theory of Automata, MRI Symposia Series*, 12:529–561, 1962.
4. J.-M. Champarnaud and F. Coulon. Theoretical study and implementation of the canonical automaton. Technical Report AIA 2003.03, LIFAR, Université de Rouen, 2003.
5. J.-M. Champarnaud, F. Coulon, and T. Paranthoën. Compact and fast algorithms for regular expression search. Technical Report AIA 2003.01, LIFAR, Université de Rouen, 2003.
6. J. Daciuk, B.-W. Watson, and R.-E. Watson. Incremental construction of minimal acyclic finite state automata and transducers. In L. Karttunen, editor, *FSMNLP'98*, pages 48–55. Association for Computational Linguistics, Somerset, New Jersey, 1998.
7. John E. Hopcroft. An $n \log n$ algorithm for minimizing the states in a finite automaton. In Z. Kohavi, editor, *The Theory of Machines and Computations*, pages 189–196. Academic Press, 1971.
8. H. Hunt, D. Rosenkrantz, and T. Szymanski. On the equivalence, containment and covering problems for the regular and context-free languages. *J. Comput. System Sci.*, 12:222–268, 1976.
9. L. Ilie and S. Yu. Algorithms for computing small NFAs. In K. Diks and W. Rytter, editors, *Lecture Notes in Computer Science*, volume 2420, pages 328–340. Springer, 2002.
10. J. Jaja. *An introduction to parallel algorithms*. Addison-Wesley, 1992.

11. T. Jiang and B. Ravikumar. Minimal NFA problems are hard. *SIAM J. Comput. Vol 22, No 6*, pages 1117–1141, 1993.
12. T. Kameda and P. Weiner. On the state minimization of nondeterministic finite automata. *IEEE Trans. Comp.*, C(19):617–627, 1970.
13. H. Sengoku. Minimization of nondeterministic finite automata. Master's thesis, Kyoto University, 1992.
14. S. Wu and U. Manber. Fast text searching algorithm allowing errors. In *Communication of the ACM*, 31, pages 83–91, October 1992.

# Tile Rewriting Grammars*

Stefano Crespi Reghizzi** and Matteo Pradella

DEI – Politecnico di Milano and
CNR IEIIT-MI
Piazza Leonardo da Vinci, 32
I-20133 Milano, Italy
{crespi,pradella}@elet.polimi.it

**Abstract.** Past proposals for applying to pictures or 2D languages the generative grammar approach do not match in our opinion the elegance and descriptive adequacy that made Context Free grammars so successful for 1D languages. In a renewed attempt, a model named Tile Rewriting Grammar is introduced combining the rewriting rules with the Tiling System of Giammaresi and Restivo which define the family of Recognizable 2D languages. The new grammars have isometric rewriting rules which for string languages are equivalent to CF rules. TRG have the capacity to generate a sort of 2D analogues of Dyck languages. Closure properties of TRG are proved for some basic operations. TRG strictly include TS as well as the context-free picture grammars of Matz.

## 1 Introduction

In the past several proposals have been made for applying to pictures or 2D languages the generative grammar approach but in our opinion none of them matches the elegance and descriptive adequacy that made Context Free (CF) grammars so successful for 1D i.e. string languages. A picture is a rectangular array of symbols (the pixels) from a finite alphabet.

A survey of formal definition models for picture languages is [3] where different approaches are compared and related: tiling systems, cellular automata, and grammars. The latter had been surveyed in more detail by [6]. Classical 2D grammars can be grouped in two categories[1] called *matrix* and *array* grammars respectively.

The matrix grammars, introduced by A. Rosenfeld, impose the constraint that the left and right parts of a rewriting rule must be isometric arrays; this condition overcomes the inherent problem of "shearing" which pops up while substituting a subarray in a host array.

---

* Work partially supported by MIUR, Progetto *Linguaggi formali e automi, teoria e applicazioni.*
** Lecturer of Formal Languages, Università della Svizzera Italiana.
[1] Leaving aside the graph grammar models because they generate graphs, not 2D matrices.

Z. Ésik and Z. Fülöp (Eds.): DLT 2003, LNCS 2710, pp. 206–217, 2003.
© Springer-Verlag Berlin Heidelberg 2003

Siromoney's array grammars are parallel-sequential in nature, in the sense that first a horizontal string of nonterminals is derived sequentially, using the horizontal productions; and then the vertical derivations proceed in parallel, applying a set of vertical productions. Several variations have been made, for instance [1]. A particular case are the two dimensional right-linear grammars studied in [3].

More recently Matz [5] has proposed a new approach relying on the notion of row and column concatenation and their closures. A production of his *context-free picture grammars* is similar to a classical 1D Context Free (CF) production, with the difference that the right part is a 2D regular expression. The shearing difficulty does not arise because, say, row concatenation is a partial operation which is only defined on matrices of identical width.

Exploring a different course, our new model is introduced, named *Tile Rewriting Grammar* (TRG), intuitively combining Rosenfeld's isometric rewriting rules with the Tiling System (TS) of Giammaresi and Restivo [2]. The latter defines the family of Recognizable 2D languages (the same as those accepted by so called *on-line tessellation automata* of Inoue and Nakamura [4]).

A TRG production is a schema having to the left a nonterminal symbol and to the right a local 2D language over terminals and nonterminals; that is the right part is specified by a set of fixed size tiles.

Similar to Rosenfeld's matrix grammars, the shearing problem is solved by a isometric constraint on the left and right parts, but a TRG production is not compelled to have a fixed rectangular size. The left part denotes a rectangle of any size filled with the same nonterminal. Whatever size the left part takes, the same size is assigned to the right part. Viewing rewriting from right to left, the effect is to replace each pixel of the right picture by the same nonterminal. To make this simple idea effective, it is necessary to impose a tree partial order on the areas which are rewritten, thus moving from the syntax tree of string languages to some sort of well nested prisms. A simple nonterminal marking device implements the partial ordering.

To our knowledge this approach is novel and is able to generate an interesting gamut of pictures, grids, spirals, and in particular a language of nested frames which is in some way the analogue of a Dyck language.

Some formal properties of the new model are here proved. Derivations can be computed in a canonical order. The family is closed with respect to some basic operations: row/column concatenation and closures, union, and rotation.

Comparing with other families, TRGs' generative capacity is stronger than that of Tiling Systems, as shown by the non-TS language made by the vertical concatenation of two specularly symmetrical squares. TRG grammars are proved more powerful than Matz's CF picture grammars.

In Sect. 2 we gradually and informally introduce the isometric productions and marking device for string languages. Then we list the basic definitions for picture languages, and we define our model and show the closure properties. In Sect. 3 the family of TRG pictures is compared with the other families.

## 2    Tile Rewriting Grammars

### 2.1    Introducing One Dimensional Isometric Grammars

For a gradual presentation we first sketch the TRG model for string grammars, showing it is tantamount to a notational variant of classical CF grammars. The essential features are that rewriting rules are isometric and of unbounded length. These properties set the ground for exploiting the model in two dimensions in the next section: because isometric rules matching arrays of unbounded size are exempt from the problems (shearing and narrow scope) encountered by early attempts at using phrase structure grammars for pictures.

Consider a CF grammar $G = (\Sigma, N, S, R)$ in Chomsky normal form with rules of the forms $A \to b$ and $A \to BC$, where $b$ is a terminal and $A, B, C$ are nonterminals.

For a derivation

$$S \overset{*}{\Rightarrow}_G uAv \overset{*}{\Rightarrow}_G uzv \in \Sigma^+, A \in N$$

we associate to each nonterminal an attribute, the length of the terminal string generated, by writing

$$(S, n) \overset{*}{\Rightarrow}_G u(A, j)v \overset{*}{\Rightarrow}_G uzv \in \Sigma^+$$

where $n = |uzv|$ and $j = |z|$.

An isometric grammar $G_I = (\Sigma, N, S, R_I)$ consists of a countable set of rules corresponding to the rules of $G$ as specified in the table:

| CF rule | Isometric rule |
|---|---|
| $A \to b$ | $\underline{A} \to b$ |
| $A \to BC$ | $\underline{AA} \to \underline{B}\ \underline{C}$ |
| | $\underline{AAA} \to \underline{BB}\ \underline{C}$ |
| | $\underline{AAA} \to \underline{B}\ \underline{CC}$ |
| | $\cdots$ |
| | $\underline{A^n} \to \underline{B^p}\ \underline{C^q}, \forall n, p, q : n \geq 2, n = p + q$ |

An underbraced string must be rewritten all at once, so that the derivations of a isometric grammar are in step by step correspondence with CF derivations. More precisely, it is clear that the CF derivation

$$(S, m) \overset{*}{\Rightarrow}_G u(A, n)v \Rightarrow_G u(B, p)(C, q)v \overset{*}{\Rightarrow}_G uzv \in \Sigma^m, |z| = n = p + q$$

exists iff the isometric derivation exists

$$\underline{S^m} \overset{*}{\Rightarrow}_{G_I} u\,\underline{A^n}\,v \Rightarrow_{G_I} u\,\underline{B^p}\,\underline{C^q}\,v \overset{*}{\Rightarrow}_{G_I} uzv$$

Therefore the isometric grammar is just a notational variant of the CF grammar and defines the same language.

The semigraphical representation using underbraces has to be abandoned for a symbolic notation, that will be more convenient for the 2D case. Symbols in the right part of a rule are marked by natural indices, qualified as *update* indices and used to denote underbraces, so that the CF rule $A \to BC$ produces the set of isometric rules

$$A^n \to B_1{}^p C_2{}^q, \ \forall n, p, q : n \geq 2, n = p + q \tag{1}$$

All nonterminals which were grouped by an underbrace are now marked with the same distinct index, called *area* index. An area index updating rule is added to a derivation step.

Initially the axiom has area index 0, and the derivation starts from $(S_0)^m$. Assume a string $\gamma = \alpha(A_i)^n \beta$ deriving from $(S_0)^m$ is such that the nonterminals that would be underbraced carry the same index $i$, which differs from all other indices. Moreover neither $\alpha$ nor $\beta$ contain $A_i$, that is $(A_i)^n$ is the largest substring containing $A_i$. Let $\mu(\gamma)$ be the maximum area index present in $\gamma$. Then when a derivation step is performed, by applying the rewriting rule (1), the indices of the right hand side symbols are incremented by $\mu(\gamma)$:

$$\alpha(A_i)^n \beta \Rightarrow \alpha(B_r)^p (C_s)^q \beta$$

$$n = p + q, \ r = 1 + \mu(\gamma), \ s = 2 + \mu(\gamma)$$

As a consequence, the area indices of the resulting string preserve the divisions.

*Example 1.* Consider the following CF grammar: $S \to SS; S \to AB; A \to a; B \to b$. We can write it as follows:

$$S^n \to (S_1)^p (S_2)^q; \ S^n \to (A_1)^p (B_2)^q, \ \forall n, p, q : n = p + q \geq 2$$
$$A \to a_0; \ B \to b_0$$

Notice that we use update index 0 for terminals.
The CF derivation (with sizes):

$$(S, 4) \Rightarrow (S, 2)(S, 2) \Rightarrow (A, 1)(B, 1)(S, 2) \overset{2}{\Rightarrow} abS \Rightarrow ab(A, 1)(B, 1) \overset{2}{\Rightarrow} abab.$$

corresponds to the derivation with indexed symbols:

$$(S_0)^4 \Rightarrow (S_1)^2 (S_2)^2 \Rightarrow A_3 B_4 (S_2)^2 \overset{2}{\Rightarrow} a_3 b_4 (S_2)^2 \Rightarrow a_3 b_4 A_5 B_6 \overset{2}{\Rightarrow} a_3 b_4 a_5 b_6.$$

Since the right hand side of (1) is a locally testable language over $\{B_1, C_1\}$, the set of rules will be defined in closed form by listing the substrings of length two (or tiles) which may occur:

$$A \to \{B_1 C_2\}; \ A \to \{B_1 B_1, B_1 C_2\}; \ A \to \{B_1 C_2, C_2 C_2\};$$
$$A \to \{B_1 B_1, B_1 C_2, C_2 C_2\}$$

Notice that by the isometric hypothesis the left hand side is assumed to match the length of the right hand side, and its length is not specified.

Although this technique is clumsy for string languages, it permits to do without 2D bracketing in picture languages.

## 2.2   Basic Definitions

We list the preliminary notation and definitions, most of them as well as the omitted ones are in [3].

**Definition 1.** *For a terminal alphabet* $\Sigma$, *the set of pictures is* $\Sigma^{**}$. *For* $h, k \geq 1$, $\Sigma^{(h,k)}$ *denotes the set of pictures of size* $(h, k)$ *(i.e.* $|p| = (h, k)$*):*

$$p \in \Sigma^{(h,k)} \iff p = \begin{matrix} p(1,1) & \ldots & p(1,k) \\ \vdots & \ddots & \vdots \\ p(h,1) & \ldots & p(h,k) \end{matrix}$$

   *Row and column concatenations are denoted* $\ominus$ *and* $\oslash$, *respectively.* $p \ominus q$ *is defined iff* $p$ *and* $q$ *have the same number of columns; the resulting picture is the vertical juxtaposition of* $p$ *over* $q$. $p^{k\ominus}$ *is the vertical juxtaposition of* $k$ *copies of* $p$; $p^{*\ominus}$ *is the corresponding closure.* $\oslash, {}^{k\oslash}, {}^{*\oslash}$ *are the column analogous.*

**Definition 2.** *The (vertical) mirror image and the (clockwise) rotation of a picture* $p$ *(with* $|p| = (h, k)$*), respectively, are defined as follows:*

$$Mirror(p) = \begin{matrix} p(h,1) & \ldots & p(h,k) \\ \vdots & \ddots & \vdots \\ p(1,1) & \ldots & p(1,k) \end{matrix} \quad ; \quad p^R = \begin{matrix} p(h,1) & \ldots & p(1,1) \\ \vdots & \ddots & \vdots \\ p(h,k) & \ldots & p(1,k) \end{matrix}$$

**Definition 3.** *Let* $p \in \Sigma^{(h,k)}, q \in \Sigma^{(h',k')}, h' \leq h, k' \leq k$. $q$ *is called a subpicture (or* window*) of* $p$ *at position* $(i, j)$ *(also written* $q \trianglelefteq_{(i,j)} p$*) iff:*

$$\forall i', j'(i' \leq h' \wedge j' \leq k' \wedge i < h - h' \wedge j < k - k' \Rightarrow p(i + i', j + j') = q(i', j'))$$

   *We will use the shortcut* $q \trianglelefteq p$, *for* $\exists i, j (q \trianglelefteq_{(i,j)} p)$.

**Definition 4.** *For a picture* $p \in H^{**}$ *the set of subpictures with size* $(h, k)$ *is:*

$$B_{h,k}(p) = \{q \in H^{(h,k)} \mid q \trianglelefteq p\}$$

   *We assume* $B_{1,k}$ *to be only defined on* $\Sigma^{(1,*)}$ *(horizontal string), and* $B_{h,1}$ *on* $\Sigma^{(*,1)}$ *(vertical string).*

**Definition 5.** *Consider a set* $\omega$ *of pictures* $t \in \Sigma^{(i,j)}$. *The* locally testable language *defined by* $\omega$ *(called* $LOC(\omega)$*) is the set of pictures* $p \in \Sigma^{**}$ *such that* $B_{i,j}(p) = \omega$.
   *Moreover the* locally testable language in the strict sense *defined by* $\omega$ *(written* $LOCss(\omega)$*) is the set of pictures* $p \in \Sigma^{**}$ *such that* $B_{i,j}(p) \subseteq \omega$

**Definition 6.** Substitution. *If $p, q, q'$ are pictures, $q \trianglelefteq_{(i,j)} p$, and $q, q'$ have the same size, then $p[q'/q_{(i,j)}]$ denotes the picture obtained by replacing the occurrence of $q$ at position $(i, j)$ in $p$ with $q'$.*

*For simplicity, we shall drop the subscript $(i, j)$ when there is a single occurrence of $q$ in $p$.*

The main definition follows.

**Definition 7.** *A* Tile Rewriting Grammar (TRG) *is a tuple $(\Sigma, N, S, R)$, where $\Sigma$ is the terminal alphabet, $N$ is a set of nonterminal symbols, $S \in N$ is the starting symbol, $R$ is a set of rules.*

*An* indexed symbol *is an element from $I = (\Sigma \times \{0\}) \cup (N \times \mathbb{N}_{\geq 1})$.*

*$R$ may contain two kinds of rules:*

**Type 1:**  *$A \rightarrow t$, where $A \in N$, $t \in I^{(h,k)}$, with $h, k > 0$;*
**Type 2:**  *$A \rightarrow \omega$, where $A \in N$ and $\omega \subseteq \{t \mid t \in I^{(i,j)}\}$, with $i, j > 0$.*

Notice that type 1 is not a special case of type 2. Moreover, the update index used for terminals is 0, while indices for nonterminals are greater than 0.

Intuitively a rule of type 1 is intended to match a subpicture of small bounded size, identical to the right part $t$. A rule of type 2 matches any subpicture of any size which can be tiled using *all* the elements $t$ of the tile set $\omega$.

We define a procedure to be used in the derivation step.

**Definition 8.** Rewrite procedure. *Consider a TRG $G = (\Sigma, N, S, R)$, a rule $\rho \in R$, and a picture $p \in ((\Sigma \cup N) \times \mathbb{N})^{**}$. Then its* maximum index *(written $\mu(p)$) is defined as: $\mu(p) = Max\{k \mid p(i, j) = (\chi, k), \chi \in \Sigma \cup N\}$.*

**Rewrite** *(in $\rho, p$; out $q$)*
**1:** *Find a maximal[2] $r \in (A, k)^{**}$, where $A$ is the left part of $\rho$, such that $r \trianglelefteq_{(m,n)} p$, for some $m, n$. $r$ is called the* application area.
**2:** *If $\rho = A \rightarrow \omega$ (i.e. $\rho$ is a type 2 rule), then choose a picture $s \in LOC(\omega)$, with $|s| = |r|$.*
*Otherwise if $\rho = A \rightarrow t$ (type 1), set $s := t$, the right part of $\rho$.*
**3:** *Let $s'$ be the picture defined by $\forall i, j : s(i, j) = (\chi, l), \chi \in \Sigma \cup N \Rightarrow s'(i, j) = (\chi, l + \mu(p))$. That is, $s'$ is computed incrementing the indices in $s$ by $\mu(p)$.*
**4:** *Return $q = p[s'/r_{(m,n)}]$.*
**end.**

**Definition 9.** *A* derivation in one step *is a relation*

$$\Rightarrow_G \subseteq ((\Sigma \cup N) \times \mathbb{N})^{(h,k)} \times ((\Sigma \cup N) \times \mathbb{N})^{(h,k)}$$

*Then $p \Rightarrow_G q$ iff there exists a rule $\rho \in R$ such that $Rewrite(\rho, p, q)$.*
*We say that $q$ is* derivable from $p$ in $n$ step, *in symbols $p \stackrel{n}{\Rightarrow}_G q$, iff $p = q$, when $n = 0$, or there is a picture $r \in ((\Sigma \cup N) \times \mathbb{N})^{(h,k)}$ such that $p \stackrel{n-1}{\Rightarrow}_G r$ and $r \Rightarrow_G q$. As usual $p \stackrel{*}{\Rightarrow}_G q$ says that $p \stackrel{n}{\Rightarrow}_G q$, for some $n \geq 0$.*

---

[2] i.e. it is not the case that $\exists q \neq r$ such that $q \in (A, k)^{**}$ and $r \trianglelefteq q$.

**Definition 10.** *The picture language defined by a TRG $G$ (written $L(G)$) is the set of $p \in \Sigma^{**}$ such that, if $|p| = (h,k)$, then $(S,0)^{(h,k)} \overset{*}{\Rightarrow}_G p|_1$, where $|_1$ denotes the projection on $\Sigma$ of the elements in $\Sigma \times \mathbb{N}$. For short we write $S \overset{*}{\Rightarrow}_G p$.*

The following obvious statements may be viewed as a 2D formulation of well known properties of 1D CF derivations.

Let $p_1 \Rightarrow p_2 \Rightarrow \ldots \Rightarrow p_n$ be a derivation, and $\alpha(p_1), \alpha(p_2), \ldots \alpha(p_{n-1})$ the corresponding application areas.

**Disjointness of application areas:** For any $p_i, p_j, i < j$, either $\alpha(p_i) \trianglelefteq \alpha(p_j)$, or $\alpha(p_i)$, $\alpha(p_j)$ are disjoint.

**Canonical derivation:** Let $c_i = (x_i, y_i)$ be the coordinates of the top leftmost corner of the application area $\alpha(p_i)$. The previous derivation is *lexicographic* iff $i < j$ implies $c_i \leq_{lex} c_j$ (where $\leq_{lex}$ is the usual lexicographic order). Then

$$L(G) \equiv \{p \mid S \overset{*}{\Rightarrow}_G p \text{ and } \overset{*}{\Rightarrow}_G \text{ is a lexicographic derivation}\}$$

To simplify the notation, in the sequel we will use subscripts for indexed symbols (e.g. $S_1$ instead of $(S,1)$ ). Moreover, we will drop the index when possible, for terminals and in rules where all nonterminals have the same index.

To illustrate we present two examples.

*Example 2. Chinese boxes.* $G = (\Sigma, N, S, R)$, where $\Sigma = \{\ulcorner, \urcorner, \llcorner, \lrcorner, o\}$, $N = \{S\}$, and $R$ consists of the rules:

$$S \to \begin{array}{c} \ulcorner \urcorner \\ \llcorner \lrcorner \end{array}; \quad S \to \left\{ \begin{array}{llllllll} \ulcorner o & o\,S & o\,S & S\,S & o\,o & S\,S & o\,\urcorner & S\,o & S\,o \\ o\,S, & \llcorner o\,, & o\,S, & o\,o\,, & S\,S, & S\,S, & S\,o\,, & S\,o\,, & o\,\lrcorner \end{array} \right\}$$

(where indices have been dropped).
A picture in $L(G)$ is:

$$\begin{array}{cccccc} \ulcorner & o & o & o & o & \urcorner \\ o & \ulcorner & o & o & \urcorner & o \\ o & o & \ulcorner & \urcorner & o & o \\ o & o & \llcorner & \lrcorner & o & o \\ o & \llcorner & o & o & \lrcorner & o \\ \llcorner & o & o & o & o & \lrcorner \end{array}$$

For convenience, we will often specify a set of tiles by a sample picture exhibiting the tiles as its subpictures (using $B_{h,k}$ - see Definition 4). We write $|$ to separate alternative right parts of rules with the same left part (analogously to string grammars). Therefore, the previous grammar becomes:

$$S \to \begin{array}{c} \ulcorner \urcorner \\ \llcorner \lrcorner \end{array} \;\middle|\; B_{2,2} \begin{pmatrix} \ulcorner & o & o & \urcorner \\ o & S & S & o \\ o & S & S & o \\ \llcorner & o & o & \lrcorner \end{pmatrix}$$

*Example 3. 2D Dyck analogue.* The next language $L_{box}$, a superset of Chinese boxes, can be defined by a sort of cancellation rule. But since terminals cannot be cancelled without shearing the picture, we replace them by a character $b$, the blank or background.

**Empty frame:** Let $k \geq 0$. An *empty frame* is a picture defined by the regular expression: $(\ulcorner \oplus (\circ)^k \oplus \urcorner) \ominus (\circ \oplus b^{k \oplus} \oplus \circ)^{k \ominus} \ominus (\llcorner \oplus (\circ)^k \oplus \lrcorner)$, i.e. a box containing just $b$'s.

**Cancellation:** The *cancellation* of an empty frame $p$ is the picture $del(p)$ obtained by applying the projection $del(x) = b, x \in \Sigma \cup \{b\}$.

A picture $p$ is in $L_{box}$ iff by repeatedly applying $del$ on subpictures which are empty frames, an empty frame is obtained, as in the picture:

```
┌ o o o o ┐┌ o o ┐
o ┌ ┐┌ ┐ o o ┌ ┐ o
o └ ┘└ ┘ o o └ ┘ o
└ o o o o ┘└ o o ┘
```

This time we need to use indices in the productions of the grammar:

$$S \to B_{2,2} \begin{pmatrix} S_1 \ S_1 \ S_2 \ S_2 \\ S_1 \ S_1 \ S_2 \ S_2 \end{pmatrix} \mid B_{2,2} \begin{pmatrix} S_1 \ S_1 \\ S_1 \ S_1 \\ S_2 \ S_2 \\ S_2 \ S_2 \end{pmatrix}$$

To illustrate we list the derivation steps of the previous picture:

$$
\begin{array}{l}
S_0 \ S_0 \ S_0 \ S_0 \ S_0 \ S_0 \ S_0 \ S_0 \ S_0 \ S_0 \\
S_0 \ S_0 \ S_0 \ S_0 \ S_0 \ S_0 \ S_0 \ S_0 \ S_0 \ S_0 \\
S_0 \ S_0 \ S_0 \ S_0 \ S_0 \ S_0 \ S_0 \ S_0 \ S_0 \ S_0 \\
S_0 \ S_0 \ S_0 \ S_0 \ S_0 \ S_0 \ S_0 \ S_0 \ S_0 \ S_0
\end{array}
\Rightarrow_G
\begin{array}{l}
S_1 \ S_1 \ S_1 \ S_1 \ S_1 \ S_1 \ S_2 \ S_2 \ S_2 \ S_2 \\
S_1 \ S_1 \ S_1 \ S_1 \ S_1 \ S_1 \ S_2 \ S_2 \ S_2 \ S_2 \\
S_1 \ S_1 \ S_1 \ S_1 \ S_1 \ S_1 \ S_2 \ S_2 \ S_2 \ S_2 \\
S_1 \ S_1 \ S_1 \ S_1 \ S_1 \ S_1 \ S_2 \ S_2 \ S_2 \ S_2
\end{array}
\overset{2}{\Rightarrow}_G
$$

```
┌ o  o  o  o ┐┌ o  o ┐
o S₃ S₃ S₃ S₃ o o S₄ S₄ o
o S₃ S₃ S₃ S₃ o o S₄ S₄ o
└ o  o  o  o ┘└ o  o ┘
```
$\Rightarrow_G$
```
┌ o  o  o  o ┐┌ o  o ┐
o S₅ S₅ S₆ S₆ o o S₄ S₄ o
o S₅ S₅ S₆ S₆ o o S₄ S₄ o
└ o  o  o  o ┘└ o  o ┘
```
$\Rightarrow_G$

```
┌ o  o  o ┐┌ o  o ┐
o ┌ ┐ S₆ S₆ o o S₄ S₄ o
o └ ┘ S₆ S₆ o o S₄ S₄ o
└ o  o  o ┘└ o  o ┘
```
$\overset{2}{\Rightarrow}_G$
```
┌ o o o o ┐┌ o o ┐
o ┌ ┐┌ ┐ o o ┌ ┐ o
o └ ┘└ ┘ o o └ ┘ o
└ o o o o ┘└ o o ┘
```

Although this language can be viewed as a 2D analogue of a Dyck's string language, variations are possible and we do not claim the same algebraic properties as in 1D.

## 2.3   Closure Properties

For simplicity, in the following theorem we suppose that $L(G_1), L(G_2)$ contain pictures of size at least $(2,2)$.

**Theorem 1.** *The family $\mathcal{L}(TRG)$ is closed under union, column/row concatenation, column/row closure operations, rotation, and alphabetical mapping (or projection).*

*Proof.* Consider two grammars $G_1 = (\Sigma, N_1, A, R_1)$ and $G_2 = (\Sigma, N_2, B, R_2)$. Suppose for simplicity that $N_1 \cap N_2 = \emptyset$, and $S \notin N_1 \cup N_2$. Then it is easy to show that the grammar $G = (\Sigma, N_1 \cup N_2 \cup \{S\}, S, R_1 \cup R_2 \cup R)$, where

**Union $\cup$:**

$$R = \left\{ S \to B_{2,2} \begin{pmatrix} A\ A \\ A\ A \end{pmatrix},\ S \to B_{2,2} \begin{pmatrix} B\ B \\ B\ B \end{pmatrix} \right\}$$

is such that $L(G) = L(G_1) \cup L(G_2)$.

**Concatenation $①/⊖$:**

$$R = \left\{ S \to B_{2,2} \begin{pmatrix} A\ A\ B\ B \\ A\ A\ B\ B \end{pmatrix} \right\}$$

is such that $L(G) = L(G_1) ① L(G_2)$. The row concatenation case is analogous.

**Closures $*①/*⊖$:**

$$G = (\Sigma, N_1 \cup \{S\}, S, R_1 \cup R)$$

where

$$R = \left\{ S \to B_{2,2} \begin{pmatrix} S_1\ S_1\ S_2\ S_2 \\ S_1\ S_1\ S_2\ S_2 \end{pmatrix} \right\}$$

is such that $L(G) = L(G_1)^{*①}$. The row closure case is analogous.

**Rotation $R$:** Construct the grammar $G = (\Sigma, N, A, R')$, where $R'$ is such that, if $B \to t \in R_1$ is a type 1 rule, then $B \to t^R$ is in $R'$; if $B \to \omega \in R_1$ is a type 2 rule, then $B \to \omega'$ is in $R'$, with $t \in \omega$ implies $t^R \in \omega'$. It is easy to verify that $L(G) = L(G_1)^R$.

**Projection $\pi$:** Consider a grammar $G = (\Sigma_1, N, S, R)$ and a projection $\pi : \Sigma_1 \to \Sigma_2$. It is possible to build a grammar $G' = (\Sigma_2, N', S, R')$, such that $L(G') = \pi(L(G))$. Indeed, let $\Sigma_1'$ be a set of new nonterminals corresponding to elements in $\Sigma_1$, then $N' = N \cup \Sigma_1'$; $R' = \phi(R) \cup R''$, where $\phi : \Sigma_1 \times \{0\} \to \Sigma_1' \times \{k\}$ is the alphabetical mapping $\phi(a) = a'_k$, where $k$ is a fixed unused updating index, and it is naturally extended to TRG rules. Moreover, we need the additional projection rules $R''$ (where $v = \pi(a)$):

$$a' \to v_0 \mid \left\{ \begin{matrix} v_0\ v_0 \\ v_0\ v_0 \end{matrix} \right\} \mid \{ v_0\ v_0 \} \mid \left\{ \begin{matrix} v_0 \\ v_0 \end{matrix} \right\}$$

## 3    Comparison with Other Models

We first compare with Tiling Systems, then with Matz's CF grammars.

**Theorem 2.**

$$\mathcal{L}(LOCss) \subseteq \mathcal{L}(TRG)$$

*Proof.* Consider a local two-dimensional language over $\Sigma$ defined by the set of allowed blocks $\Theta$.

Let $\Theta_0 = \{\begin{smallmatrix} x_0 & y_0 \\ z_0 & w_0 \end{smallmatrix} \mid \begin{smallmatrix} x & y \\ z & w \end{smallmatrix} \in \Theta\}$, then an equivalent TRG is $G = (\Sigma, \{S\}, S, R)$, where $R$ is the set $\{S \to \theta \mid \theta \subseteq \Theta_0\}$.

**Theorem 3.**

$$\mathcal{L}(TS) \subseteq \mathcal{L}(TRG)$$

*Proof.* This result is a consequence of Theorems 1, 2, and the fact that $\mathcal{L}(TS)$ is the closure of $\mathcal{L}(LOCss)$ with respect to projection.

The following strict inclusion is an immediate consequence of the fact that, for 1D languages, $\mathcal{L}(TS) \subset \mathcal{L}(CF)$, and $\mathcal{L}(TRG) = \mathcal{L}(CF)$. But we prefer to prove it by exhibiting an interesting picture language, made by the vertical concatenation of two specularly symmetrical rectangles.

**Theorem 4.**

$$\mathcal{L}(TS) \neq \mathcal{L}(TRG)$$

*Proof.* Let $\Sigma = \{a, b\}$. Consider the language $L = \{p \mid p = s \ominus Mirror(s) \text{ and } s \in \Sigma^{(h,k)}, h > 1, k \geq 1\}$. We prove that $L \notin \mathcal{L}(TS)$ using a technique very similar to that of Theorem 7.5 in [3].

Consider the grammar $G$:

$$A \to \begin{array}{c|c} a & b \\ a & b \end{array} \mid B_{2,1} \begin{pmatrix} a \\ A \\ A \\ a \end{pmatrix} \mid B_{2,1} \begin{pmatrix} b \\ A \\ A \\ b \end{pmatrix}$$

$$S \to B_{2,2} \begin{pmatrix} A\,S\,S \\ A\,S\,S \end{pmatrix} \mid B_{2,1} \begin{pmatrix} A \\ A \end{pmatrix}$$

Without proof, it is easy to see that $L(G) = L$.

We prove by contradiction that $L \notin \mathcal{L}(TS)$. Suppose that $L \in \mathcal{L}(TS)$, therefore $L$ is a projection of a local language $L'$ defined over some alphabet $\Gamma$. Let $\sigma = |\Sigma|$ and $\gamma = |\Gamma|$, with $\sigma \leq \gamma$. For an integer $n$, let:

$$L_n = \{p \mid p = s \ominus Mirror(s) \text{ and } |s| = (n, n)\}.$$

Clearly, $|L_n| = \sigma^{n^2}$. Let $L'_n$ be the set of pictures in $L'$ over $\Gamma$ whose projections are in $L_n$. By choice of $\gamma$ and by construction of $L_n$ there are at most $\gamma^n$ possibilities for the $n$-th and $(n+1)$-th rows in the pictures of $L'_n$, because this is the number of mirrored stripe pictures of size $(2, n)$ on $\Gamma$.

For $n$ sufficiently large $\sigma^{n^2} \geq \gamma^n$. Therefore, for such $n$, there will be two different pictures $p = s_p \ominus Mirror(s_p), q = s_q \ominus Mirror(s_q)$ such that the corresponding $p' = s'_p \ominus s''_p, q' = s'_q \ominus s''_q$ have the same $n$-th and $(n+1)$-th rows. This implies that, by definition of local language, pictures $v' = s'_p \ominus s''_q, w' = s'_q \ominus s''_p$ belong to $L'_n$, too. Therefore, pictures $\pi(v') = s_p \ominus Mirror(s_q)$, and $\pi(w') = s_q \ominus Mirror(s_p)$ belong to $L_n$. But this is a contradiction.

The other family of languages to be compared are a different generalisation of CF grammars in two dimensions, Matz's CF Picture Grammars $(CFPG)$[5]. These grammars are syntactically very similar to 1D CFs. The main difference is that their right parts use $\oplus, \ominus$ operators. Nonterminals denote unbound rectangular pictures. Derivation is analogous to 1D, but the resulting regular expression may or may not define a picture (e.g. $a \oplus (b \ominus b)$ does not generate any picture).

**Theorem 5.**

$$\mathcal{L}(CFPG) \subseteq \mathcal{L}(TRG)$$

*Hint of the proof.* Consider now a Matz's CFPG grammar in Chomsky Normal Form. It may contain three types of rules: $A \to B \oplus C$; $A \to B \ominus C$; $A \to a$. Then, $A \to B \oplus C$ corresponds to the following TRG rules:

$$A \to B_{2,2} \begin{pmatrix} B\,B\,C\,C \\ B\,B\,C\,C \end{pmatrix} \mid B_{2,2} \begin{pmatrix} B\,C\,C \\ B\,C\,C \end{pmatrix} \mid B_{2,2} \begin{pmatrix} B\,B\,C \\ B\,B\,C \end{pmatrix} \mid$$

$$B\,C \mid B_{1,2} \begin{pmatrix} B\,B\,C\,C \end{pmatrix} \mid B_{1,2} \begin{pmatrix} B\,C\,C \end{pmatrix} \mid B_{1,2} \begin{pmatrix} B\,B\,C \end{pmatrix}$$

Notice that a CFPG production $A \to B \oplus B$, with two copies of the same nonterminal $B$, imposes the use of indices in the corresponding TRG productions:

$$A \to B_{2,2} \begin{pmatrix} B_1\,B_1\,B_2\,B_2 \\ B_1\,B_1\,B_2\,B_2 \end{pmatrix} \mid \ldots$$

The $\ominus$ case is analogous, while $A \to a$ is trivial.

**Theorem 6.**

$$\mathcal{L}(CFPG) \neq \mathcal{L}(TRG)$$

*Proof.* It is a consequence of Theorems 3, 4, and 5, and the fact that $\mathcal{L}(TS) \not\subseteq \mathcal{L}(CFPG)$, as reported in [5].

An example of a TRG but not CFPG language is the following. We know from [5] that the language which consists of a thin cross of $b$'s on a field of $a$'s is not in $\mathcal{L}(CFPG)$. It is easy to show that the following TRG defines the "cross" language:

$$S \to B_{2,2} \begin{pmatrix} B\,B\,A\,A \\ B\,B\,A\,A \\ C\,C\,D\,D \\ C\,C\,D\,D \end{pmatrix}; \quad B \to B_{2,2} \begin{pmatrix} a\,a \\ a\,a \\ b\,b \end{pmatrix}; \quad A \to B_{2,2} \begin{pmatrix} b\,a\,a \\ b\,a\,a \\ b\,b\,b \end{pmatrix}$$

$$C \to B_{2,2} \begin{pmatrix} a\,a \\ a\,a \end{pmatrix}; \quad D \to B_{2,2} \begin{pmatrix} b\,a\,a \\ b\,a\,a \end{pmatrix}$$

# 4    Conclusions

The new TRG grammar model extends the context-free string grammars to two dimensions. As the application area of a rewriting rule is an unbounded subpicture, we use tiling to specify it. In a derivation the application areas rewritten are partially ordered by the subpicture relation, as in a syntax tree. In three dimensions a derivation can be depicted as a set of prisms whose bases are the application areas of the productions, the analogue of the syntax tree for strings. The Dyck strings then become the picture language of well nested boxes, in one of several possible variations.

The expressive power of TRG is greater than of two previous models: the Tiling Systems of [2] which define 2D recognizable languages; and Matz's context free picture grammars, another generalisation of context free grammars. More work is needed to compare with other grammar families and to assess the suitability of TRG for practical applications.

The analogy with CF string grammars raises to the educated formal linguists many obvious questions, such as the formulation of a pumping lemma, of Chomsky-Schutzenberger theorem, etc.

It would be also interesting to define a type 1 TRG model in the sense of Chomsky's hierarchy.

**Acknowledgement.** We thank Antonio Restivo for calling our attention to the problem of inventing a class of grammars suitable to define "2D Dyck languages". We thank Alessandra Cherubini and Pierluigi San Pietro for their comments, and the anonymous referees for helpful suggestions.

# References

1. Henning Fernau and Rudolf Freund. Bounded parallelism in array grammars used for character recognition. In Petra Perner, Patrick Wang, and Azriel Rosenfeld, editors, *Advances in Structural and Syntactical Pattern Recognition (Proceedings of the SSPR'96)*, volume 1121, pages 40–49. Springer-Verlag, 1996.
2. Dora Giammarresi and Antonio Restivo. Recognizable picture languages. *International Journal Pattern Recognition and Artificial Intelligence*, 6(2-3):241–256, 1992. Special Issue on *Parallel Image Processing*.
3. Dora Giammarresi and Antonio Restivo. Two-dimensional languages. In Arto Salomaa and Grzegorz Rozenberg, editors, *Handbook of Formal Languages*, volume 3, Beyond Words, pages 215–267. Springer-Verlag, Berlin, 1997.
4. Katsushi Inoue and Akira Nakamura. Some properties of two-dimensional on-line tessellation acceptors. *Information Sciences*, 13:95–121, 1977.
5. Oliver Matz. Regular expressions and context-free grammars for picture languages. In *14th Annual Symposium on Theoretical Aspects of Computer Science*, volume 1200 of *Lecture Notes in Computer Science*, pages 283–294, Lübeck, Germany, 27 February–March 1 1997. Springer-Verlag.
6. Rani Siromoney. Advances in Array Languages. In Hartmut Ehrig, Manfred Nagl, Grzegorz Rozenberg, and Azriel Rosenfeld, editors, *Proc. 3rd Int. Workshop on Graph-Grammars and Their Application to Computer Science*, volume 291 of *Lecture Notes in Computer Science*, pages 549–563. Springer-Verlag, 1987.

# Distributed Pushdown Automata Systems: Computational Power

Erzsébet Csuhaj-Varjú[1,*], Victor Mitrana[2,**], and György Vaszil[1]

[1] Computer and Automation Research Institute, Hungarian Academy of Sciences
Kende u. 13-17, 1111 Budapest, Hungary
{csuhaj,vaszil}@sztaki.hu
[2] Faculty of Mathematics and Computer Science, University of Bucharest
Str. Academiei 14, 70109, Bucharest, Romania
and
Research Group in Mathematical Linguistics, Rovira i Virgili University
Pça. Imperial Tàrraco 1, 43005 Tarragona, Spain
vmi@fll.urv.es

**Abstract.** We introduce distributed pushdown automata systems consisting of several pushdown automata which work in turn on the input string placed on a common one-way input tape. The work of the components is based on protocols and strategies similar to those that cooperating distributed grammar systems use. We investigate the computational power of these mechanisms under different protocols for activating components and two ways of accepting the input string: with empty stacks or with final states which means that all components have empty stacks or are in final states, respectively, when the input string was completely read.

## 1 Introduction

Distributed systems are recently in the focus of interest in computer science. One of the main goals, when designing these systems, is to increase the computational power of the components by cooperation and communication and/or to decrease the complexity of different tasks by distribution and parallelism. For this purpose, various models have been proposed and intensively investigated, among them we find the formal language theoretic paradigm called *grammar system* [3, 5]. A grammar system is a finite set of grammars which cooperate and communicate in deriving words of a common language. Two main architectures have been distinguished in the area, cooperating distributed (CD) grammar systems [2] and parallel communicating (PC) grammar systems [11]. In CD grammar systems the grammars generate a common sentential form in turn, according to a specific protocol, while in PC grammar systems they generate their own sentential form in parallel and communicate these strings to each other by requests.

\* Work supported in part by a grant from the NATO Scientific Committee in Spain
\*\* Work supported by the Centre of Excellence in Information Technology, Computer Science and Control, ICA1-CT-2000-70025, HUN-TING project, WP5.

Z. Ésik and Z. Fülöp (Eds.): DLT 2003, LNCS 2710, pp. 218–229, 2003.

Grammar systems theory has been explored extensively in the last twelve years, see the monograph [3] and more recently the chapter [5] in [12]. The obtained results demonstrated that cooperation and communication increases the power of the individual components: large language classes were described by systems of very simple grammars belonging to grammar classes with weak computational power.

In spite of the notable development in the area of generative tools, only a very little has been done with respect to automata systems realizing similar architectures and working under similar strategies. These investigations are of particular interest from several points of view: they might lead to a comparison between the power of distributed generative and accepting devices, and might give information on the boundaries of describing language classes in terms of automata systems. In the following, we briefly summarize the previous work done in this area.

In [6] special types of multi-stack pushdown automata were introduced. These mechanisms are usual multi-stack pushdown automata whose stacks cooperate in the accepting process under strategies borrowed from CD grammar systems. However, they cannot be seen as the automata counterpart of CD grammar systems.

Concerning the automata counterpart of PC grammar systems, in [4], PC pushdown automata systems communicating by stacks are introduced. PC finite automata systems communicating by states were introduced in [10] where it was shown that all variants can be simulated by multi-head finite automata [7]. Many incomparability results between these PC finite automata systems and the corresponding PC regular grammar systems were reported. Some undecidability results were proved in [8]. A survey of PC finite automata systems can be found in [9].

In this paper we define and investigate systems of pushdown automata with working modes very close to that of CD grammar systems. These systems, called distributed pushdown automata systems (DPAS, for short), consist of several pushdown automata working in turn on the input string under different protocols, similar to those under which CD grammar systems work. A distributed pushdown automata system has a common one-way input tape, one reading head, and several central units. Each central unit has its own finite set of states and accesses the topmost symbol of its own pushdown memory. At any moment, only one central unit is active, the others are "frozen". When active, the central unit can also read the current input symbol by means of the common reading head. The activation of a component means that the central unit of that component takes control over the reading head. The protocols for activating components are similar to those used by CD grammar systems, namely, the active automaton must perform moves of exactly $k$ steps, at least $k$ steps, at most $k$ steps, where $k \geq 1$, or the automaton must continue its work as long as it is able to perform a move. This latter way of functioning is called the $t$-mode of computation. There are two ways in which these systems can accept input strings: with empty stacks or with final states, meaning that all components have empty stacks or are in final states, respectively, when the input string has been completely read.

In this paper we examine the computational power of DPASs. We prove that in the $=$ 1-mode, the $\geq$ 1-mode, and the $\leq k$-mode of computation, systems with $n$ components describe exactly the class of the $n$-shuffles of context-free languages, while in the $= k$-mode and the $\geq k$-mode of computation the class of accepted languages are included in the non-erasing homomorphic images of the $n$-shuffles of context-free languages. On the other hand, DPASs recognize context-sensitive languages only. With the exception of the languages accepted in the $t$-mode of computation with final states, all languages recognized by DPASs under the above computation modes and both ways of accepting are semilinear. Thus, language classes of these systems significantly differ from the language classes of context-free CD grammar systems, since the latter ones in the $=$ 1-mode, the $\geq$ 1-mode, and the $\leq k$-mode of derivation determine the context-free language class, in the $k$-mode or $\geq k$-mode of derivation, $k \geq 2$, they describe a subclass of matrix languages, and in the $t$-mode of derivation they identify the class of $ET0L$ languages.

## 2    Basic Notions

We assume that the reader is familiar with the basic concepts of automata and formal language theory; for further details we refer to [12].

An alphabet is a finite and nonempty set of symbols. Any sequence of symbols from an alphabet $V$ is called a string (a word) over $V$. For an alphabet $V$, we denote by $V^*$ the free monoid generated by $V$ under the operation of concatenation; the empty string is denoted by $\varepsilon$ and the semigroup $V^* - \{\varepsilon\}$ is denoted by $V^+$. The length of $x \in V^*$ is denoted by $|x|$, and $|x|_a$ denotes the number of occurrences of the symbol $a$ in $x$. A subset of $V^*$ is called a language over $V$. The family of context-free and context-sensitive languages is denoted by $CF$ and $CS$, respectively.

We shall also denote by $Rec_X(A)$ the language accepted by a pushdown automaton $A$ with final state if $X = f$, or with empty stack if $X = \varepsilon$. We note that pushdown automata characterize the class of context-free languages in both modes of acceptance. A similar machine characterization of context-sensitive languages is obtained by nondeterministic Turing machines with a linear space bound, that is, **NSPACE**$(n) = CS$.

For a given alphabet $V = \{a_1, a_2, \ldots, a_n\}$, $n \geq 1$, and a string $x \in V^*$, we define the *Parikh mapping* of $x$ by the vector $\psi_V(x) = (|x|_{a_1}, |x|_{a_2}, \ldots, |x|_{a_n})$. The subscript $V$ is omitted when it is clear from the context. The Parikh mapping is extended to languages $L \subseteq V^*$ by $\psi(L) = \{\psi(x) \mid x \in L\}$. It is well-known that the Parikh mapping of any context-free language is a semilinear set.

Now we recall the definition of the *shuffle operation* which we shall use frequently in the sequel, and which has been intensively investigated in formal language and concurrency theory. Formally, this operation is defined recursively on the words over an alphabet $V$ as follows:

$$ш(\varepsilon, x) = ш(x, \varepsilon) = \{x\}, \text{ for any } x \in V^*$$
$$ш(ax, by) = \{a\}ш(x, by) \cup \{b\}ш(ax, y), \text{ for all } a, b \in V, \ x, y \in V^*.$$

The operation is extended to languages by $ш(L_1, L_2) = \bigcup_{x \in L_1, y \in L_2} ш(x, y)$, and to have $k$ arguments, with $k > 2$, by

$$ш_k(x_1, x_2, \ldots, x_k) = ш(ш_{k-1}(x_1, x_2, \ldots, x_{k-1}), \{x_k\}),$$

where $x_1, \ldots, x_k \in V^*$. Similarly, $ш_k$ can be extended to languages in the following way:

$$ш_k(L_1, L_2, \ldots, L_k) = \bigcup_{x_i \in L_i, 1 \le i \le k} ш_k(x_1, x_2, \ldots, x_k).$$

For a family of languages $F$, we denote

$$Shuf_n(F) = \{ш_n(L_1, L_2, \ldots, L_n) \mid L_i \in F, 1 \le i \le n\}.$$

Now we give the definition of the main concept of the paper.

A *distributed pushdown automata system* (a DPAS, in short) of degree $n$, where $n \ge 1$, is a construction

$$\mathcal{A} = (V, A_1, A_2, \ldots, A_n),$$

where $V$ is an alphabet, and for each $i$, $1 \le i \le n$, $A_i = (Q_i, V, \Gamma_i, f_i, q_i, Z_i, F_i)$ is a pushdown automaton with the set of states $Q_i$, initial state $q_i \in Q_i$, alphabet of input symbols $V$, alphabet of pushdown symbols $\Gamma_i$, initial contents of the pushdown memory $Z_i \in \Gamma_i$, set of final states $F_i \subseteq Q_i$, and transition mapping $f_i$ from $Q_i \times V \cup \{\varepsilon\} \times \Gamma_i$ into the finite subsets of $Q_i \times \Gamma_i^*$. We refer to the automaton $A_i$, $1 \le i \le n$, as the $i$th *component* of $\mathcal{A}$.

An *instantaneous description* (ID) of a distributed pushdown automata system as above is a $2n + 1$-tuple $(x, s_1, \alpha_1, s_2, \alpha_2, \ldots, s_n, \alpha_n)$, where $x \in V^*$ is the part of the input string to be read, $s_i$, $1 \le i \le n$, is the current state of automaton $A_i$, and $\alpha_i \in \Gamma_i^*$ is the contents of the pushdown memory of the same automaton. We sometimes also refer to an instantaneous description of a DPAS as a *configuration* of the system.

A *one step move* of $\mathcal{A}$ performed by the component $A_i$, $1 \le i \le n$, is defined by the binary relation $\vdash_i$ on all IDs of $\mathcal{A}$ in the following way:

$$(ax, s_1, \alpha_1, s_2, \alpha_2, \ldots, s_i, \alpha_i, \ldots, s_n, \alpha_n) \vdash_i (x, s_1, \alpha_1, s_2, \alpha_2, \ldots, r, \beta, \ldots, s_n, \alpha_n)$$

if and only if $(r, \delta) \in f_i(s_i, a, A)$, where $a \in V \cup \{\varepsilon\}$, $\alpha_i = A\gamma$, and $\beta = \delta\gamma$.

Let $\vdash_i^k$ denotes the $k$-th power of $\vdash_i$ for some $k \ge 1$, and $\vdash_i^*$ denotes the reflexive and transitive closure of $\vdash_i$. Let now $C_1, C_2$ be two IDs of a distributed pushdown automata system. We say that $C_1$ directly derives $C_2$

- by a *move consisting of $k$ steps*, denoted by $C_1 \vdash_{\mathcal{A}}^{=k} C_2$, if and only if $C_1 \vdash_i^k C_2$ for some $1 \le i \le n$,

- by a *move consisting of at least k steps*, denoted by $C_1 \vdash_{\mathcal{A}}^{\geq k} C_2$, if and only if $C_1 \vdash_i^l C_2$ for some $l \geq k$ and $1 \leq i \leq n$,
- by a *move consisting of at most k steps*, denoted by $C_1 \vdash_{\mathcal{A}}^{\leq k} C_2$, if and only if $C_1 \vdash_i^l C_2$ for some $l \leq k$ and $1 \leq i \leq n$,
- by a *move representing a sequence of steps that cannot be continued*, denoted by $C_1 \vdash_{\mathcal{A}}^t C_2$, if and only if $C_1 \vdash_i^* C_2$ for some $1 \leq i \leq n$, and there is no $C'$ with $C_2 \vdash_i C'$.

Let us denote the derivation modes defined above as

$$M = \{t\} \cup \{\leq k, = k, \geq k \mid k \geq 1\}.$$

The *language* accepted (or recognized) by a DPAS $\mathcal{A}$ with *final states* in the computation mode $Y \in M$ is defined by

$$Rec_f(\mathcal{A}, Y) = \{w \in V^* \mid (w, q_1, Z_1, q_2, Z_2, \ldots, q_n, Z_n) \vdash_{\mathcal{A}}^{Y*}$$
$$(\varepsilon, s_1^{(m)}, \alpha_1^{(m)}, s_2^{(m)}, \alpha_2^{(m)}, \ldots, s_n^{(m)}, \alpha_n^{(m)}) \text{ with}$$
$$s_i^{(m)} \in F_i, \ 1 \leq i \leq n\}.$$

If we take $\alpha_i^{(m)} = \varepsilon$ and discard the condition $s_i^{(m)} \in F_i$, for all $1 \leq i \leq n$, then we obtain the definition of the *language* accepted by a DPAS $\mathcal{A}$ as above with *empty stacks* in the computation mode $Y \in M$ and we denote it by $Rec_\varepsilon(\mathcal{A}, Y)$.

We illustrate the above notions through an example which will also be useful in the sequel.

**Example 1.** *Let $\mathcal{A}$ be a DPAS of degree two over the alphabet $V = \{a, b, c, d\}$, having components $A_1$ and $A_2$ with the transition mappings $f_1$, $f_2$, as follows:*

$$
\begin{array}{ll}
f_1(q_1, a, Z_1) = \{(s_1, aZ_1)\} & f_2(q_2, b, Z_2) = \{(s_2, bZ_2)\} \\
f_1(s_1, a, a) = \{(s_1, aa)\} & f_2(s_2, b, b) = \{(s_2, bb)\} \\
f_1(s_1, c, a) = \{(p_1, \varepsilon)\} & f_2(s_2, d, b) = \{(p_2, \varepsilon)\} \\
f_1(p_1, c, a) = \{(p_1, \varepsilon)\} & f_2(p_2, d, b) = \{(p_2, \varepsilon)\} \\
f_1(p_1, \varepsilon, Z_1) = \{(s_f, Z_1)\} & f_2(p_2, \varepsilon, Z_2) = \{(s_f, Z_2)\} \\
f_1(s_f, X, Z_1) = \{(s_e, Z_1)\}, X \in \{b, c\} & f_2(s_f, Y, Z_2) = \{(s_e, Z_2)\}, Y \in \{a, c\} \\
f_1(p_1, b, X) = \{(s_e, X)\}, X \in \{a, Z_1\} & f_2(r, a, Y) = \{(s_e, Y)\}, r \in \{s_2, p_2\}, \\
 & \qquad Y \in \{b, Z_2\} \\
f_1(p_1, d, X) = \{(s_e, X)\}, X \in \{a, Z_1\} & f_2(p_2, c, Y) = \{(s_e, Y)\}, Y \in \{b, Z_2\}.
\end{array}
$$

*Let $q_1, Z_1$ and $q_2, Z_2$ be the initial state and the initial stack contents of components $A_1$ and $A_2$, respectively, and let $s_f$ be the only final state for both components. The language accepted by the system with final states in the computation mode $t$ is $Rec_f(\mathcal{A}, t) = \{a^n b^m c^n d^m \mid n, m \geq 1\}$, which is a non-context-free language.*

The *family of languages* accepted by distributed pushdown automata systems of degree $n$, $n \geq 1$, with final states or empty stacks in the computation mode $Y \in M$ is denoted by $\mathcal{L}_f(DPAS, Y, n)$ or $\mathcal{L}_\varepsilon(DPAS, Y, n)$, respectively.

## 3   Computational Power

Context-free CD grammar systems working in the $Y$-mode of derivation, $Y \in \{= 1, \geq 1\} \cup \{\leq k \mid k \geq 1\}$, are exactly as powerful as the context-free grammars [2]. A natural question is the following: What is the computational power of distributed pushdown automata systems under the corresponding computation modes? The first result shows that the accepted language class differs from the class of context-free languages.

**Proposition 1.** $\mathcal{L}_X(DPAS, Y, n) = Shuf_n(CF)$, for any $X \in \{f, \varepsilon\}$, $Y \in \{= 1, \geq 1\} \cup \{\leq k \mid k \geq 1\}$, $n \geq 2$.

*Proof.* Let us consider first the computation mode $= 1$. It can immediately be seen that for any DPAS $\mathcal{A} = (V, A_1, A_2, \ldots, A_n)$ the equality $Rec_X(\mathcal{A}, = 1) = \amalg_n(Rec_X(A_1), Rec_X(A_2), \ldots, Rec_X(A_n))$ holds for both $X \in \{f, \varepsilon\}$. Conversely, if for a given $n$-tuple of context-free languages $(L_1, L_2, \ldots, L_n)$ over an alphabet $V$ we construct an $n$-tuple of pushdown automata $(A_1, \ldots, A_n)$ such that $Rec_X(A_i) = L_i$, $1 \leq i \leq n$, then the language accepted by the distributed pushdown automata system $\mathcal{A} = (V, A_1, A_2, \ldots, A_n)$ in the $= 1$-mode of computation is $\amalg_n(L_1, L_2, \ldots, L_n)$.

Since there is no restriction concerning the order of the components of a DPAS following each other under the computation process, any language that can be obtained in the $\geq 1$-mode or the $\leq k$-mode of computation with $k \geq 1$ can be accepted in the $= 1$-mode of computation as well, and conversely.   $\square$

The following statement gives useful information on the form of the languages of distributed pushdown automata systems recognized in the $t$-mode, the $= k$-mode, and the $\geq k$-mode of computation for $k \geq 2$.

**Lemma 2.** *For any DPAS $\mathcal{A}$ of degree $n$, where $n \geq 2$, and for any $Y \in \{t\} \cup \{= k, \geq k \mid k \geq 2\}$, there are context-free languages $L_1, L_2, \ldots, L_n$ such that*

$$L_{\sigma(1)} L_{\sigma(2)} \ldots L_{\sigma(n)} \subseteq Rec_\varepsilon(\mathcal{A}, Y) \subseteq \amalg_n(L_1, L_2, \ldots, L_n)$$

*for any permutation $\sigma$ of $\{1, 2, \ldots, n\}$.*

*Proof.* Let $\mathcal{A} = (V, A_1, A_2, \ldots, A_n)$, with $n \geq 2$, be a DPAS. For $Y = t$, if we consider $L_i = Rec_\varepsilon(A_i)$, $1 \leq i \leq n$, then any word in $L_{\sigma(1)} L_{\sigma(2)} \ldots L_{\sigma(n)}$ is in $Rec_\varepsilon(\mathcal{A}, t)$, and $Rec_\varepsilon(\mathcal{A}, t)$ is in $\amalg_n(L_1, L_2, \ldots, L_n)$, thus the statement holds.

For $Y \in \{= k \mid k \geq 2\}$, we take as $L_i$, $1 \leq i \leq n$, the language recognized by $A_i$ with empty stack in such a way that each word in $L_i$ is accepted by a computation consisting of a multiple of $k$ number of steps. Actually, this language is $Rec_\varepsilon(\mathcal{A}_i, = k)$, where $\mathcal{A}_i = (V, A_i)$. Notice that $L_i$ is a context-free language. Then, the languages $L_1, \ldots, L_n$ satisfy the requirements of the statement.

A similar reasoning works for $Y \in \{\geq k \mid k \geq 2\}$ with the context-free languages $L_i = Rec_\varepsilon(\mathcal{A}_i, \geq k)$ and $\mathcal{A}_i = (V, A_i)$, $1 \leq i \leq n$.   $\square$

As for pushdown automata, the relationship between acceptance with empty stacks and with final states is a natural question also for distributed pushdown automata systems. We shall see that the ways of acceptance are not equally powerful in the $t$-mode of computation.

**Proposition 3.**

1. $\mathcal{L}_\varepsilon(DPAS, Y, n) = \mathcal{L}_f(DPAS, Y, n)$ for all $Y \in M \setminus \{t\}$ and $n \geq 1$.
2. $\mathcal{L}_\varepsilon(DPAS, t, n) \subset \mathcal{L}_f(DPAS, t, n)$ for all $n \geq 2$.

*Proof.* By Proposition 1, to prove the first statement it is sufficient to show that it holds for the $\geq k$ and $= k$ derivation modes, and this can easily be derived from the corresponding proofs in the case of pushdown automata. The inclusion in the second statement can also be demonstrated in a similar manner, while the properness of this inclusion follows from Example 1 and Lemma 2.     □

Now we continue by investigating the relationship of the language classes accepted in the different computation modes. Similarly to context-free CD grammar systems, the $= k$ and the $\geq k$ modes, for $k \geq 2$, increase the power of systems having at least two components.

**Proposition 4.** $\mathcal{L}_X(DPAS, Y, n) \subset \mathcal{L}_\varepsilon(DPAS, Z, n)$ for $X \in \{f, \varepsilon\}$, $Y \in \{= 1, \geq 1\} \cup \{\leq k \mid k \geq 1\}$, $Z \in \{= k, \geq k \mid k \geq 2\}$ and $n \geq 2$.

*Proof.* We first show that $\mathcal{L}_X(DPAS, Y, n) \subseteq \mathcal{L}_\varepsilon(DPAS, Z, n)$, for $n \geq 2$. By Proposition 1, it is sufficient to show that $Shuf_n(CF) \subseteq \mathcal{L}_\varepsilon(DPAS, Z, n)$. For given context-free languages $L_1, L_2, \ldots, L_n$, we construct $n$ pushdown automata $A_1, A_2, \ldots, A_n$ such that $L_i = Rec_\varepsilon(A_i)$, $1 \leq i \leq n$. By standard techniques, we can modify these automata in such a way that for each move of the original automaton, the new automaton performs exactly $k$ (for the $= k$-mode), or at least $k$ (for the $\geq k$-mode), steps. Then we form from these modified automata DPASs which accept with empty stacks, in the considered computation mode from $Z$, the language $⧢_n(L_1, L_2, \ldots, L_n)$.

To prove that the inclusion is proper, let us consider the non-context-free language $E = h(⧢(\{a^n b^n \mid n \geq 1\}, \{c^m d^m \mid m \geq 1\}))$, where $h(x) = x^k$, for $x \in \{a, b, c, d\}$. This language is in $\mathcal{L}_\varepsilon(DPAS, = k, 2) \cap \mathcal{L}_\varepsilon(DPAS, \geq k, 2)$, we leave the construction to the reader. We show that this language cannot be expressed as $⧢_p(E_1, E_2, \ldots, E_p)$ for any context-free languages $E_1, E_2, \ldots, E_p$, with $p \geq 2$. To see this, assume the contrary, namely $E = ⧢_p(E_1, E_2, \ldots, E_p)$. Let $w_i \in E_i$, $w_i \neq \varepsilon$, $1 \leq i \leq p$, such that $w_1 w_2 \ldots w_p \in E$. Furthermore, let $w_1 = a\alpha$, where $\alpha \in \{a, b, c, d\}^*$. Then, there exists $2 \leq t \leq p$ such that $w_t = e\beta$, for some $e \in \{a, b, c, d\}$, $\beta \in \{a, b, c, d\}^*$ and $w_1 w_2 \ldots w_t \notin a^+$.

Let $t$ be the minimal number with this property. If $w_t = e\beta$ with $e \neq a$, then the string $ae\alpha\beta w_2 \ldots w_{t-1} w_{t+1} \ldots w_p$ obviously cannot be in $E$. If $w_t = a^r\beta$, for some $r \geq 1$ where the first letter of $\beta$ is different from $a$, then we distinguish two cases. If $r$ is a multiple of $k$, then the word $aa^r\beta\alpha w_2 \ldots w_{t-1} w_{t+1} \ldots w_p$ cannot be in $E$. If $r$ is not a multiple of $k$, then the word $w_t w_1 \ldots w_{t-1} w_{t+1} \ldots w_p$ is not in $E$.

In conclusion, $Shuf_n(CF) = \mathcal{L}_X(DPAS, Y, n) \subset \mathcal{L}_\varepsilon(DPAS, Z, n)$, for $Y, Z$ as above, and $n \geq 2$. □

We do not know whether $Shuf_n(CF) \subset \mathcal{L}_X(DPAS, t, m)$ holds for any $X = \{\varepsilon, f\}$ and $n, m \geq 2$. On the other hand, we have the following proposition.

**Proposition 5.** $\mathcal{L}_\varepsilon(DPAS, t, 2) \setminus \mathcal{L}_f(DPAS, = k, n) \neq \emptyset$ for any $k \geq 1$ and $n \geq 1$.

*Proof.* For a given $k \geq 2$, let us consider the language

$$L = h(\sqcup\!\sqcup(\{a^n b^n \mid n \geq 1\}, \{c^m d^m \mid m \geq 1\})),$$

where $h$ is a homomorphism defined by $h(x) = x^{2k}$ for $x \in \{a, b, c, d\}$. Assume that $L$ can be accepted by a distributed pushdown automata system $\mathcal{A}$ in the $= k$-mode of computation. Let $w \in L$, and let us consider the process of accepting $w$. Since $w$ is a sequence of $2k$ long blocks of identical symbols, there is a position in each of these blocks where an active (or just activated) component starts its $k$-step long transition sequence and during this sequence reads a nonempty block of at most $k$ identical symbols. If modify $w$ by replacing with each other such sequences $x^+$ and $y^+$, $x, y \in \{a, b, c, d\}$, $x \neq y$, containing at most $k$ identical symbols, then we obtain a word which is still accepted by $\mathcal{A}$, but not in $L$ any more. This is a contradiction, so $L$ cannot be accepted by $\mathcal{A}$ in the $= k$-mode of computation.

However, $L$ is accepted with empty stacks in the $t$-mode of computation by the DPAS of degree 2 having the transition mapping defined as follows:

$$f_1(q_1, a, Z_1) = \{([q_1, 1], aZ_1)\} \quad f_2(q_2, c, Z_2) = \{([q_2, 1], cZ_2)\}$$
$$f_1([q_1, i], a, a) = \{([q_1, j], aa)\} \quad f_2([q_2, i], c, c) = \{([q_2, j], cc)\}$$

where $j = i + 1$, if $i \neq 2k$ or $j = 1$, if $i = 2k$,

$$f_1([q_1, i], X, Y) = \{(q_e^1, Y)\} \quad f_2([q_2, i], X', Y') = \{(q_e^2, Y)\},$$

where $X \in \{b, c, d\}, Y \in \{a, Z_1\}, X' \in \{a, b, d\}, Y' \in \{c, Z_2\}, i \neq 2k$,

$$f_1([q_1', 2k], \varepsilon, Z_1) = \{(q_f^1, \varepsilon)\} \quad f_2([q_2', 2k], \varepsilon, Z_2) = \{(q_f^2, \varepsilon)\}$$
$$f_1([q_1', i], b, a) = \{([q_1', j], \varepsilon)\} \quad f_2([q_2', i], d, c) = \{([q_2', j], \varepsilon)\},$$

where $j = i + 1$, if $i \neq 2k$ or $j = 1$, if $i = 2k$,

$$f_1([q_1, 2k], b, a) = \{([q_1', 1], \varepsilon)\} \quad f_2([q_2, 2k], d, c) = \{([q_2', 1], \varepsilon)\}$$
$$f_1([q_1', i], X, Y) = \{(q_e^1, Y)\} \quad f_2([q_2', i], X', Y') = \{(q_e^2, Y)\},$$

where $X \in \{a, c, d\}, Y \in \{a, Z_1\}, X' \in \{a, b, c\}, Y' \in \{c, Z_2\}, i \neq 2k$.

To see this, consider the following:

1. No component can be disabled before it made a multiple of $2k$ moves, except the case when that component entered an error state, $q_e^1$ or $q_e^2$. For each move among these ones the same symbol is effectively read.

2. After reading a symbol $b$ or a symbol $d$, a further reading of a symbol $a$ or $c$, respectively, blocks the automata system.

3. Each component checks by empty stack the equality between the number of occurrences of $a$ and $b$ or $c$ and $d$, respectively.

By these considerations it follows that $L$ is accepted by the distributed automata system, $\mathcal{A}$, above.     □

Although we are not able to precisely identify the class of languages recognized by distributed pushdown automata systems in the $= k$-mode or the $\geq k$-mode of computation, the next theorem describes the relation of these languages to the shuffles of context-free languages.

**Theorem 6.**

*1. For each $L \subseteq V^*$, $L \in \mathcal{L}_X(DPAS, = k, n)$, where $k \geq 2, n \geq 1$, there exist an alphabet $V'$, a homomorphism $h : V' \to V^*$ with $1 \leq |h(a)| \leq k$ for any $a \in V'$, and a language $L' \subseteq V'^*$ such that $h(L') = L$ and $L' \in \mathcal{L}_X(DPAS, = 1, n)$, $X \in \{\varepsilon, f\}$, $n \geq 1$.*

*2. For each $L \subseteq V^*$, $L \in \mathcal{L}_X(DPAS, \geq k, n)$, where $k \geq 2, n \geq 1$, there exist an alphabet $V'$, a homomorphism $h : V' \to V^*$ with $1 \leq |h(a)| \leq 2k - 1$ for any $a \in V'$, and a language $L' \subseteq V'^*$ such that $h(L') = L$ and $L' \in \mathcal{L}_X(DPAS, = 1, n)$, $X \in \{\varepsilon, f\}$, $n \geq 1$.*

*Proof.* We start with the first statement. Let $\mathcal{A} = (V, A_1, A_2, \ldots, A_n)$ be a distributed pushdown automata system of degree $n \geq 1$ with components $A_i = (Q_i, V, \Gamma_i, f_i, q_i, Z_i, F_i), 1 \leq i \leq n$, and let $L$ be the language accepted by $\mathcal{A}$ in the $= k$-mode of computation. We construct a distributed pushdown automata system $\mathcal{A}' = (V', A'_1, A'_2, \ldots, A'_n)$ with $A'_i = (Q'_i, V, \Gamma'_i, f'_i, q'_i, Z'_i, F_i)$ that accepts in the $= 1$-mode of computation a language $L' \in V'^*$ with $h(L') = L$.

The idea of the construction is the following: The number of all possible $k$ step moves that can follow a configuration of a component of $\mathcal{A}$ is finite, and all of them can be constructed based on the state, the next (at most) $k$ symbols of the input, and the (at most) $k$ topmost stack symbols in the starting configuration of the $k$ steps. Thus, we can encode this information into the states, input letters, and the transition mapping of the new automata such that each $k$ step move of component $A_i$ in $\mathcal{A}$ can be simulated by a specific one step move of component $A'_i$ of DPAS $\mathcal{A}', 1 \leq i \leq n$.

Now we construct the components of $\mathcal{A}'$. Let $V' = \{[a_1 \ldots a_j] \mid a_i \in V, 1 \leq i \leq j \leq k\}$, and let us define $h : V' \to V^*$ with $h([a_1 \ldots a_s]) = a_1 \ldots a_s$, for $1 \leq s \leq k$. Let $Q'_i = Q_i \cup \{q'_i\} \cup \{(<q, y_1, \ldots, y_p >\mid y_t \in \Gamma_i, 1 \leq t \leq p \leq k, q \in Q_i\}$, and $\Gamma'_i = \Gamma_i \cup \{Z'_i\}$. We define the transition mapping $f'_i$ of component $A'_i$, $1 \leq i \leq n$, as follows. For each $k$ step move performed by $A_i$, starting in state $q$ with the topmost stack symbols $y_1 \ldots y_p$, effectively used, $y_j \in \Gamma_i, 1 \leq j \leq p \leq k$, reading the string $a_1 a_2 \ldots a_l$ on the input tape (this string may be empty), ending in state $s$, and replacing the string of the topmost $p$ stack symbols above with $\beta \in \Gamma_i^*$, formally written as $(q, a_1 \ldots a_l, y_1 \ldots y_p) \vdash_i^{=k} (s, \varepsilon, \beta)$, we define the following sequence of transitions:

$$(<q, y_1 >, \varepsilon) \in f'_i(q, \varepsilon, y_1), y_1 \neq Z'_i,$$

$$(< q, y_1, \ldots, y_j >, \varepsilon) \in f_i'(< q, y_1, \ldots, y_{j-1} >, \varepsilon, y_j), \text{ for } 2 \leq j \leq p, y_j \neq Z_i',$$
$$(s, \beta A) \in f_i'(< q, y_1, \ldots, y_p >, [a_1 \ldots a_l], A), A \in \Gamma_i', \text{ if } a_1 a_2 \ldots a_l \neq \varepsilon,$$
$$(s, \beta A) \in f_i'(< q, y_1, \ldots, y_p >, \varepsilon, A), A \in \Gamma_i', \text{ if } a_1 a_2 \ldots a_l = \varepsilon.$$

Furthermore, for simulating the start of a computation in $A_i$ we add transition $(q_i, Z_i Z_i') \in f_i'(q_i', \varepsilon, Z_i')$, and to guarantee the acceptance with empty stacks we consider $(q, \varepsilon) \in f_i'(q, \varepsilon, Z_i')$ for any $q \neq q_i'$, where $q \in Q_i'$. Note that the sets of final states are the same for $A_i$ and $A_i'$, $1 \leq i \leq n$.

Now we describe the work of $\mathcal{A}'$. First, $A_i'$ being in a state $q$ ($A_i$ being in the same state) starts to store the $p$ topmost symbols of its stack in the intermediate states, symbol by symbol, provided that they are the $p$ topmost symbols of the stack of $A_i$. During this process, its reading head does not move. When $A_i'$ finishes this process, it reads a symbol $[a_1 \ldots a_l]$ from the input tape, enters state $s$ and writes $\beta$ on the top of its stack. This last move actually simulates the $k$ step move of $A_i$. If $A_i'$ interrupts its work before the last move is performed, and another component, say $A_j'$, starts its work, then the simulation still remains correct, since only the last move of the process reads symbols from the input. Thus, $L = Rec_X(\mathcal{A}, = k) = h(Rec_X(\mathcal{A}', = 1))$ clearly holds.

For the $\geq k$-mode, we can modify the above proof as follows: In addition to the transitions simulating the $k$ step moves performed by $A_i$, we construct all the transitions which simulate the moves performed by this automaton in $k + 1, \ldots, 2k - 1$ steps. Then, the new automaton $A_i'$ simulates the work of automaton $A_i$ in the $\geq k$-mode of computation. $\qquad \square$

While CD grammar systems are able to generate both semilinear and non-semilinear languages in the $t$-mode of derivation, DPASs with empty stacks accept only semilinear languages, as we see from Proposition 1 and Lemma 2. By Proposition 3, the same holds for final state acceptance in the computation modes different from $t$. Although we do not know if $\mathcal{L}_f(DPAS, t, n)$ contains only semilinear languages, we can prove the following statement.

**Theorem 7.** $\mathcal{L}_X(DPAS, Y, p) \subseteq \mathbf{NSPACE}(n) = CS$ for any $X \in \{f, \varepsilon\}$, $p \geq 1$, and $Y \in M$.

*Proof.* For $X \in \{f, \varepsilon\}$ and $Y \in \{= 1, \geq 1\} \cup \{\leq k \mid k \geq 1\}$ the statement follows from Proposition 1, since the languages of $Shuf_p(CF)$ are context-sensitive for any $p \geq 1$. For $X \in \{f, \varepsilon\}$ and $Y \in \{= k, \geq k \mid k \geq 2\}$ the statement follows from Theorem 6, since the context-sensitive language class is closed under non-erasing homomorphisms.

For $Y = t$, we give a proof for $X = f$ only, then by Proposition 3, the case of $X = \varepsilon$ follows. Let $\mathcal{A} = (V, A_1, A_2, \ldots, A_p)$ be a DPAS of degree $p \geq 1$ with $A_i = (Q_i, V, \Gamma_i, f_i, q_i, Z_i, F_i)$, $1 \leq i \leq p$. In the first phase we transform each component $A_i$ of $\mathcal{A}$ into a new component $A_i' = (Q_i', V, \Gamma_i', f_i', q_i, Z_i', F_i')$ which never removes the topmost symbol of its pushdown memory in an $\varepsilon$-move. The construction of $A_i'$ is done as follows:

$$Q_i' = Q_i \cup \{s_f^i\}, \quad F_i' = F_i \cup \{s_f^i\}, \quad \Gamma_i' = \{Z_i'\} \cup \{[q, X, r] \mid q, r \in Q_i, X \in \Gamma_i\}.$$

We construct recursively the following subset $\Lambda_i$ of $\Gamma'_i$:

$$\Lambda_{i,0} = \{[s, A, t] \mid (t, \varepsilon) \in f_i(s, \varepsilon, A)\},$$
$$\Lambda_{i,j+1} = \Lambda_{i,j} \cup \{[s, A, t] \mid (r, B_1 B_2 \ldots B_m) \in f_i(s, \varepsilon, A), m \geq 1, \text{ and there are}$$
$$\text{states } r_1, \ldots, r_{m-1} \text{ such that all symbols } [r, B_1, r_1], [r_1, B_2, r_2], \ldots,$$
$$[r_{m-1}, B_m, t] \in \Lambda_{i,j}\}, \text{ for all } j \geq 0.$$

Let now $\Lambda_i = \Lambda_{i,j}$ such that $\Lambda_{i,j} = \Lambda_{i,j+1}$. We define the transition mapping of $A'_i$ by:

$$f'_i(q_i, \varepsilon, Z'_i) = \{(q_i, [q_i, Z_i, s]) \mid s \in Q_i\} \cup \{(s^i_f, Z'_i) \mid [q_i, Z_i, s] \in \Lambda_i, s \in F_i\},$$
$$f'_i(q, \varepsilon, [q, A, s]) = \{(t, g_i([t, B_1, t_1] \ldots [t_{m-1}, B_m, s])) \mid (t, B_1 B_2 \ldots B_m) \in$$
$$f_i(q, \varepsilon, A), t_1, t_2, \ldots, t_{m-1} \in Q_i, m \geq 1\} \setminus \{(r, \varepsilon) \mid r \in Q_i\},$$
$$f'_i(q, a, [q, A, s]) = \{(t, g_i([t, B_1, t_1] \ldots [t_{m-1}, B_m, s])) \mid (t, B_1 B_2 \ldots B_m) \in$$
$$f_i(q, a, A), t_1, t_2, \ldots, t_{m-1} \in Q_i, m \geq 1\}, a \in V,$$

where $g_i$ is a finite substitution defined by

$$g_i([r, B, r']) = \begin{cases} [r, B, r'], & \text{if } [r, B, r'] \notin \Lambda_i \\ \{\varepsilon, [r, B, r']\}, & \text{if } [r, B, r'] \in \Lambda_i. \end{cases}$$

It is not difficult to see that $Rec_f(\mathcal{A}, t) = Rec_f(\mathcal{A}', t)$, where $\mathcal{A}' = (V, A'_1, A'_2, \ldots, A'_p)$.

Now, given $\mathcal{A}'$ we construct an on-line Turing machine with $p$ storage tapes which works as follows:

- The read-only input tape contains the string of length $n$ which is to be analyzed.
- The states are $(p+1)$-tuples formed by the states of the components of $\mathcal{A}'$ and a number between 1 and $p$ indicating which component is active.
- The $j$-th storage tape stores the prefix of length $n$ (the topmost $n$ symbols) of the contents of the pushdown memory of component $j$. By our construction, the rest of this memory cannot contribute to the accepting process.
- This machine accepts the input string when the string was completely read and all the components of the current state are final states in the corresponding automata.

At each step, the machine uses effectively only one storage tape, namely the tape corresponding to the active component of $\mathcal{A}'$, which can be updated according to the transition mapping of the active component. If the contents of a storage tape tends to become longer than $n$, then the $n$ topmost symbols are shifted to the beginning of the tape over the symbols existing there. By the construction of $\mathcal{A}'$, the deleted symbols cannot contribute to the accepting process, the prefix of length $n$ of the contents of each pushdown memory is sufficient for accepting or rejecting the input string. Therefore, $Rec_f(\mathcal{A}', t) \in \mathbf{NSPACE}(n)$ which concludes the proof.    $\square$

# 4   Final Remarks

In this paper we introduced distributed pushdown automata systems and examined their computational power in some basic computational modes. Many questions and problems have remained open; we plan to return to them in the near future. We finish the paper by pointing out three of them:

1. Does the relation $Shuf_n(CF) \subseteq \mathcal{L}_X(DPAS, t, n)$ for any $X \in \{f, \varepsilon\}$, $n \geq 2$ hold? We suspect a negative answer.

2. Are the families $\mathcal{L}_\varepsilon(DPAS, t, n)$ and $\mathcal{L}_\varepsilon(DPAS, = k, n)$ for any $k, n \geq 2$ incomparable? We suspect an affirmative answer.

3. Are all the hierarchies $\mathcal{L}_X(DPAS, Y, n) \subseteq \mathcal{L}_X(DPAS, Y, n + 1)$ infinite? A partial answer (affirmative) is given in [1] for $Y \in \{= 1, \geq 1\} \cup \{\leq k \mid k \geq 1\}$.

# References

1. M.H. ter Beek, E. Csuhaj-Varjú, V. Mitrana, Teams of pushdown automata, submitted.

2. E. Csuhaj-Varjú, J. Dassow, On cooperating distributed grammar systems, *J. Inform. Process. Cybern., EIK*, 26 (1990), 49–63.

3. E. Csuhaj-Varjú, J. Dassow, J. Kelemen, Gh. Păun, *Grammar Systems. A grammatical approach to distribution and cooperation*, Gordon and Breach, 1994.

4. E. Csuhaj-Varjú, C. Martín-Vide, V. Mitrana, Gy. Vaszil, Parallel communicating pushdown automata systems, *Intern. J. Found. Comp. Sci.*, 11:4 (2000), 633–650.

5. J. Dassow, Gh. Păun, G. Rozenberg, Grammar systems, in vol. 2 of [12].

6. J. Dassow, V. Mitrana, "Stack cooperation in multi-stack pushdown automata", *J. Comput. System Sci.* 58 (1999), 611–621.

7. O. H. Ibarra, One two-way multihead automata, *J. Comput. System Sci.* 7 (1973), 28–36.

8. C. Martín-Vide, V. Mitrana, Some undecidable problems for parallel communicating finite automata systems, *Inform. Proc. Letters*, 77:5-6 (2001), 239–245.

9. C. Martín-Vide, V. Mitrana, Parallel communicating automata systems. A survey. Invited article in *Korean Journal of Computational and Applied Mathematics*, 7:2 (2000), 237–257.

10. C. Martín-Vide, A. Mateescu, V. Mitrana, Parallel finite automata systems communicating by states, *Intern. J. Found. Comp. Sci.*, 13:5 (2002), 733–749.

11. Gh. Păun, L. Sântean, Parallel communicating grammar systems: the regular case, *Ann. Univ. Bucharest, Ser. Matem.-Inform.* 38 (1989), 55–63.

12. G. Rozenberg, A. Salomaa, *Handbook of Formal Languages*, Springer-Verlag, Berlin, vol. 1–3, 1997.

# On Well Quasi-orders on Languages[*]

Flavio D'Alessandro[1] and Stefano Varricchio[2]

[1] Dipartimento di Matematica, Università di Roma "La Sapienza"
Piazzale Aldo Moro 2, 00185 Roma, Italy
dalessan@mat.uniroma1.it
http://mat.uniroma1.it/people/dalessandro

[2] Dipartimento di Matematica, Università di Roma "Tor Vergata"
via della Ricerca Scientifica, 00133 Roma, Italy
varricch@mat.uniroma2.it
http://mat.uniroma2.it/~varricch

**Abstract.** Let $G$ be a context-free grammar and let $L$ be the language of all the words derived from any variable of $G$. We prove the following generalization of Higman's theorem: any division order on $L$ is a well quasi-order on $L$. We also give applications of this result to some quasi-orders associated with unitary grammars.

## 1 Introduction

A *quasi-order* on a set $S$ is called a *well quasi-order* (*wqo*) if every nonempty subset $X$ of $S$ has at least one minimal element in $X$ but no more than a finite number of (non-equivalent) minimal elements.

Well quasi-orders have been widely investigated in the past. In [9] Higman gives a very general theorem on division orders in abstract algebras that in the case of semigroups becomes: *Let $S$ be a semigroup quasi-ordered by a division order $\leq$. If there exists a generating set of $S$ well quasi-ordered by $\leq$, then $S$ will also be so.* From this one derives that the *subsequence ordering* in free monoids is a wqo.

In [12] Kruskal extends Higman's result, proving that some embeddings on finite trees are well quasi-orders. In the last years many papers have been devoted to the applications of wqo's to formal language theory. The most important result is a generalization of the famous Myhill-Nerode theorem on regular languages. In [6] Ehrenfeucht et al. proved that a language is regular if and only if it is upward-closed with respect to a monotone well quasi-order. From this result many regularity conditions have been derived (see for instance [2,3,4,5]).

In [6] unavoidable sets of words are characterized in terms of the wqo property of a suitable unitary grammar: a set $I$ is unavoidable if and only if the derivation relation $\Rightarrow_I^*$ of the unitary semi-Thue system associated with the finite set $I \subseteq A^+$ is a wqo. An extension of the previous result has been given by Haussler in [8], considering set of words which are *subsequence unavoidable*.

---

[*] This work was partially supported by MIUR project *"Linguaggi formali e automi: teoria e applicazioni"*.

Z. Ésik and Z. Fülöp (Eds.): DLT 2003, LNCS 2710, pp. 230–241, 2003.
© Springer-Verlag Berlin Heidelberg 2003

In [11] some extensions of Higman and Kruskal's theorem to regular languages and rational trees have been given. Further applications of the wqo theory to formal languages are given in [7,10].

In this paper we give a new generalization of Higman's theorem. First of all we give the notion of *division order* on a language $L$: a quasi order $\leq$ on $A^*$ is called a *division order* on $L$ if it is monotone and for any $u, v \in L$ if $u$ is factor of $v$ then $u \leq v$. When $L$ is the whole free monoid $A^*$ this notion is equivalent to the classical one, but, in general, a quasi-order on $A^*$ could be a division order on a set $L$ and not on $A^*$. Then, given a context-free grammar $G$ with set of variables $V = \{X_1, X_2, \ldots, X_n\}$, let $L_i$ be the language of the words generated setting $X_i$ as start symbol and let $L = \bigcup_{i=1}^n L_i$. Our main theorem states that any division order on $L$ is a well quasi-order on $L$. In particular, if $L$ is a context-free language generated by a grammar with only one variable, then any division order on $L$ is a wqo on $L$. This generalizes Higman's theorem on finitely generated free monoids, since for any finite alphabet $A$, the set $A^*$ can be generated by a context-free grammar having only one variable.

In the second part of the paper we study the wqo property in relation to some quasi-orders associated with unitary grammars. Let $I$ be a finite set of words and let $\Rightarrow_I^*$ be the derivation relation associated with the semi-Thue system

$$\{\epsilon \rightarrow u, \ u \in I\}.$$

One can also consider the relation $\vdash_I^*$ as the transitive and reflexive closure of $\vdash_I$ where $v \vdash_I w$ if

$$v = v_1 v_2 \cdots v_{n+1},$$

$$w = v_1 a_1 v_2 a_2 \cdots v_n a_n v_{n+1},$$

where the $a_i$'s are letters, and $a_1 a_2 \cdots a_n \in I$.

We set $L_I^\epsilon = \{w \in A^* \mid \epsilon \Rightarrow_I^* w\}$, $L_{\vdash_I}^\epsilon = \{w \in A^* \mid \epsilon \vdash_I^* w\}$ and prove that

- There exists a finite set $I$ such that $\Rightarrow_I^*$ is not a wqo on $L_I^\epsilon$.
- There exists a finite set $I$ such that $\vdash_I^*$ is not a wqo on $L_{\vdash_I}^\epsilon$.
- For any finite set $I$ the relation $\vdash_I^*$ is a wqo on $L_I^\epsilon$.

Finally we observe that for any finite set $I$, the relation $\vdash_I^*$ is a division order on the language $L_{\vdash_I}^\epsilon$. Therefore, our main theorem does not hold in general on an arbitrary language. On the other hand the language $L_{\vdash_I}^\epsilon$ is not context-free.

## 2    Preliminaries

The main notions and results concerning quasi-orders and languages are shortly recalled in this section. Let $A$ be a finite *alphabet* and $A^*$ the free monoid generated by $A$. The elements of $A$ are usually called *letters* and those of $A^*$ *words*. The identity of $A^*$ is denoted $\epsilon$ and called the *empty word*.

A word $w \in A^*$ can be written uniquely as a sequence of letters as $w = a_1 a_2 \cdots a_n$, with $a_i \in A$, $1 \leq i \leq n$, $n > 0$. The integer $n$ is called the *length* of $w$ and denoted $|w|$. For all $a \in A$, $|w|_a$ denotes the number of occurrences of the letter $a$ in $w$. Let $w \in A^*$. The word $u \in A^*$ is a *factor* of $w$ if there exist $p, q \in A^*$ such that $w = puq$. If $w = uq$, for some $q \in A^*$ (resp. $w = pu$, for some $p \in A^*$), then $u$ is called a *prefix* (resp. a *suffix*) of $w$. The set of all prefixes (resp. suffixes, factors) of $w$ is denoted $Pref(w)$ (resp. $Suf(w)$, $F(w)$).

A subset $L$ of $A^*$ is called a *language*. If $L$ is a language of $A^*$, then $\mathrm{alph}(L)$ is the smallest subset $B$ of $A$ such that $L \subseteq B^*$. A binary relation $\leq$ on a set $S$ is a *quasi-order* (qo) if $\leq$ is reflexive and transitive. Moreover, if $\leq$ is symmetric, then $\leq$ is an equivalence relation. The meet $\leq \cap \leq^{-1}$ is an equivalence relation $\sim$ and the quotient of $S$ by $\sim$ is a *poset* (partially ordered set).

An element $s \in X \subseteq S$ is *minimal* in $X$ with respect to $\leq$ if, for every $x \in X$, $x \leq s$ implies $x \sim s$. For $s, t \in S$ if $s \leq t$ and $s$ is not equivalent to $t$ mod $\sim$, then we set $s < t$. A part $X$ of $S$ is *upper-closed*, or simply *closed*, with respect to $\leq$ if the following condition is satisfied:

$$\text{if } x \in X \text{ and } x \leq y \text{ then } y \in X.$$

We shall denote by $\mathrm{Cl}(X)$ the *closure* of $X$,

$$\mathrm{Cl}(X) = \{s \in S \mid \exists\, x \in X \text{ such that } x \leq s\},$$

so that $X$ is closed if and only if $X = \mathrm{Cl}(X)$. For any $X \subseteq S$ one has $X \subseteq \mathrm{Cl}(X)$. Moreover, if $Y \subseteq X$, then $\mathrm{Cl}(Y) \subseteq \mathrm{Cl}(X)$. A closed set $X$ is called *finitely generated* if there exists a finite subset $F$ of $X$ such that $\mathrm{Cl}(F) = X$.

A quasi-order in $S$ is called a *well quasi-order* (wqo) if every non-empty subset $X$ of $S$ has at least one minimal element but no more than a finite number of (non-equivalent) minimal elements. We say that a set $S$ is *well quasi-ordered* (wqo) by $\leq$, if $\leq$ is a well quasi-order on $S$.

There exists several conditions which characterize the concept of well quasi-order and that can be assumed as equivalent definitions (cf. [5]).

**Theorem 1.** *Let $S$ be a set quasi-ordered by $\leq$. The following conditions are equivalent:*

i. *$\leq$ is a well quasi-order;*
ii. *the ascending chain condition holds for the closed subsets of $S$;*
iii. *every infinite sequence of elements of $S$ has an infinite ascending subsequence;*
iv. *if $s_1, s_2, \ldots, s_n, \ldots$ is an infinite sequence of elements of $S$, then there exist integers $i, j$ such that $i < j$ and $s_i \leq s_j$;*
v. *there exists neither an infinite strictly descending sequence in $S$ (i.e. $\leq$ is well founded), nor an infinity of mutually incomparable elements of $S$;*
vi. *$S$ has the finite basis property, i.e. every closed subset $S$ is finitely generated.*

Let $\sigma = \{s_i\}_{i \geq 1}$ be an infinite sequence of elements of $S$. Then $\sigma$ is called *good* if it satisfies condition (iv) of Theorem 1 and it is called *bad* otherwise, that is, for all integers $i, j$ such that $i < j$, $s_i \not\leq s_j$. It is worth noting that, by condition (iv) above, a useful technique to prove that $\leq$ is a wqo on $S$ is to prove that no bad sequence exists in $S$.

If $\rho$ and $\sigma$ are two relations on sets $S$ and $T$ respectively, then the direct product $\rho \otimes \sigma$ is the relation on $S \times T$ defined as

$$(a, b) \; \rho \otimes \sigma \; (c, d) \iff a \, \rho \, c \text{ and } b \, \sigma \, d.$$

The following lemma is well known (see [5], Ch. 6).

**Lemma 1.** *The following conditions hold:*
*1) Every subset of a wqo set is wqo;*
*2) If $S$ and $T$ are wqo by $\leq_S$ and $\leq_T$ respectively, then $S \times T$ is wqo by $\leq_S \otimes \leq_T$.*

Let us now suppose that the set $S$ is a semigroup. Let $S^1 = S$ if $S$ is a monoid, otherwise $S^1$ is the monoid obtained by adding the identity to $S$.

**Definition 1.** *A quasi-order $\leq$ in a semigroup $S$ is* monotone on the right (on the left) *if for all $x_1, x_2, y \in S$*

$$x_1 \leq x_2 \text{ implies } x_1 y \leq x_2 y \; (y x_1 \leq y x_2).$$

*A quasi-order is* monotone *if it is monotone on the right and on the left.*

**Definition 2.** *A quasi-order $\leq$ in a semigroup $S$ is a* division order *if it is monotone and, for all $s \in S$ and $x, y \in S^1$*

$$s \leq xsy.$$

The ordering by division in abstract algebras was studied by Higman [9] who proved a general theorem that in the case of semigroups becomes:

**Theorem 2.** *Let $S$ be a semigroup quasi-ordered by a division order $\leq$. If there exists a generating set of $S$ well quasi-ordered by $\leq$ then so will be $S$.*

If $n$ is a positive integer, then the set of all positive integers less or equal than $n$ is denoted $[n]$. If $f$ is a map then $\mathrm{Im}(f)$ denotes the set of images of $f$.

## 3  Main Result

We now prove our main result. For this purpose, it is useful to give some preliminary definitions and results. We assume the reader to be familiar with the basic theory of context–free languages. It is useful to recall few elements of the vocabulary (cf. [1]).

A *context-free grammar* is a triplet $G = (V, \, A, \, P)$ where $V$ and $A$ are finite sets of *variables* and *terminals*, respectively. $P$ is the set of *productions*: each element of $P$ is of the form $X \to u$ with $X \in V$ and $u \in \{V \cup A\}^*$.

The relation $\Rightarrow_G$, simply denoted by $\Rightarrow$, is the binary relation on the set $\{V \cup A\}^*$ defined as: $w_1 \Rightarrow w_2$ if and only if $w_1 = w' X w''$, $w_2 = w' u w''$ where $X \to u$ is a production of $G$ and $w', w'' \in \{V \cup A\}^*$. The relation $\Rightarrow^*$ is the reflexive and transitive closure of $\Rightarrow$. Let $V = \{X_1, X_2, \ldots, X_n\}$. For every $i = 1, \ldots, n$, the language generated by $X_i$ is $L(X_i) = \{u \in A^* \mid X_i \Rightarrow^* u\}$. We shall adopt the convention to denote $L(X_i)$ by $L_i$ whenever no ambiguity or confusion arises.

**Definition 3.** *Let $\leq$ be a quasi-order on $A^*$. Then $\leq$ is said to be compatible with $G$ if the following condition holds:*

*for every production of $G$ of the kind $X_i \longrightarrow u_1 Y_1 u_2 Y_2 \cdots u_m Y_m u_{m+1}$, where, $u_k \in A^*$, for $k = 1, \ldots, m+1$, and $Y_k \in V$, $k = 1, \ldots, m$, one has:*

$$x_k \leq u_1 x_1 u_2 x_2 \cdots u_m x_m u_{m+1},$$

*for any choice of $x_i \in L(Y_i)$, for $i = 1, \ldots, m$ and for any $k \in \{1, \ldots, m\}$.*

The following result holds.

**Proposition 1.** *If $\leq$ is a monotone quasi-order compatible with $G$, then $\leq$ is a wqo on $L = \bigcup_{i=1}^{n} L_i$.*

*Proof.* In this proof, for the sake of simplicity, we assume that the grammar $G$ does not contain neither unitary productions nor $\epsilon$-productions. By contradiction, deny the claim of the proposition. Hence there exists a bad sequence in $L$. Select $v_1 \in L$ such that $v_1$ is the first term of a bad sequence in $L$ and its length $|v_1|$ is as small as possible. Then select a word $v_2 \in L$ such that $v_1$, $v_2$ (in that order) are the first two terms of a bad sequence in $L$ and $|v_2|$ is as small as possible. Then select a word $v_3 \in L$ such that $v_1$, $v_2$, $v_3$ (in that order) are the first three terms of a bad sequence in $L$ and $|v_3|$ is as small as possible. Assuming the Axiom of Choice, this process yields a bad sequence $\gamma = \{v_i\}_{i \geq 1}$ in $L$. This sequence is minimal in the following sense: let $\alpha = \{z_i\}_{i \geq 1}$ be a bad sequence of $L$ and let $k$ be a positive integer such that, for $i = 1, \ldots, k$, $z_i = v_i$. Then $|v_{k+1}| \leq |z_{k+1}|$.

Since $P$ is finite, we may consider a subsequence $\sigma = \{v_{i_\ell}\}_{i_\ell \geq 1}$ of the sequence above, which satisfies the following property:

$$\forall \ell \geq 1, \ X_k \Rightarrow p \Rightarrow^* v_{i_\ell}, \tag{1}$$

where $X_k \to p$ is a production and $p = u_1 Y_1 u_2 Y_2 \cdots u_m Y_m u_{m+1}$. By the sake of simplicity, let us rename the terms of $\sigma$ as: for every $\ell \geq 1$, $w_\ell = v_{i_\ell}$. Hence, by (1), for every $\ell \geq 1$, one has

$$w_\ell = u_1 x_1^\ell u_2 x_2^\ell \cdots u_m x_m^\ell u_{m+1}, \text{ with}$$

$$x_1^\ell \in L(Y_1), \ x_2^\ell \in L(Y_2), \ \ldots, \ x_m^\ell \in L(Y_m).$$

For every $j = 1, \ldots, m$, set $F_j = \{x_j^i\}_{i \geq 1}$.

The following claim is crucial.

**Claim.** *For every $j = 1, \ldots, m$, $F_j$ is wqo by $\leq$.*

**Proof of the Claim:** By contradiction, let $j$ be a positive integer with $1 \leq j \leq m$ such that $F_j$ is not wqo by $\leq$. Let $\tau = \{y_i\}_{i \geq 1}$ be a bad sequence in $F_j$.

We first observe that, for all $i \geq 1$, there exists a positive integer $g(i)$ such that $y_i = x_j^{g(i)}$. Without loss of generality we may assume that for every $i \geq 1$, $g(i) \geq g(1)$. Indeed, if the above condition is not satisfied one can consider a subsequence of $\tau$ satisfying this property.

Consider now the sequence

$$v_1, \; v_2, \; \ldots, \; v_{i_{g(1)}-1}, \; y_1, \; y_2, \; \ldots, \; y_i \; \ldots$$

By construction, every term of the sequence above belongs to $L$. Moreover one easily proves the latter sequence is bad. Since $\gamma$ and $\{y_i\}_{i\geq 1}$ are bad sequences in $L$, this amounts to show that for $h, k$, $1 \leq h \leq i_{g(1)} - 1$, $k \geq 1$, one has $v_h \not\leq y_k$. Indeed, suppose $v_h \leq y_k$. Since $y_k = x_j^{g(k)}$, then $v_h \leq x_j^{g(k)}$. Since for every $\ell = 1, \ldots, m$, $x_\ell^{g(k)} \in L(Y_\ell)$, the fact that $\leq$ is compatible with $G$ entails

$$x_j^{g(k)} \; \leq \; u_1 x_1^{g(k)} u_2 \cdots u_m x_m^{g(k)} u_{m+1} = w_{g(k)} = v_{i_{g(k)}}.$$

Hence $v_h \leq v_{i_{g(k)}}$. Since $g(1) \leq g(k)$, one has $h < i_{g(1)} \leq i_{g(k)}$ and this contradicts that $\gamma$ is bad. Hence $v_h \not\leq y_k$.

Now we observe that $y_1$ is a proper factor of $w_{g(1)} = v_{i_{g(1)}}$, since the grammar does not contain neither unitary productions nor $\epsilon$-productions. Thus $|y_1| < |v_{i_{g(1)}}|$ and this contradicts that $\gamma$ is minimal. Hence, no bad sequence in $F_j$ exists so $F_j$ is wqo by $\leq$.   ◇

Let $\mathcal{F} = F_1 \times F_2 \times \cdots \times F_j \times \cdots \times F_m$. By condition (2) of Lemma 1 and the claim above, one has the set $\mathcal{F}$ is wqo by the canonical extension of $\leq$ on $\mathcal{F}$. Consider now the sequence of $\mathcal{F}$ defined as

$$\{ (x_1^i, \; x_2^i, \; x_3^i, \; \ldots, \; x_m^i) \}_{i\geq 1}.$$

Since $\mathcal{F}$ is wqo, the latter sequence is good so there exist two positive integers $i$, $j$ such that $i < j$ and, for every $\ell = 1, \ldots, m$, $x_\ell^i \leq x_\ell^j$. The previous condition and the monotonicity of $\leq$ entails $w_i \leq w_j$. The latter contradicts that $\gamma$ is bad. This proves that $L$ is wqo by $\leq$.

If the grammar $G$ contains either unitary productions or $\epsilon$-productions, the proof is almost the same. One has only to consider minimal bad sequences, assuming as a parameter the minimal length of a derivation of a word.

$\square$

The corollary below immediately follows from condition (1) of Lemma 1 and Proposition 1.

**Corollary 1.** *Let $G = (V, A, P)$ be a context-free grammar where $V = \{X_1, X_2, \ldots, X_n\}$. If $\leq$ is a monotone quasi-order compatible with $G$, then $L_i$ is wqo by $\leq$ for every $i = 1, \ldots, n$.*

The following notion is a natural extension of that of division order in the free monoid.

**Definition 4.** *Let $L \subseteq A^*$ be a language and let $\leq$ be a quasi-order. Then $\leq$ is a division order on $L$ if $\leq$ is monotone and the following condition holds:*

$$u \leq xuy \text{ for every } u \in L, \; x, \; y \in A^* \text{ with } xuy \in L.$$

The following theorem holds.

**Theorem 3.** *Let $G = (V, A, P)$ be a context-free grammar and, according to the previous notation, let $L = \bigcup_{i=1}^{n} L_i$ be the union of all languages generated by $G$. If $\leq$ is a division order on $L$, then $\leq$ is a well quasi-order on $L$.*

*Proof.* It is easily checked that $\leq$ is compatible with $G$. Indeed, let $X_i \rightarrow p$ be a production of $G$. Suppose $p = u_1 Y_1 \cdots u_m Y_m u_{m+1}$ with $u_i \in A^*$, for $i = 1, \ldots, m+1$ and $Y_i \in V$, for $i = 1, \ldots, m$. Let $x_i \in L(Y_i)$ for every $i = 1, \ldots, m$. Hence $u_1 x_1 \cdots u_m x_m u_{m+1} \in L$. Since $\leq$ is a division order on $L$, one has

$$x_i \leq (u_1 x_1 \cdots x_{i-1} u_i) x_i (u_{i+1} x_{i+1} \cdots u_m x_m u_{m+1}),$$

for every $i = 1, \ldots, m$.

Then the result follows from Proposition 1. □

## 4    Well Quasi-orders and Unitary Grammars

We now prove an interesting corollary of Proposition 1 concerning unitary semi-Thue systems. Following [5], we recall that a *rewriting system*, or *semi-Thue system* on an alphabet $A$ is a pair $(A, \pi)$ where $\pi$ is a binary relation on $A^*$. Any pair of words $(p, q) \in \pi$ is called a *production* and denoted by $p \rightarrow q$. Let us denote by $\Rightarrow_\pi$ the derivation relation of $\pi$, that is, for $u, v \in A^*$, $u \Rightarrow_\pi v$ if and only if

$$\exists\, (p, q) \in \pi \text{ and } \exists\, h, k \in A^* \text{ such that } u = hpk, \quad v = hqk.$$

The *derivation relation* $\Rightarrow_\pi^*$ is the transitive and reflexive closure of $\Rightarrow_\pi$. One easily verifies that $\Rightarrow_\pi^*$ is a monotone quasi-order on $A^*$.

A semi-Thue system is called *unitary* if $\pi$ is a finite set of productions of the kind

$$\epsilon \rightarrow u, \ u \in I, \ I \subseteq A^*.$$

Such a system is then determined by the finite set $I \subseteq A^*$. The derivation relation of it and its transitive and reflexive closure are denoted by $\Rightarrow_I$ (or, simply, $\Rightarrow$) and $\Rightarrow_I^*$ (or, simply, $\Rightarrow^*$), respectively. We set $L_I^\epsilon = \{u \in A^* \mid \epsilon \Rightarrow^* u\}$.

The following Lemma states that a unitary semi-Thue system may be simulated by a suitable context-free grammar and it belongs to the folklore.

**Definition 5.** *Let $I$ be a finite subset of $A^*$. Let $G_I = (V, A, P)$ be the context-free grammar where $V = \{X\}$, $A = \text{alph}(I)$ and $P$ is the set of productions defined as:*

- $X \longrightarrow \epsilon$,
- *for every $u = a_1 \cdots a_n \in I$, where $a_i \in A$, $1 \leq i \leq n$,*

$$X \longrightarrow X a_1 X a_2 X \cdots X a_n X.$$

**Lemma 2.** *Let $I$ be a finite subset of $A^*$. Then $L(G_I) = L(X) = L_I^\epsilon$.*

Let $I$ be a finite subset of $A^*$. Then we denote by $\vdash_I$ the binary relation of $A^*$ defined as: for every $u, v \in A^*$, $u \vdash_I v$ if

$$u = u_1 u_2 \cdots u_{n+1},$$

$$v = u_1 a_1 u_2 a_2 \cdots u_n a_n u_{n+1},$$

with $u_i \in A^*$, $a_i \in A$, and $a_1 \cdots a_n \in I$.

The relation $\vdash_I^*$ is the transitive and reflexive closure of $\vdash_I$. One easily verifies that $\vdash_\pi^*$ is a monotone quasi-order on $A^*$. Moreover $L_{\vdash_I}^\epsilon$ denotes the set of all words derived from the empty word by applying $\vdash_I^*$, that is

$$L_{\vdash_I}^\epsilon = \{ u \in A^* \mid \epsilon \vdash_I^* u \}.$$

Generally $\Rightarrow_I^*$ is not a wqo on $L_I^\epsilon$. In fact let $A = \{a, b, c\}$, $I = \{ab, c\}$, and consider the sequence $\sigma = \{acb, aacbb, aaacbbb, \ldots, a^n cb^n \ldots\}$. It is easy to see that the elements of $\sigma$ are pairwise incomparable with respect to $\Rightarrow_I^*$, so that $\sigma$ is bad. We observe that $\sigma$ is not bad with respect to $\vdash_I^*$. Indeed for any $n, m$, $n \leq m$, one has $a^n cb^n \vdash_I^* a^m cb^m$.

The following theorem holds.

**Theorem 4.** *Let $I$ be a finite set of words. Then $\vdash_I^*$ is wqo on $L_I^\epsilon$.*

*Proof.* First we prove that $\vdash_I^*$ is compatible with the grammar $G_I$. According to the definition of $G_I$ and by Lemma 2, the task amounts to show that, for every $v = a_1 \cdots a_m \in I$ and, for every $u_1, \ldots, u_{m+1} \in L_I^\epsilon$, one has, for $i = 1, \ldots m+1$,

$$u_i \vdash_I^* u_1 a_1 u_2 a_2 \cdots u_m a_m u_{m+1}.$$

Since $\Rightarrow_I^*$ is monotone and $u_i \in L_I^\epsilon$, for every $i = 1, \ldots, m+1$, one has $u_i \Rightarrow_I^* u_1 \cdots u_{m+1}$. So $u_i \vdash_I^* u_1 \cdots u_{m+1} \vdash_I^* u_1 a_1 \cdots a_m u_{m+1}$. Finally, the claim follows from Proposition 1 and Lemma 2. □

We prove that, for a suitable finite set $I$ of words over a finite alphabet, the quasi-order $\vdash_I^*$ is not a wqo on $L_{\vdash_I}^\epsilon$. For this purpose, let $A = \{a, b, c, d\}$ be a four-letter alphabet and let $\bar{A} = \{\bar{a}, \bar{b}, \bar{c}, \bar{d}\}$ be a disjoint copy of $A$. Let $\tilde{A} = A \cup \bar{A}$ and let $I = \{a\bar{a}, b\bar{b}, c\bar{c}, d\bar{d}\}$.

Now consider the sequence $\{S_n\}_{n \geq 1}$ of words of $\tilde{A}^*$ defined as: for every $n \geq 1$,

$$S_n = adb\bar{b}c\bar{c}\bar{a}(a\bar{d}dc\bar{c}c\bar{c}\bar{a})^n a\bar{d}b\bar{b}\bar{a}.$$

The following result holds.

**Proposition 2.** *$\{S_n\}_{n \geq 1}$ is a bad sequence in $\tilde{A}^*$ with respect to $\vdash_I^*$. Hence $\vdash_I^*$ is not wqo on $\tilde{A}^*$. In particular, $\vdash_I^*$ is not wqo on $L_{\vdash_I}^\epsilon$.*

*Remark 1.* We observe that one can easily prove that $\vdash_I^*$ is a division order on $L_{\vdash_I}^\epsilon$. Therefore, if one drops the hypothesis on the structure of $L$, Theorem 3 does not hold any more. On the other hands the language $L_{\vdash_I}^\epsilon$ is not context-free.

In order to prove Proposition 2, we need some preliminary definitions and lemmas.

**Lemma 3.** *Let* $u \in L_{\vdash_I}^\epsilon$. *For every* $p \in Pref(u)$ *and* $x \in A$, $|p|_{\bar{x}} \leq |p|_x$.

*Proof.* $u \in L_{\vdash_I}^\epsilon$ implies $\epsilon \vdash_I^k u$, for some $k \geq 0$. By induction on $k$, one easily derives the assertion. $\qquad\square$

The following definitions will be used later.

**Definition 6.** *Let* $u = a_1 \cdots a_n$ *and* $v = b_1 \cdots b_m$ *be two words over* $\tilde{A}$ *with* $n \leq m$. *An embedding of* $u$ *in* $v$ *is a map* $f : [n] \longrightarrow [m]$ *such that* $f$ *is increasing and, for every* $i = 1, \ldots, n$, $a_i = b_{f(i)}$.

**Definition 7.** *Let* $u, v \in \tilde{A}^*$ *and let* $f$ *be an embedding of* $u$ *in* $v$. *Let* $v = b_1 \cdots b_m$. *Then* $\langle v - u \rangle_f$ *is the subword of* $v$ *defined as*

$$\langle v - u \rangle_f = b_{i_1} \cdots b_{i_\ell} \text{ where, for every } k = 1, \ldots \ell,$$

$$i_k \notin \mathrm{Im}(f).$$

*The word* $\langle v - u \rangle_f$ *is called the* difference *of* $v$ *and* $u$ *with respect to* $f$.

It is useful to remark that $\langle v - u \rangle_f$ is obtained from $v$ by deleting, one by one, all the letters of $u$ according to $f$.

*Example 1.* Let $u = a\bar{a}$ and $v = ab\bar{a}\bar{b}a\bar{a}$. Let $f$ and $g$ be two embeddings of $u$ in $v$ defined respectively as: $f(1) = 1$, $f(2) = 3$, and $g(1) = 5$, $g(2) = 6$. Then we have $\langle v - u \rangle_f = b\bar{b}a\bar{a}$ and $\langle v - u \rangle_g = ab\bar{a}\bar{b}$.

*Remark 2.* A word $u$ is a subsequence of $v$ if and only if there exists an embedding of $u$ in $v$.

*Remark 3.* An embedding $f$ of $u$ in $v$ is uniquely determined by two factorizations of $u$ and $v$ of the form

$$u = u_1 u_2 \cdots u_n, \qquad v = v_1 u_1 v_2 u_2 \cdots v_n u_n v_{n+1}$$

with $u_i, v_i \in \tilde{A}^*$.

In the sequel, according to the latter remark, $\langle v - u \rangle_f$ may be written as

$$\langle v - u \rangle_f = v_1 v_2 \cdots v_n v_{n+1}.$$

**Lemma 4.** *Let* $u, v \in L_{\vdash_I}^\epsilon$ *such that* $u \vdash_I^* v$. *Then there exists an embedding* $f$ *of* $u$ *in* $v$ *such that*

$$\langle v - u \rangle_f \in L_{\vdash_I}^\epsilon.$$

*Proof.* By induction on $k \geq 0$ such that $u \vdash_I^k v$. If $k = 0$, then $u = v$ so $\langle v - u \rangle_f = \epsilon \in L_{\vdash_I}^\epsilon$. Suppose $k = 1$. Thus $u = u_1 u_2 u_3$ and $v = u_1 x u_2 \bar{x} u_3$ where $x \in A$ and $u_1 u_2 u_3 \in L_{\vdash_I}^\epsilon$. Hence $\langle v - u \rangle_f = x\bar{x} \in L_{\vdash_I}^\epsilon$. The basis of the induction is proved.

Let us prove the induction step. Suppose $u \vdash_I^{k+1} v$ with $k \geq 1$. Then there exists $w \in L_{\vdash_I}^\epsilon$ such that $u \vdash_I^k w$ and $w \vdash_I v$. By the induction hypothesis, there exists an embedding $f$ of $u$ in $w$ such that $\langle w - u \rangle_f \in L_{\vdash_I}^\epsilon$. Suppose $u = a_1 \cdots a_n$ and $w = u_1 a_1 u_2 a_2 \cdots u_i a_i \cdots u_n a_n u_{n+1}$ with $a_i \in \tilde{A}$, $u_i \in \tilde{A}^*$. Hence $\langle w - u \rangle_f = u_1 u_2 \cdots u_{n+1} \in L_{\vdash_I}^\epsilon$. Since $w \vdash_I v$, suppose that

$$v = u_1 a_1 u_2 a_2 \cdots u_i x \cdots u_j \bar{x} \cdots u_n a_n u_{n+1},$$

with $x \in A$ (the other cases determined by different positions of $x$ and $\bar{x}$ are treated similarly). From the latter condition, one easily sees that $f$ may be extended to an embedding $g$ of $u$ in $v$ such that

$$\langle v - u \rangle_g = u_1 u_2 \cdots u_i x \cdots u_j \bar{x} \cdots u_n u_{n+1}.$$

$\square$

**Lemma 5.** *For every* $m, n \geq 1$,
  *(1)* $S_n \in L_{\vdash_I}^\epsilon$;
  *(2)* $S_n \in F(S_m)$ *if and only if* $n = m$;
  *(3) Suppose* $n \leq m$. *Let* $Q = adb\bar{b}c\bar{c}\bar{a}(a\bar{d}dc\bar{c}\bar{c}\bar{a})^n a\bar{d}$. *Then* $Q \in Pref(S_n) \cap Pref(S_m)$.

*Proof.* By induction on $n$, condition (1) is easily proved. Conditions (2) and (3) immediately follow from the structure of words of $\{S_n\}_{n \geq 1}$. $\square$

**Lemma 6.** *Let* $n, m$ *be positive integers such that* $n \leq m$. *If* $S_n \vdash_I^* S_m$ *then* $S_n = S_m$.

*Proof.* Let $n \leq m$ be positive integers. Then

$$S_n = adb\bar{b}c\bar{c}\bar{a}(a\bar{d}dc\bar{c}\bar{c}\bar{a})^n a\bar{d}b\bar{b}\bar{a} \text{ and}$$

$$S_m = adb\bar{b}c\bar{c}\bar{a}(a\bar{d}dc\bar{c}\bar{c}\bar{a})^n (a\bar{d}dc\bar{c}\bar{c}\bar{a})^k a\bar{d}b\bar{b}\bar{a}, \text{ with } k \geq 0.$$

By Lemma 4, the hypothesis $S_n \vdash_I^* S_m$ implies there exists an embedding $f$ of $S_n$ in $S_m$ such that $\langle S_m - S_n \rangle_f \in L_{\vdash_I}^\epsilon$.

We now prove the following claim.

**Claim.** *The following conditions hold:*
  1) $\forall i = 1, \ldots, 9 + 8n$, $f(i) = i$. In particular, by condition (3) of Lemma 5, $f$ is the identity on the common prefix $Q = adb\bar{b}c\bar{c}\bar{a}(a\bar{d}dc\bar{c}\bar{c}\bar{a})^n a\bar{d}$ of $S_n$ and $S_m$.
  2) $f(|S_n| - i) = |S_m| - i$, for $i = 0, 1, 2$.

**Proof of the Claim:** First we observe that, for all $n \geq 1$, $b\bar{b}$ occurs exactly twice as a factor of $S_n$. This immediately entails condition (2) and $f(i) = i$ for all $i = 1, \ldots, 4$.

The proof of condition (1) is divided into the following two steps.

**Step 1.** *Let $i$ be a positive integer such that $i \leq 9 + 8n$. If $a_i \in \{a, \bar{a}, d, \bar{d}\}$, then $f(i) = i$.*

We first observe that, for all $i$ such that $4 \leq i \leq 9 + 8n$, one has:

- If $a_i = d$ (resp. $a_i = \bar{d}$) then $i = 10 + 8\ell$ (resp. $i = 9 + 8\ell$), with $\ell \geq 0$;
- If $a_i = a$ (resp. $a_i = \bar{a}$) then $i = 8(\ell + 1)$ (resp. $i = 8(\ell + 1) - 1$), with $\ell \geq 0$.

Now we prove Step 1 by induction on $\ell \geq 0$. One easily checks that $f(2) = 2$ yields $f(9) = 9$. Indeed, if $f(9) > 9$ then $\langle S_m - S_n \rangle_f = v'v''$, with $v'$, $v'' \in \tilde{A}^*$ and $|v'|_{\bar{d}} = 1 > |v'|_d = 0$. By Lemma 3, $\langle S_m - S_n \rangle_f \notin L_{\vdash_I}^\epsilon$ which contradicts the choice of $f$. Hence $f(9) = 9$. This entails $f(7) = 7$ and $f(8) = 8$.

By using a similar argument, conditions $f(10) = 10$ and $f(15) = 15$ follow from $f(8) = 8$. The basis of the induction is proved.

Let us prove the induction step. Let $i = 10 + 8(\ell - 1)$. Then $a_i = d$ and, by induction hypothesis, $f(i) = i$. This yields $f(9 + 8\ell) = 9 + 8\ell$. Indeed, otherwise, $\langle S_m - S_n \rangle_f = v'v''$, with $v'$, $v'' \in \tilde{A}^*$ and $|v'|_{\bar{d}} = 1 > |v'|_d = 0$. As before, $\langle S_m - S_n \rangle_f \notin L_{\vdash_I}^\epsilon$ which contradicts the choice of $f$. Hence $f(9 + 8\ell) = 9 + 8\ell$ which entails $f(8\ell) = 8\ell$. This proves Step 1.

**Step 2.** *Let $i$ be a positive integer such that $i \leq 9 + 8n$. If $a_i \in \{c, \bar{c}\}$, then $f(i) = i$.*

First we observe that every occurrence of $c\bar{c}$ in $S_n$ is a factor of an occurrence of $db\bar{b}c\bar{c}\bar{a}$ or $dc\bar{c}c\bar{c}\bar{a}$. Let us consider the second case (the first is similarly treated). Set $dc\bar{c}c\bar{c}\bar{a} = a_i \cdots a_{i+5}$ with $i \geq 1$. By Step 1, $f(i) = i$ and $f(i + 5) = i + 5$ which immediately entails $f(i + \ell) = i + \ell$, for $\ell = 1, \ldots, 4$. This proves Step 2. Finally, Condition (1) follows from Step 1 and Step 2. ◇

Suppose now $k > 0$. Then the previous claim implies

$$\langle S_m - S_n \rangle_f = dc\bar{c}c\bar{c}\bar{a}(a\bar{d}dc\bar{c}c\bar{c}\bar{a})^{k-1}a\bar{d}.$$

Let $p = dc\bar{c}c\bar{c}\bar{a}$. Since $p \in Pref(\langle S_m - S_n \rangle_f)$ and $|p|_{\bar{a}} > |p|_a$, Lemma 3 implies $\langle S_m - S_n \rangle_f \notin L_{\vdash_I}^\epsilon$. Hence the case $n < m$ is not possible. This proves the Lemma. □

**Proof of Proposition 2:** By contradiction, deny. Thus there exist $n, m \geq 1$ such that $n < m$ and $S_n \vdash_I^* S_m$. By Lemma 6, $S_n = S_m$. Hence, by condition (2) of Lemma 5, $n = m$ which is a contradiction. This proves that the sequence $\{S_n\}_{n \geq 1}$ is bad. □

# References

1. J. Berstel, *Transductions and Context-Free Languages*. Teubner, Stuttgart, 1979.
2. D.P. Bovet and S. Varricchio, On the regularity of languages on a binary alphabet generated by copying systems. *Information Processing Letters* **44**, 119–123 (1992).
3. A. de Luca and S. Varricchio, Some regularity conditions based on well quasi-orders. *Lecture Notes in Computer Science*, Vol. 583, pp. 356–371, Springer-Verlag, Berlin, 1992.
4. A. de Luca and S. Varricchio, Well quasi-orders and regular languages. *Acta Informatica* **31**, 539–557 (1994).

5. A. de Luca and S. Varricchio, *Finiteness and regularity in semigroups and formal languages*. EATCS Monographs on Theoretical Computer Science. Springer, Berlin, 1999.
6. A. Ehrenfeucht, D. Haussler, and G. Rozenberg, On regularity of context-free languages. *Theoretical Computer Science* **27**, 311–332 (1983).
7. T. Harju and L. Ilie, On well quasi orders of words and the confluence property. *Theoretical Computer Science* **200**, 205–224 (1998).
8. D. Haussler, Another generalization of Higman's well quasi-order result on $\Sigma^*$. *Discrete Mathematics* **57**, 237–243 (1985).
9. G. H. Higman, Ordering by divisibility in abstract algebras. *Proc. London Math. Soc.* **3**, 326–336 (1952).
10. L. Ilie and A. Salomaa, On well quasi orders of free monoids. *Theoretical Computer Science* **204**, 131–152 (1998).
11. B. Intrigila and S.Varricchio, On the generalization of Higman and Kruskal's theorems to regular languages and rational trees. *Acta Informatica* **36**, 817–835 (2000).
12. J. Kruskal, The theory of well-quasi-ordering: a frequently discovered concept. *J. Combin. Theory, Ser. A*, **13**, 297–305 (1972).

# Frequency of Symbol Occurrences in Simple Non-primitive Stochastic Models*

Diego de Falco, Massimiliano Goldwurm, and Violetta Lonati

Università degli Studi di Milano
Dipartimento di Scienze dell'Informazione
via Comelico 39, 20135 Milano, Italy
{defalco,goldwurm,lonati}@dsi.unimi.it

**Abstract.** We study the random variable $Y_n$ representing the number of occurrences of a given symbol in a word of length $n$ generated at random. The stochastic model we assume is a simple non-ergodic model defined by the product of two primitive rational formal series, which form two distinct ergodic components. We obtain asymptotic evaluations for the mean and the variance of $Y_n$ and its limit distribution. It turns out that there are two main cases: if one component is dominant and non-degenerate we get a Gaussian limit distribution; if the two components are equipotent and have different leading terms of the mean, we get a uniform limit distribution. Other particular limit distributions are obtained in the case of a degenerate dominant component and in the equipotent case when the leading terms of the expectation values are equal.

## 1 Introduction

The analysis of the frequency of pattern occurrences in a long string of symbols, usually called *text*, is a classical problem that is of interest in several research areas of computer science and molecular biology. In computer science for instance it has been studied in connection with the design of algorithms for approximate pattern-matching [13,16] and the analysis of problems of code synchronization [10]. The problem is particularly relevant in molecular biology to study properties of DNA sequences and for gene recognition [20]. For instance, biological informations can be obtained from unexpected frequencies of special deviant motifs in a DNA text [7,9,15]. Moreover, the frequency problems in a probabilistic framework are studied in [12,1,17,14]. In this context a set of one or more patterns is given and the text is randomly generated by a memoryless source (also called *Bernoulli model*) or a Markovian source (the *Markovian model*) where the probability of a symbol in any position only depends on the previous occurrence.

A more general approach, developed in the area of automata and formal languages, is recently proposed in [3,4], where the pattern is reduced to a single symbol and the text is randomly generated according to a stochastic model

---

* This work has been supported by the Project M.I.U.R. COFIN "Formal languages and automata: theory and applications".

Z. Ésik and Z. Fülöp (Eds.): DLT 2003, LNCS 2710, pp. 242–253, 2003.

defined by a rational formal series in two non-commutative variables. We recall that there are well-known linear time algorithms to generate a random word in such a model [6] which we call the *rational model* in this work. The frequency problem in this model is also related to the ambiguity of rational grammars and to the asymptotic form of the coefficients of rational and algebraic formal series studied for instance in [21,18].

It is proved that the symbol frequency problem in the rational model includes, as a special case, the general frequency problem of regular patterns in the Markovian model (studied in [14]) and it is also known that the two models are not equivalent [3]. The symbol frequency problem in the rational model is studied in [3,4] in the ergodic case, i.e. when the matrix associated with the rational formal series (counting the transitions between states) is primitive. Under this hypothesis, asymptotic expressions for the mean and the variance of the statistics under investigation are obtained, together with their limit distributions expressed in the form of both central and local limit theorems [3,4].

In this work we study the symbol frequency problem in the rational model in a simple non-ergodic case, that is when the rational series defining the stochastic source is the product of two primitive rational series. This case is rather representative of a more general situation where the matrix associated with the rational model has two primitive components. We obtain asymptotic evaluations for the mean and the variance of the number of symbol occurrences and its limit distribution. It turns out that there are two main cases. In the dominant case the main eigenvalue associated with one component is strictly greater than the main eigenvalue associated with the other. In the equipotent case these two eigenvalues are equal.

If one component is dominant and does not degenerate[1], the main terms of mean and variance are determined by such a component and we get a Gaussian limit distribution. We also determine the limit distribution when there exists a dominant degenerate component. Apparently, this has a large variety of possible forms depending even on the other (non-main) eigenvalues of the secondary component and including the geometric law in some simple cases.

If the two components are equipotent and have different leading terms of the mean, then the variance is of a quadratic order showing there is not a concentration phenomenon around the average value of our statistics. In this case we get a uniform limit distribution between the constants of the leading terms of the expected values associated with the two components.

However, in the equipotent case, if the leading terms of the two means are equal then the variance reduces to a linear order of growth and we have again a concentration phenomenon. In this case the limit distribution depends on the main terms of the variances associated with the two components: if they are equal we obtain a Gaussian limit distribution again; if they are different we obtain a limit distribution defined by a mixture of Gaussian random variables of mean 0 and variance uniformly distributed in a given interval.

---

[1] i.e., considering the series of the dominant component, both symbols of the alphabet appear in some words with non-null coefficient.

The main contribution of these results is related to the non-ergodic hypothesis. To our knowledge, the pattern frequency problem in the Markovian model is usually studied in the literature under ergodic hypothesis and Gaussian limit distributions are generally obtained. On the contrary, here we get in many cases limit distributions quite different from the Gaussian one.

We think our analysis is significant also from a methodical point of view: we adapt methods and ideas introduced to deal with the Markovian model to a more general stochastic model, the rational one, which seems to be the natural setting for these techniques.

Due to space constraints, in this paper all proofs are omitted. They can be found in [5] and rely on singularity analysis of the bivariate generating functions associated with the statistics under investigation.

The computations described in our examples are executed by using MATHE-MATICA [22].

## 2    Preliminary Notions

### 2.1    Perron–Frobenius Theory

The Perron–Frobenius theory is a well-known subject widely studied in the literature (see for instance [19]). To recall its main results we first establish some notation. For every pair of matrices $T = [T_{ij}]$, $S = [S_{ij}]$, the expression $T > S$ means that $T_{ij} > S_{ij}$ for every pair of indices $i, j$. As usual, we consider any vector $v$ as a column vector and denote by $v'$ the corresponding row vector. We recall that a nonnegative matrix $T$ is called *primitive* if there exists $m \in \mathbb{N}$ such that $T^m > 0$. The main properties of such matrices are given by the following theorem [19, Sect.1].

**Theorem 1 (Perron–Frobenius).** *Let $T$ be a primitive nonnegative matrix. There exists an eigenvalue $\lambda$ of $T$ (called Perron–Frobenius eigenvalue of $T$) such that:*

1. *$\lambda$ is real and positive;*
2. *with $\lambda$ we can associate strictly positive left and right eigenvectors;*
3. *$|\nu| < \lambda$ for every eigenvalue $\nu \neq \lambda$;*
4. *if $0 \leq C \leq T$ and $\gamma$ is an eigenvalue of $C$, then $|\gamma| \leq \lambda$; moreover $|\gamma| = \lambda$ implies $C = T$;*
5. *$\lambda$ is a simple root of the characteristic polynomial of $T$.*

### 2.2    Moments and Limit Distribution of a Discrete Random Variable

Let $X$ be an integer valued random variable (r.v.), such that $\Pr\{X = k\} = p_k$ for every $k \in \mathbb{N}$. Consider its moment generating function $\Psi_X(z) = \sum_{k \in \mathbb{N}} p_k e^{zk}$; then the first two moments of $X$ can be computed by

$$\mathbb{E}(X) = \Psi_X'(0) , \qquad \mathbb{E}(X^2) = \Psi_X''(0) . \tag{1}$$

Moreover, the characteristic function of $X$ is defined by

$$\Phi_X(t) = \mathbb{E}(e^{itX}) = \Psi_X(it)$$

$\Phi_X$ is always well-defined for every $t \in \mathbb{R}$, it is periodic of period $2\pi$ and it completely characterizes the r.v. $X$. Moreover it represents the classical tool to prove convergence in distribution: a sequence of random variables $\{X_n\}_n$ converges to a r.v. $X$ in distribution[2] if and only if $\Phi_{X_n}(t)$ tends to $\Phi_X(t)$ for every $t \in \mathbb{R}$. Several forms of the central limit theorem are classically proved in this way [8].

## 3 The Rational Stochastic Model

Here we define the rational stochastic model. According to [2] a formal series in the non-commutative variables $a, b$, with coefficients in the semiring $\mathbb{R}_+$ of non-negative real numbers, is a function $r : \{a, b\}^* \longrightarrow \mathbb{R}_+$. For any word $w \in \{a, b\}^*$, we denote by $(r, w)$ the value of $r$ at $w$ and a series $r$ is usually represented as a sum in the form

$$r = \sum_{w \in \{a,b\}^*} (r, w) w$$

The set of all such series is denoted by $\mathbb{R}_+\langle\!\langle a, b \rangle\!\rangle$. It is well-known that $\mathbb{R}_+\langle\!\langle a, b \rangle\!\rangle$ forms a semiring with respect to the traditional operation of sum and Cauchy product.

Now, given $r \in \mathbb{R}_+\langle\!\langle a, b \rangle\!\rangle$, we can define a stochastic model as follows. Consider a positive $n \in \mathbb{N}$ such that $(r, w) \neq 0$ for some string $w \in \{a, b\}^*$ of length $n$. For every integer $0 \le k \le n$ set

$$\varphi_k^{(n)} = \sum_{|w|=n, |w|_a=k} (r, w)$$

and define the random variable (r.v.) $Y_n$ such that

$$\Pr\{Y_n = k\} = \frac{\varphi_k^{(n)}}{\sum_{j=0}^{n} \varphi_j^{(n)}}.$$

Roughly speaking, $Y_n$ represents the number of occurrences of $a$ in a word of length $n$ randomly generated according to the stochastic model defined by $r$. This model is of particular interest in the case of rational series. We recall that a series $r \in \mathbb{R}_+\langle\!\langle a, b \rangle\!\rangle$ is said to be *rational* if for some integer $m > 0$ there exists a monoid morphism $\mu : \{a, b\}^* \longrightarrow \mathbb{R}_+^{m \times m}$, a pair of (column) vectors $\xi, \eta \in \mathbb{R}_+^m$ such that $(r, w) = \xi' \mu(w) \eta$ for every $w \in \{a, b\}^+$. The triple $(\xi, \mu, \eta)$ is called *linear representation* of $r$.

---

[2] I.e. $\lim_{n \to \infty} F_{X_n}(\tau) = F_X(\tau)$ for every point $\tau \in \mathbb{R}$ of continuity for $F_X$, where $F_{X_n}(\tau) = \Pr\{X_n \le \tau\}$ and $F_X(\tau) = \Pr\{X \le \tau\}$.

We say that $\{Y_n\}$ is defined in a *rational* stochastic model if the associated series $r$ is rational. It turns out that classical probabilistic models as the Bernoulli or the Markov processes, frequently used to study the number of occurrences of regular patterns in random words [12,14], are special cases of rational stochastic models [3].

## 4   The Primitive Case

The asymptotic behaviour of $Y_n$ is studied in [3,4] when $r$ is rational and admits a primitive linear representation, i.e. a linear representation $(\xi, \mu, \eta)$ such that the matrix $M = A + B$ is primitive, where $A = \mu(a)$ and $B = \mu(b)$. Under this hypothesis, let $\lambda$ be the Perron–Frobenius eigenvalue of $M$; from Theorem 1, one can prove that, for each $n \in \mathbb{N}$,

$$M^n = \lambda^n \left(uv' + C(n)\right)$$

where $C(n)$ is a real matrix such that, for some $c > 0$ and $0 \le \varepsilon < 1$, $|C(n)_{ij}| \le c\varepsilon^n$ (for any $i, j$ and all $n$ large enough) and $v'$ and $u$ are strictly positive left and right eigenvectors of $M/\lambda$ corresponding to the eigenvalue 1, normed so that $v'u = 1$. Moreover, the matrix $C = \sum_{n=0}^{\infty} C(n)$ is well-defined and $v'C = Cu = 0$.

Using these properties, it is proved in [3] that the mean and the variance of $Y_n$ satisfy the relations

$$\mathbb{E}(Y_n) = \beta n + \frac{\delta}{\alpha} + \mathrm{O}\left(\varepsilon^n\right), \qquad \mathbb{V}ar(Y_n) = \gamma n + \mathrm{O}(1) \tag{2}$$

where $\alpha, \beta, \gamma$ and $\delta$ are constants defined by

$$\beta = \frac{v'Au}{\lambda}, \qquad \gamma = \beta - \beta^2 + 2\frac{v'ACAu}{\lambda^2}$$

$$\alpha = (\xi'u)(v'\eta), \qquad \delta = \left(\xi'C\frac{A}{\lambda}u\right)(v'\eta) + (\xi'u)\left(v'\frac{A}{\lambda}C\eta\right).$$

Notice that $B = 0$ implies $\beta = 1$ and $\gamma = \delta = 0$, while $A = 0$ implies $\beta = \gamma = \delta = 0$; on the other side, if $A \ne 0 \ne B$ then $\beta > 0$ and it turns out that also $\gamma > 0$.

Relations (2) is proved from (1) observing that $\Psi_{Y_n}(z) = \frac{h_n(z)}{h_n(0)}$, where

$$h_n(z) = \sum_{k=0}^{n} \varphi_k^{(n)} e^{zk} = \xi'(Ae^z + B)^n \eta,$$

and studying the asymptotic behaviour of $h_n(0)$, $h_n'(0)$ and $h_n''(0)$. This analysis is essentially based on Theorem 1 and on a sort of simple differential calculus for matrices.

Finally, the characteristic function $\Phi_{Y_n}(t) = \frac{h_n(it)}{h_n(0)}$ is used in [3] to prove that, if $M$ is primitive and $A \neq 0 \neq B$, then the distribution of $Y_n$ approximates a normal distribution, i.e. for every $x \in \mathbb{R}$

$$\lim_{n \longrightarrow +\infty} \Pr\left\{ \frac{Y_n - \beta n}{\sqrt{\gamma n}} \leq x \right\} = \frac{1}{\sqrt{2\pi}} \int_{-\infty}^{x} e^{-\frac{t^2}{2}} dt .$$

## 5   The Product Model

Given two primitive linear representations $(\xi_1, \mu_1, \eta_1)$ and $(\xi_2, \mu_2, \eta_2)$ over the alphabet $\{a, b\}$, let $r$ be the formal series defined by

$$(r, w) = \sum_{w=xy} [\xi_1' \mu_1(x) \eta_1] \cdot [\xi_2' \mu_2(y) \eta_2]$$

for every $w \in \{a, b\}^*$. It turns out that $r$ admits a linear representation $(\xi, \mu, \eta)$ given by

$$\xi = \begin{pmatrix} \xi_1 \\ 0 \end{pmatrix}, \quad \mu(x) = \begin{pmatrix} \mu_1(x) & \eta_1 \xi_2' \mu_2(x) \\ 0 & \mu_2(x) \end{pmatrix}, \quad \eta = \begin{pmatrix} \eta_1 \xi_2' \eta_2 \\ \eta_2 \end{pmatrix} \quad (3)$$

Using the notation introduced in the previous section, from now on we refer the terms $M$, $A$, $B$ and $h_n(z)$ to the product series $r$. To avoid trivial cases, throughout this work we assume $A \neq 0 \neq B$. We also use the obvious extension of appending indices 1 and 2 to the values associated with the linear representation $(\xi_1, \mu_1, \eta_1)$ and $(\xi_2, \mu_2, \eta_2)$, respectively. Thus, for each $i = 1, 2$, the values $Y_n^{(i)}$, $M_i$, $\lambda_i$, $A_i$, $B_i$, $h_n^{(i)}(z)$, $\beta_i$, $\gamma_i$ are well-defined and associated with the linear representation $(\xi_i, \mu_i, \eta_i)$.

From the decomposition (3) it is easy to see that $h_n(z)$ is given by

$$h_n(z) = \sum_{i=0}^{n} \xi_1'(A_1 e^z + B_1)^i \eta_1 \xi_2'(A_2 e^z + B_2)^{n-i} \eta_2 = \sum_{i=0}^{n} h_i^{(1)}(z) h_{n-i}^{(2)}(z) \quad (4)$$

which is the convolution of $h_n^{(1)}(z)$ and $h_n^{(2)}(z)$. Since $(\xi_1, \mu_1, \eta_1)$ and $(\xi_2, \mu_2, \eta_2)$ are primitive, we can consider the Perron-Frobenius eigenvalues $\lambda_1, \lambda_2$ of $M_1$ and $M_2$, respectively. The properties of $Y_n$ now depend on whether these two values are distinct or equal. In the first case the rational representation associated with the largest one determines the main characteristics of $Y_n$. We say that $(\xi_i, \mu_i, \eta_i)$ is the *dominant* component if $\lambda_1 \neq \lambda_2$ and $\lambda_i = \max\{\lambda_1, \lambda_2\}$. On the contrary, if $\lambda_1 = \lambda_2$, both components give a contribution to the asymptotic behaviour of $Y_n$ and hence we say they are *equipotent*.

## 6   Main Results

In this section we summarize the main results concerning the product model. We consider separately the case $\lambda_1 > \lambda_2$ (the case $\lambda_1 < \lambda_2$ is symmetric) and the case $\lambda_1 = \lambda_2$. In both cases, we first determine asymptotic expressions for mean and variance of $Y_n$ and then we study its limit distribution.

## 6.1  Dominant Case

Using (4) and the results of the primitive case, one can determine asymptotic expressions for $h_n(0)$ and its derivatives, which yield the following theorem.

**Theorem 2.** *Assume $\lambda_1 > \lambda_2$. Then the following statements hold:*
   *i) if $A_1 \neq 0 \neq B_1$, then $\mathbb{E}(Y_n) = \beta_1 n + O(1)$ and $\mathrm{Var}(Y_n) = \gamma_1 n + O(1)$;*
   *ii) if $A_1 \neq 0$ and $B_1 = 0$, then $\mathbb{E}(Y_n) = n + O(1)$ and $\mathrm{Var}(Y_n) = c_1 + O(\varepsilon^n)$;*
   *iii) if $A_1 = 0$ and $B_1 \neq 0$, then $\mathbb{E}(Y_n) = c_2 + O(\varepsilon^n)$ and $\mathrm{Var}(Y_n) = c_3 + O(\varepsilon^n)$;*
*where $\beta_1 > 0$, $\gamma_1 > 0$ and $c_i$ and $\varepsilon$ are constants such that $c_i > 0$ and $|\varepsilon| < 1$.*

As far as the limit distribution of $\{Y_n\}$ is concerned, if the dominant component does not degenerate (i.e. $A_1 \neq 0 \neq B_1$) the analysis is similar to the primitive case and gives rise to a Gaussian limit distribution [3]. On the contrary, if the dominant component degenerates, the limit distribution may assume different forms, depending on the second component. In both cases the proof is based on the analysis of the characteristic function of $Y_n$.

**Theorem 3.** *Let $\lambda_1 > \lambda_2$. If $A_1 \neq 0 \neq B_1$ then $\frac{Y_n - \beta_1 n}{\sqrt{\gamma_1 n}}$ converges in distribution to the normal random variable of mean 0 and variance 1.*

If either $A_1 = 0$ or $B_1 = 0$ then $\gamma_1 = 0$ and the previous theorem does not hold.

**Theorem 4.** *Let $\lambda_1 > \lambda_2$. If $A_1 = 0$, then the random variables $Y_n$ converges in distribution to the random variable $Z$ of characteristic function*

$$\Phi_Z(t) = \frac{\xi_2'(\lambda_1 I - A_2 e^{it} - B_2)^{-1} \eta_2}{\xi_2'(\lambda_1 I - M_2)^{-1} \eta_2} \tag{5}$$

*If $B_1 = 0$, then the random variables $n - Y_n$ converges in distribution to the random variable $W$ of characteristic function*

$$\Phi_W(t) = \frac{\xi_2'(\lambda_1 I - A_2 - B_2 e^{it})^{-1} \eta_2}{\xi_2'(\lambda_1 I - M_2)^{-1} \eta_2}.$$

Some comments on the random variables $Z$ and $W$ are now necessary. First observe that, when the matrices $M_2$, $A_2$ and $B_2$ have size 1, $Z$ and $W$ are geometric random variables. Indeed, in this case $M_2 = A_2 + B_2 = \lambda_2 < \lambda_1$ and we get

$$\Phi_Z(t) = \frac{1 - \frac{A_2}{\lambda_1 - B_2}}{1 - \frac{A_2}{\lambda_1 - B_2} e^{it}} \qquad \text{and} \qquad \Phi_W(t) = \frac{1 - \frac{B_2}{\lambda_1 - A_2}}{1 - \frac{B_2}{\lambda_1 - A_2} e^{it}}$$

which are the characteristic functions of geometric random variables of parameter $\frac{A_2}{\lambda_1 - B_2}$ and $\frac{B_2}{\lambda_1 - A_2}$ respectively. However, the range of possible behaviours of these random variables is much larger than what these examples show. To see this fact consider the function $\Phi_Z(t)$ in (5); it can be expressed in the form

$$\Phi_Z(t) = \sum_{j=0}^{\infty} \frac{\xi_2' (M_2/\lambda_2)^j \eta_2 \cdot (\lambda_2/\lambda_1)^j}{\sum_{i=0}^{\infty} \xi_2' (M_2/\lambda_2)^i \eta_2 \cdot (\lambda_2/\lambda_1)^i} \Phi_{Y_j^{(2)}}(t)$$

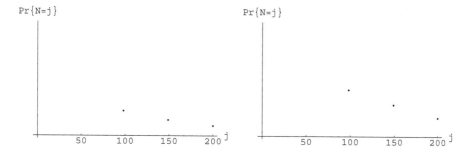

**Fig. 1.** Probability law of the random variable $N$ defined in (6), for $j = 0, 1, \ldots, 200$. In the first picture we compare the case $\mu = 0.00001$ and $\mu = -0.89$. In the second one we compare the case $\mu = 0.00001$ and $\mu = +0.89$.

and hence it describes the random variable $Y_N^{(2)}$, where $N$ is the random variable defined by the law

$$\Pr\{N = j\} = \frac{\xi_2' \, (M_2/\lambda_2)^j \, \eta_2 \cdot (\lambda_2/\lambda_1)^j}{\sum_{i=0}^{\infty} \xi_2' \, (M_2/\lambda_2)^i \, \eta_2 \cdot (\lambda_2/\lambda_1)^i} \, . \tag{6}$$

If $B_2 = 0$ then by (5) $Z$ reduces to $N$, and an example of the rich range of its possible forms is shown by considering the case where $(A_1 = 0 = B_2)$ $\lambda_1 = 1.009$, $\lambda_2 = 1$, and the second component is represented by a generic $(2 \times 2)$-matrix whose eigenvalues are 1 and $\mu$ such that $-1 < \mu < 1$. In this case, since the two main eigenvalues have similar values, the behaviour of $\Pr\{N = j\}$ for small $j$ depends on the second component and in particular on its smallest eigenvalue $\mu$. In Fig. 1 we plot the probability law of $N$ defined in (6) for $j = 0, 1, \ldots, 200$ in three cases: $\mu = -0.89$, $\mu = 0.00001$ and $\mu = 0.89$; the first picture compares the curves in the cases $\mu = -0.89$ and $\mu = 0.00001$, while the second picture compares the curves when $\mu = 0.00001$ and $\mu = 0.89$. Note that in the second case, when $\mu$ is almost null, we find a distribution similar to a (lengthy) geometric law while, for $\mu = -0.89$ and $\mu = 0.89$, we get a quite different behaviour which approximates the previous one for large values of $j$.

## 6.2   Equipotent Components

In this section we consider the random variable $Y_n$ assuming $\lambda_1 = \lambda_2$. Under this hypothesis two main subcases arise, depending on whether $\beta_1$ and $\beta_2$ are equal. First, we present the following theorem concerning the mean and the variance of $Y_n$, which can be obtained from equation (4) by singularity analysis.

**Theorem 5.** *If* $\lambda_1 = \lambda_2$, *then the mean and the variance of the random variable* $Y_n$ *are given by*

$$\mathbb{E}(Y_n) = \frac{\beta_1 + \beta_2}{2} n + O(1)$$

$$\mathrm{Var}(Y_n) = \begin{cases} \dfrac{(\beta_1 - \beta_2)^2}{12} n^2 + O(n) & \text{if } \beta_1 \neq \beta_2 \\ \dfrac{\gamma_1 + \gamma_2}{2} n + O(1) & \text{if } \beta_1 = \beta_2 \end{cases}$$

In the case $\beta_1 \neq \beta_2$ it is clear from the previous theorem that the variance is of a quadratic order. Hence, by using the characteristic function of $Y_n/n$ one can prove the following result.

**Theorem 6.** *If* $\lambda_1 = \lambda_2$ *and* $\beta_1 \neq \beta_2$ *then* $Y_n/n$ *converges in law to a random variable having uniform distribution in the interval* $[\min\{\beta_1, \beta_2\}, \max\{\beta_1, \beta_2\}]$.

If $\beta_1 = \beta_2$ then, since $A \neq 0 \neq B$, we have $A_i \neq 0 \neq B_i$ for $i = 1$ or $i = 2$. This implies $\gamma_i \neq 0$ and hence the variance is linear in $n$ (see Theorem 5). In this case we get a concentration phenomenon of $Y_n$ around its mean and we get two different limit distributions according to whether $\gamma_1 = \gamma_2$ or not. In the following, $\beta$ and $\gamma$ are defined by

$$\beta = \beta_1 = \beta_2, \qquad \gamma = \frac{\gamma_1 + \gamma_2}{2}.$$

First, we consider the case where the main terms of the variances are equal.

**Theorem 7.** *If* $\lambda_1 = \lambda_2$, $\beta_1 = \beta_2$ *and* $\gamma_1 = \gamma_2$ *then* $\frac{Y_n - \beta n}{\sqrt{\gamma n}}$ *converges in distribution to the normal random variable of mean* $0$ *and variance* $1$.

At last, we consider the case where the main terms of the variances are not equal.

**Theorem 8.** *If* $\lambda_1 = \lambda_2$, $\beta_1 = \beta_2$ *and* $\gamma_1 \neq \gamma_2$ *then* $\frac{Y_n - \beta n}{\sqrt{\gamma n}}$ *converges in distribution to the random variable of characteristic function*

$$\Phi(t) = \frac{2 \left( e^{-\frac{\gamma_2}{2\gamma} t^2} - e^{-\frac{\gamma_1}{2\gamma} t^2} \right) \gamma}{(\gamma_1 - \gamma_2) t^2} \tag{7}$$

One can prove that the probability density corresponding to the characteristic function (7) is a mixture of Gaussian densities of mean 0, with variances uniformly distributed in the interval with extremes $\frac{\gamma}{\gamma_1}$ and $\frac{\gamma}{\gamma_2}$. Indeed, it is easy to see that

$$\Phi(t) = \frac{1}{\left( \frac{\gamma_2}{\gamma} - \frac{\gamma_1}{\gamma} \right)} \int_{\frac{\gamma_1}{\gamma}}^{\frac{\gamma_2}{\gamma}} e^{-\frac{1}{2} v t^2} dv$$

In Fig. 2 we illustrate the form of the limit distributions obtained in this section (i.e. when $\lambda_1 = \lambda_2$ and $\beta_1 = \beta_2$). We represent the density of the random variable having characteristic function (7), for different values of the ratio $p = \gamma_2/\gamma_1$. When $p$ approaches 1, the curve tends to a Gaussian density according to Theorem 7; if $\gamma_2$ is much greater than $\gamma_1$, then we find a density with a cuspid in the origin corresponding to Theorem 8.

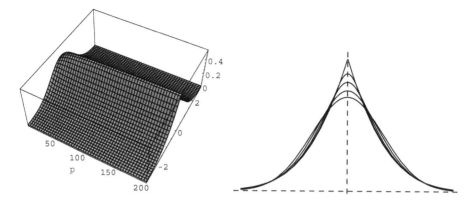

**Fig. 2.** The first picture represents the density of the random variable having character-istic function (7), according to the parameter $p = \gamma_2/\gamma_1$. The second picture represents some sections obtained for $p = 1.0001, 5, 15, 50, 20000$.

## 7  Examples

In this section we present an example which compares the limit distributions obtained in the non-degenerate dominant case and in the equipotent case, with different leading terms of the mean associated with each component.

This example is based on the automata represented in Fig. 3. They define two ergodic components with matrices $M_i$, $A_i$, $B_i$, $i = 1, 2$. The arrays $\xi_i$ and $\eta_i$ are given by the values included in the states.

Multiplying the matrices $A_i$ and $B_i$ (for $i = 1, 2$) by suitable factors, it is possible to build a family of primitive linear representations where we may have $\lambda_1 = \lambda_2$ or $\lambda_1 \neq \lambda_2$. In all cases, it turns out that $\beta_1 = 0.146447$ and $\beta_2 = 0.733333$ (and hence $\beta_1 \neq \beta_2$).

Figure 4 illustrates the probability function of the random variable $Y_{50}$ in three different cases. If $\lambda_1 = 2$ and $\lambda_2 = 1$ we find a normal density of mean asymptotic to $50 \, \beta_1$. If $\lambda_1 = 1$ and $\lambda_2 = 2$ we have a normal density of mean

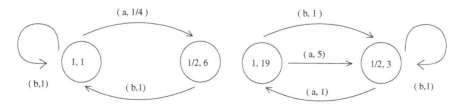

**Fig. 3.** Two weighted finite automata over the alphabet $\{a, b\}$, defining two primitive linear representations $(\xi_i, \mu_i, \eta_i)$, $i = 1, 2$. The matrices $A_i = \mu_i(a)$ and $B_i = \mu_i(b)$ are defined by the labels associated with transitions in the pictures. The values of the components of the arrays $\xi_i$ and $\eta_i$ are included in the corresponding states.

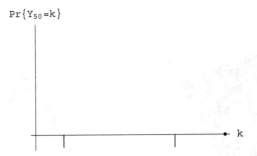

**Fig. 4.** Probability functions of $Y_{50}$, corresponding to a formal series derived from the automata of Fig. 3 with weighted expanded by a constant factor, in the case where $(\lambda_1, \lambda_2)$ are equal to (2,1), (1,2) and (1,1). The vertical bars have abscissas $50\beta_1$ and $50\beta_2$, respectively.

asymptotic to $50\beta_2$. Both situations correspond to Theorem 3. If $\lambda_1 = \lambda_2 = 1$, we recognize the convergence to the uniform distribution in the interval $[50\beta_1, 50\beta_2]$ according to Theorem 6.

# References

1. E. A. Bender and F. Kochman. The distribution of subword counts is usually normal. *European Journal of Combinatorics*, 14:265–275, 1993.
2. J. Berstel and C. Reutenauer. *Rational series and their languages*, Springer-Verlag, New York – Heidelberg – Berlin, 1988.
3. A. Bertoni, C. Choffrut, M. Goldwurm, and V. Lonati. On the number of occurrences of a symbol in words of regular languages. *Rapporto Interno n. 274-02*, Dipartimento di Scienze dell'Informazione, Università degli Studi di Milano, February 2002 (to appear in TCS).
4. A. Bertoni, C. Choffrut, M. Goldwurm, and V. Lonati. The symbol-periodicity of irreducible finite automata. *Rapporto Interno n. 277-02*, Dipartimento di Scienze dell'Informazione, Università degli Studi di Milano, April 2002 (available at http://homes.dsi.unimi.it/~goldwurm/home.html).
5. D. de Falco, M. Goldwurm, and V. Lonati. Frequency of symbol occurrences in simple non-primitive stochastic models. *Rapporto Interno n. 287-03*, Dipartimento di Scienze dell'Informazione, Università degli Studi di Milano, February 2003 (available at http://homes.dsi.unimi.it/~goldwurm/home.html).
6. A. Denise. Génération aléatoire uniforme de mots de langages rationnels. *Theoretical Computer Science*, 159:43–63, 1996.
7. J. Fickett. Recognition of protein coding regions in DNA sequences. *Nucleic Acid Res*, 10:5303–5318, 1982.
8. P. Flajolet and R. Sedgewick. The average case analysis of algorithms: multivariate asymptotics and limit distributions. *Rapport de recherche n. 3162*, INRIA Rocquencourt, May 1997.
9. M.S. Gelfand. Prediction of function in DNA sequence analysis. *J. Comput. Biol.*, 2:87–117, 1995.

10. L.J. Guibas and A. M. Odlyzko. Maximal prefix-synchronized codes. *SIAM J. Appl. Math.*, 35:401–418, 1978.

11. L.J. Guibas and A. M. Odlyzko. Periods in strings. *Journal of Combinatorial Theory. Series A*, 30:19–43, 1981.

12. L.J. Guibas and A. M. Odlyzko. String overlaps, pattern matching, and nontransitive games. *Journal of Combinatorial Theory. Series A*, 30(2):183–208, 1981.

13. P. Jokinen and E. Ukkonen. Two algorithms for approximate string matching in static texts *Proc. MFCS 91*, Lecture Notes in Computer Science, vol. n.520, Springer, 240–248, 1991.

14. P. Nicodeme, B. Salvy, and P. Flajolet. Motif statistics. In *Proceedings of the 7th ESA*, J. Nešetril editor. Lecture Notes in Computer Science, vol. n.1643, Springer, 1999, 194–211.

15. B. Prum, F. Rudolphe and E. Turckheim. Finding words with unexpected frequencies in deoxyribonucleic acid sequence. *J. Roy. Statist. Soc. Ser. B*, 57: 205–220, 1995.

16. M. Régnier and W. Szpankowski. On the approximate pattern occurrence in a text. *Proc. Sequence '97*, Positano, 1997.

17. M. Régnier and W. Szpankowski. On pattern frequency occurrences in a Markovian sequence. *Algorithmica*, 22 (4):621–649, 1998.

18. C. Reutenauer. *Propriétés arithmétiques et topologiques de séries rationnelles en variables non commutatives*, These Sc. Maths, Doctorat troisieme cycle, Université Paris VI, 1977.

19. E. Seneta. *Non-negative matrices and Markov chains*, Springer–Verlag, New York Heidelberg Berlin, 1981.

20. M. Waterman. *Introduction to computational biology*, Chapman & Hall, New York, 1995.

21. K. Wich. Sublinear ambiguity. In *Proceedings of the 25th MFCS*, M. Nielsen and B. Rovan editors. Lecture Notes in Computer Science, vol. n.1893, Springer, 2000, 690–698.

22. S. Wolfram. *The Mathematica book* Fourth Edition, Wolfram Media – Cambridge University Press, 1999.

# On Enumeration of Müller Automata

Michael Domaratzki*

School of Computing, Queen's University
Kingston, ON K7L 3N6 Canada
domaratz@cs.queensu.ca

**Abstract.** In this paper, we consider the problem of enumeration of Müller automata with a given number of states. Given a Müller automata, its acceptance table $\mathcal{F}$ is admissible if, for each element $f \in \mathcal{F}$, there exists an infinite word whose set of states visited infinitely often is exactly $f$.
We consider acceptance tables in Müller automata which are never admissible, regardless of the choice of transition function $\delta$. We apply the results to enumeration of Müller automata by number of states.

## 1 Introduction

Müller automata are a model of automata accepting infinite words in which acceptance is determined by a table of states, which is a subset of $2^Q$, where $Q$ is the set of states (for definitions, see Sect. 2). An infinite word $w$ is accepted if and only if the set of states visited infinitely often by the automata when reading $w$ is a member of the table.

Recently, there has been renewed interest in the enumeration of finite automata (see, e.g., [1,2,3,4]). However, the problem of enumerating finite automata which accept infinite words has not been previously considered. In this paper, we consider the problem of enumerating Müller automata with a given number of states. This problem is interesting because of the substantial difference in the acceptance condition between Müller automata and automata accepting finite words.

To obtain bounds on number of Müller automata, we consider all subsets of $2^Q$ for a fixed set $Q$, and examine those elements $\mathcal{F}$ for which there exists some $f \in \mathcal{F}$ such that, regardless of our choice of $\delta$, no infinite word exists which visits exactly the states in $f$ infinitely often. We call such $\mathcal{F}$ *strongly inadmissible*. We show that for $|Q| = n$, the number of strongly inadmissible tables is $2^{2^n - 1}(1 + \lambda(n))$ for some function $\lambda$ for which $\lambda(n) \to 0$ as $n \to \infty$.

Note that the measure of number of states is not a typical measure of the descriptional complexity of Müller automata. In particular, the size of the acceptance table of the automaton is a more natural measure in terms of descriptional complexity (for a Müller automaton $M = (Q, \Sigma, \delta, q_0, \mathcal{F})$, Pin and Perrin [5, Sect. 10, Chap. 1] use the size $\max(|Q|, |\delta|, |\mathcal{F}|)$ in their discussion of size trade-offs between Müller and other automata accepting infinite words). However, we

---

* Research supported in part by an NSERC PGS-B graduate scholarship.

employ the measure of number of states since it is a more natural measure for the enumerative problem; we note that enumerating Müller automata by number of states and size of acceptance table is an interesting unexamined research problem.

## 2    Preliminaries

We review some basic notions of infinite words and automata acting on them. For any undefined notions, please see Pin and Perrin [5] or Thomas [6]. Let $\Sigma$ be our finite alphabet. A right infinite word over $\Sigma$ is an infinite sequence of symbols from $\Sigma$. Formally, an infinite word $u$ over $\Sigma$ is a function from $\mathbb{N}$ to $\Sigma$: the $i$-th letter of $u$ is $u(i)$. We often denote $u(i)$ by $u_i$. The set of all right infinite words over $\Sigma$ is denoted by $\Sigma^\omega$.

A Müller automaton is a five-tuple $M = (Q, \Sigma, \delta, q_0, \mathcal{F})$, where $Q$ is the set of states, $\Sigma$ is the input alphabet, $\delta : Q \times \Sigma \to Q$ is the transition function, $q_0 \in Q$ is the start state and $\mathcal{F} \subseteq 2^Q$ is the acceptance table.

To define acceptance of words by Müller automata, let $w \in \Sigma^\omega$ be an infinite word and $\boldsymbol{q}_w \in Q^\omega$ be the infinite path through $M$ defined by $w$ as follows: $\boldsymbol{q}_w = q_0 q_1 q_2 \ldots$ where $q_i = \delta(q_{i-1}, w_i)$ for all $i \geq 1$. Let $\text{Inf}(\boldsymbol{q}_w) \in 2^Q$ denote the set of states which occur infinitely often in $\boldsymbol{q}_w$. Then $w$ is accepted by the Müller automaton $M$ if $\text{Inf}(\boldsymbol{q}_w) \in \mathcal{F}$. The language accepted by a Müller automaton $M$ is simply the set of all infinite words accepted by $M$.

Given an Müller automaton $M = (Q, \Sigma, \delta, q_0, \mathcal{F})$, an element $f \in \mathcal{F}$ is said to be admissible if there exists an infinite path $p$ starting at $q_0$ such that $\text{Inf}(p) = f$. By extension, we say that the table $\mathcal{F}$ is admissible with respect to an automaton $M$ if all its elements are admissible. We simply say that a table $\mathcal{F}$ is admissible if there exists a Müller automaton $M = (Q, \Sigma, \delta, q_0, \mathcal{F})$ that makes $\mathcal{F}$ admissible.

Let $Q$ be a set of states. If, for a fixed input alphabet $\Sigma$, there is no Müller automata $M = (Q, \Sigma, \delta, q_0, F)$ such that $\mathcal{F}$ is admissible with respect to $M$, we say that $\mathcal{F}$ is $\Sigma$-strongly inadmissible (or simply strongly admissible if $\Sigma$ is understood) on $Q$. Clearly, whether or not a table is strongly inadmissible depends only on $|\Sigma|$ and $|Q|$. If $|\Sigma| = k$ and $|Q| = n$, then we will also say that $\mathcal{F}$ is $(k, n)$-strongly inadmissible. In what follows, we assume that for any table $\mathcal{F}, \emptyset \notin \mathcal{F}$.

Let $n \geq 1$ be a natural number. We denote by $[n]$ the set $\{1, 2, \ldots, n\}$.

## 3    Admissible and Inadmissible Tables

Fix an alphabet $\Sigma$ with $|\Sigma| > 1$. We first ask the following question: Of the $2^{2^{|Q|}-1}$ possible acceptance tables on the set $Q$ of states, how many are admissible (or strongly inadmissible)? We begin with a few observations which are easily proven:

**Fact 1.** *Let $M = (Q, \Sigma, \delta, q_0, \mathcal{F})$. Then $\mathcal{F}$ is admissible iff for all $f \in \mathcal{F}$, and for all $q \in f$, the following conditions hold:*

(a) there exists a word $w \in \Sigma^*$ such that $\delta(q, w) = q$ and the sequence $q = q_{i_1} \vdash q_{i_2} \vdash \cdots \vdash q_{i_j} = q$ of states encountered while reading $w$ contains each state of $f$ at least once, and no state which is not in $f$.

(b) $q$ is reachable from $q_0$.

**Fact 2.** *If $\mathcal{F}$ is $(k, n)$-strongly inadmissible, then $\mathcal{F}'$ is $(k, n)$-strongly inadmissible for all $\mathcal{F}' \supseteq \mathcal{F}$.*

First, we note that if we have more letters than states, no table is $\Sigma$-strongly inadmissible.

**Lemma 1.** *Let $\Sigma, Q$ be fixed. If $|\Sigma| \geq |Q|$ then there are no $\Sigma$-strongly inadmissible tables on $Q$.*

*Proof.* The proof is obvious: Let $Q = \{q_0, \ldots, q_{n-1}\}$ and $\Sigma = \{a_0, a_1, \ldots, a_{k-1}\}$ with $k \geq n$. Then we set $\delta(q_i, a_j) = q_j$ for all $0 \leq i, j \leq n - 1$. Consider any subset $\{q_{i_1}, q_{i_2}, \ldots, q_{i_\ell}\} \subseteq Q$. Then the word $a_{i_2} a_{i_3} \cdots a_{i_\ell} a_{i_1}$ will take us through the following sequence of states $q_{i_1} \vdash q_{i_2} \vdash \cdots \vdash q_{i_\ell} \vdash q_{i_1}$. Thus, by Fact 1, any possible nonempty $\mathcal{F} \subseteq 2^Q$ is admissible. □

Thus, our results on strong inadmissibility will require that $|\Sigma| < |Q|$.

## 3.1   Enumerating Strongly Inadmissible Tables

We begin with a result which allows us to characterize strongly inadmissible tables.

**Theorem 3.** *Let $\Sigma$ be an alphabet with $|\Sigma| = k \geq 2$ and $Q$ be a set of states with $|Q| = n$. Let $S \subseteq Q$ be an arbitrary non-empty set of states. Let $\mathcal{F} \subseteq 2^Q$. If there exists a subset $\{f_1, f_2, \ldots, f_\ell\} \subseteq \mathcal{F}$ such that $\ell \geq (k - 1)|S| + 2$ and*

(a) $f_i \supseteq S$ for all $1 \leq i \leq \ell$; and
(b) $f_i \cap f_j = S$ for all $1 \leq i < j \leq \ell$,

*then $\mathcal{F}$ is strongly inadmissible.*

*Proof.* Assume that $\mathcal{F}$ satisfies the above conditions. Let $M = (Q, \Sigma, \delta, q_0, \mathcal{F})$ be a Müller automaton such that $\mathcal{F}$ is admissible.

First, we establish that for all $s \in S$, there exists a letter $a \in \Sigma$ such that $\delta(s, a) \in S$.

Assume not, that is, there exists some $s \in S$ such that $\delta(s, a) \notin S$ for all $a \in \Sigma$. Now, consider each $f_i$ for $1 \leq i \leq \ell$. Since $s \in S \subseteq f_i$, there must be a path starting at $s$ and passing through all the elements of $f_i$. Since no letter $a$ of $\Sigma$ satisfies $\delta(s, a) \in S$, we must have that the path through $f_i$ starting at $s$ must pass through some state of $f_i \setminus S$ next. Call that state $q_i$. Now, for all $1 \leq i < i' \leq \ell$, $q_i \neq q_{i'}$ since $(f_i \setminus S) \cap (f_{i'} \setminus S) = \emptyset$. Thus, there are $\ell$ states which are reachable from $s$. As there are only $k$ transitions leaving $s$ (as $\delta$ is deterministic), we have that $\ell \leq k$; this implies

$$(k - 1)|S| + 2 \leq k. \tag{1}$$

For $|S| = 1$, this implies $k + 1 \le k$. Otherwise (recall $S \ne \emptyset$), $1 - |S| < 0$ and thus (1) implies

$$k \le \frac{2 - |S|}{1 - |S|}.$$

This contradicts $k \ge 2$, since $(2 - |S|)/(1 - |S|) \le 1$ for all $|S| > 1$. Thus, for all $s \in S$, there is some $a \in \Sigma$ such that $\delta(s, a) \in S$.

Since each of the $f_i$ are distinct, there is at most one index $i_0$ with $1 \le i_0 \le \ell$ such that $f_i = S$. Now, for all $f_i$ with $1 \le i \le \ell$ and $i \ne i_0$, there is some $q_i \in f_i$ such that $q_i \notin S$. Furthermore, since $f_i \cap f_j = S$ for $i \ne j$, we have that $q_i \ne q_j$. Assume without loss of generality that for each $q_i$, there is some $s_i \in S$ such that $\delta(s_i, b_i) = q_i$ for some $b_i \in \Sigma$. Indeed, if no such $s_i$ existed for this choice of $q_i \in f_i \setminus S$, we choose another state in $f_i \setminus S$. Certainly, for some $q_i \in f_i \setminus S$, a state $s_i \in S$ exists such that $\delta(s_i, b_i) \in f_i \setminus S$, otherwise there could be no path which cycled through all the states of $f_i$ (including $S$, since $S \subseteq f_i$) and no others, and $\mathcal{F}$ would not be admissible.

Thus, for each $i$ with $1 \le i \le \ell$ ($i \ne i_0$) there is some $(s_i, b_i) \in S \times \Sigma$ such that $\delta(s_i, b_i) \in f_i \setminus S$. Now, each $(s_i, b_i)$ must be distinct, since their image is in $f_i \setminus S$, and $(f_i \setminus S) \cap (f_j \setminus S) = \emptyset$.

Thus, we have at least $\ell - 1 + |S|$ transitions defined on states of $S$: $|S|$ transitions of the form $\delta(s, a) \in S$ for each $s \in S$, and at least $\ell - 1$ transitions of the form $\delta(s, b) \in f_i \setminus S$ for some $s \in S$ and $1 \le i \le \ell$, with one possible exception, $i_0$. Thus, since there are $k|S|$ transitions for the states in $S$, we must have that

$$\ell - 1 + |S| \le k|S|$$

or

$$\ell \le (k - 1)|S| + 1.$$

But this contradicts our choice of $\ell$. Thus, $\mathcal{F}$ is strongly inadmissible.    □

We now show that $(k - 1)|S| + 2$ is best possible in Theorem 3, provided that $|Q \setminus S|$ is large enough:

**Lemma 2.** *Let $k \ge 2$, $n \ge 3$. Let $|Q| = n$ and $|\Sigma| = k$. Then for all $S \subseteq Q$ with $S \ne \emptyset$ and $k|S| \le n$, there exists a table $\mathcal{F} \subseteq 2^Q$ such that there exists a subset $\{f_1, f_2, \ldots, f_\ell\} \subseteq \mathcal{F}$ with $\ell = (k - 1)|S| + 1$ and*

*(a) $f_i \supseteq S$ for all $1 \le i \le \ell$; and*
*(b) $f_i \cap f_j = S$ for all $1 \le i < j \le \ell$.*

*Furthermore, $\mathcal{F}$ is admissible.*

*Proof.* Let $|S| = j$, $S = \{s_1, s_2, \ldots, s_j\}$, and $Q \setminus S = \{q_{j+1}, \ldots, q_n\}$. As $kj \le n$, we have that $\ell - 1 = (k - 1)j \le n - j$. Thus, we define $\mathcal{F}$ as $\mathcal{F} = \{f_1, f_2, \ldots, f_\ell\}$ with

$$f_i = S \cup \{q_{j+i}\}$$

for $1 \le i \le \ell - 1$ and $f_\ell = S$.

Let $\Sigma = \{a_i : 1 \leq i \leq k\}$. We define $\delta : Q \times \Sigma \to Q$ as follows: $\delta(s_i, a_1) = s_{i+1}$ for all $1 \leq i \leq j$, where the addition is taken modulo $j$. This leaves the letters $a_i$ for $2 \leq i \leq k$ which we can assign as we wish. This gives $(k-1)|S|$ transitions which can go from states in $S$ to states not in $S$. In particular, as $\ell - 1 = (k-1)|S|$, for each state $q_{j+i}$ for $1 \leq i \leq \ell - 1$, we can pick a pair $(s_{m_{j+i}}, a_{m_{j+i}}) \in S \times (\Sigma \setminus \{a_1\})$ such that $\delta(s_{m_{j+i}}, a_{m_{j+i}}) = q_{j+i}$ for all $1 \leq i \leq \ell - 1$. In this way, we define $\delta$ on $S$. For each $q_{j+i}$, we define $\delta(q_{j+i}, a_1) = s_{m_{j+i}}$, so that there is a letter which takes us from state $q_{j+i}$ to the state $s_{m_{j+i}}$ in $S$ from which we can reach $q_{j+i}$.

Thus, if we define $\delta$ arbitrarily on the remaining state-letter pairs, and let the initial state be any state in $S$, then it is easy to verify that the resulting DFA $M = (Q, \Sigma, \delta, q_0, \mathcal{F})$ makes $\mathcal{F}$ admissible. □

We can easily show that there exists strongly inadmissible tables which do not satisfy the conditions of Theorem 3. For instance, consider $|\Sigma| = 2$ and

$$\mathcal{F} = \{\{1\}, \{2\}, \{1, 2\}, \{1, 2, 3\}\}.$$

It must be the case that any transition function admitting $\mathcal{F}$ must loop on states 1 and 2. Thus, states 1 and 2 each have exactly one transition remaining. Let $i \to j$ indicate that there exists some $a \in \Sigma$ such that $\delta(i, a) = j$. If the remaining transition leaving 1 is set to be $1 \to 2$, then we have two choices: we can either set $2 \to 3$ or $2 \to 1$.

If we choose $2 \to 3$, we cannot have a loop on the set $\{1, 2\}$, since all transitions on those states have been assigned and no such loop exists. If we choose $2 \to 1$, then we cannot have a loop on the set $\{1, 2, 3\}$, since 3 is not reachable from 2 or 1. Thus, we cannot set $1 \to 2$.

Thus, we must have $1 \to 3$. But in this case, $\{1, 2\}$ cannot be looped upon. Thus, there is no assignment of the transitions such that $\mathcal{F}$ is admissible, and thus the conditions of Theorem 3 are sufficient but not necessary.

Thus, we can now use Theorem 3 to give a lower bound on the number of $(2, n)$-strongly inadmissible tables.

**Theorem 4.** Let $n \geq 3$. Then there are at least $2^{2^n - 1}(1 + o(1))$ $(2, n)$-strongly inadmissible tables.

*Proof.* Let $Q = [n]$. Let $1 \leq i < j < k \leq n$. Then we note that $\mathcal{F} = \{\{i\}, \{i, j\}, \{i, k\}\}$ is $(2, n)$-strongly inadmissible, as it satisfies the conditions of Theorem 3.

We now give a lower bound on number of ways we can choose a table $\mathcal{F}'$ which is a superset of $\mathcal{F}$ for some choice of $i \neq j \neq k$. To this end, we define sets $S_\ell(i, j, k) \subseteq 2^{2^Q}$ for all $1 \leq i < j < k \leq n$, to satisfy the following: For all $T \in S_\ell(i, j, k)$,

(a) $\{\{i\}, \{i, j\}, \{i, k\}\} \subseteq T$;
(b) $\{i'\} \notin T$ for all $1 \leq i' < i$;
(c) $\{i, j'\} \notin T$ for all $i < j' < j$;
(d) $\{i, k'\} \notin T$ for all $j < k' < k$.

By (a), each $T \in S_\ell(i, j, k)$ is $(2, n)$-strongly inadmissible. Conditions (b)–(d) are required to avoid double counting between different sets of the form $S_\ell(i, j, k)$.

In fact, it is easy to see that, if $(i, j, k) \neq (i', j', k')$ and $1 \leq i < j < k \leq n$ and $1 \leq i' < j' < k' \leq n$ then $S_\ell(i, j, k) \cap S_\ell(i', j', k') = \emptyset$.

Now, we calculate $|S_\ell(i, j, k)|$ for $1 \leq i < j < k \leq n$. We count the number of subsets $s \subseteq Q$ such that the inclusion or exclusion of $s$ in each member $T$ in $S_\ell(i, j, k)$ is fixed by our choice of $i, j, k$. We note that

(i)  All the sets $\{i'\}$ for $1 \leq i' < i$ are excluded from $T$, whereas $\{i\} \in T$ is necessary. Thus, $i$ singletons are fixed with respect to elements of $S_\ell(i, j, k)$.
(ii) All the sets $\{i, j'\}$ for $i < j' < j$ are excluded from $T$, while $\{i, j\} \in T$. This fixes $j - i$ subsets of size 2.
(iii) All the sets $\{i, k'\}$ for $j < k' < k$ are excluded from $T$, while $\{i, k\} \in T$. Thus $k - j$ additional subsets of size 2 are fixed as elements/non-elements of $T$.

Summing the number of sets in (i)–(iii), we find that $k$ subsets are fixed. Thus, any of the other $2^n - 1 - k$ (recall that we do not consider tables $\mathcal{F}$ with $\emptyset \in \mathcal{F}$) elements may or may not be added to an arbitrary element $T$ of $S_\ell(i, j, k)$. Thus

$$|S_\ell(i, j, k)| = 2^{2^n - 1 - k}.$$

We now consider

$$S_\ell = \bigcup_{i=1}^{n-2} \bigcup_{j=i+1}^{n-1} \bigcup_{k=j+1}^{n} S_\ell(i, j, k).$$

Each element of $S_\ell$ is $(2, n)$-strongly inadmissible. Further, since the $S_\ell(i, j, k)$ are mutually disjoint,

$$|S_\ell| = \sum_{i=1}^{n-2} \sum_{j=i+1}^{n-1} \sum_{k=j+1}^{n} |S_\ell(i, j, k)|$$

$$= \sum_{i=1}^{n-2} \sum_{j=i+1}^{n-1} \sum_{k=j+1}^{n} 2^{2^n - 1 - k}$$

We then note that

$$|S_\ell| = \sum_{i=1}^{n-2} \sum_{j=i+1}^{n-1} \sum_{k=j+1}^{n} 2^{2^n - 1 - k}$$

$$= 2^{2^n - 1} \sum_{i=1}^{n-2} \sum_{j=i+1}^{n-1} \sum_{k=j+1}^{n} 2^{-k}$$

$$= 2^{2^n - 1} \sum_{i=1}^{n-2} \left( \frac{1}{2^i} - \frac{1}{2^{n-1}} - \frac{n - i - 1}{2^n} \right)$$

$$= 2^{2^n - 1} \left( 1 - \frac{1}{2^{n-2}} - \frac{n - 2}{2^{n-1}} - \frac{(n - 1)(n - 2)}{2^{n+1}} \right)$$

Thus, if

$$\lambda(n) = -\frac{1}{2^{n-2}} - \frac{n-2}{2^{n-1}} - \frac{(n-1)(n-2)}{2^{n+1}}$$

then $\lambda(n) \to 0$ as $n \to \infty$ and there are at least $2^{2^n-1}(1 + \lambda(n))$ $(2, n)$-strongly inadmissible tables of size $n$.                                                                         □

Note that if we repeat Lemma 4 for $\Sigma$ with $|\Sigma| = k \geq 2$, we can consider tables containing

$$\{\{i\}, \{i, j_1\}, \{i, j_2\}, \ldots, \{i, j_k\}\}$$

for some $1 \leq i < j_1 < j_2 < \cdots < j_k \leq n$. Repeating the above analysis, we may prove the following:

**Lemma 3.** *Let $k \geq 2$. Then for all $n \geq k + 1$, there exist at least*

$$2^{2^n-1}\left(1 - \frac{\sum_{m=0}^{k}\binom{n}{m}}{2^n}\right) \tag{2}$$

$(k, n)$-*strongly inadmissible tables.*

Note that Lemma 3 reflects our intuition that if $|\Sigma| = |Q|$, then there are no $\Sigma$-strongly inadmissible tables over $Q$. Indeed, if we take $n = k$ in (2), then we get zero $(n, n)$-strongly inadmissible tables.

## 3.2 Upper Bound on $k$-Admissibility

Let $\Sigma = \{a_1, a_2, \ldots, a_k\}$ for some $k \geq 2$. We now give a lower bound on the number of admissible tables. Consider $Q = [n]$ and $\delta$ given by

$$\delta(i, a_j) = i + j \pmod n \tag{3}$$

for all $i \in Q$ and $a_j \in \Sigma$.

We define $S_k \subseteq 2^Q$ to be the set of all subsets of $Q$ which can be cycled upon under $\delta : Q \times \Sigma \to Q$ given by (3). Thus, for any non-empty $T \subseteq S_k$, $T$ will be an admissible table. Consequently, $2^{|S_k|}$ will give us a lower bound on the number of admissible tables for a state set of size $n$ and alphabet of size $k$.

To estimate the size of $S_k$, we will construct an explicit set $U_k$. We first consider the $\lfloor \frac{n}{k} \rfloor$ "blocks" of $k$ successive states. In particular, let $B_i = \{ki + 1, ki + 2, \ldots, k(i+1)\}$ for $0 \leq i \leq \lfloor \frac{n}{k} \rfloor - 1$.

For each $B_i$, we claim that, for each of the $2^{k-1}$ subsets of $T_i \subseteq \{ki + 2, \ldots, k(i+1)\}$, we can define a word $w_i$ which visits $ki + 1$, each of the elements of $T_i$ in ascending order, followed by $k(i + 1) + 1$ (the first element of $B_{i+1}$).

In fact, since there are edges from each element $x \in B_i$ to each $y \in B_i$ with $y > x$, (as well as the first element of $B_{i+1}$) we can easily define $w_i$ to be the word which takes us from each $x \in T_i$ to the next $y \in T_i$ with $y > x$, or the first element of $B_{i+1}$ if no such $y$ exists.

Thus, for each choice of $T_i$ for $0 \leq i \leq \lfloor \frac{n}{k} \rfloor - 1$, the corresponding element of $U_k$ is the union of (a) the sets $T_i$ and (b) the elements $ki+1$ for all $0 \leq i \leq \lfloor \frac{n}{k} \rfloor - 1$.

However, we also get another $2^{n-k\lfloor n/k \rfloor - 1}$ choices (if $k$ does not divide $n$; 1 choice if $k|n$) for possible subsets of the states $k\lfloor n/k \rfloor + 1, \ldots, n$, which can be reached in the same manner as subsets of $B_i$. Call this subset $T_0$.

Thus, choosing a subset $T_i$ for each $0 \leq i \leq \lfloor \frac{n}{k} \rfloor - 1$, as well as a subset $T_0$, we get that $|U_k| \geq 2^{\lfloor n/k \rfloor (k-1)} 2^{n-k\lfloor n/k \rfloor - 1} = 2^{n-\lfloor n/k \rfloor - 1}$.

We can improve this estimation further as follows. Consider rotating the previous subsets $B_i$ by some factor $j$, $1 \leq j \leq k$, in the following sense: Let $B_i^{(j)} = \{ki + j, ki + 2, \ldots, k(i+1) + j - 1\}$ (modulo $n$ if necessary).

Then we can repeat the previous arguments with each of the $k - 1$ other choices of $j$, with one minor difference. Let $U_k^{(j)}$ be those chosen with rotation $j$ for $1 \leq j \leq k$. We construct $U_k^{(j)}$ in the same manner as $U_k$, with sets $T_i^{(j)}$ at each stage. However, to avoid double counting, we insist that for all $2 \leq j \leq k$, there exists some $i$ such that $T_i^{(j)} \subseteq B_i^{(j)}$ is chosen to be empty: $T_i^{(j)} = \emptyset$. Thus, we get that each element will contain at least one "gap" between $ki + j$ and $k(i+1) + j$ for some choice of $i$. This will ensure that $U_k^{(j_1)} \cap U_k^{(j_2)}$ will be empty if $j_1 \neq j_2$. Note that for $j = 1$, $U_k^{(1)}$ does not require a gap; that is, $U_k^{(1)} = U_k$.

Now, arguing similarly to $U_k$, we get that if $2 \leq j \leq k$,

$$|U_k^{(j)}| \geq \lfloor \frac{n}{k} \rfloor (2^{k-1})^{\lfloor n/k \rfloor - 1} 2^{n-k\lfloor n/k \rfloor - 1}$$

$$= \lfloor \frac{n}{k} \rfloor 2^{n-\lfloor n/k \rfloor - k}.$$

The factor of $\lfloor \frac{n}{k} \rfloor$ comes from the choice of $i$ for which $T_i^{(j)} = \emptyset$. For the remaining $\lfloor n/k \rfloor - 1$ $B_i^{(j)}$, we get our factor of $2^{k-1}$ as before.

Thus, we now let $U_k' = \bigcup_{j=1}^{k} U_k^{(j)}$. This yields

$$|U_k'| = 2^{n-\lfloor n/k \rfloor - 1} + (k-1)\lfloor \frac{n}{k} \rfloor 2^{n-\lfloor n/k \rfloor - k}$$

$$= 2^{n-\lfloor n/k \rfloor - 1} \left(1 + \frac{\lfloor \frac{n}{k} \rfloor (k-1)}{2^{k-1}}\right)$$

Thus, as $|S_k| \geq |U_k'|$, we have established the following result:

**Theorem 5.** *Let $k \geq 2$. For $n > k$, there are*

$$\Omega(2^{2^{n-\lfloor n/k \rfloor - 1}(1+(\lfloor n/k \rfloor (k-1))/2^{k-1})})$$

*$k$-admissible tables on $Q$ with $|Q| = n$.*

Note that if $n = k$, this yields a lower bound of $2^{2^n(1/4+o(1))}$ admissible tables, which reflects our knowledge that there are no $(n, n)$-strongly inadmissible tables.

Figure 1 gives counts and estimates of $(k, n)$-strongly inadmissible tables for state set size $n = 1, 2, 3, 4$ and $k = 2, 3$.

| $k \setminus n$ | 1 | 2 | 3 | 4 |
|---|---|---|---|---|
| 2 | 0 | 0 | 44 | $\geq 22067$ |
| 3 | 0 | 0 | 0 | $\geq 7000$ |

**Fig. 1.** Number of $(k, n)$-strongly inadmissible tables for $n \leq 4$ states, $k = 2, 3$

### 3.3  $(1, n)$-Strongly Inadmissible Tables

Until now, we have not considered the case of a unary alphabet; i.e., $|\Sigma| = 1$. This case is trivial, however, we include it for completeness.

We note that in any $M = (Q, \{a\}, \delta, q_0, \mathcal{F})$, there can be at most one loop which is reachable from the start state, since there is at most one labeled edge exiting each state in $Q$. Thus, since we can chose any non-empty $S \subseteq Q$ to be in the single loop, any table of the form $\{S\}$ is admissible. Further, these are the only $(1, n)$-admissible tables.

Thus, we note the following theorem:

**Theorem 6.** *There are exactly $2^{2^n - 1} - 2^n$ $(1, n)$-strongly inadmissible tables.*

## 4    Enumeration of Müller Automata

We now turn to enumeration of Müller automata. One problem in enumerating deterministic Müller automata is that, like nondeterministic automata which accept finite words, minimal-state deterministic Müller automata are not unique. To see this, we can consider an example given by Staiger [7, Ex. 2, p. 447], depicted in Fig. 2. With tables $\mathcal{F}_1 = \{\{1\}, \{2\}\}$, $\mathcal{F}_2 = \{\{1\}\}$, $\mathcal{F}_3 = \{\{1\}\}$,

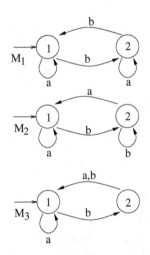

**Fig. 2.** Three Müller automata accepting $(a + b)^* a^\omega$

respectively, each automata accepts the language $(a + b)^*a^\omega$ (We can get two more automata accepting the same language by changing the initial state from 1 to 2 on both $M_2$ and $M_3$).

In the following, we establish some rough upper and lower bounds on the number of $\omega$-languages accepted by Müller automata with $n$ states.

## 4.1  Upper Bound

Recall that an automata $M = (Q, \Sigma, \delta, q_0, \mathcal{F})$ is *initially connected* if for all $q \in Q$, there is some word $w \in \Sigma^*$ such that $\delta(q_0, w) = q$. That is, every state is reachable from the initial state.

We first recall the function $C_k(n)$, which gives the number of pairwise non-isomorphic initially connected finite automata on $n$ states over a $k$-letter alphabet without any final states (Robinson [8] and Liskovets [9] give a recurrence relation for $C_k(n)$). Then we have the following results, due to Robinson [8] (see discussion and corrections in Domaratzki *et al.* [3]; see also Korshunov [10, p. 50]).

**Lemma 4.** *For all $k \geq 2$, there exists a positive constant $\gamma_k$ such that*

$$C_k(n) = n^{kn}\gamma_k^{n(1+o(1))}.$$

The following gives an asymptotic formula for the number of non-isomorphic initially-connect finite automata, without labeled states. The factor of $(n-1)!$ comes from the relabeling of all non-initial states.

**Lemma 5.** *For all $k \geq 2$, $n \geq 1$, the number of unlabeled pairwise non-isomorphic initially connected finite automata on $n$ states over a $k$-letter alphabet without final states is $C_k(n)/(n-1)!$.*

Let $f_k^{(\omega)}(n)$ be the number of distinct $\omega$-regular languages accepted by a Müller DFA with $n$ states over a $k$-letter alphabet. To give an upper bound on $f_k^{(\omega)}(n)$, we can simply take $C_k(n)/(n-1)!$ multiplied by our upper bound on the number of admissible tables on $n$ states over a $k$-letter alphabet. Thus, we can give the following corollary of Lemmas 4, 5 and 3:

**Corollary 1.** *For all $n \geq 3$ and $k \geq 2$,*

$$f_k^{(\omega)}(n) \leq \frac{C_k(n)}{(n-1)!} 2^{2^n-1} \frac{\sum_{m=0}^{k} \binom{n}{m}}{2^n}.$$

*Thus,*

$$f_k^{(\omega)}(n) \leq \frac{n^{kn}}{(n-1)!} \gamma_k^{n(1+o(1))} 2^{2^n-1} \frac{\sum_{m=0}^{k} \binom{n}{m}}{2^n}.$$

## 4.2  Lower Bound

To give a lower bound on the number of $\omega$-languages accepted by deterministic Müller automata with $n$ states, we construct a set of Müller automata, each recognizing a distinct $\omega$-language. We use the construction of the set $U_k$ from Sect. 3.2. Let $\Sigma = \{a_1, a_2, \ldots, a_k\}$. Let $Q = [n]$ and $\delta : Q \times \Sigma \to Q$ be given as in Sect. 3.2.

We define a size measure $|| \cdot || : \Sigma^* \to \mathbb{N}$ as follows:

$$||\epsilon|| = 0$$
$$||a_i w|| = i + ||w|| \quad \forall a_i \in \Sigma, w \in \Sigma^*$$

Then for all $f \in U_k$, there exists some word which visits the elements of $f$ in ascending order, let us denote this word by $w_f$. Further, for any $s \subseteq U_k$, let $w(s) = \{w_f : f \in s\}$.

Note that, if $f \in U_k$, regardless of $|w_f|$, $||w_f|| = n$. This is since the letter $a_i$ will shift us forward from state $j$ to $j + i$, by definition of $\delta$. We claim that for any distinct $f, g \in U_k$, the words $w_f, w_g$ satisfy

$$w_f^\omega \neq w_g^\omega.$$

Assume not, and that $|w_f| \leq |w_g|$. Then $w_g$ has, as prefix, $w_f$. But then as $w_f \neq w_g$, we must have $||w_g|| > n$. This contradicts the definition of $w_g$.

For any $s \subseteq U_k$, define $M_s = (Q, \Sigma, \delta, 1, s)$. We claim that if $s, t$ are distinct subsets of $U_k$, $L(M_s) \neq L(M_t)$. To see this, note that there is some word $w_f^\omega$ which is in exactly one of $L(M_s)$ and $L(M_t)$, corresponding to the set $f \subseteq Q$ which is in exactly one of $s$ and $t$. Thus, all non-empty subsets of $U_k$ give a distinct $\omega$-regular language, and we have established the following:

**Theorem 7.** *Let $n$ and $k \geq 2$ be integers with $n \geq k + 1$. Then*

$$f_k^{(\omega)}(n) \geq 2^{2^{n - \lfloor n/k \rfloor - 1} - 1}.$$

We note that the results of Theorem 7 and Corollary 1 are not asymptotically matching.

**Acknowledgments.** Kai Salomaa made many comments and corrections on drafts of this paper. I am also very grateful to the anonymous referees, who helped me with many comments and corrections, especially with corrections to Lemma 3 and Theorem 6.

## References

1. Câmpeanu, C., Păun, A.: The number of similarity relations and the number of minimal deterministic finite cover automata. In Champarnaud, J.M., Maurel, D., eds.: Pre-Proceedings of the Seventh International Conference on Implementation and Application of Automata. (2002) 71–80

2. Domaratzki, M.: Improved bounds on the number of automata accepting finite languages. In Ito, M., Toyama, M., eds.: DLT 2002: Developments in Language Theory, Sixth International Conference. (2002) 159–181

3. Domaratzki, M., Kisman, D., Shallit, J.: On the number of distinct languages accepted by finite automata with $n$ states. To appear, J. Automata, Languages and Combinatorics **7** (2002) 469–486

4. Liskovets, V.: Exact enumeration of acyclic automata. To appear, Formal Power Series and Algebraic Combinatorics 2003 (2003)

5. Perrin, D., Pin, J.E.: Infinite Words. Available electronically at http://www.liafa.jussieu.fr/~jep/Resumes/InfiniteWords.html (2003)

6. Thomas, W.: Automata on infinite objects. In van Leeuwen, J., ed.: Handbook of Theoretical Computer Science. Elsevier (1990) 133–192

7. Staiger, L.: Finite-state $\omega$-languages. J. Comp. Sys. Sci. **37** (1983) 434–448

8. Robinson, R.W.: Counting strongly connected finite automata. In Alavi, Y., Chartrand, G., Lesniak, L., Lick, D., Wall, C., eds.: Graph Theory with Applications to Algorithms and Computer Science. (1985) 671–685

9. Liskovets, V.: The number of connected initial automata. Cybernetics **5** (1969) 259–262

10. Korshunov, A.: Enumeration of finite automata (in Russian). Problemy Kibernetiki **34** (1978) 5–82

# Branching Grammars: A Generalization of ET0L Systems

Frank Drewes[1] and Joost Engelfriet[2]

[1] Department of Computing Science, Umeå University
S–901 87 Umeå, Sweden
drewes@cs.umu.se
[2] Department of Computer Science, Leiden University, P.O. Box 9612
NL-2300 RA Leiden, The Netherlands
engelfri@liacs.nl

**Abstract.** Generalizing ET0L systems, we introduce branching synchronization grammars with nested tables. Branching synchronization grammars with tables of nesting depth $n$ have the same string- and tree-generating power as $n$-fold compositions of top-down tree transducers.

## 1   Introduction

Context-free Chomsky grammars and context-free Lindenmayer systems (0L systems) certainly belong to the most well-studied types of grammars in formal language theory. They are both general enough to describe interesting languages and simple enough to guarantee nice properties (see [RS97]). The well-known ET0L systems by Rozenberg [Roz73] generalize both context-free grammars and 0L systems. Their rules are organized in tables, yielding a global synchronization mechanism: In every derivation step, a table is chosen and all nonterminals in the sentential form are replaced in parallel, using rules from the chosen table.[1]

In this paper, we propose an extension of ET0L systems, called branching synchronization grammars with nested tables (branching grammars, for short), in which pairs of nonterminals can, but need not be synchronized with each other. In a branching grammar of nesting depth $n$ (where $n \in \mathbb{N}$), the tables are organized in a tree-like hierarchy of depth $n$. We exploit this hierarchy to obtain $n$ increasingly strong levels of synchronization. A pair of nonterminals in a sentential form may be synchronized at any of these levels, or not at all.

Our main result characterizes the language generating power of branching grammars in terms of output languages of top-down tree transducers (td transducers). The investigation of the latter was initiated in the late sixties and early seventies by Rounds and Thatcher [Rou70,Tha70,Tha73], and later continued by many others. See the books by Gécseg and Steinby [GS84] (or the survey article [GS97]) and Fülöp and Vogler [FV98].

---

[1] We shall speak of terminals and nonterminals because our definition of ET0L systems in Sect. 2 makes this distinction.

Z. Ésik and Z. Fülöp (Eds.): DLT 2003, LNCS 2710, pp. 266–278, 2003.

It is known from [Eng76] that the ET0L languages are precisely the yields of output languages of td transducers having a monadic input alphabet. Intuitively, the states of the transducer are the nonterminals of the ET0L system and the input symbols represent the tables. The synchronization results from the copying of subtrees of the input tree that takes place in a computation of the transducer.

We turn from monadic input trees to arbitrary ones and from td transducers to compositions of td transducers. Our main result states that branching grammars of nesting depth $n$ have exactly the same language generating power as compositions of $n$ td transducers (which are known to constitute a strict hierarchy [Eng82]). We obtain this result by considering branching *tree* grammars, which are branching grammars in which the right-hand sides of rules are trees, i.e., strings having a term-like structure, with the nonterminals at the leaves. We note that Vágvölgyi defined the $n$-synchronized R-transducer in [Vág86] and showed that they realize the composition of $n$ top-down tree transducers. However, we have not been able to discover a direct relationship between these transducers and branching tree grammars.

Apart from being of independent theoretical interest, we expect the main result to be useful for applications in which only the language generating power of compositions of td transducers is needed, while the tree transformations as such are of minor interest (see, e.g., [Dre00,Dre01]). In such cases it is appropriate to describe the desired language by means of a single grammatical device rather than by a cumbersome and (from the algorithmic point of view) inefficient $n$ stage process using $n$ td transducers. Furthermore, our characterization can be used to prove new results about output languages of td transducers. We exemplify this by pointing out two such results in the conclusion.

The paper is structured as follows. In Sect. 2 we introduce branching grammars and discuss some of their basic properties. In Sect. 3 we define branching tree grammars, recall regular tree grammars and td transducers, and present a result about total td transducers needed later. Section 4 is devoted to our main result. A short conclusion in Sect. 5 finishes the paper.

Due to lack of space, the results are not proved in detail. Interested readers are referred to [DE02] for full proofs and additional background as well as examples.

## 2   Branching Grammars

In this section, we introduce branching grammars. However, let us first enumerate some of the terminology and notation we use.

We denote the set of natural numbers (including 0) by $\mathbb{N}$. For $n \in \mathbb{N}$, $[n]$ denotes $\{1, \ldots, n\}$. As usual, $X^*$ denotes the set of all strings over a set $X$. The length of $s \in X^*$ is denoted by $|s|$, and the empty string by $\lambda$. The concatenation of two strings $s$ and $t$ is denoted $st$, and $s^n$ denotes the $n$-fold concatenation $s \cdots s$. Somewhat ambiguously, we let $X^n$ denote the set of all $n$-tuples $(x_1, \ldots, x_n)$ over elements $x_i \in X$. For $\gamma = (x_1, \ldots, x_n) \in X^n$ and $x \in X$, $\gamma.x = (x_1, \ldots, x_n, x)$, and for $k \in \{0, \ldots, n\}$, $\mathrm{first}_k(\gamma) = (x_1, \ldots, x_k)$ restricts $\gamma$ to its first $k$ components.

In spite of the ambiguity above, we do not identify $n$-tuples with strings of length $n$. In particular, we will consider the set $(X^n)^*$ of strings of $n$-tuples. Thus, the string $(x,y,y)(x,y,y)(x,x,x)(y,y,y)$ of 3-tuples has length 4 and differs from the string $xyyxyyxxxyyy$ of length 12.

Branching grammars generalize ET0L systems, whose definition we recall now. An *ET0L system* is a tuple $G = (N, T, J, R, S)$ where $N$ and $T$ are disjoint alphabets of *nonterminals* resp. *terminals*, $J \neq \emptyset$ is an alphabet of *table symbols*, $R$ is the *table specification*, which assigns to every $\tau \in J$ a finite set $R(\tau)$ of rules $A \to \zeta$ such that $A \in N$ and $\zeta \in (N \cup T)^*$, and $S \in N$ is the *initial nonterminal*.

Consider strings $\xi_1, \xi_2 \in (N \cup T)^*$ where $\xi_1 = w_0 A_1 w_1 \cdots A_h w_h$ for some $h \in \mathbb{N}$, $w_0, \ldots, w_h \in T^*$, and $A_1, \ldots, A_h \in N$. There is a derivation step $\xi_1 \Rightarrow \xi_2$ if some $R(\tau)$ ($\tau \in J$) contains rules $A_1 \to \zeta_1, \ldots, A_h \to \zeta_h$ with $\xi_2 = w_0 \zeta_1 w_1 \cdots \zeta_h w_h$. Note the parallel mode of derivation which is typical for Lindenmayer systems. In each derivation step all nonterminals are replaced in parallel. Furthermore, all rules applied in a step must be taken from the same $R(\tau)$, called a *table*. The language generated by $G$ is $L(G) = \{w \in T^* \mid S \Rightarrow^* w\}$. The class of all such languages is denoted by ET0L. This class includes the context-free languages because $G$ generates the same language as the context-free grammar $(N, T, R(\tau), S)$ if $J$ is a singleton $\{\tau\}$.

Branching grammars generalize the concept of synchronization of ET0L systems by introducing different levels of synchronization, so that the synchronization need not be equally strong for all pairs of nonterminals in a given sentential form. The idea is to have $n$ possible levels of synchronization instead of just one. In a (parallel) derivation step, the synchronization between descendants of two nonterminals can be retained at the same level or released to a lower level. This is accomplished by replacing $J$ with $J^n$ in the table specification (i.e., tables are now specified by $n$-tuples of table symbols), and by augmenting the nonterminals in the right-hand sides of rules with elements of $I^n$ (i.e., with $n$-tuples of synchronization symbols). As will be seen below, the latter determines whether or not two occurrences of nonterminals are synchronized at a given level.

**Definition 1 (branching grammar).** Let $n \in \mathbb{N}$. A *grammar with branching synchronization and with nested tables of depth $n$* (*branching grammar*, for short) is a tuple $G = (N, T, I, J, R, S)$ where

- $N$ and $T$ are disjoint alphabets of *nonterminals* resp. *terminals*,
- $I$ and $J$ are nonempty alphabets of *synchronization* resp. *table symbols*,
- $R$, the *table specification*, assigns to every $\tau \in J^n$ a finite set $R(\tau)$ of rules $A \to \zeta$ such that $A \in N$ and $\zeta \in ((N \times I^n) \cup T)^*$, and
- $S \in N$ is the *initial nonterminal*.

The number $n$ is the *nesting depth* (or just *depth*) of $G$.

A set of rules $R(\tau)$, $\tau \in J^n$, is called a *table of $G$*. When there is no risk of confusion we may also call $\tau$ a table, and we may use $\tau$ to refer to $R(\tau)$. Intuitively, the table specification organizes the tables in a tree-like way. This gives rise to the following inductive definition of *supertables*. The supertables at nesting depth $n$ are the actual tables $R(\tau)$ where $\tau \in J^n$. Those at nesting depth $k$, $0 \le k < n$, are obtained by taking the union of the supertables at nesting depth $k + 1$: $R(\tau) = \bigcup_{j \in J} R(\tau.j) = \bigcup \{R(\tau') \mid \tau' \in J^n, \text{first}_k(\tau') = \tau\}$ for all $\tau \in J^k$. In particular, $R() = \bigcup_{\tau \in J^n} R(\tau)$ is the unique supertable at depth 0, consisting of all rules of $G$.

An element of $(I^n)^*$, i.e., a string of $n$-tuples of synchronization symbols, is called a *synchronization string*. An element of $\text{SN}_G = N \times (I^n)^*$, consisting of a nonterminal and a synchronization string, is called a *synchronized nonterminal*. The initial synchronized nonterminal of $G$ is $(S, \lambda)$.

Derivations will yield sentential forms $\xi = w_0(A_1, \varphi_1)w_1 \cdots (A_h, \varphi_h)w_h$ where $h \in \mathbb{N}$, $w_0, \ldots, w_h \in T^*$, $(A_1, \varphi_1), \ldots, (A_h, \varphi_h) \in \text{SN}_G$, and $|\varphi_1| = \cdots = |\varphi_h|$. As a convention, denoting a string $\xi$ in this manner is from now on always meant to imply that $h \in \mathbb{N}$, $w_0, \ldots, w_h \in T^*$, and $(A_1, \varphi_1), \ldots, (A_h, \varphi_h) \in \text{SN}_G$, where all synchronization strings $\varphi_1, \ldots, \varphi_h$ have equal length. Note that this convention applies to the right-hand sides of rules as well, which are written as $v_0(B_1, \alpha_1)v_1 \cdots (B_l, \alpha_l)v_l$. In this case, $|\alpha_1| = \cdots = |\alpha_l| = 1$, of course.

In a derivation, synchronization strings are accumulated in the nonterminals[2] of a sentential form. To formalize this, we first define an auxiliary step relation. For every $(A, \varphi) \in \text{SN}_G$ and every rule $r = A \to v_0(B_1, \alpha_1)v_1 \cdots (B_l, \alpha_l)v_l$

$$(A, \varphi) \Rightarrow_r v_0(B_1, \varphi\alpha_1)v_1 \cdots (B_l, \varphi\alpha_l)v_l.$$

Before we turn to the definition of a general derivation step, where a string contains several terminals and nonterminals (Definition 2), let us discuss how synchronization works. Suppose a sentential form contains synchronized nonterminals $(A, \varphi)$ and $(B, \psi)$. Since derivations start with $(S, \lambda)$ and are fully parallel, the accumulated synchronization strings $\varphi$ and $\psi$ have the same length, namely the length $m$ of the derivation. Thus, $\varphi = \alpha_1 \cdots \alpha_m$ and $\psi = \beta_1 \cdots \beta_m$ for certain $\alpha_1, \beta_1, \ldots, \alpha_m, \beta_m \in I^n$. Writing the $\alpha_i$ and $\beta_i$ as column vectors we can therefore view $\varphi$ and $\psi$ as $n \times m$-matrices of synchronization symbols:

$$\varphi = \begin{matrix} \alpha_{1,1} & \cdots & \alpha_{1,m} \\ \vdots & \ddots & \vdots \\ \alpha_{n,1} & \cdots & \alpha_{n,m} \end{matrix} \qquad \psi = \begin{matrix} \beta_{1,1} & \cdots & \beta_{1,m} \\ \vdots & \ddots & \vdots \\ \beta_{n,1} & \cdots & \beta_{n,m} \end{matrix}$$

where $\alpha_i = (\alpha_{1,i}, \ldots, \alpha_{n,i})$ and $\beta_i = (\beta_{1,i}, \ldots, \beta_{n,i})$ for $i \in [m]$. Now, the number $k$ of rows, counted from the top, up to which both matrices are equal determines how tightly $(A, \varphi)$ and $(B, \psi)$ are synchronized. We call this number their *level of synchronization*. The rules to be applied to them must be taken from the same supertable at depth $k$ but can be chosen independently within one such supertable. The highest level of synchronization is given by $k = n$, i.e., $\varphi = \psi$, in which case the rules must be taken from the same table. Thus, nesting depth $n$ leads to $n$ increasingly strong levels of synchronization (not counting level 0).

---

[2] We will often speak of nonterminals instead of synchronized nonterminals if there is no danger of confusion.

To formalize this, suppose we are given a set $X$. We extend $\text{first}_k$ to strings of $n$-tuples in the usual way: $\text{first}_k(s) = \text{first}_k(\gamma_1) \cdots \text{first}_k(\gamma_m)$ for all $s = \gamma_1 \cdots \gamma_m \in (X^n)^*$, mapping $(X^n)^*$ to $(X^k)^*$. With respect to the matrix notation used above this means that $\text{first}_k$ cuts off rows $k+1, \ldots, n$. Now, for $s, s' \in (X^n)^*$ with $|s| = |s'|$ let

$$\text{level}(s, s') = \max\{k \in \{0, \ldots, n\} \mid \text{first}_k(s) = \text{first}_k(s')\}.$$

Note that $\text{level}(s, s) = n$ and that $\text{level}(s, s') = n$ implies $s = s'$.

Thus, for synchronized nonterminals $(A, \varphi)$ and $(B, \psi)$, $\text{level}(\varphi, \psi)$ yields their level of synchronization, as discussed informally above. Regarding tables $\tau, \tau' \in J^n$, $\text{first}_k(\tau) = \text{first}_k(\tau')$ means that $R(\tau)$ and $R(\tau')$ are included in the same supertable at depth $k$. Hence, $\text{level}(\tau, \tau')$ yields the depth of their least common supertable.

We can now formalize derivations of branching grammars.

**Definition 2 (derivation and generated language).** Consider a branching grammar $G = (N, T, I, J, R, S)$ of depth $n$ and strings $\xi_1, \xi_2 \in (\text{SN}_G \cup T)^*$ where $\xi_1 = w_0(A_1, \varphi_1)w_1 \cdots (A_h, \varphi_h)w_h$. There is a *derivation step* $\xi_1 \Rightarrow_G \xi_2$ (or simply $\xi_1 \Rightarrow \xi_2$) if there are $\tau_1, \ldots, \tau_h \in J^n$ and $r_1 \in R(\tau_1), \ldots, r_h \in R(\tau_h)$ such that

(i)  $\xi_2 = w_0\zeta_1 w_1 \cdots \zeta_h w_h$ for $\zeta_1, \ldots, \zeta_h$ with $(A_j, \varphi_j) \Rightarrow_{r_j} \zeta_j$ for all $j \in [h]$ and
(ii) $\text{level}(\tau_i, \tau_j) \geq \text{level}(\varphi_i, \varphi_j)$ for all $i, j \in [h]$.

The *language generated by* $G$ is $L(G) = \{w \in T^* \mid (S, \lambda) \Rightarrow^* w\}$. The class of all languages generated by branching grammars of nesting depth $n$ is denoted by $\text{BS}_n$, and $\text{BS} = \bigcup_{n \in \mathbb{N}} \text{BS}_n$ (where BS stands for *branching synchronization*).

Note that, similar to the case of ET0L systems, derivation steps of a branching grammar are fully parallel. Of course, the choice of $r_1, \ldots, r_h$ in the definition above does not depend on the actual synchronization symbols which appear in $\varphi_1, \ldots, \varphi_h$. The only value that matters is $\text{level}(\varphi_i, \varphi_j)$—an (injective) renaming or permutation of synchronization symbols does not affect the language generated by a branching grammar. A similar remark holds for the symbols in $J$.

Generalizing the remark that context-free grammars correspond to ET0L systems with only one table, a context-free grammar may be considered to be a branching grammar (of any depth $\geq 0$) with only one table symbol, i.e., $|J| = 1$. Furthermore, if a branching grammar has depth 0 then $I$ and $J$ do not play any role because $I^0 = \{()\} = J^0$. Thus, $\text{BS}_0$ is the class of context-free languages. Moreover, an ET0L system may be seen as a branching grammar (of any depth $\geq 1$) with one synchronization symbol, i.e., $|I| = 1$. Hence, $\text{ET0L} \subseteq \text{BS}_1$. It can be shown by means of results of Skyum [Sky76] and Engelfriet, Rozenberg, and Slutzki [ERS80] that this inclusion is in fact proper.

It follows from our main result, together with the results of [Eng82], that the classes $(\text{BS}_n)_{n \in \mathbb{N}}$ constitute an infinite hierarchy which is strict on each level. An example in $\text{BS}_2 \setminus \text{BS}_1$ (known from [Eng82]) is the language $L_0 \subseteq \{a, b, \$\}^*$ defined as $L_0 = \{w\$w \mid w \in \{a, b\}^* \text{ and } \#_a(w) = 2^n \text{ for some } n \in \mathbb{N}\}$, where $\#_a(w)$ counts the number of occurrences of $a$ in $w$.

To generate $L_0$ by a branching grammar of depth 2 we use synchronization symbols $0, 1$, table symbols $0, 1, 2$, and nonterminals $S$ (the initial nonterminal), $A$, and $B$. The tables are the following (where we denote a synchronized nonterminal $(A, \varphi)$ by $A\langle\varphi\rangle$ and $\varphi$ as a column vector in order to enhance readability):

$$R(0,0) = \{ \ S \to A\langle{}^0_0\rangle \$ A\langle{}^0_0\rangle \ \}, \qquad R(2,0) = \{ \ A \to B\langle{}^0_0\rangle a B\langle{}^0_1\rangle \ \},$$
$$R(1,0) = \{ \ A \to A\langle{}^0_0\rangle A\langle{}^0_1\rangle \ \}, \qquad R(2,1) = \{ \ B \to b B\langle{}^0_0\rangle \ \},$$
$$R(2,2) = \{ \ B \to \lambda \ \}.$$

The table $R(0,0)$ is applied exactly once, namely in the first derivation step. The two resulting nonterminals on both sides of the $\$$ are then synchronized with each other at level 2. During the derivation this property carries over to the pairwise corresponding descendants of these two nonterminals, guaranteeing that the same table (i.e., the same rule) is applied to them and hence the same string is generated on both sides of the $\$$. Now consider the prefix to the left of the $\$$ and disregard the other half. Clearly, in every sentential form all nonterminals in this part of the string are synchronized with each other at level 1. Thus, in each step the same supertable must be applied to all nonterminals. Hence, the derivation corresponds to a derivation in an ET0L system with tables $\{A \to AA\}$ and $\{A \to BaB, B \to bB, B \to \lambda\}$. Consequently, a derivation first duplicates all $A$'s $n$ times, yielding $A^{2^n}$. When $R(2)$ is applied the first time, each $A$ is replaced with $BaB$. Finally, each $B$ can produce an arbitrary number of $b$'s.

As the example shows (for a simple case), the set of nonterminals of a sentential form is partitioned into groups of mutually synchronized nonterminals on each of the levels $1, \ldots, n$. Two nonterminals $(A, \varphi)$ and $(B, \psi)$ belong to the same group at level $k$ if $\mathrm{first}_k(\varphi) = \mathrm{first}_k(\psi)$. Hence, the partition into groups at level $k + 1$ is a refinement of the one at level $k$. In a derivation step, a group splits if nonterminals in the right-hand sides are provided with different tuples of synchronization symbols—the synchronization branches. Note that each group at level $n$ (consisting of all nonterminals with the same synchronization string) branches into at most $|I^n|$ new groups of descendants at level $n$.

In the remainder of this section, we discuss some important special cases.

**Definition 3.** Let $G = (N, T, I, J, R, S)$ be a branching grammar of depth $n$ and let $Q \subseteq R()$ be a subset of its set of rules.

1. The set of rules $Q$ is *total* if every nonterminal is the left-hand side of some rule in $Q$ and *deterministic* if it does not contain distinct rules with the same left-hand side. The grammar $G$ is *total* (*deterministic*) if every nonempty table $R(\tau)$, $\tau \in J^n$, is total (deterministic, respectively).
2. The grammar $G$ is *terminable* if the following holds: for every $\xi \in (\mathrm{SN}_G \cup T)^*$ with $(S, \lambda) \Rightarrow^+ \xi$ there exists some $w \in T^*$ such that $\xi \Rightarrow^* w$.

The example above (generating the language $L_0$) is deterministic and terminable, but not total. In fact, none of its tables is total.

What happens if we increase the depth $n$ of a branching grammar $G$ by one and replace every nonempty table by all its deterministic subtables without synchronizing any nonterminals at the new level? We obtain a deterministic branching grammar $G'$ of depth $n + 1$ generating the same language.

**Lemma 4.** For every branching grammar $G$ of depth $n$ there is a deterministic branching grammar $G'$ of depth $n+1$ such that $L(G') = L(G)$. The construction preserves totality and terminability.

*Proof sketch.* Change every rule $r = A \to v_0(B_1, \alpha_1)v_1 \cdots (B_l, \alpha_l)v_l$ of $G$ into the rule $r' = A \to v_0(B_1, \alpha_1.1)v_1 \cdots (B_l, \alpha_l.l)v_l$, assuming w.l.o.g. that $[l] \subseteq I$, and define $R'(\tau.j) = \{r' \mid r \in Q_j\}$ where $Q_1, \ldots, Q_k$ are all (total, if $G$ is total) deterministic subsets of $R(\tau)$, assuming w.l.o.g. that $[k] \subseteq J$.                    ☐

The lemma also holds for branching tree grammars, defined in the next section, because the construction does not affect the structure of right-hand sides.

## 3    Tree Grammars and Tree Transducers

A *ranked alphabet* is an alphabet $\Sigma$ such that every symbol $a \in \Sigma$ is given a *rank* in $\mathbb{N}$. For every $k \in \mathbb{N}$, the set of symbols of rank $k$ in $\Sigma$ is denoted by $\Sigma_k$. For a ranked alphabet $\Sigma$ and a set $A$ disjoint with $\Sigma$, the set $T_\Sigma(A)$ of trees over $\Sigma$ and $A$ is the smallest subset $T$ of $(\Sigma \cup A)^*$ such that $A \subseteq T$ and, for all $f \in \Sigma_k$ ($k \in \mathbb{N}$) and $t_1, \ldots, t_k \in T$, $f\, t_1 \cdots t_k \in T$. We usually write $f[t_1, \ldots, t_k]$ instead of $f\, t_1 \cdots t_k$. The set $T_\Sigma(\emptyset)$ of trees over $\Sigma$ is denoted by $T_\Sigma$. A subset of $T_\Sigma$ is called a *tree language*. Given a set $T$ of trees, $\Sigma(T)$ denotes the set of all trees $f[t_1, \ldots, t_k]$ where $f \in \Sigma_k$ and $t_1, \ldots, t_k \in T$ for some $k \in \mathbb{N}$. The *yield* of a tree $t$ is the string $\text{yield}(t)$ of all leaves of $t$, read from left to right. For a class $\mathcal{L}$ of tree languages, $\text{yield}(\mathcal{L}) = \{\text{yield}(L) \mid L \in \mathcal{L}\}$.

A branching grammar $G = (N, \Sigma, I, J, R, S)$ is a *branching tree grammar* if its terminal alphabet $\Sigma$ is ranked and the right-hand side of each rule is a tree in $T_\Sigma(N \times I^n)$. Obviously, this implies that every sentential form is a tree in $T_\Sigma(SN_G)$. In particular, $L(G) \subseteq T_\Sigma$. We denote by BST ($\text{BST}_n$) the class of tree languages generated by branching tree grammars (of nesting depth $n$). The reader is referred to [DE02] for examples. There, we discuss a family of tree languages $L_0, L_1, \ldots$ generated by branching tree grammars of depth $0, 1, \ldots$.

Let us now recall the definition of regular tree grammars. A *regular tree grammar* is a quadruple $G = (N, \Sigma, R, S)$ where $N$ is the alphabet of nonterminals, $\Sigma$ is the ranked alphabet of terminals, $S \in N$ is the initial nonterminal, and $R$ is a finite set of rules $A \to \zeta$ where $A \in N$ and $\zeta \in T_\Sigma(N)$. Hence, $G$ is a context-free Chomsky grammar with the additional requirements that the terminal alphabet $\Sigma$ is ranked and every right-hand side of a rule in $R$ is a tree in $T_\Sigma(N)$. The derivation relation $\Rightarrow$ and the generated language $L(G)$, which is called a *regular tree language*, are defined as usual. The class of all regular tree languages is denoted by REGT. Obviously, regular tree grammars can be identified with branching tree grammars of depth 0, i.e., $\text{REGT} = \text{BST}_0$.

It is well known that the context-free languages are the yields of the regular tree languages. A similar result relates $\text{BS}_n$ and $\text{BST}_n$ (with a similar proof).

**Lemma 5.** For every $n \in \mathbb{N}$, $\mathrm{BS}_n = \mathrm{yield}(\mathrm{BST}_n)$.

To recall top-down tree transducers, we fix a countably infinite set $X = \{x_1, x_2, \ldots\}$ of pairwise distinct symbols of rank 0, called variables. For $k \in \mathbb{N}$, $X_k$ denotes $\{x_1, \ldots, x_k\}$.

**Definition 6 (td transducer).** A *top-down tree transducer* (td transducer, for short) is a tuple $td = (\Sigma, \Sigma', Q, R, q_0)$ where $\Sigma$ and $\Sigma'$ are ranked alphabets of *input* resp. *output symbols*, $Q$ is a ranked alphabet of *states* of rank 1, $q_0 \in Q$ is the *initial state*, and $R$ is a finite set of rules of the form $q[f[x_1, \ldots, x_k]] \to \zeta$ where $q \in Q$, $f \in \Sigma_k$ for some $k \in \mathbb{N}$, and $\zeta \in \mathrm{T}_{\Sigma'}(Q(X_k))$.

Thus, a td transducer is a special term rewrite system. Its derivation relation $\xi_1 \Rightarrow_{td} \xi_2$ (or simply $\xi_1 \Rightarrow \xi_2$) is defined as usual for trees $\xi_1, \xi_2 \in \mathrm{T}_{\Sigma'}(Q(\mathrm{T}_\Sigma))$. The *top-down tree transduction (td transduction) computed by* $td$ is given by $td(s) = \{s' \in \mathrm{T}_{\Sigma'} \mid q_0[s] \Rightarrow^* s'\}$ for all $s \in \mathrm{T}_\Sigma$.

The left-hand side $q[f[x_1, \ldots, x_k]]$ of a rule is briefly denoted by $qf$ and the rule is said to be a $qf$-rule. A td transducer $td = (\Sigma, \Sigma', Q, R, q_0)$ is *total* (*deterministic*) if $R$ contains at least (resp. at most) one $qf$-rule for every $q \in Q$ and $f \in \Sigma$. It is a *finite state relabelling* if the right-hand side of every $qf$-rule, where $f \in \Sigma_k$, has the form $g[q_1[x_1], \ldots, q_k[x_k]]$ (and thus does not change the structure of input trees). The set of all td transductions is denoted by TD. Moreover, tTD, dTD, and RELAB denote the sets of total td transductions, deterministic td transductions, and finite state relabellings, respectively.

Given a class $\mathcal{C}$ of tree transductions (i.e., a class of binary relations on trees such as TD) and a class $\mathcal{L}$ of tree languages, we denote by $\mathcal{C}(\mathcal{L})$ the class of all tree languages of the form $tr(L)$ where $tr \in \mathcal{C}$ and $L \in \mathcal{L}$. This construction can be iterated: $\mathcal{C}^0(\mathcal{L}) = \mathcal{L}$ and, for $n \in \mathbb{N}$, $\mathcal{C}^{n+1}(\mathcal{L}) = \mathcal{C}(\mathcal{C}^n(\mathcal{L}))$.

One can show that $\mathrm{tTD}(\mathcal{L}) = \mathrm{TD}(\mathcal{L})$ for all classes $\mathcal{L}$ of input tree languages that are closed under RELAB (see [DE02]). To apply this to the classes that are of interest here, note that $\mathrm{RELAB}(\mathrm{REGT}) \subseteq \mathrm{REGT}$ [GS84, Corollary III.6.6], and that TD is closed under composition with finite state relabellings [GS84, Theorem III.3.15]. By induction on $n$ we thus get the following theorem, which is used to present our main result.

**Theorem 7.** $\mathrm{TD}^n(\mathrm{REGT}) = \mathrm{tTD}^n(\mathrm{REGT})$ for all $n \in \mathbb{N}$.

## 4   The Main Result

In this section, we present the main result of this paper. We first show that $\mathrm{TD}^n(\mathrm{REGT}) \subseteq \mathrm{BST}_n$ for all $n \in \mathbb{N}$. It is convenient (and interesting in itself) to prove that these languages can even be generated by branching tree grammars that are total and terminable. The proof will be by induction on $n$. Lemma 4 makes it possible to turn a total terminable branching tree grammar of depth $n$ into a total terminable branching tree grammar of depth $n+1$ that generates the same language and is deterministic in addition. Furthermore, Theorem 7 allows us to restrict our attention to total td transducers. Exploiting this, it basically remains to prove the following lemma.

**Lemma 8.** Let $G$ be a total deterministic terminable branching tree grammar, and let $td$ be a total td transducer. Then a total terminable branching tree grammar $G'$ of the same nesting depth can be constructed such that $L(G') = td(L(G))$. If $td$ is deterministic, then so is $G'$.

In the proof of the lemma (which we omit), $G'$ is constructed in a rather standard way—the right-hand side of a new rule is obtained by "running" $td$ on the right-hand side of the original rule. Terminability of $G$ is needed because $td$ can delete parts of a right-hand side, and totality of $td$ is needed to show that $G'$ is terminable. We use the lemma to prove the first direction of our main result.

**Theorem 9.** Let $n \in \mathbb{N}$. For every language $L \in TD^n(REGT)$ there is a total terminable branching tree grammar $G$ of nesting depth $n$ with $L(G) = L$.

*Proof.* We proceed by induction on $n$. For $n = 0$ it has been observed earlier that $BST_0 = REGT = TD^0(REGT)$. Furthermore, total and terminable regular tree grammars are a well-known normal form of regular tree grammars.

Now, let $n > 0$. By Theorem 7, $L = td(L_0)$ for some $L_0 \in tTD^{n-1}(REGT)$ and $td \in tTD$. By the induction hypothesis, $L_0 = L(G_0)$ for a total terminable branching tree grammar $G_0$ of depth $n - 1$. Thus, Lemma 4 yields a total, terminable, and deterministic branching tree grammar $G_0'$ of depth $n$ with $L(G_0') = L_0$. Hence, by Lemma 8 there is a total terminable branching tree grammar $G$ of depth $n$ such that $L(G) = td(L_0)$, as claimed.    $\square$

We now turn to the other inclusion: $BST_n \subseteq TD^n(REGT)$ for all $n \in \mathbb{N}$. Since $BST_0 = REGT$, this holds for $n = 0$. Thus, let $n \geq 1$ and consider an arbitrary but fixed branching tree grammar $G = (N, \Sigma, I, J, R, S)$ of depth $n \geq 1$. Without loss of generality, let $I = \{0, \ldots, d-1\}$ for some $d \in \mathbb{N}$.

The translation of $G$ to a composition of $n$ td transducers (applied to a regular tree language) makes use of the notion of *synchronization trees* to be defined below. The set of synchronization trees depends only on $I$, $J$, and $n$. Intuitively, the synchronization trees represent all correctly synchronized choices of tables in the derivations of $G$. We will need $n - 1$ td transducers to generate the set of synchronization trees for the given $I$, $J$, and $n$. Then, another td transducer $td_G$, whose states are the nonterminals of $G$ and whose rules are obtained from those of $G$ in a rather direct manner, exploits the information in the synchronization trees (which it takes as input), in order to generate the language $L(G)$.

Let 'num' denote the bijection num: $I^n \to [d^n]$ given by num$(i_1, \ldots, i_n) = 1 + \sum_{j=1}^n i_j \cdot d^{n-j}$ for all $i_1, \ldots, i_n \in I$. Thus, 'num' simply interprets $(i_1, \ldots, i_n)$ as a natural number written in base-$d$ notation and adds 1. Furthermore, let $\Sigma_{\langle n \rangle}$ denote the ranked alphabet consisting of a symbol $\perp$ of rank 0 and all $\tau \in J^n$ viewed as symbols of rank $d^n$. Thus, the internal symbols of a tree $s \in T_{\Sigma_{\langle n \rangle}}$ are tables of nesting depth $n$, and every such symbol has as many children as there are possibilities to branch out in the synchronization. Among others, this can be exploited in order to address the nodes of $s$ by synchronization strings $\varphi \in (I^n)^*$. For this, we define the set internal$(s) \subseteq (I^n)^*$ of *internal nodes* of $s$ and the table $s(\varphi)$ at such a node $\varphi$ inductively, as follows. If $s = \perp$ then internal$(s) = \emptyset$. If $s = \tau[s_1, \ldots, s_{d^n}]$ for some $\tau \in J^n$ and $s_1, \ldots, s_{d^n} \in T_{\Sigma_{\langle n \rangle}}$, then internal$(s) = \{\lambda\} \cup \{\alpha\varphi \mid \alpha \in I^n, \varphi \in \text{internal}(s_{\text{num}(\alpha)})\}$, $s(\lambda) = \tau$, and $s(\alpha\varphi) = s_{\text{num}(\alpha)}(\varphi)$ for all $\alpha \in I^n$ and all $\varphi \in \text{internal}(s_{\text{num}(\alpha)})$.

**Definition 10 (synchronization tree).** The set of *n-synchronization trees* is the set SYNC$_n$ of all trees $s \in T_{\Sigma_{\langle n \rangle}}$ such that level$(s(\varphi), s(\varphi')) \geq$ level$(\varphi, \varphi')$ for all $\varphi, \varphi' \in \text{internal}(s)$ of equal length.

Thus, the requirement is that the tables at nodes $\varphi, \varphi'$ fulfill condition (ii) of Definition 2. We now show that SYNC$_n$ can be generated by a composition of $n - 1$ td transducers applied to a regular tree language.

**Theorem 11.** SYNC$_n \in \text{TD}^{n-1}(\text{REGT})$.

*Proof sketch.* The proof is by induction on $n$. For $n = 1$ it suffices to notice that SYNC$_1 = T_{\Sigma_{\langle n \rangle}}$. Indeed, if $n = 1$ in Definition 10 then level$(\varphi, \varphi') = 1$ means $\varphi = \varphi'$ and thus level$(s(\varphi), s(\varphi')) = 1$, which shows that all trees in $T_{\Sigma_{\langle 1 \rangle}}$ are 1-synchronization trees.

Let $n \geq 1$ and assume that SYNC$_n \in \text{TD}^{n-1}(\text{REGT})$. We construct a td transducer $td = (\Sigma_{\langle n \rangle}, \Sigma_{\langle n+1 \rangle}, \{q\}, R, q)$ whose application to SYNC$_n$ yields SYNC$_{n+1}$. The set $R$ consists of all rules

$$q\tau \to \tau.j[\underbrace{q[x_1], \ldots, q[x_1]}_{d \text{ times}}, \ldots, \underbrace{q[x_{d^n}], \ldots, q[x_{d^n}]}_{d \text{ times}}]$$

and $q\perp \to \perp$, where $\tau \in J^n$ and $j \in J$, as well as the rule $q\perp \to \perp$.

Intuitively, $td$ copies the subtrees of the input tree (representing a correct choice of tables at the levels $1, \ldots, n$ of synchronization) $d$ times, and adds nondeterministically an arbitrary choice of tables at level $n + 1$.    □

Now, we define the td transducer $td_G$ which completes the implementation of the branching tree grammar $G = (N, \Sigma, I, J, R, S)$ by a series of $n$ td transducers. Let $td_G = (\Sigma_{\langle n \rangle}, \Sigma, N, R', S)$ where $R'$ consists of all rules

$$A\tau \to v_0 B_1[x_{\text{num}(\alpha_1)}]v_1 \cdots B_l[x_{\text{num}(\alpha_l)}]v_l$$

such that $\tau \in J^n$ and $A \to v_0(B_1, \alpha_1)v_1 \cdots (B_l, \alpha_l)v_l$ is a rule in $R(\tau)$. There are no rules in $R'$ with left-hand side $A \perp$. Note that if $G$ is deterministic, then so is $td_G$. To show that $td_G(\text{SYNC}_n) = L(G)$, we must establish a correspondence between derivations in $G$ and synchronization trees. Let $(A, \varphi) \in \text{SN}_G$ and $s \in$ SYNC$_n$. A derivation in $G$ starting with $(A, \varphi)$ is said to be *s-synchronized* if, for every step $w_0(A_1, \varphi\varphi_1)w_1 \cdots (A_h, \varphi\varphi_h)w_h \Rightarrow w_0\zeta_1 w_1 \cdots \zeta_h w_h$ of this derivation using tables $\tau_1, \ldots, \tau_h$, it holds for all $i \in [h]$ that $\varphi_i \in \text{internal}(s)$ and $\tau_i = s(\varphi_i)$. Now, the following lemma can be shown in a rather straightforward manner.

**Lemma 12.** For all $(A, \varphi) \in \text{SN}_G$, $s \in$ SYNC$_n$, and $s' \in T_\Sigma$, $A[s] \to^*_{td_G} s'$ if and only if there is an *s*-synchronized derivation $(A, \varphi) \Rightarrow^*_G s'$. Furthermore, taking $(A, \varphi) = (S, \lambda)$, every derivation $(S, \lambda) \Rightarrow^*_G s'$ with $s' \in T_\Sigma$ is *s*-synchronized for some $s \in$ SYNC$_n$.

To see how to construct $s$ in the last part of the lemma, consider a derivation $D = (\xi_1 \Rightarrow_G \xi_2 \Rightarrow_G \cdots \Rightarrow_G \xi_m)$ where $\xi_1 = (S, \lambda)$. Let internal$(s)$ be the (finite, prefix-closed) set of all $\varphi \in (I^n)^*$ such that $(A, \varphi)$ occurs in $\xi_i$ for some $A \in N$ and $i \in [m-1]$. Furthermore, if $\tau$ is the table applied to this occurrence of $(A, \varphi)$ in the derivation step $\xi_i \Rightarrow_G \xi_{i+1}$, define $s(\varphi) = \tau$. By Definition 2(ii) this construction of $s$ is consistent and $s$ is indeed a synchronization tree. Note that $D$ is $s$-synchronized by the definition of $s$.

Lemma 12 yields the desired equality $L(G) = td_G(\mathrm{SYNC}_n)$. Combining this with Theorem 11 and Theorem 9, we get the main result of this paper (using Lemma 5 for the second part).

**Theorem 13.** For $n \in \mathbb{N}$, $\mathrm{BST}_n = \mathrm{TD}^n(\mathrm{REGT})$ and $\mathrm{BS}_n = \mathrm{yield}(\mathrm{TD}^n(\mathrm{REGT}))$.

The branching tree grammar of Theorem 9 is total and terminable. This yields a normal form for branching grammars.

**Theorem 14.** For every branching (tree) grammar $G$ there is a total terminable branching (tree) grammar $G'$ of the same nesting depth such that $L(G') = L(G)$. If $G$ is deterministic, then so is $G'$.

Using the results obtained so far, one can also characterize the languages generated by deterministic branching grammars. We will use $\mathrm{dBS}_n$ to denote the class of languages generated by deterministic branching grammars of depth $n$, and similarly $\mathrm{dBST}_n$ for the corresponding tree languages.

**Theorem 15.** For every $n \geq 1$, $\mathrm{dBST}_n = \mathrm{dTD}(\mathrm{TD}^{n-1}(\mathrm{REGT}))$ and $\mathrm{dBS}_n = \mathrm{yield}(\mathrm{dTD}(\mathrm{TD}^{n-1}(\mathrm{REGT})))$.

## 5    Conclusion

We have proved that the classes $\mathrm{yield}(\mathrm{TD}^n(\mathrm{REGT}))$ and $\mathrm{TD}^n(\mathrm{REGT})$ are generated by branching (tree) grammars of nesting depth $n$, for all $n \in \mathbb{N}$. This establishes for the first time a suitable grammatical device for the generation of these well-known language classes. This is useful for applications. Moreover, the main result can be used to prove new results regarding $\mathrm{yield}(\mathrm{TD}^n(\mathrm{REGT}))$ and $\mathrm{TD}^n(\mathrm{REGT})$. We exemplify this by formulating two such results.

The first result is that for every $n \geq 1$ there is a single tree language $K_{n-1} \in \mathrm{TD}^{n-1}(\mathrm{REGT})$ such that $\mathrm{TD}^n(\mathrm{REGT}) = \mathrm{TD}(\{K_{n-1}\})$. In other words, the tree languages in the class $\mathrm{TD}(\mathrm{TD}^{n-1}(\mathrm{REGT}))$ can be generated by top-down tree transducers from just one element of $\mathrm{TD}^{n-1}(\mathrm{REGT})$. The second result is that for every $n$ there is a single language $L_n \in \mathrm{yield}(\mathrm{TD}^n(\mathrm{REGT}))$ such that $\mathrm{yield}(\mathrm{TD}^n(\mathrm{REGT})) = \mathrm{FST}(\{L_n\})$, where FST is the class of finite state transductions. In other words, $\mathrm{yield}(\mathrm{TD}^n(\mathrm{REGT}))$ which is known to be a full AFL [Bak78, Theorem 13], is a full *principal* AFL.

The tree language $K_{n-1}$ mentioned above is, in fact, the language $\mathrm{SYNC}_n$ of $n$-synchronization trees with $I$ and $J$ both equal to $\{0, 1\}$ (recall from the previous section that $\mathrm{SYNC}_n$ depends on $I$ and $J$). The inclusion $\mathrm{TD}(\{\mathrm{SYNC}_n\}) \subseteq$

$\mathrm{TD}^n(\mathrm{REGT})$ is immediate from Theorem 11. To prove that $\mathrm{TD}^n(\mathrm{REGT}) \subseteq \mathrm{TD}(\{\mathrm{SYNC}_n\})$ with $I = J = \{0, 1\}$, it follows from the results of the previous section that it suffices to show that every tree language in $\mathrm{BST}_n$ can be generated by a branching tree grammar with $I = J = \{0, 1\}$. This implies the existence of the language $L_n$ mentioned above, applying known results from AFL and AFA theory to the CTPD transducers of [ERS80] (which are closely related to td transducers).

We omit the proofs, but state the resulting theorem.

**Theorem 16.** (1) For every $n \geq 1$, $\mathrm{TD}^n(\mathrm{REGT}) = \mathrm{TD}(\{\mathrm{SYNC}_n\})$.
(2) For every $n \geq 0$, $\mathrm{yield}(\mathrm{TD}^n(\mathrm{REGT}))$ is a full principal AFL.

**Acknowledgement.** We thank the referees for some helpful comments.

# References

[Bak78]   Brenda S. Baker. Tree transducers and tree languages. *Information and Control*, 37:241–266, 1978.

[DE02]    Frank Drewes and Joost Engelfriet. Branching Synchronization Grammars with Nested Tables. Technical Report UMINF 02.22, Umeå University, 2002.

[Dre00]   Frank Drewes. Tree-based picture generation. *Theoretical Computer Science*, 246:1–51, 2000.

[Dre01]   Frank Drewes. Tree-based generation of languages of fractals. *Theoretical Computer Science*, 262:377–414, 2001.

[Eng76]   Joost Engelfriet. Surface tree languages and parallel derivation trees. *Theoretical Computer Science*, 2:9–27, 1976.

[Eng82]   Joost Engelfriet. Three hierarchies of transducers. *Mathematical Systems Theory*, 15:95–125, 1982.

[ERS80]   Joost Engelfriet, Grzegorz Rozenberg, and Giora Slutzki. Tree transducers, L systems, and two-way machines. *Journal of Computer and System Sciences*, 20:150–202, 1980.

[FV98]    Zoltán Fülöp and Heiko Vogler. *Syntax-Directed Semantics: Formal Models Based on Tree Transducers*. Springer, 1998.

[GS84]    Ferenc Gécseg and Magnus Steinby. *Tree Automata*. Akadémiai Kiadó, Budapest, 1984.

[GS97]    Ferenc Gécseg and Magnus Steinby. Tree languages. In G. Rozenberg and A. Salomaa, editors, *Handbook of Formal Languages. Vol. III: Beyond Words*, chapter 1, pages 1–68. Springer, 1997.

[Rou70]   William C. Rounds. Mappings and grammars on trees. *Mathematical Systems Theory*, 4:257–287, 1970.

[Roz73]   Grzegorz Rozenberg. Extension of tabled 0L systems and languages. *International Journal of Computer and Information Sciences*, 2:311–334, 1973.

[RS97]    Grzegorz Rozenberg and Arto Salomaa, editors. *Handbook of Formal Languages*, volume 1–3. Springer, 1997.

[Sky76]   Sven Skyum. Decomposition theorems for various kinds of languages parallel in nature. *SIAM Journal of Computing*, 5:284–296, 1976.

[Tha70]   James W. Thatcher. Generalized[2] sequential machine maps. *Journal of Computer and System Sciences*, 4:339–367, 1970.

[Tha73]    James W. Thatcher. Tree automata: an informal survey. In A.V. Aho, editor, *Currents in the Theory of Computing*, pages 143–172. Prentice Hall, 1973.

[Vág86]    Sándor Vágvölgyi. On compositions of root-to-frontier tree transformations. *Acta Cybernetica*, 7:443–480, 1986.

# Learning a Regular Tree Language from a Teacher

Frank Drewes and Johanna Högberg

Department of Computing Science, Umeå University
S–901 87 Umeå, Sweden
{drewes,johanna}@cs.umu.se

**Abstract.** We generalize an inference algorithm by Angluin, that learns a regular string language from a "minimally adequate teacher", to regular tree languages. This improves a similar algorithm proposed by Sakakibara. In particular, we show how our algorithm can be used to avoid dead states, thus answering a question by Sakakibara.

## 1 Introduction

Grammatical inference addresses the problem of algorithmically "learning" a language for which no explicit grammatical description is available. Here, learning means to construct an appropriate grammar (or some other type of formal device for language description) for the unknown language from examples or similar information.

An interesting setting in which a regular language can be learned in polynomial time was proposed by Angluin in [Ang87]. In this approach the learning algorithm, called the learner, has access to a "minimally adequate teacher" who can answer two types of queries. The teacher can check whether a given string is an element of the unknown language $U$. Furthermore, given some finite automaton $A$, she can check whether $A$ correctly recognizes $U$. If not, she will return a counterexample—a string that is erroneously accepted or rejected by $A$.

This situation is rather natural (though of course idealized). One may imagine the teacher to be a human expert who tries to convey her knowledge to the learner. Even if she does not know the correct automata-theoretic description of $U$ she might be able to check proposed examples and solutions using her domain-specific knowledge and intuition.

Regular tree languages are a well-known generalization of regular string languages. Nearly all classical results for regular string languages carry over to regular tree languages. In particular, there is a convenient type of language acceptor for this class of languages, namely the finite tree automaton (fta, see [TW68]). There exists a nice relationship between regular tree languages and context-free string languages: The latter are precisely the yields of the former (where the yield of a tree is the string of leaves of this tree, read from left to right). Intuitively, the tree language consists of the derivation trees of strings in the string language. In fact, for this it suffices to consider so-called skeletal

Z. Ésik and Z. Fülöp (Eds.): DLT 2003, LNCS 2710, pp. 279–291, 2003.

trees, in which internal nodes are unlabeled [LJ78]. Such a tree reveals only the syntactic structure of the string but not the concrete rules generating it.

Motivated by these connections, Sakakibara [Sak90] extended the algorithm proposed by Angluin to skeletal regular tree languages. In this way, context-free string languages can be learned in polynomial time, provided that the teacher is able to check (skeletal) trees and fta's (instead of strings and ordinary finite automata), yielding skeletal trees as counterexamples.

In this paper, we propose an alternative extension of the algorithm by Angluin, which generalizes and improves the one in [Sak90]. Since our algorithm, restricted to skeletal tree languages, can be applied to context-free string languages in exactly the same way as the one by Sakakibara (i.e., one can simply substitute our algorithm for the one by Sakakibara to improve the learning algorithm for context free string languages in [Sak90]), we focus on regular tree languages throughout. We consider regular tree languages in general, instead of restricting our attention to skeletal ones. This generalization is rather straightforward, but not unimportant since regular tree languages are not only useful as a syntactic basis for context-free string grammars, but also for, e.g., graph and picture generating devices (see, e.g., [Eng94,Dre00]). Thus, future work may use the results presented in this paper in order to devise learning algorithms for graph and picture languages.

Algorithmically, we keep the basic approach by Angluin, but modify and extend it in two respects. From the point of view of complexity, a deficiency of the algorithms in [Ang87,Sak90] is that both are quite sensitive to the size of counterexamples provided by the teacher. If the teacher selects counterexamples that are larger than necessary, the runtime of the algorithms is increased considerably. This problem is especially serious in the tree case because a tree may have exponentially many subtrees relative to its height. The approach proposed here uses the counterexamples provided by the teacher in a new way in order to overcome this problem.

The second improvement attacks the problem of "dead states". As pointed out in [Sak90], for languages that occur in practice, the constructed fta will typically have a rather sparse transition table. However, the basic learning algorithm is not able to benefit from this fact because it constructs only total tree automata, i.e., tree automata with a total transition function. Thus, transitions have to be specified (and computed) even in those cases where a dead state is involved. Fortunately, the way in which we exploit counterexamples makes it possible to modify the algorithm in order to avoid dead states, thus resulting in a partial transition function.

The paper is structured as follows. In the next section, we collect some basic notions regarding trees and tree automata. In Section 3 we present our algorithm for total fta's. In Section 4, we discuss how this algorithm can be modified in order to avoid dead states. In this short version, detailed proofs are omitted. They will be given in the long version.

## 2    Trees and Tree Automata

The set of natural numbers (including 0) is denoted by $\mathbb{N}$. For $n \in \mathbb{N}$, $[n]$ denotes the set $\{1, \ldots, n\}$. The cardinality of a set $S$ is denoted by $|S|$. Given a function $f \colon A \to B$, we denote the canonical extension of $f$ to subsets of $A$ by $f$ as well, i.e., $f(S) = \{f(a) \mid a \in S\}$ for all $S \subseteq A$. We use the symbol '$\equiv$' in order to compare logical statements, i.e., $A \equiv B$ if $A$ and $B$ are both true or both false.

A *ranked alphabet* is a finite set of ranked symbols, where a ranked symbol is a pair $(f, k)$ consisting of a symbol $f$ and a rank $k \in \mathbb{N}$. Somewhat ambiguously, we shall usually identify $(f, k)$ with $f$ and say that $k$ is the rank of the symbol $f$. By $\Sigma_{(k)}$ ($k \in \mathbb{N}$) we denote the set of all symbols in $\Sigma$ of rank $k$. The set $\mathrm{T}_\Sigma$ of all trees over $\Sigma$ is defined inductively, as usual: It is the smallest set of formal expressions such that $f[t_1, \cdots, t_k] \in \mathrm{T}_\Sigma$ for every $f \in \Sigma_{(k)}$ and all $t_1, \ldots, t_k \in \mathrm{T}_\Sigma$. The trees $t_1, \ldots, t_k$ are said to be the *direct subtrees* of the tree. The set *subtrees*$(t)$ of subtrees of a tree $t$ consists of $t$ itself and all subtrees of its direct subtrees. Given a set $T$ of trees, $\Sigma(T)$ denotes the set of all trees of the form $f[t_1, \ldots, t_k]$ such that $f \in \Sigma_{(k)}$ for some $k \in \mathbb{N}$ and $t_1, \ldots, t_k \in T$. A subset of $\mathrm{T}_\Sigma$ is called a *tree language*. As usual, the size $|t|$ of a tree $t$ is given by $|t| = 1 + \sum_{i=1}^{k} |t_i|$ for every tree $t = f[t_1, \ldots, t_k]$.

Let $\square$ be a special symbol of rank 0. A tree $c \in \mathrm{T}_{\Sigma \cup \{\square\}}$ in which $\square$ occurs exactly once is called a *context* (over $\Sigma$). The set of all contexts over $\Sigma$ is denoted by $\mathrm{C}_\Sigma$. For $c \in \mathrm{C}_\Sigma$ and a tree $s$, we denote by $c[\![s]\!]$ the tree obtained by substituting $s$ for the unique occurrence of $\square$ in $c$. The *depth* of a context $c \in \mathrm{C}_\Sigma$ is the length of the path from the root to the unique occurrence of $\square$ in $c$. Formally, $depth(\square) = 0$ and if $c = c'[\![f[s_1, \ldots, s_i, \square, s_{i+1}, \ldots, s_k]]\!]$ for some $c' \in \mathrm{C}_\Sigma$, $f \in \Sigma_{(k+1)}$ ($k \in \mathbb{N}$), and $s_1, \ldots, s_k \in \mathrm{T}_\Sigma$, then $depth(c) = depth(c') + 1$.

**Definition 1 (finite-state tree automaton).**    A (total and deterministic) bottom-up finite-state tree automaton (fta, for short) is a tuple $A = (\Sigma, Q, \delta, F)$ where $\Sigma$ is the ranked input alphabet, $Q$ is the finite set of states, $\delta$ is the transition function assigning to every $f \in \Sigma_{(k)}$ ($k \in \mathbb{N}$) and all $q_1, \ldots, q_k \in Q$ a state $\delta(q_1 \cdots q_k, f) \in Q$, and $F \subseteq Q$ is the set of accepting (or final) states.

The transition function extends to trees, yielding a function $\delta \colon \mathrm{T}_\Sigma \to Q$ in the obvious way: if $t = f[t_1, \ldots, t_k] \in \mathrm{T}_\Sigma$ then $\delta(t) = \delta(\delta(t_1) \cdots \delta(t_k), f)$. The set of trees accepted by $A$ is $L(A) = \{t \in \mathrm{T}_\Sigma \mid \delta(t) \in F\}$. Such a tree language is called a *regular tree language* and $A$ is said to *recognize* $L(A)$.

Given an fta $A = (\Sigma, Q, \delta, F)$, it follows by an obvious induction from the definition of $\delta$ that any occurrence of a subtree $s$ in a tree $t$ may be replaced with any tree $s'$ without affecting $\delta(t)$, provided that $\delta(s') = \delta(s)$.

**Lemma 2.** Let $A = (\Sigma, Q, \delta, F)$ be an fta. For all contexts $c \in \mathrm{C}_\Sigma$ and all trees $s, s' \in \mathrm{T}_\Sigma$ with $\delta(s) = \delta(s')$ it holds that $\delta(c[\![s]\!]) = \delta(c[\![s']\!])$.

It is well known (see, e.g., [Bra68]) that the Myhill-Nerode theorem carries over to regular tree languages. For this, let $L \subseteq \mathrm{T}_\Sigma$. Given two trees $s, s' \in \mathrm{T}_\Sigma$, let $s \sim_L s'$ if and only if, for every context $c \in \mathrm{C}_\Sigma$, either both of $c[\![s]\!]$ and $c[\![s']\!]$

are in $L$ or none of them is. Obviously, $\sim_L$ is an equivalence relation on $T_\Sigma$. The equivalence class containing $s \in T_\Sigma$ is denoted by $[s]_L$. The *index* of $L$ is the cardinality of $\{[s]_L \mid s \in T_\Sigma\}$, i.e., the number of equivalence classes of $\sim_L$. Note that the index of $L$ may be infinite. For later use, we note here the following obvious lemma.

**Lemma 3.** Let $\Sigma$ be a ranked alphabet, $L \subseteq T_\Sigma$ a tree language, and $s, s' \in T_\Sigma$.

(1) If $s = f[s_1, \ldots, s_k]$, then $[s]_L$ is uniquely determined by $f$ and $[s_1]_L, \ldots, [s_k]_L$.
(2) For all contexts $c \in C_\Sigma$, if $c[\![s]\!] \in L \not\equiv c[\![s']\!] \in L$ then $s \not\sim_L s'$.

For every fta $A$, $L(A)$ is of finite index since, by Lemma 2, $\delta(s) = \delta(s')$ implies $s \sim_{L(A)} s'$. Conversely, if a tree language $L$ is of finite index, we can easily build an fta recognizing $L$. We shall denote this fta by $A_L$ in the following. Its states are the equivalence classes of $\sim_L$. Given some $f \in \Sigma_{(k)}$ and states $[s_1]_L, \ldots, [s_k]_L$, we define the transition function by $\delta_L([s_1]_L, \ldots, [s_k]_L, f) = [s]_L$ where $s = f[s_1, \ldots, s_k]$. An equivalence class $[s]_L$ is an element of the set $F_L$ of final states of $A_L$ if $s \in L$ (which, by the definition of $\sim_L$, is the case if and only if $[s]_L \subseteq L$). The fta $A_L$ is a very special one. As one can easily show, it is the unique minimal fta recognizing $L$, up to a bijective renaming of states.

**Lemma 4.** Let $\Sigma$ be a ranked alphabet and $L \subseteq T_\Sigma$ a tree language. Then $L$ is of finite index if and only if $L$ is regular. In this case, $A_L$ is the unique minimal fta recognizing $L$ (up to a bijective renaming of states).

## 3    Learning a Regular Tree Language

In this section, we present the first version of our learning algorithm, called the *learner*. For the rest of the paper, let us fix an arbitrary ranked alphabet $\Sigma$, and let $r$ be the maximum rank of symbols in $\Sigma$. We assume that $\Sigma_{(0)} \neq \emptyset$ (i.e., $T_\Sigma \neq \emptyset$). The aim is to learn an unknown regular tree language $U \subseteq T_\Sigma$ of index $I \in \mathbb{N}$. Here, learning means to construct an fta recognizing $U$ (which will, in fact, turn out to be $A_U$). Following the approach of a "minimally adequate teacher" by Angluin [Ang87], the learner may use the help of a teacher who is able to perform two tasks:

1. Given some tree $t \in T_\Sigma$, the teacher will check whether or not $t \in U$.
2. Given an fta $A$ with less than $I$ states, a tree *counterexample*$(A) \in (U \setminus L(A)) \cup (L(A) \setminus U)$ in the symmetric difference of $U$ and $L(A)$ will be returned.

Note that we assume, for the sake of simplicity, that the index $I$ of $U$ is given to the learner as a parameter. As this is only used as a termination criterion and does not affect the computation as such, it is equivalent to the assumption of Angluin that the teacher, when asked for a counterexample, first decides whether $A = A_U$ (see [Ang87]). For complexity considerations, the teacher is assumed to answer both types of questions in constant time, i.e., we do not take into account the complexity of the teacher.

Regarding its basic structure, our algorithm is similar to the one proposed by Angluin for regular string languages, which was already extended to (skeletal) regular tree languages by Sakakibara [Sak90]. The difference is that the learner proposed below makes use of counterexamples in a more sophisticated way. This avoids the need to work with a huge number of large trees if the teacher returns unnecessarily large counterexamples.

At any stage of the computation, the learner will maintain finite sets of trees $S \subseteq \mathrm{T}_\Sigma$ and contexts $C \subseteq C_\Sigma$ satisfying certain conditions. Intuitively, one may imagine that the algorithm builds a table whose rows are indexed by the elements of $S \cup \Sigma(S)$ and whose columns are indexed by the elements of $C$. The cell in row $s$ and column $c$ contains an observation—a truth value indicating whether $c[\![s]\!] \in U$. Note that this is just a useful mental image; implementations need not build the table explicitly, or can represent the information it contains in a different way. In principle, the teacher can be asked whenever the value of some table cell is required, which means that only $S$ and $C$ need to be maintained explicitly. Of course, maintaining a table may be useful from a practical point of view in order to avoid excessive inquiries. For this reason, and in order to support intuition, we shall therefore adopt the terminology of [Ang87,Sak90] and call the pair $(S, C)$ an *observation table*. The precise definition of observation tables reads as follows.

**Definition 5 (observation table).** Let $S$ be a finite subset of $\mathrm{T}_\Sigma$ and let $C$ be a finite nonempty subset of $C_\Sigma$. Then $T = (S, C)$ is called an *observation table* if the following additional conditions are satisfied.

1. For every tree $f[s_1, \dots, s_k] \in S$, the trees $s_1, \dots, s_k$ are in $S$ as well—we say that $S$ is *subtree closed*.
2. For every context $c_0$ of the form $c[\![f[s_1, \dots, s_{i-1}, \Box, s_{i+1}, \dots, s_k]]\!] \in C$, the context $c$ is in $C$ as well and $s_1, \dots, s_{i-1}, s_{i+1}, \dots, s_k$ are in $S$—we say that $C$ is *generalization closed* resp. *S-composed*. In the situation above, $c$ is called the *generalization* of $c_0$.

The main algorithmic idea behind the learner is to collect, in $S$, representatives of the equivalence classes of $\sim_U$ and, in $C$, contexts witnessing that these representatives belong to different equivalence classes. In contrast to the algorithms of [Ang87,Sak90], the one presented below adds to $S$ only trees that are pairwise nonequivalent. Trees in $S$ belong to different equivalence classes if the observed behaviours recorded in the observation table differ. Of course, the converse does not necessarily hold as the learner may not yet have discovered the right contexts.

In order to continue, we need some additional terminology and notation. Given an observation table $T = (S, C)$ and a tree $s \in S \cup \Sigma(S)$, we denote by $obs_T(s)$ the observed behaviour of $s$.[1] Formally, $obs_T(s)$ denotes the function

---

[1] For this, Angluin and Sakakibara used the notation $row_T(s)$.

$obs \colon C \to \{true, false\}$ such that, for all $c \in C$, $obs(c) = true$ if and only if $c[\![s]\!] \in U$. We now define two central properties of observation tables, which generalize the respective properties in [Ang87] and are similar to those used in [Sak90].

**Definition 6 (*closed* and *consistent* observation tables).** Let $T = (S, C)$ be an observation table. We say that $T$ is *closed* if $obs_T(\Sigma(S)) \subseteq obs_T(S)$, and that $T$ is *consistent* if the following condition is satisfied: For all $f \in \Sigma_{(k)}$ ($k \in \mathbb{N}$) and all $s_1, \ldots, s_k, s'_1, \ldots, s'_k \in S$, if $obs_T(s_i) = obs_T(s'_i)$ for all $i \in [k]$ then $obs_T(f[s_1, \ldots, s_k]) = obs_T(f[s'_1, \ldots, s'_k])$.

In other words, $T$ is closed if the observed behaviour of every element of $\Sigma(S)$ can already be observed among the behaviours of elements of $S$, and $T$ is consistent if the observed behaviour of every tree in $S \cup \Sigma(S)$ is uniquely determined by the observations made for its direct subtrees.

From a closed and consistent observation table $T = (S, C)$, we can synthesize an fta $A_T = (\Sigma, Q_T, \delta_T, F_T)$. Its set of states is $Q_T = \{obs_T(s) \mid s \in S\}$. The transition function is given by $\delta_T(obs_T(s_1) \cdots obs_T(s_k), f) = obs_T(f[s_1, \ldots, s_k])$ for all $f \in \Sigma_{(k)}$ ($k \in \mathbb{N}$) and $s_1, \ldots, s_k \in S$, and $F_T = \{obs_T(s) \mid s \in S \cap U\}$. It is easy to check that $\delta_T$ is well defined and that $A_T$ can be constructed in time linear in the size of its transition table, which yields the following lemma.

**Lemma 7.** For every closed and consistent observation table $T = (S, C)$, the fta $A_T$ can be synthesized in time $O(|S|^r |C|)$.

The next two lemmas are similar to the respective ones in [Ang87,Sak90].

**Lemma 8.** Let $T = (S, C)$ be a closed and consistent observation table. Then $\delta_T(s) = obs_T(s)$ for all $s \in S \cup \Sigma(S)$.

**Lemma 9.** Let $T = (S, C)$ be a closed and consistent observation table. For all trees $s \in S \cup \Sigma(S)$ and all contexts $c \in C$, $A_T$ accepts $c[\![s]\!]$ if and only if $c[\![s]\!] \in U$. Moreover, $A_T$ is the unique minimal fta with this property (up to a bijective renaming of states).

We now describe the learner. It maintains an observation table which it builds in $m \le 2I$ steps, applying in each step one of the three procedures CLOSURE, RESOLVE, and EXTEND. Its overall structure is this one (similar to [Sak90] and directly extending the one of [Ang87]):

```
T = (S, C)  := ({a}, {□}) for some arbitrary a ∈ Σ_(0) ;
while |{obs_T(s) | s ∈ S}| < I do
      if T is not closed then T  := CLOSURE(T)
      else if T is not consistent then T  := RESOLVE(T)
      else T  := EXTEND(T)
end while;
return A_T
```

Thus, the learner starts with an initial observation table $T_1 = (\{a\}, \{\Box\})$, where $a$ is an arbitrary symbol of rank 0. Each execution of the loop body (i.e., of CLOSURE, RESOLVE, or EXTEND), yields a new observation table. In the following discussion, we will denote the table obtained after $l - 1$ steps ($l \in [m]$) by $T_l$ and its two components by $S_l$ resp. $C_l$. Although we omit the proofs due to lack of space, the reader may easily check that the procedures CLOSURE, RESOLVE, and EXTEND described below, guarantee that each of the tables $T_l$ ($l \in [m]$) satisfies the following conditions:

(A) $T_l$ is indeed an observation table,
(B) for all distinct trees $s, s' \in S_l$, $s \not\sim_U s'$,
(C) $|S_l| + |C_l| = l + 1$, and
(D) the number of contexts in $C_l$ does not exceed the number of behaviours observed for trees in $S_l$, i.e. $|C_l| \leq |obs_{T_l}(S_l)|$.

The observation table $T_1$ obviously satisfies (A)–(D). Hence, it suffices to note that the procedures CLOSURE, RESOLVE, and EXTEND given next preserve these properties. The computation of CLOSURE($T$) relies on the precondition that $T = (S, C)$ is not closed. It simply searches for a tree $s \in \Sigma(S)$ such that $obs_T(s) \notin obs_T(S)$ and adds it to $S$.

procedure CLOSURE($T$) where $T = (S, C)$
    find $s \in \Sigma(S)$ such that $obs_T(s) \notin obs_T(S)$;
    return $(S \cup \{s\}, C)$;

Note that several invokations may be necessary in order to obtain a closed table. In order to define RESOLVE($T$), we first need to discuss an easy observation. If $T$ is not consistent, then there are trees $f[s_1, \ldots, s_k], f[t_1, \ldots, t_k] \in \Sigma(S_l)$ such that $obs(f[s_1, \ldots, s_k]) \neq obs(f[t_1, \ldots, t_k])$ although $obs(s_i) = obs(t_i)$ for all $i \in [k]$. Consequently, there exists some $i \in [k]$ such that

$$obs(f[t_1, \ldots, t_{i-1}, s_i, \ldots, s_k]) \neq obs(f[t_1, \ldots, t_i, s_{i+1}, \ldots, s_k])$$

since otherwise $obs(f[s_1, \ldots, s_k]) = \cdots = obs(f[t_1, \ldots, t_k])$ when $i$ runs from 1 to $k$. In other words, setting $c = f[t_1, \ldots, t_{i-1}, \Box, s_{i+1}, \ldots, s_k]$, $s = s_i$, and $s' = t_i$ we obtain trees $c[s], c[s'] \in \Sigma(S)$ with $s, s' \in S$, such that $obs(s) = obs(s')$ but $obs(c[s]) \neq obs(c[s'])$. Moreover, it holds that $depth(c) = 1$. The procedure RESOLVE($T$) below exploits this, using the precondition that $T$ is closed but not consistent.

procedure RESOLVE($T$) where $T = (S, C)$
    find $c[s], c[s'] \in \Sigma(S)$ where $s, s' \in S$ and $depth(c) = 1$ such that
        $obs_T(c[s]) \neq obs_T(c[s'])$ and $obs_T(s) = obs_T(s')$;
    find $t, t' \in S$ such that
        $obs_T(t) = obs_T(c[s])$ and $obs_T(t') = obs_T(c[s'])$;
    find $c' \in C$ such that $obs_T(t)(c') \neq obs_T(t')(c')$;
    return $(S, C \cup \{c'[c]\})$;

As explained in the previous paragraph, the procedure will find appropriate $s$, $s'$, and $c$. The fact that $t$ and $t'$ exist follows from the closedness of $T$, while the existence of $c'$ is a consequence of $obs_T(t) = obs_T(c[\![s]\!]) \neq obs_T(c[\![s']\!]) = obs_T(t')$. Note that the generalization of $c'[\![c]\!]$ equals $c'$. Hence, $C \cup c[\![c']\!]$, is generalization closed and $S$-composed if $C$ is.

To compute $\text{EXTEND}(T)$ we synthesize $A_T$ and ask the teacher for a counterexample $t$. In principle, we could then simply add $t$ to $S$. However, if the teacher returns unnecessarily large counterexamples, this procedure would add trees of potentially unbounded size to $S$. What is more, in order to keep $S$ subtree closed, all subtrees of $t$ would have to be added to $S$ as well, potentially resulting in a huge table. In order to avoid this, we extract from $t$, by means of repeated substitutions of subtrees, a counterexample $s$ with the property that $S \cup \{s\}$ is subtree-closed. In this way, only a single tree need to be added to $S$, which makes sure that there will never be more than $I$ trees in $S$. The procedure $\text{EXTEND}$ synthesizes the automaton $A_T$ in order to obtain a counterexample from the teacher which it passes to $\text{EXTRACT}$. $\text{EXTRACT}$ calls itself recursively, passing as a parameter the counterexample. With each call the counterexample is modified to weed out superfluous information.

```
procedure EXTEND(T) where T = (S,C)
    A_T := synthesize(T);
    return EXTRACT(T, counterexample(A_T));

procedure EXTRACT(T,t) where T = (S,C)
    choose c ∈ C_Σ and s ∈ subtrees(t) ∩ Σ(S) \ S such that t = c[[s]];
    if there exists s' ∈ S such that
        obs_T(s') = obs_T(s) and t ∈ U ≡ c[[s']] ∈ U then
        return EXTRACT(T, c[[s']]);
    else
        return (S ∪ {s}, C)
    end if
```

Let us discuss this briefly. The procedure first locates some subtree $s$ of $t$ which is an element of $\Sigma(S) \setminus S$. Such a subtree $s$ must exist in every counterexample because, by Lemma 9, no counterexample can be an element of $\Sigma(S)$. In particular, $t \notin S$, which implies that it must contain a subtree in $\Sigma(S) \setminus S$.

The procedure now attempts to replace $s$ with a tree $s' \in S$, in such a way that also the resulting tree $c[\![s']\!]$ is a counterexample. In order to verify that $c[\![s']\!]$ is indeed a counterexample if $obs_T(s') = obs_T(s)$ and $t \in U \equiv c[\![s']\!] \in U$, we reason as follows. By Lemma 8 we have $\delta_{A_T}(s') = obs_T(s') = obs_T(s) = \delta_{A_T}(s)$ and thus, by Lemma 2, $t = c[\![s]\!] \in L(A_T)$ if and only if $c[\![s']\!] \in L(A_T)$. But $t$ is a counterexample, so $t \in L(A_T)$ if and only if $t \notin U$. Together with the two equivalences above this means that $c[\![s']\!] \in L(A_T)$ if and only if $c[\![s']\!] \notin U$, proving that $c[\![s']\!]$ is a counterexample as well.

If it is not possible to replace $s$ with some suitable tree $s' \in S$, the procedure returns $S \cup \{s\}$. In each recursive call the number of occurrences of subtrees which are not elements of $S$ is reduced by at least one. Consequently, the procedure terminates after less than $|t|$ recursive calls.

To see that property (B) is retained, suppose $T$ has that property. We have to argue that $s' \not\sim_U s$ for all $s' \in S$. This is clear if $obs_T(s') \neq obs_T(s)$. Otherwise, the counterexample of the last recursive call satisfies $t \in U \not\equiv c[\![s']\!] \in U$, where $t = c[\![s]\!]$. By Lemma 3, we thus have $s \not\sim_U s'$ in this case, too.

This ends the description of the learner. We now show that it terminates and returns the desired fta.

**Theorem 10.** The learner returns $A_U$ after less than $2I$ loop executions.

*Proof.* According to (D) there cannot be more contexts in $C_l$ than there are trees in $S_l$, for all $l \in [m]$. Since (C) states that $|C_l| + |S_l| = l + 1$, this means $|S_l| > l/2$. We also know that the learner halts when $S_l$ has $I$ elements (by (B)), so we conclude that it will halt before $l = 2I$.

Now, let $A_{T_m}$ be the returned automaton. Then $T_m$ is a closed, consistent observation table and $A_{T_m}$ is the unique minimal automaton such that, for all $s \in S_m$ and $c \in C_m$, $c[\![s]\!] \in L(A_{T_m})$ if and only if $c[\![s]\!] \in U$ (see Lemma 9). However, $A_U$ has the same property and the same number of states, so $A_{T_m} = A_U$ up to a bijective renaming of states.                                                                    □

Let us now discuss the time complexity of the learner. As previously mentioned, the size of the observation table generated by the learner is bounded by the index of the unknown language. However, there is no restriction on the size of the counterexamples. Even if the procedure EXTEND avoids adding these large trees and their subtrees to $S$, the running time of EXTEND itself still depends on the input from the teacher. This cannot be avoided since any algorithm would at least have to inspect the counterexamples returned by the teacher. Hence, we have to use both $I$ and the maximum size of the counterexamples as parameters when describing the complexity of the learner.

By Theorem 10 the learner will perform less than $2I$ iterations of the main loop. For the running time of CLOSURE and RESOLVE we shall now derive upper bounds expressed in terms of $|S|$ and $|C|$, where $T = (S, C)$ is the table they process. Furthermore, we express the running time of EXTEND in terms of $|S|$, $|C|$, and $|t|$, where $t$ is the counterexample supplied by the teacher.

To compute CLOSURE($T$) we will have to consider each pair in $S \times \Sigma(S)$ and compare the observed behaviours of the two trees (in a worst case situation). This task can obviously be carried out in $O(|S|^{r+1}|C|)$ operations. To compute RESOLVE($T$) we first find the triple $s$, $s'$, $c$ presented in the description of the procedure. A straightforward implementation could examine every tree $f[s_1, \ldots, s_k] \in \Sigma(S) \setminus S$, for each value of $i \in [k]$ building the context $c = f[s_1, \ldots, s_{i-1}, \Box, s_{i+1}, \ldots, s_k]$ and checking whether there exists a tree $s' \in S$ such that $obs_T(s') = obs_T(s)$ but $obs_T(c[\![s']\!]) \neq obs_T(c[\![s]\!])$. This search requires $O(|S|^{r+1}|C|)$ operations. Once $s$, $s'$, and $c$ have been found, $t$ and $t'$ can be discovered in time $O(|S||C|)$ and $c'$ can be located in $O(|C|)$ steps. Hence, RESOLVE runs in time $O(|S|^{r+1}|C|)$.

To estimate the running time of EXTEND($T$), recall that $A_T$ can be constructed in time $O(|S|^r|C|)$ (Lemma 7). For the complexity of EXTRACT($T,t$), note first that there can be at most $h \leq |t|$ calls, with counterexamples $t_1, \ldots, t_h$ being passed to them (where $t_1 = t$). In the $i$th call we first have to search $t_i$ for some occurrence of a subtree $s$ in $\Sigma(S) \setminus S$. For this, we check the subtrees of $t_i$ in a bottom-up fashion until we find one which is not an element of $S$. As the chosen $s$ is replaced with a tree $s' \in S$ and thus need not be considered in the following recursive calls, the overall time needed for this step, taking the sum over all recursive calls, is $O(|t||S|)$.

Next, we have to check in each recursive call whether some $s' \in S$ satisfies the condition "$obs_T(s') = obs_T(s)$ and $t \in U \equiv c[\![s']\!] \in U$", which takes $O(|S||C|)$ operations, summing up to $O(|t||S||C|)$ over all recursive calls. Altogether, EXTEND($T,t$) thus runs in time $O(|S|^r|C| + |t||S||C|)$.

Using these estimations, it is easy to prove the following theorem.

**Theorem 11.** Let $c_{max}$ the size of the largest counterexample returned by the teacher during the execution of the learner. Then the running time of the learner is bounded by $O((I^r + c_{max})I^3)$.

Note that polynomial runtime is obtained only if the maximum rank $r$ of symbols in $\Sigma$ is fixed. If we consider $r$ to be part of the input, an exponential behaviour is obtained. However, thanks to the use of the procedure EXTRACT, the term in the runtime estimation which depends exponentially on $r$ is independent of $c_{max}$. In other words, the negative effects of large ranks and large counterexamples do not amplify each other.

The learner will ask for at most $I$ counterexamples and make a maximum of $O(I^{r+1} + c_{max}I)$ membership queries, where the first summand is the number of entries in the observation table it has to maintain and the second summand is the number of calls made to EXTRACT.

## 4    Avoiding Dead States

As remarked by Sakakibara in [Sak90], one deficiency of the learner constructed in the previous section is that it constructs a total fta, and hence has to specify $\delta$ even for tuples of states containing or yielding the dead state (assuming that $A_U$ contains a dead state). In most practical applications, the table describing the transition function will be sparse in the sense that there are only relatively few $\delta(q_1 \cdots q_k, f)$ which do not result in the dead state. For example, sets of derivation trees of programming languages or (subsets of) natural languages usually have this property. Hence, it would be useful if one could avoid producing dead states, working with partial fta's instead of total ones. We shall in the following discuss how this can be done, allthough we will only address correctness issues informally. However, from the presentation below the correctness of the constructions as well as the improvement with respect to complexity for almost all practical applications of the algorithm should be more or less obvious.

Consider an fta $A = (\Sigma, Q, \delta, F)$. A tree $s \in T_\Sigma$ is said to be *malformed* (with respect to $A$) if $c[\![s]\!] \notin L(A)$ for all contexts $c \in C_\Sigma$. In particular, $s \notin A$. The state $\delta(s)$ is then called a *dead state*. Note that malformed trees are necessarily equivalent. Hence, a minimal fta cannot contain more than one dead state. Suppose we use $A$ in order to check whether a tree $t$ is an element of $L(A)$. Working bottom up, we may reject $t$ as soon as (and if) a dead state occurs in the computation since this means that we have discovered a subtree $s$ of $t$ which is malformed. This means that we can turn $A$ into an equivalent *partial* fta $A' = (\Sigma, Q', \delta', F)$, where $Q'$ is obtained from $Q$ by removing all dead states. The transition function $\delta'$ is defined by $\delta'(q_1 \cdots q_k, f) = \delta(q_1 \cdots q_k, f)$ for all $q_1, \ldots, q_k \in Q'$ and $f \in \Sigma_{(k)}$ unless $\delta(q_1 \cdots q_k, f)$ is a dead state in $A$, in which case $\delta'(q_1 \cdots q_k, f)$ is undefined. In this way, the extension of $\delta'$ to $T_\Sigma$ becomes a partial function. The set of trees accepted by $A'$ is the set of all trees $t \in T_\Sigma$ such that $\delta'(t)$ is defined and an element of $F$.

The basic idea behind the modification of the learner is that a candidate $s$ for inclusion in $S$ (found by CLOSURE or EXTEND) for which all observations are negative is tentatively classified as malformed. It is then not immediately added to $S$ but put aside. The corresponding state $obs_T(s)$ is considered to be a dead state and is neither considered in the definitions of table closedness and consistency, nor when generating $A_T$. Later on, it will either be added to $S$, namely if it turns out that it is actually not malformed, or it will be discarded. In the following paragraphs, we discuss this in more detail.

An observation table is now a triple $T = (S, C, M)$, where $(S, C)$ is an observation table in the sense of Definition 5 and $M$ is either undefined (we write $M = \bot$) or is a single tree in $\Sigma(S) \setminus S$ such that (a) $c[\![M]\!] \notin U$ for all $c \in C$ and (b) $obs_T(s) \neq obs_T(M)$ for all $s \in S$ (where $obs_T = obs_{(S,C)}$). Thus, the tree $M$ which is suspected to be malformed gives rise to negative observations only, and there is no tree in $S$ that gives rise to the same observations. The table $T$ is closed if $\{obs_T(s) \mid s \in \Sigma(S)\} \subseteq \{obs_T(s) \mid s \in S \cup \{M\}\}$, where we define $S \cup \{M\} = S$ if $M = \bot$. Thus, $T$ is closed in the old sense, except for $obs_T(M)$ which need not be represented by the trees in $S$. The definition of consistency is not affected.

We synthesize $A_T$ as before, but let $\delta(obs_T(s_1) \cdots obs_T(s_k), f)$ be undefined if $obs_T(f[s_1, \ldots, s_k]) = obs_T(M)$. Then, Lemmas 8 and 9 carry over to the new definitions without difficulty.

The initial table is $T_1 = (\{a\}, \{\Box\}, \bot)$ if $a \in U$ and $T_1 = (\emptyset, \{\Box\}, a)$ otherwise. The learner stops and returns $A_T$ as soon as $|\{obs_T(s) \mid s \in S \cup \{M\}\}| = n$. The procedures CLOSURE, RESOLVE, and EXTEND are modified as follows.

**procedure** CLOSURE($T$) **where** $T = (S, C, M)$
    **find** $s \in \Sigma(S)$ **such that** $obs_T(s) \notin obs_T(S \cup \{M\})$;
    **if** $c[\![M]\!]\neg \in U$ for all $c \in C$ **then return** $(S, C, s)$
    **else return** $(S \cup \{s\}, C, \bot)$;

Thus, the tree $s$ found is classified as being malformed if all its observations are negative throughout. Note that this cannot happen if $M \neq \bot$, due to the requirement $obs_T(s) \notin obs_T(S \cup \{M\})$.

The procedure RESOLVE can be kept unchanged except for the last statement $\texttt{return } (S, C \cup \{c'[\![c]\!]\})$, since the new context may prove that $M$ is not malformed, in which case it has to be moved to $S$. Hence, the last line of RESOLVE must be changed as follows:

$\texttt{if } M \neq \bot \texttt{ and } c'[\![c[\![M]\!]]\!] \in U \texttt{ then return } (S \cup \{M\}, C \cup \{c'[\![c]\!]\}, \bot)$
$\texttt{else return } (S, C \cup \{c'[\![c]\!]\}, M);$

Finally, EXTEND is adapted by adding two lines:

$\texttt{procedure EXTEND}(T, t) \texttt{ where } T = (S, C, M)$
$\quad \texttt{choose } c \in C_\Sigma \texttt{ and } s \in subtrees(t) \cap \Sigma(S) \setminus S \texttt{ such that } t = c[\![s]\!];$
$\quad \texttt{if } M \neq \bot \texttt{ and } obs_T(s) = obs_T(M) \texttt{ then return } (S \cup \{s\}, C, \bot);$
$\quad \texttt{if there exists } s' \in S \texttt{ such that}$
$\quad\quad obs_T(s') = obs_T(s) \texttt{ and } t \in U \equiv c[\![s']\!] \in U \texttt{ then}$
$\quad\quad \texttt{return EXTRACT}(T, c[\![s']\!]);$
$\quad \texttt{else}$
$\quad\quad \texttt{if } c[\![s]\!]^\neg \in U \texttt{ for all } c \in C \texttt{ then return } (S, C, s)$
$\quad\quad \texttt{return } (S \cup \{s\}, C, M)$
$\quad \texttt{end if}$

The important modification is that we immediately insert $s$ into $S$ and discard $M$ if the observations for both are the same. To see that this is correct, notice the following. We know that $t$ is a counterexample. Furthermore, $A_T$ rejects all trees which contain a subtree $s \in \Sigma(S)$ with $obs_T(s) = obs_T(M)$, as such trees are considered to be malformed. In particular, $A_T$ rejects $t$. As $t$ is a counterexample, this proves that $t \in U$ and hence $s$ cannot be malformed (despite the fact that all observations that can be made so far are negative).

**Acknowledgment.** We thank the referees for useful comments. Some valuable suggestions, which could not be implemented due to time and space restrictions, will be considered in the full version.

# References

[Ang87]  Dana Angluin. Learning regular sets from queries and counterexamples. *Information and Computation*, 75:87–106, 1987.

[Bra68]  W.S. Brainerd. The minimalization of tree automata. *Information and Computation*, 13:484–491, 1968.

[Dre00]  Frank Drewes. Tree-based picture generation. *Theoretical Computer Science*, 246:1–51, 2000.

[Eng94]  Joost Engelfriet. Graph grammars and tree transducers. In S. Tison, editor, *Proc. CAAP 94*, volume 787 of *Lecture Notes in Computer Science*, pages 15–37. Springer, 1994.

[LJ78]  L.S. Levy and A.K. Joshi. Skeletal structural descriptions. *Information and Control*, 39:192–211, 1978.

[Sak90]   Yasubumi Sakakibara. Learning context-free grammars from structural data in polynomial time. *Theoretical Computer Science*, 76:223–242, 1990.

[TW68]   James W. Thatcher and Jesse B. Wright. Generalized finite automata theory with an application to a decision-problem of second-order logic. *Mathematical Systems Theory*, 2:57–81, 1968.

# On Three Classes of Automata-Like P Systems

Rudolf Freund[1], Carlos Martín-Vide[2], Adam Obtułowicz[3], and
Gheorghe Păun[*4 2]

[1] Department of Computer Science, Technische Universität Wien
Karlsplatz 13, A-1040 Wien, Austria
rudi@emcc.at
[2] Research Group on Mathematical Linguistics, Rovira i Virgili University
Pl. Imperial Tárraco 1, 43005 Tarragona, Spain
cmv@astor.urv.es
[3] Institute of Mathematics of the Polish Academy of Sciences
Śniadeckich 8, PO Box 137, 00-950 Warsaw, Poland
adamo@impan.gov.pl
[4] Institute of Mathematics of the Romanian Academy
PO Box 1-764, 70700 Bucureşti, Romania
gpaun@imar.ro,gp@astor.urv.es

**Abstract.** We investigate the three classes of accepting P systems considered so far, namely the P automata of Csuhaj-Varjú, Vaszil [3], their variant introduced by Madhu, Krithivasan [10], and the related machinery of Freund, Oswald [5]. All three variants of automata-like P systems are based on symport/antiport rules. For slight variants of the first two classes we prove that any recursively enumerable language can be recognized by systems with only two membranes (this considerably improves the result from [3], where systems with seven membranes were proved to be universal). We also introduce the initial mode of accepting strings (the strings are introduced into the system, symbol by symbol, at the beginning of a computation), and we briefly investigate this mode for the three classes of automata, especially for languages over a one-letter alphabet. Some open problems are formulated, too.

## 1   Introduction

Membrane computing is a branch of natural computing which aims to investigate distributed parallel computing devices that are abstracted from cell functioning. We refer to [13] and to the web page \%http://psystems.disco.unimib.it for details and for references. In short, in the compartments of a membrane structure we place multisets of symbol objects which evolve by means of local rules. Applying the rules in a non-deterministic, maximally parallel way, the system passes from one configuration to another one, thereby performing a computation. Only halting computations produce a result, which consists of the objects present in a specified output membrane.

[*] Work done in the framework of the Contract No ICA1-CT-2000-70024 between IM-PAN, Warsaw, Poland, and the European Community and under the Fifth Framework Programme, project "MolCoNet" IST-2001-32008

Z. Ésik and Z. Fülöp (Eds.): DLT 2003, LNCS 2710, pp. 292–303, 2003.
© Springer-Verlag Berlin Heidelberg 2003

An interesting class of P systems is that based on purely communicative rules, using so-called symport/antiport rules, [12]. Specifically, we use rules for transferring objects through membranes corresponding to the symport and antiport known in biology: symport is the process of passing two molecules through a membrane in the same direction, while antiport refers to the passage of two molecules in opposite directions. In mathematical (generalized) terms, a symport rule is of the form $(x, in)$ or $(x, out)$, where $x$ is a string representing the multiset of objects which enter, respectively exit, the membrane, while an antiport rule is of the form $(x, out; y, in)$, with $x$ indicating the objects which exit the membrane at the same time when the objects indicated by $y$ enter the membrane. Such rules can also be used in a conditional manner. For instance, a symport rule written in the form $(x, in)|_z$ is used to introduce the elements of $x$ in a membrane $i$ only if the elements of $z$ are already present in membrane $i$; we say that $z$ acts as a *promoter*.

As sketched above, a P system is used to *generate* a set of numbers or of strings. In the present paper we deal with the automata-like approach, using a P system as an acceptor, more precisely, as a string acceptor. The idea is not new, it was already explored for the first time in [3], and then in [10] and [5].

In [3] the so-called P automata are introduced, as membrane systems using conditional symport rules of the form $(y, in)|_x$, with a very important difference in defining the computation steps: the multisets are processed in a sequential manner, exactly one rule is applied in each region of the system. This is a feature very useful in programming computations in such systems. On the other hand, note the strong restriction that all rules use the command *in*, hence the communication is done in a one-way manner, top-down. Modulo a projection of sequences of multisets to strings (consisting only of symbols from a given terminal alphabet), these devices are shown to be universal in [3]: any recursively enumerable language can be recognized by a P automaton (with at least seven membranes). Actually, some further features are present in the systems from [3], but we do not consider them here. We improve the result from [3]: two membranes suffice (and this is an optimal result, because one-membrane systems can never halt after having started a computation).

A variant of the devices from [3] was proposed in [10], this time closer to the automata style: one considers both objects and states, and rules of the form $(qy, in)|_{px}$, where $p, q$ are states and $x, y$ are mutisets of objects. Each region has associated only one state object. A multiset which enters the system during a computation is accepted if the system halts in a final state. In [10], systems of this type with two membranes are proved to accept all recursively enumerable sets of natural numbers. We extend here this result to languages, even for systems with restricted forms of rules.

Note the similarities and differences of the systems of the two types above with the multiset automata from [2], where one uses multiset rewriting-like rules of the form $px \rightarrow qy$. It is also worth mentioning that the conditional symport rules are a particular form of "boundary rules", as introduced in [1], where rules of the form $y[_i x \rightarrow [_i xy$ were considered, with the obvious meaning that $y$ enters region $i$ provided that $x$ is already there. Thus, our universality results extend also to systems from [1].

A similar type of string accepting devices was considered in [5], but with a simpler (standard) definition: take usual P systems with symport/antiport rules, and accept the multiset of terminal symbols taken from the environment during a halting computation. The universality is again obtained, this time by systems with only one membrane, using antiport rules only.

In the present paper we also consider a restricted version of the devices of all these three types, which brings accepting P systems closer to the way usual automata work and which we call *initial*: the symbols of a string are introduced into the system one by one, in the first steps of the computation; further symbols can be introduced, at the same time or later; the string is accepted if the computation eventually halts.

For the devices from [3] and for a simplified variant of systems from [10] where however we allow several state objects to be present in the same region, we prove that any one-letter recursively enumerable language can be recognized in the initial mode by systems with only two membranes, and with only two objects present in each symport rule.

## 2   Prerequisites

We use the standard formal language theory notation and terminology. In particular, by $|x|$ we denote the length of the word $x$ over $V$. For $x \in V^*$ and $U \subseteq V$, by $|x|_U$ we denote the number of occurrences of symbols from $U$ in $x$. For any family of languages $FL$, by $1FL$ we denote the family of one-letter languages from $FL$ and by $NFL$ the family of length sets of languages in $FL$.

Several times in this paper we will make use of representing a string over $V = \{a_1, \ldots, a_k\}$, by its value in base $k+1$. Specifically, if $w = a_{i_1} a_{i_2} \ldots a_{i_n}$, $1 \leq i_j \leq k$ for all $1 \leq j \leq n$, then $val_{k+1}(w) = i_1 \cdot (k+1)^{n-1} + \ldots + i_{n-1} \cdot (k+1) + i_n$. Note that $val_{k+1}(a_{i_1} a_{i_2} \ldots a_{i_n}) = val_{k+1}(a_{i_1} \ldots a_{i_{n-1}}) \cdot (k+1) + i_n$.

A universal computational model very useful in our framework is the model of register machines (see [11] for some original definitions, [9] for variants called counter automata, and [7], [8] for definitions like that we use in this paper).

An *n-register machine* is a construct $M = (n, R, q_s, q_h)$, where:

- $n$ is the number of registers,
- $R$ is a set of labelled instructions of the form $q_1 : (op(r), q_2, q_3)$, where $op(r)$ is an operation on register $r$ of $M$, and $q_1, q_2, q_3$ are labels from a given set $lab(M)$ (the labels are associated in a one-to-one manner to the instructions),
- $q_s$ is the initial/start label, and
- $q_h$ is the final label.

The machine is capable of the following instructions:

- $(A(r), q_2, q_3)$: add 1 to the contents of register $r$ and proceed to any of the instructions (labelled with) $q_2$ and $q_3$.

- $(S(r), q_2, q_3)$: if register $r$ is not empty, then subtract 1 from its contents and go to instruction $q_2$, otherwise proceed to instruction $q_3$.
- *halt*: stop the machine. This instruction can only be assigned to the final label $q_h$.

A register machine $M$ is said to compute a partial function $f : \mathbf{N} \longrightarrow \mathbf{N}$ if, starting with any number $m$ in register 1 and with instruction $q_s$, $M$ halts in the final label $q_h$ with register 1 containing $f(m)$; if the final label cannot be reached, then $f(m)$ remains undefined.

A register machine can also recognize an input number $m \in \mathbf{N}$ placed in register 1: $m$ is accepted if the machine stops by the *halt* instruction. If the machine does not halt, then the analysis is not successful.

Register machines (with a small number of registers, but this fact is not of interest for what follows) are known to be computationally universal, equal in power to Turing machines. In our proofs, we will make use of the following two propositions.

**Lemma 1.** *Each recursively enumerable set of natural numbers can be recognized by a register machine.*

**Lemma 2.** *For any recursively enumerable language $L \subseteq T^*$, with $\mathrm{card}(T) = k$, there exists a register machine $M$ recognizing $L$ in the following way: for every $w \in T^*$, $w \in L$ if and only if $M$ halts when started with $val_{k+1}(w)$ in its first register; in the halting configuration, all registers of the machine are empty.*

## 3   P Systems with Symport/Antiport

We start by recalling the "standard" definition of a P system with symport/antiport, as introduced in [12] (hence used to generate natural numbers).

A *P system with symport/antiport rules* is a construct $\Pi$ of the form

$$\Pi = (V, T, \mu, w_1, \ldots, w_m, E, R_1, \ldots, R_m, i_o),$$

where: (1) $V$ is an alphabet of *objects*, (2) $T \subseteq V$ is the terminal alphabet; (3) $\mu$ is a membrane structure (with the membranes labelled by natural numbers $1, \ldots, m$ in a one-to-one manner); (4) $w_1, \ldots, w_m$ are multisets over $V$ associated with the regions $1, \ldots, m$ of $\mu$; (5) $E \subseteq V$ is the set of objects which are supposed to appear in an arbitrarily large number of copies in the environment; (6) $R_1, \ldots, R_m$ are finite sets of symport and antiport rules associated with the membranes $1, \ldots, m$; a symport rule is of the form $(x, in)$ or $(x, out)$, where $x \in V^+$ (with the meaning that the objects specified by $x$ enter, respectively exit, the membrane), and an antiport rule is of the form $(x, out; y, in)$, where $x, y \in V^+$, which means that the multiset $x$ is sent out of the membrane and $y$ is taken into the membrane region from the surrounding region; (7) $i_o$ is the label of the output membrane, an elementary one in $\mu$.

Starting from the *initial configuration*, which consists of $w_1, \ldots, w_m$ in the membrane structure $\mu$, the system passes from one configuration to another one by applying the rules from each $R_i$ in a non-deterministic and maximally parallel way. (The environment is supposed to be inexhaustible, at each moment all objects from $E$ are available in any number of copies we need.) A sequence of transitions is called a *computation*; a computation is *successful* if and only if it halts.

With a successful computation we associate a *result*, in the form of the number of objects from $T$ present in membrane $i_0$ in the halting configuration. The set of all such numbers computed (we also say *generated*) by $\Pi$ is denoted by $N(\Pi)$, and the family of all sets $N(\Pi)$, computed by P systems with at most $m$ membranes, with symport rules $(x, in), (x, out)$ with $|x| \leq r$, and antiport rules $(x, out; y, in)$ with $|x|, |y| \leq t$ is denoted by $NOP_m(sym_r, anti_t)$, $m \geq 1$ and $r, t \geq 0$.

Proofs of the following results, which improve previously known results in this area, can be found in [9], [6].

**Theorem 1.** $NRE = NOP_m(sym_r, anti_t)$, for $(m, r, t) \in \{(1, 1, 2), (3, 2, 0), (2, 3, 0)\}$.

With the symport/antiport rules we can also associate *promoters*, in the form $(x, in)|_z, (x, out)|_z, (x, out; y, in)|_z$, where $z$ is a multiset of objects. Such a rule, associated with membrane $i$, is applied only if $z$ is present in the region of membrane $i$. The use of promoters $z$ of length at most $u$ is indicated by adding $p_u$ in front of $sym, anti$, respectively, in the notation of families $NOP_m(sym_r, anti_t)$. The uncontrolled case corresponds to having $p_0$.

A language accepting version of these systems was considered in [5]: A string $w$ over the alphabet $T$ is recognized by the analysing P system $\Pi$ if and only if there is a successful computation of $\Pi$ such that the sequence of terminal symbols taken from the environment during the computation is exactly $w$. If more than one terminal symbol is taken from the environment in one step, then any permutation of these symbols constitutes a valid subword of the input string. The language of all strings $w \in T^*$ recognized by $\Pi$ is denoted by $A(\Pi)$. (In this case the output membrane $i_o$ plays no role, hence it can be omitted.)

In order to recognize strings in a way closer to automata style, we consider a restricted mode of accepting strings, called *initial*: take a string $x = x(1)x(2)\ldots x(n)$, with $x(i) \in T, 1 \leq i \leq n$; in the steps $1, 2, \ldots, n$ of a computation we place one copy of $x(1), x(2), \ldots, x(n)$, respectively, in the environment (together with the symbols of $E$); in each step $1, 2, \ldots, n$ we request that the symbol $x(1), x(2), \ldots, x(n)$, respectively, is introduced into the system (by a symport or an antiport rule, alone or together with other symbols); after exhausting the string $x$, the computation may continue, maybe introducing further symbols into the system, but without allowing the symbols of $x$ to leave the system; if the computation eventually halts, then the string $x$ is recognized. If the system halts before ending the string, or at some step $i$ the symbol $x(i)$ is not taken from the environment, then the string is not accepted. The language of strings accepted by $\Pi$ in the initial mode is denoted by $A_I(\Pi)$.

It is important to note that in the initial mode of using a P system we can use, say, a rule of the form $(a, in)$ for introducing a symbol of the analysed string into the system: $a$ is not necessarily an element of $E$, it is only present on "the tape". Contrast this with the non-initial case, when such a rule will cause the computation to continue forever: if $a$ is present in the environment, then it is present in arbitrarily many copies, thus the rule $(a, in)$ can be used infinitely often, because the environment is inexhaustible.

The family of all languages $A_I(\Pi)$, accepted by systems $\Pi$ with at most $m$ membranes, with symport rules $(x, in)$, $(x, out)$ with $|x| \leq r$, and antiport rules $(x, out; y, in)$ with $|x|, |y| \leq t$, is denoted by $A_I LP_m(sym_r, anti_t)$, $m \geq 1$, $r, t \geq 0$.

Throughout the rest of the paper we make use of the following convention:

**Convention.** When comparing two devices generating/accepting languages we ignore the empty string and when comparing two devices computing/accepting natural numbers we ignore the number 0.

## 4    The Universality of P Automata

Based on the model of analysing P systems as defined above we now consider a simpler variant of the devices investigated in [3] (especially, we do not use the concept of final states), i.e., we take constructs of the form $\Pi = (V, T, \mu, w_1, \ldots, w_m, E, R_1, \ldots, R_m)$, with all components as above, but with the rules from the sets $R_i, 1 \leq i \leq m$, being of the form $(y, in)|_x$. Several rules can have the same promoter multiset $x$ or can share objects from their promoting multisets. Moreover, objects which appear in the promoter $x$ of a rule associated with a membrane $i$ may also appear in the "moving multiset" $y$ of a rule associated with a membrane placed inside membrane $i$.

However, the most important difference compared with the general case of (analysing) P systems with symport/antiport rules as definded before is that in each step, in each region, at most one rule is used; if *at least* one rule can be used, then *exactly one* of them is chosen and used.

As for the analysing P systems defined in the previous section, a string $w$ over $T^*$ is accepted if its symbols are taken from the environment during a halting computation, in the order they appear in $w$. By $A(\Pi)$ we denote the language of strings accepted by $\Pi$ in the arbitrary mode and by $A_I(\Pi)$ the language of strings accepted in the initial mode. The obtained families of languages are denoted by $ALP_m(p_u sym_r, owts)$ and $A_I LP_m(p_u sym_r, owts)$, respectively; the meaning of all parameters is as usual, while the indication $owts$ refers to the specific form (one-way communication, using only top-down symport rules) and working mode (sequential, one rule per region) of the systems we use.

With our notation, in [3] it is proved that $RE = ALP_7(p_4 sym_6, owts)$. We are going to improve the result from [3] from the point of view of all parameters (number of membranes, size of promoters, size of moved multisets) for languages over arbitrary alphabets. However, we first consider the simpler case of one-letter alphabets also looking for the power of the initial mode for introducing the string in the system.

**Theorem 2.** $1RE = 1ALP_2(p_2sym_2, owts) = 1A_ILP_2(p_2sym_2, owts)$.

*Proof.* We use Lemma 1. Let $L \subseteq \{a_{n+1}\}^*$ be a language whose length set is recognized by a register machine $M = (n, R, q_s, q_h)$. We construct the register machine $M' = (n + 1, R', \bar{q}_s, q_h)$ with $lab(M') = lab(M) \cup \{\bar{q}_s, p_1\}$ and

$$R' = R \cup \{\bar{q}_s : (S(n + 1), p_1, q_s), \ p_1 : (A(1), \bar{q}_s, \bar{q}_s)\}.$$

The register machine $M'$ starts by moving the contents of its register $n + 1$ into register 1 and after that it simulates the machine $M$.

Starting from $M'$ we construct the P system

$$\Pi = (V, \{a_{n+1}\}, [_1[_2 \ ]_2]_1, b, \lambda, V, R_1, R_2),$$
$$V = \{a_j \mid 1 \le j \le n + 1\} \cup \{q, q' \mid q \in lab(M')\} \cup \{b, c\},$$
$$R_1 = \{(a_{n+1}, in)|_b, \ (a_{n+1}\bar{q}_s, in)|_b, \ (c, in)|_{a_{n+1}}, \ (c, in)|_{\bar{q}_s q_h}, \ (c, in)|_c\}$$
$$\cup \{(q_2 a_r, in)|_{q_1}, \ (q_3 a_r, in)|_{q_1} \mid \text{for } q_1 : (A(r), q_2, q_3) \in R'\}$$
$$\cup \{(q_2, in)|_{q_1 a_r}, \ (q'_3, in)|_{q_1},$$
$$(c, in)|_{q'_3}, \ (q_3, in)|_{q_1 q'_3} \mid \text{for } q_1 : (S(r), q_2, q_3) \in R'\},$$
$$R_2 = \{(q, in) \mid q \in lab(R') \setminus (\{\bar{q}_s\} \cup \{q_1 \mid \text{for } q_1 : (S(r), q_2, q_3) \in R'\})\}$$
$$\cup \{(\bar{q}_s b, in)\} \cup \{(q_1 a_r, in), \ (q_1 q'_3, in) \mid \text{for } q_1 : (S(r), q_2, q_3) \in R'\}.$$

We start with the object $b$ in region 1. As long as $b$ is here, in each step one symbol $a_{n+1}$ is introduced into the system by the rule $(a_{n+1}, in)|_b$. At any moment we can also use the rule $(a_{n+1}\bar{q}_s, in)|_b$. It introduces the label $\bar{q}_s$ in the system, hence at the next step $\bar{q}_s b$ will enter membrane 2, and simultaneously from the set $R_1$ we can either use again one of the rules $(a_{n+1}, in)|_b$, $(a_{n+1}\bar{q}_s, in)|_b$, or we can start simulating a computation in $M$.

If we use the rule $(a_{n+1}, in)|_b$, then, because $\bar{q}_s$ is "lost" in region 2, the only rule which, in the next step, can be used in region 1 is $(c, in)|_{a_{n+1}}$, and then the computation continues forever by means of the rule $(c, in)|_c$. If we use the rule $(a_{n+1}\bar{q}_s, in)|_b$, then the new copy of $\bar{q}_s$ will remain forever in region 1 of the system (note that we do not have a rule of the form $(\bar{q}_s, in)$ in $R_2$, while the unique copy of $b$ was already introduced in membrane 2). Because $\bar{q}_s$ is present, the system will start to simulate the work of $M'$ (see the details below). If the computation in $M$ never stops, then the computation in $\Pi$ never stops and the string is not recognized. If the computation in $M'$ stops, the label $q_h$ is introduced, but then the computation in $\Pi$ again continues forever: in the presence of $\bar{q}_s q_h$, we can bring $c$ into the system, and then we can use forever the rule $(c, in)|_c$ from $R_1$.

Therefore, in order to have a halting computation, after introducing the first $\bar{q}_s$ into the system, we have to start simulating a computation in $M'$; as we have noted above (see the passing from $M$ to $M'$), this means the simulation of a computation in $M$, after moving the contents of register $n + 1$ into register 1. Therefore, the contents of register $n + 1$ is only once emptied and after that we only work on registers $1, \ldots, n$.

Thus, any string $a_{n+1}^m, m \geq 1$, can be introduced into the system, with $\bar{q}_s$ also present in region 1, and no copy of $a_{n+1}$ will ever be taken from the environment afterwards in a halting computation. Hence, the arbitrary and the initial modes of working are identical for $\Pi$.

The simulation of the instructions of $M'$ is done as follows. After bringing $\bar{q}_s$ into the system, this object goes into region 2 together with $b$, at the same time promoting a rule from $R_1$. In general, let us assume that we have a label $q_1$ in region 1.

In the case of an add-instruction $q_1 : (A(r), q_2, q_3)$, the simulation is simply done by bringing inside one of the labels $q_2, q_3$, together with a copy of $a_r$, by means of rules $(q_2 a_r, in)|_{q_1}$ and $(q_3 a_r, in)|_{q_1}$ of $R_1$, whereas $q_1$ itself goes to region 2 by $(q_1, in)$ in $R_2$.

Assume that in region 1 we have the label $q_1$ associated with the subtract-instruction $q_1 : (S(r), q_2, q_3)$. The rule $(q_1 a_r, in) \in R_2$ can be used or not, depending on the existence of at least one copy of $a_r$, the same with the rule $(q_2, in)|_{q_1 a_r} \in R_1$, while $(q_3', in)|_{q_1} \in R_1$ can always be used. Thus, we distinguish two cases:

*Case* (i): At least one copy of $a_r$ is present in region 1. Then $q_1$ is introduced in region 2 together with a copy of $a_r$, and from $R_1$ we use either one of the rules $(q_2, in)|_{q_1 a_r}, (q_3', in)|_{q_1}$. If the first one is used, this is the correct simulation of the instruction $q_1 : (S(r), q_2, q_3)$ in the case when the subtraction is possible. If we use the second rule from $R_1$, then the object $q_3'$ enters region 1 and remains here. At the next step, the only applicable rule from $R_1$ is $(c, in)|_{q_3'}$, and the computation will continue forever. This is again consistent with the work of $M'$, where the continuation with $q_3$ is not correct if the subtraction is possible.

*Case* (ii): If no copy of $a_r$ is present in region 1, then $q_1$ remains in region 1, and the rule $(q_2, in)|_{q_1 a_r}$ cannot be used. Thus, we have to use the rule $(q_3', in)|_{q_1}$. Because both $q_1$ and $q_3'$ are present, we may use the rule $(q_3, in)|_{q_1 q_3'} \in R_1$, which brings inside the label $q_3$; at the same time, $q_1$ and $q_3'$ enter region 2. We have correctly simulated the instruction $q_1 : (S(r), q_2, q_3)$ in the case when the subtraction is not possible. Instead of the rule $(q_3, in)|_{q_1 q_3'} \in R_1$, we can also use the rule $(c, in)|_{q_3'}$, but this will lead to a non-halting computation.

Therefore, the computation continues without introducing the trap object $c$ only if the instructions of $M'$ are correctly simulated. Clearly, we stop only if the number $m$ is recognized by $M'$ (hence by $M$), and $a_{n+1}^m$ is thus accepted by $\Pi$. Consequently, $A_I(\Pi) = A(\Pi) = L$, and the proof is complete.    □

In the case of the arbitrary mode, this result can be extended to arbitrary alphabets.

**Theorem 3.** $RE = ALP_2(p_2 sym_2, owts)$.

*Proof.* This time we use Lemma 2. Consider a language $L \subseteq V^*$, for some $V = \{b_1, \ldots, b_k\}$, which is accepted by a register machine $M = (n, R, q_s, q_h)$.

We now construct a P system $\Pi$, of degree 2, which works as follows. Starting in the initial configuration, a symbol $b_j$ is introduced into region 1. Then $j$ copies of $a_{n+1}$ are introduced, too. Assume that we also have some number $\alpha$ of copies of $a_{n+2}$ present in the system. We multiply this number $\alpha$ with $k+1$, passing from the $\alpha$ copies of $a_{n+2}$ to $k\alpha$ copies of $a_{n+1}$. In this way, we have $k\alpha + j$ copies of $a_{n+1}$, which corresponds to the value in base $k+1$ of the string we have introduced into the system so far. After completing this step, we change all objects $a_{n+1}$ to objects $a_{n+2}$ and we continue. At any step, we can also switch to simulating the work of $M$ over the (value in base $k+1$ of the) string already introduced into the system. To this aim, we first change all objects $a_{n+2}$ to objects $a_1$ and introduce the label $q_s$ in the system.

The introduction of $b_j$ and of $j$ copies of $a_{n+1}$ is done by rules in the following way: Assume that in the skin region we have the object $q_{s,any}$. Then, for all $j = 1, 2, \ldots, k$, in $R_1$ we introduce

$$(b_j q_{s,j}, in)|_{q_{s,any}}, \ (q_{s,j,1} a_{n+1}, in)|_{q_{s,j}}, \text{ and}$$
$$(q_{s,j,t+1} a_{n+1}, in)|_{q_{s,j,t}} \text{ for all } 1 \leq t \leq j-1,$$

and in $R_2$ we introduce

$$(q_{s,any}, in) \text{ and } (q_{s,j,t}, in) \text{ for all } 1 \leq t \leq j.$$

The "state objects" $q$ (with subscripts) are immediately sent into region 2, hence they correctly control the process and then "disappear". When $q_{s,j,j}$ is introduced, we pass to the multiplication by $k+1$ of the available copies of $a_{n+2}$ as described above. This can be done by a register machine $M_j = (3, R_j, q_{s,j,j}, q_{s,any})$, with the following instructions:

$$q_{s,j,j} : (S(n+2), q_{j,1}, q'_j),$$
$$q_{j,t} : (A(n+1), q_{j,t+1}, q_{j,t+1}), \text{ for all } 1 \leq t \leq k,$$
$$q_{j,k+1} : (A(n+1), q_{s,j,j}, q_{s,j,j}),$$
$$q'_j : (S(n+1), q''_j, q'''_j),$$
$$q''_j : (A(n+2), q'_j, q'_j),$$
$$q'''_j : (A(n+3), q_{s,any}, q_{move}).$$

The three registers of $M_j$ are numbered with $n+1, n+2$, and $n+3$, because $M_j$ will be integrated in a large machine, with $n+3$ registers. The last instruction $q'''_j : (A(n+3), q_{s,any}, q_{move})$ allows us to start the input of another symbol $b_j$ by $q_{s,any}$ or, by $q_{move}$, to switch to the register machine $M_{move}$ which changes all symbols $a_{n+2}$ to $a_1$, clears register $n+3$, and ends in the label $q_s$ (which is the initial label of $M$); $M_{move}$ uses the following instructions:

$$q_{move} : (S(n+2), q'_{move}, q''_{move}),$$
$$q'_{move} : (A(1), q_{move}, q_{move}),$$
$$q''_{move} : (S(n+3), q''_{move}, q_s).$$

Now, consider the procedure of passing from the add- and subtract-instructions of a register machine to conditional symport rules as used in the proof of Theorem 2; the obtained rules are placed in two sets, $R_1$ and $R_2$. We apply this procedure for all instructions of the register machines $M_j$, $1 \leq j \leq k$, and $M_{move}$, as specified above. We obtain a "subsystem" of $\Pi$ which introduces a string $w \in \{b_1, \ldots, b_k\}^*$ into the skin membrane, and at the same time produces $val_{k+1}(w)$ copies of $a_1$ there; at the end of this process, the object $q_s$ is also placed in the skin region. Therefore, the work of $M$ now can be simulated as in the proof of Theorem 2. In total, we get a system $\Pi$ which accepts exactly the strings which are accepted by $M$, hence $A(\Pi) = L$. □

Combinatoric calculations show that the result in Theorem 2 cannot be extended to languages over arbitrary alphabets with more than one letter in the case of the initial mode:

**Theorem 4.** For all $m, u, r \geq 1$, $A_I LP_m(p_u sym_r, owts) \neq RE$.

*Proof.* Assume that in a P system recognizing a given language over the terminal alphabet $\Sigma$ (containing $t \geq 2$ symbols) in total we have $k$ different objects, $k \geq t$. In each step, at most $r$ (new) objects can enter the skin membrane. After an input of $n$ terminal letters, i.e., after $n$ steps in the initial mode, there can be at most $nr + c$ objects in the whole system, where $c$ is the number of initial objects in the initial configuration. In each of the $m$ membrane regions, the number of occurrences of one of the $k$ different objects can take a value from 0 to $nr + c$; hence, in total at most $(nr + c + 1)^{km}$ different configurations can be obtained. On the other hand, the number of different strings of length $n$ is $t^n$, which number exceeds the number $(nr + c + 1)^{km}$ for numbers $n$ being large enough. Hence, for sufficiently large numbers $n$, no P system working in the given $owts$ mode can distinguish all the initial inputs of length $n$ representing the strings of length $n$, which proves that P automata working in the $owts$ mode and in the initial mode are not computationally universal. □

## 5    The Universality of P Automata with States

In this section, we now consider the variant of P automata as introduced in [10], in a simplified version. Specifically, we deal with systems of the form $\Pi = (V, K, T, \mu, w_1, \ldots, w_m, R_1, \ldots, R_m)$, where all components are as usual, $K \subseteq V \setminus T$ is the set of *states*, and the rules from sets $R_i, 1 \leq i \leq m$, are of the form $(qy, in)|_{px}$, where $p, q \in K$ and $x, y \in V^*$. The rules are used in the maximally parallel manner. Using a rule $(qy, in)|_{px} \in R_i$ is possible only if $px$ is included in the multiset present in region $i$, and the effect is that the multiset $y$ is introduced in region $i$ from the surrounding region, while $p$ is replaced by $q$. Thus, in each region we can use at most as many rules as occurrences of state objects exist in that region. As usual, a string over $T$ is accepted if it is a sequence of symbols from $T$ taken into the system during a halting computation. The language of all strings accepted by $\Pi$ is denoted by $A(\Pi)$ and the language of strings accepted in the initial mode is denoted by $A_I(\Pi)$. The family of all languages $A(\Pi)$, accepted by systems $\Pi$ as above with at most $m$ membranes, using rules $(qy, in)|_{px}$ with $|qy| \leq r, |px| \leq u$, and with at most t states is denoted by $ALP_m(p_u sym_r, state_t)$.

If the strings are recognized in the initial mode, then we write $A_I LP$ instead of $ALP$.

Clearly, the automata from [10] are extensions of those from [3]: we can consider a unique state, $p$, and instead of a rule $(y, in)|_x$ we consider the rule $(py, in)|_{px}$. Thus, the following results are consequences of Theorems 2 and 3. Direct proofs can be given, too, much simpler than the proofs of Theorems 2 and 3, but we leave this as a task for the reader.

**Theorem 5.** $1RE = 1ALP_2(p_3 sym_3, state_1) = 1A_I LP_2(p_3 sym_3, state_1)$.

**Theorem 6.** $RE = ALP_2(p_3 sym_3, state_1)$.

The questions whether these results are optimal as well as whether Theorem 6 also holds for the initial mode remain as open problems.

## 6     Analysing P Systems with Symport/Antiport

Let us now consider the P systems as introduced in [5]. In the arbitrary case they can recognize all recursively enumerable languages even when using only one membrane. More precisely, the following result was proved in [5]:

**Theorem 7.** *Each language $L \in RE$ can be recognized by an analysing P system with only one membrane using antiport rules $(x, out; y, in)$ with $(|x|, |y|) \in \{(1, 2), (2, 1)\}$ only (and no symport rule).*

Thus, there is nothing else to do about the general case; what remains to do is to investigate the initial mode of accepting strings. We will return to this topic in a forthcoming paper, here we only state some preliminary results; due to lack of space the proofs are left to the reader.

First let us mention the following observation:

**Lemma 3.** *Let $Q \in NOP_m(sym_r, anti_t)$ be a set of numbers generated by a P system of the type associated with the family $NOP_m(sym_r, anti_t)$, which in the halting configuration counts the symbols from a set $T$ present in an elementary membrane $i_o$ and such that the only rules affecting the contents of membrane $i_o$ are involving elements of $T$ and are of the form $(a, in)$, $a \in T$. Then $\{b^j \mid j \in Q\} \in 1A_I LP_{m+2}(sym_{r'}, anti_t)$, where $r' = \max(r, 2)$.*

As the systems used in the proofs of the results elaborated in [9] and [6] have the properties from the lemma cited above, as a consequence of Theorem 1 and the preceding lemma we get:

**Theorem 8.** $1RE = 1A_I LP_n(sym_r, anti_t)$ *for* $(n, r, t) \in \{(3, 2, 2), (5, 2, 0), (4, 3, 0)\}$.

As a consequence of the proof given in [5] for Theorem 7, we immediately get the following results:

**Theorem 9.** $1RE = 1A_I LP_1(sym_0, anti_3) = 1A_I LP_2(sym_0, anti_2)$.

# 7 Conclusions

We have investigated analysing P systems with symport/antiport rules of the forms considered in [3], [10], and [5], recognizing strings in the modes proposed in these papers, or with a particular mode of introducing the strings into the system which we call *initial*. The results from [3], [10] were significantly improved: all recursively enumerable languages can be recognized in the arbitrary mode by systems with only two membranes, while the same result holds true also in the initial mode for languages over a one-letter alphabet. For these systems investigated in [3] and [10] as well as for the systems considered in [5], the initial mode needs further investigations.

**Acknowledgements.** We gratefully acknowledge some helpful comments of two unknown referees. Moreover, we thank Hendrik Jan Hoogeboom for pointing out Theorem 4 and its proof.

# References

1. Bernardini, F., Manca, V.: P Systems with Boundary Rules. In: [14] 107–118
2. Csuhaj-Varjú, E., Martín-Vide, C., Mitrana, V.: Multiset Automata. In: Calude, C.S., Păun, Gh., Rozenberg, G., Salomaa, A. (eds.): Multiset Processing. LNCS, Vol. 2235. Springer-Verlag, Berlin Heidelberg New York (2001) 69–84
3. Csuhaj-Varjú, E., Vaszil, G.: P Automata. In: [14] 219–233
4. Freund, R., Oswald, M.: P Systems with Activated/Prohibited Membrane Channels. In: [14] 261–269
5. Freund, R., Oswald, M.: A Short Note on Analysing P Systems. Bulletin of the EATCS 78 (October 2002) 231–236
6. Freund, R., Păun, A.: Membrane Systems with Symport/Antiport: Universality Results. In: [14] 270–287
7. Freund, R., Păun, Gh.: On the Number of Non-terminal Symbols in Graph-controlled, Programmed and Matrix Grammars. In: Margenstern, M., Rogozhin, Y. (eds.): Proc. Conf. Universal Machines and Computations, Chişinău, 2001. LNCS, Vol. 2055. Springer-Verlag, Berlin Heidelberg New York (2001) 214–225
8. Freund, R., Sosík, P.: P Systems without Priorities Are Computationally Universal. In: [14] 400–409
9. Frisco, P., Hoogeboom, H.J.: Simulating Counter Automata by P Systems with Symport/Antiport. In: [14] 288–301
10. Madhu, M., Krithivasan, K.: On a Class of P Automata, manuscript (2002)
11. Minsky, M.L.: Computation: Finite and Infinite Machines. Prentice Hall, Englewood Cliffs, New Jersey, USA (1967)
12. Păun, A., Păun, Gh.: The Power of Communication: P Systems with Symport/Antiport. New Generation Computing 20, 3 (2002) 295–306
13. Păun, Gh.: Membrane Computing: An Introduction. Springer-Verlag, Berlin Heidelberg New York (2002)
14. Păun, Gh, Rozenberg, G., Salomaa, A., Zandron, C. (eds.): Membrane Computing 2002. LNCS, Vol. 2597. Springer-Verlag, Berlin Heidelberg New York (2003)

# Computing Languages by (Bounded) Local Sets*

Dora Giammarresi

Dipartimento di Matematica, Università di Roma "Tor Vergata"
via della Ricerca Scientifica, 00133 Roma, Italy
giammarr@mat.uniroma2.it

**Abstract.** We introduce the definition of local structures as description of computations to recognize strings and characterize families of Chomsky's hierarchy in terms of projection of frontiers of local sets of structures. Then we consider particular grid structures we call *bounded-grids* and study the corresponding family of string languages by proving some closure properties and giving several examples.

## 1   Introduction

Chomsky's hierarchy of languages was defined in the fifties in terms of grammars introduced as potential models for natural languages. Despite such language classes were defined in a linguistic context, they have had their success in computer science and applied in several computational problems. Then, alternatively, the same families of languages were defined in terms of different kind of machines (finite automata, push-down automata, linear bounded automata, Turing machines, respectively) that recognize them by finite computations.

For regular languages it is also known a characterization in terms of local string languages and projections (cf. [3]). More precisely, a language $L$ is *local* if there exists a finite set $\Theta$ of strings of length 2 that correspond to all possible substrings of length 2 of the strings in $L$ To accept a string $w$ in a local language $L$, it is sufficient to perform a local test that checks, in any order, whether all its substrings of length 2 belongs to set $\Theta$. Local languages are subsets of regular languages but, it holds that, all regular languages can be obtained as projections (by alphabetic mappings) of local languages.

In this paper we generalize the definition of local sets in a way that local sets plus projections can be used to describe uniformly different families of string languages. We consider a generic *structure* to be an embedded labeled graph in which it can be identified a string, referred to as the *frontier of the structure*. Local sets of structures are defined by means of a finite set of elementary structures that represent all allowed substructures. Then, a local set of structures $S$ together with a projection $\pi$ is a description for a string language $L$ containing all the projections by $\pi$ of the frontiers of elements in $S$.

The case where structures are line-graphs with frontiers equal to the lines itself leads to define regular languages as we discussed above. As second particular

---

* This work was partially supported by MIUR project *Linguaggi formali e automi: teoria e applicazioni*.

Z. Ésik and Z. Fülöp (Eds.): DLT 2003, LNCS 2710, pp. 304–315, 2003.

case, we take as structures binary trees with frontier equal to the set of leaves (read from left to right). A local set of binary trees is defined by means of a set of trees of height one that represent all allowed subtrees. Then, it can be proved that projections of frontiers of local sets of binary trees characterize context-free languages (see [12]). As third case, we take as structures rectangular grid graphs with frontiers equal to the last row. Local sets of grids are defined as local picture languages in [5,6]. In [4,9] it is proved that context-sensitive languages can be characterized as projections of frontiers of local picture languages.

We also point out that local sets of structures and projections can be also considered by a machine perspective. We define a local machine that operates on its input string by guessing its counterimage by the projection and taking it as frontier for a local structure. Then, it builds such local structure by using the elementary structures as bricks. If it succeeds in building a structure that belongs to the local set then it accepts the input string. This means that, in general, the local sets of structures will be the "computations" to recognize/generate the languages of their frontier strings. Or, taking it from a reverse point of view, the computation of a local machine exploits geometric properties of patterns and shapes. We mention that the idea of considering pictures as computations for the strings is used also in [10] in connection with context-sensitive trasductions.

Therefore, with the help of local sets of structures and projections, we are able to provide a uniform characterization of regular, context free and context sensitive languages in terms of computational machines rather than grammars. The difference among those languages lies in the "complexity" of structures needed. One aspect of such complexity differences lies in the different topologies of structure graphs: namely, vertex degrees $\leq 2, \leq 3, \leq 4$, respectively. The other major difference is the following: while the size of a structure for either a regular or a context free language is linear with respect to its frontier, in the case of context-sensitive languages, the frontier does not give any bounds to the whole structure. This corresponds to the fact that no bounds can be given either on the length of a computation of a linear bounded automaton or on the length of a derivation of a context-sensitive grammar.

As a consequence, this characterization reveals a big gap between context-free and context-sensitive languages. This gap was also observed by R.McNaugthon, who proposed to insert in the Chomsky's hierarchy the family of *growing context-sensitive languages* as "type one-and-a-half" languages (see [11,2]).

In the second part of the paper we consider as structures what we call *bounded-grids* that are again pictures but with the frontier equal to the bottom-right semi-perimeter of the rectangular grid. Notice that, with this choice of frontier, the number of vertices of the grid is now bounded by the length of the frontier itself. We propose such bounded-grids structures as *next level* after binary tree structures in the above mentioned characterization of languages by local sets of structures and projection. We study properties of the family of string languages that are projections of frontiers of local sets of bounded-grids. We show that such languages are context-sensitive thus we refer to them as *bounded-grids context-sensitive (Bgrid-CS) languages*. Moreover we can prove that the family of Bgrid-CS languages is closed under concatenation and Kleene's star operations and under Boolean union and intersection. We also construct explicitly the local picture languages for several examples of typical context sensitive languages. Those examples show how geometric properties of patterns or shapes can be used as computations for context-sensitive languages. Moreover, describing a computation by a geometric properties of patterns can be often more intuitive than listing the derivation rules of a context-sensitive grammar are giving the transition function of a linear bounded automaton.

All the examples together with the closure properties somehow give the impression that the Bgrid-CS family is a quite big subset of context-sensitive languages. Very interestingly, family of Bgrid-CS languages coincides with a family defined in [1] by means of time-bounded grammars. In the same paper R. Book left open the problem whether such family coincides with the whole family of context sensitive languages.

We conclude the paper by discussing some related open problems and further research directions.

## 2    Preliminaries

We assume the reader familiar with basic formal language theory: we will use standard notations as, for example, in [8]. Throughout this section we implicitly assume that the alphabets do not contain symbols $\#, \$$ that will be always taken as *border symbols* for the strings. We now recall a characterization of regular string languages by means of local languages and projection (cf. [3]).

Let $\Gamma$ be a finite alphabet and let $L \subseteq \Gamma^*$ be a string language. Then $L$ is *local* if there exists a finite set $\Theta$ of strings of length 2 over $\Gamma \cup \{\#, \$\}$ that contains all allowed substrings of length 2 for the bordered strings in $L$. We write $L = \mathcal{L}(\Theta)$. Since this notion will be crucial for the rest of the paper, we give an example of local string language. Let $\Gamma = \{a, b\}$ and let $L$ be the set of all strings that begin with $a$ and do not contain two consecutive $b$'s. Then $L = \mathcal{L}(\Theta)$ with $\Theta = \{\$a, ab, aa, ba, a\#, b\#\}$. To verify that a string $w$ belongs to $L$, we scan $\$w\#$ using a window of size 2 and verify that all substrings we see through the window belong to $\Theta$. Remark that this procedure does not need to be sequential then it is not necessarily related to a particular model of machine.

Let $\Gamma$ and $\Sigma$ be two finite alphabets and let $\pi : \Gamma \longrightarrow \Sigma$ be a function to which we refer as projection. Then $\pi$ can be extended in the usual way to strings and to languages over $\Gamma$. If $L' \subseteq \Gamma^*$ then $L = \pi(L') \subseteq \Sigma^*$ will be referred to as the *projection of $L'$ by $\pi$*. Local string languages and projections are used for a characterization for regular languages in the following theorem (cf. [3]).

**Theorem 1.** *Let $L$ be a language over an alphabet $\Sigma$. Then $L$ is regular if and only if there exists a local language $L'$ over an alphabet $\Gamma$ and a projection $\pi : \Gamma \longrightarrow \Sigma$ such that $L = \pi(L')$.*

The proof comes directly from the fact that the corresponding local languages $L' \subseteq \Gamma^*$ together with the projection $\pi$ are actually an alternative way to describe a finite automaton that recognizes $L$. More precisely, the alphabet $\Gamma$ is in a one-to-one correspondence with the edges of the state-graph of the automaton. The set $\Theta$ describes the edges adjacency and thus provides a description of the state-graph. The mapping $\pi : \Gamma \rightarrow \Sigma$ gives the labelling of the edges in the state-graph. Then, the set of words of the underling local language defined by set $\Theta$ corresponds to all accepting paths (as edges sequence) in the state-graph and its projection by $\pi$ gives the language recognized by the automaton. Alternatively, we can think the local set and the projection as a description for the right-linear grammar that generates $L$.

The notion of local languages has been extended from string to picture languages in [5,6]. A picture is essentially a two-dimensional array of symbols from a finite alphabet. Given a picture $p$ of size $m \times n$, its bordered version is defined by adding symbols # all around the rectangle: the result is then a picture of size $(m + 2) \times (n + 2)$ whose first and last rows and columns contain the border symbol. A picture language $L$ over $\Gamma$ is local if it is defined by a finite set $\Theta$ of pictures of size $2 \times 2$ over $\Gamma \cup \{\#\}$ that represents all allowed subpictures of that size. In [5,6] there are several examples of local picture languages that show how geometric properties of patterns can be exploited to define squares and several other shapes and functions.

In the next section we give a more general definition of local sets and use them to define context-free and context-sensitive string languages.

## 3    Chomsky's Hierarchy by Local Sets

In the Chomsky's hierarchy, regular, context free and context sensitive languages are defined in terms of grammars that generate them. In this section we give a new characterization of the families in Chomshy's hierarchy by means of local sets and projections. This shows that what is defined by means of grammars has a meaning also in terms of topological structures.

We start with some technical definitions. Let $\Gamma$ be a finite alphabet and let $\#_1, \ldots, \#_h \notin \Gamma$ be a finite set of *border symbols*, we give the following definition.

**Definition 1.** *A structure over $\Gamma$ is an embedded labeled graph with labels in $\Gamma \cup \{\#_1, \ldots, \#_h\}$ where it is defined a  border as the set of all vertices carrying border symbols.*

Most of the definitions on graphs can be extended to structures in an obvious way. In particular, one can define a substructure of a given structure by referring to the corresponding subgraph. Border vertices are needed for the notion of *local structures* we are going to introduce. We fix one unlabeled graph (of small size): then all the structures having that one as underlying graph are referred to as *elementary structures* or, equivalently, as *structures of elementary size*.

**Definition 2.** *A set of structures $S$ over $\Gamma$ is local if there exists a finite set of elementary structures $\Theta$ over $\Gamma \cup \{\#_1, \ldots, \#_h\}$ such that, for every $s \in S$, each substructure of elementary size of $s$ belongs to $\Theta$. We write $S = \mathcal{L}(\Theta)$.*

We will use local structures to characterize family of languages and for this we need to identify a string in each structure. Let $s$ be a structure over $\Gamma$.

**Definition 3.** *The* frontier *of $s$, denoted by* fr(s)*, is a particular sequence of non-border vertices adjacent to the border.*

For each typology of structures (i.e. topology of the underlying graphs) we will fix the frontier. For the sequel, where it does not create ambiguity, we will indicate by $fr(s)$ also the string corresponding to the labels of the frontier of $s$. Then, to each set of structures $S$, we associate a corresponding language of strings $L = fr(S)$ containing all the frontiers of the structures in $S$. We will refer to $L$ as *the frontier of $S$*.

Despite we introduced structures as very general definition, for the rest of the section we will focus on three particular cases we describe below.

*1)  Lines*
The underlying graphs are lines that we assume be drawn as horizontal ones. The structures are finite lines with border at leftmost and at rightmost vertices (two different symbols are needed). Then they correspond to classical definition of *strings*. Local sets of lines are defined by using 2-vertices graphs as elementary structures and thus they correspond to the family of local string languages (cf. Sects. 2 and [3]). The frontier of a line is the line itself excluding the border.

*2)  Binary trees*
The underlying graphs are complete binary trees. The structures are finite embedded labeled trees with border symbols at the root and at the leaves (two different symbols are needed for the root and the leaves, respectively) We define local sets of trees by taking a 3-vertices binary tree as elementary structure. The frontier of a tree will be the set of vertices adjacent to the leaves.

*3)  Grids*
The underlying graphs are grids that we assume to be drawn with horizontal and vertical sides, respectively. The structures are finite embedded labeled rectangular grids with border symbols at leftmost and rightmost columns and at top and bottom rows We define local sets of grids by using a 4-vertices grid (i.e. a unit square), as elementary structure, and this corresponds exactly to the family of *local picture languages* as defined and studied in [5,6]. The frontier of a rectangular grid graph is the row next to the bottom border.

Since structures are embedded graphs then they can be implicitly identified with their representation: we assume the frontier to be a string we read in the conventional way from left to right. Moreover, for the sequel we assume that string languages and structure sets are defined on $\Sigma$ and $\Gamma$, respectively.

We now give a characterization of regular, context free and context sensitive languages of Chomsky's hierarchy by means of local sets of structures.

**Proposition 1.** *Let $L$ be a string language. Then:*

1) $L$ *is* regular *if and only if there exists a local set of lines $S$ such that $L = \pi(fr(S))$;*

*2) L is* context-free *if and only if there exists a local set of* binary trees *S such that* $L = \pi(fr(S))$;

*3) L is* context-sensitive *if and only if there exists a local set of* grids *S such that* $L = \pi(fr(S))$.

*Proof.* (Sketch)

*1)*   The proof comes directly from Theorem 1.

*2)*   The proof comes by noticing that a derivation of a context free grammar in Chomsky's Normal Form is actually a local binary tree (possibly after some minor modifications). See [12] for details.

*3)*   The proof comes from the fact that, given an accepting run of an LBA, we can take all the instantaneous configurations and write them in order one above the others. This gives a local two-dimensional array (local picture). On the other hand, given a local set of $2 \times 2$ squares, one can define a corresponding set of context-sensitive grammar rules in a way that the derivations of the grammar correspond to the local pictures. See [4,9] for the details of possible codings.

Observe that, for the case of context-free and context-sensitive languages, one can also take the projection $\pi$ as the identity function and let $\Gamma \supset \Sigma$.

In the previous section we pointed out that, in the case of regular languages, the local language $L'$ such that $\pi(L') = L$ is somehow the language of the computations for strings in $L$. This can be generalized to the local set of structures if we look at them from a machine point of view.

For each type of structures $\mathcal{T}$ (ex. lines, trees,.....) we can associate a *scanning procedure* for the structures. Then, for each $\mathcal{T}$, we can define a *local machine* as a quintuple $M = (\mathcal{T}, \Sigma, \Gamma, \Theta, \pi)$ where $\Sigma$ is the input alphabet, $\Gamma$ is the writing alphabet, $\Theta$ is a set of elementary structures and $\pi : \Gamma \to \Sigma$ is a projection. To recognize a string $w \in \Sigma^*$, a local machine $M$, does the following non-deterministic process. It rewrites $w$ as $w' \in \Gamma^*$ such that $\pi(w') = w$. Then, starting from string $w'$ taken as frontier, $M$ "builds" a structure $s$ of type $\mathcal{T}$ following the associated scanning procedure and using elements in $\Theta$ as bricks. If $M$ succeeds in building $s \in \mathcal{L}(\Theta)$, then $w$ is accepted. Informally, the action of building consists in guessing and checking that all elementary substructures belong to $\Theta$ and it can be defined by a finite state device. Notice that a local machine for line graphs corresponds to a finite automaton (see Sect. 2).

*Remark 1.* Thinking with a machine perspective, the local structure $s$ describes, and actually corresponds to, the computation to recognize the input string by a finite-state device. This implies that the size of the local structure is actually a measure of time complexity: to accept a string we need to verify that a local structure does exist. Because of the locality property, this can be done in a space at most double of the input size (trees can be built by levels from the frontier and grids can be built by rows from the bottom).

From Proposition 1, we get a characterization for regular, context-free and context-sensitive languages in terms of finite-state computation devices: the dif-

ferences among the languages is made by the "complexity" of the needed structures. It is interesting to notice that there are two main aspects in such complexity differences. The first one is the topology of the underlying graphs (vertices
degrees $\leq 2$, $\leq 3$, $\leq 4$, respectively). The other one is the size of the structures,
as number of vertices, with respect to the input string (the frontier). More precisely, excluding the border vertices, we have the following. For the line case, the
structure does not have any other vertex besides the frontier itself. For the tree
case, the structure has size at most double of the frontier. Very different is the
case of the grids where we can have any number of rows on top of the frontier.

Then, as side result, Proposition 1 reveals a big gap between context-free
and context-sensitive languages. In the next section we will introduce a kind of
bounded-grids structures that we propose as a *next level* after binary trees.

## 4    Bounded-Grids Computations

In this section we study languages that are projection of frontiers of local grids
whose size is bounded by the length of the frontier itself. We define another kind
of grid structure, we call *bounded-grids*, that differs from the one of previous
section for the frontier that is defined as the last row followed by the last column
reversed (taking the corner only once(. For example, in the structure below, the
frontier corresponds to the string $w = w_1 w_2....w_n$.

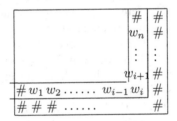

Notice that, if the string $w$ has length $n$, then the size (number of vertices) of
a local grid for $w$, is at most $(n/2)^2$: the exact size depends on the string position
in the bottom-right corner of the grid. For the rest of the section, we will always
assume the frontier defined as above. We give first the following theorem.

**Theorem 2.** *Let $L$ be a string language. If $L$ is a projection of the frontier of
a local picture language, then $L$ is context-sensitive.*

*Proof.* (Sketch) Let $L = \pi(fr(S))$ with $S = \mathcal{L}(\Theta)$ local picture language and $\Theta$
being a finite set of (2×2)-squares, or tiles. We define a linear bounded automaton
(LBA) $\mathcal{A}$ that accepts a string $w$ if and only if $w = \pi(fr(p))$ for some $p \in \mathcal{L}(\Theta)$.

The idea is to let $\mathcal{A}$ behave as a local machine for bounded-grids (see previous
section). While $\mathcal{A}$ scans and rewrites a string $w \in L$, it effectively builds a picture
$p \in \mathcal{L}(\Theta)$. The frontier of $p$ will contain a string $u$ such that $w = \pi(u)$, and all
the other positions are filled by the symbols written by $\mathcal{A}$ during its computation
to accept $w$. We distinguish between *"moving transitions"* that have the effect
of changing current head's state and position and *"rewriting transitions"* that

rewrite the symbol that is being read by the head: rewriting transitions will correspond to elements of $\Theta$.

Let $w = w_1 w_2 \ldots w_n$. In a first phase, $\mathcal{A}$ non-deterministically chooses $i$, $1 \leq i \leq n$, such that $p$ will be of size $(n{-}i{-}1 \times i)$. Then, it rewrites $w$ as $u = u_1 u_2 \ldots u_n$ such that $\pi(u) = w$, satisfying the constraint of $u$ being the frontier of $p \in \mathcal{L}(\Theta)$. That is, starting from $u_i$, $\mathcal{A}$ verifies that bottom- and right-border tiles, for $r, s = 1, \ldots, i{-}1$, $\begin{array}{|c|c|}\hline u_i & \# \\\hline \# & \# \\\hline\end{array}$, $\begin{array}{|c|c|}\hline u_r & u_{r+1} \\\hline \# & \# \\\hline\end{array}$ $\begin{array}{|c|c|}\hline \# & u_1 \\\hline \# & \# \\\hline\end{array}$, $\begin{array}{|c|c|}\hline u_s & \# \\\hline u_{s+1} & \# \\\hline\end{array}$, $\begin{array}{|c|c|}\hline \# & \# \\\hline u_n & \# \\\hline\end{array}$ are in $\Theta$. Thus, at the end of the first phase, the current string is $\#u_1 u_2 \ldots u_n \#$, while we have "built" the last two rows and last two columns of $p$ (including the borders).

Now it comes the second phase: the rewriting transitions will "fill" all the other positions of the picture $p$ according to the set of tiles $\Theta$. More precisely, $\begin{array}{|c|c|}\hline Y & B \\\hline A & X \\\hline\end{array} \in \Theta$ will correspond to the following rewriting transition: " *If A and B are to the left and to the right of X respectively, in the current string, then the LBA $\mathcal{A}$ rewrites X as Y* ". (The information about $A$ and $B$ being to the left and to the right of $X$ respectively, is assumed to be contained in the current state entered after some *moving* transitions). Moreover, to apply the above rewriting transition we need that the three symbols $AXB$ of the current string are already placed in the current picture $p$ in the following relative position: $\begin{array}{|c|c|}\hline & B \\\hline A & X \\\hline\end{array}$ . The above transition has the effect of filling the empty position in $p$ with a $Y$. Then, if $u_i$ is the BottomRight corner of $p$, then $\mathcal{A}$ can do its first rewriting only on $u_i$. After that, $u_{i-1}$ and $u_{i+1}$ can be also rewritten. For example, after applying the transitions corresponding to the tiles $\begin{array}{|c|c|}\hline X & u_{i+1} \\\hline u_{i-1} & u_i \\\hline\end{array}$, $\begin{array}{|c|c|}\hline B & X \\\hline u_{i-2} & u_{i-1} \\\hline\end{array}$, $\begin{array}{|c|c|}\hline A & u_{i+2} \\\hline X & u_{i+1} \\\hline\end{array} \in \Theta$, the constructing picture becomes:

$$\begin{array}{|c|c|c|c|c|c|}\hline \multicolumn{5}{c}{} & \vdots & \# \\\cline{6-7}\multicolumn{4}{c|}{} & & u_{i+3} & \# \\\cline{5-7}\multicolumn{3}{c|}{} & & A & u_{i+2} & \# \\\cline{4-7}\multicolumn{2}{c|}{} & & B & X & u_{i+1} & \# \\\hline \cdots & u_{i-3} & u_{i-2} & u_{i-1} & u_i & \# \\\hline \# & \# & \# & \# & \# & \# \\\hline\end{array}$$

while the current string becomes: $\#u_1 u_2 \ldots u_{i-2}\, A X B\, u_{i+2} \ldots u_n \#$. Next, $\mathcal{A}$ can rewrite symbols $u_{i-2}$ and/or $X$ and/or $u_{i+2}$. That is, the picture will be filled by diagonal waves that start from the BottomRight-corner and go towards the TopLeft-corner, according to tiles in $\Theta$.

The automaton $\mathcal{A}$ accepts $\#w\#$ when it succeeds in building the picture $p$. Last transition to be applied (the accepting transition) corresponds to a TopLeft-tile, i.e. like the following: $\begin{array}{|c|c|}\hline \# & \# \\\hline \# & A \\\hline\end{array} \in \Theta$. After the application of an accepting transition, the current string will correspond to the Left and the Top border of $p$, (i.e. all $\#$ symbols). To complete the definition of the transition function, it remains to define the *moving transitions* that put the head in the correct position for the rewriting. Observe that, once $\mathcal{A}$ fixes the position for the BottomRight corner of $p$, all the moving are the ones to build $p$ by diagonals.

An alternative proof can be given by defining a context sensitive grammar whose rules correspond to the elements of set $\Theta$.

For the sequel, languages that are projections of frontiers of picture languages will be referred to as *bounded-grids context sensitive (Bgrid-CS)* languages.

In 1971 R. Book defined *time-bounded context-sensitive languages* whose derivations (or, equivalently, accepting computations) steps are bounded by some functions $f(n)$ where $n$ is the input string length (see [1]). For the observations in Remark 1, we can affirm that the family of bounded-grids context sensitive languages coincides with the Book's family of context-sensitive languages that admits computations bounded by a quadratic function.

In the proof of the above theorem we defined a particular LBA that simulates a local machine on bounded-grids. The transitions of such LBA are completely defined by set $\Theta$ while a local picture $p \in L(\Theta)$ corresponds to an accepting computation (for the projection of its frontier string). Then a local picture it is a way to describe such computation by means of geometric patterns whose properties can be more "immediate" with respect to a table of a transition function. In particular, using the techniques in [5,6] for recognizable picture languages, it is quite easy to define local sets of pictures whose sides dimensions are in a particular relation (multiples, powers, ...). And this can be used to get properties on the frontiers of the rectangle itself: count string concatenations, mirror strings and so on. We will see explicitly some of them in the next section.

We remark that an alternative way to define bounded-grid structures could be to use squares with last row as frontier. The squareness condition can be imposed by intersecting the picture language with the language of squares on a disjoint alphabet. It can be proved that the projections of frontiers of this kind of structures give rise to the same subfamily of context-sensitive languages.

## 4.1   Examples

In this section we list different examples that show how lots of "typical" context sensitive languages are Bgrid-CS languages.

*Example 1.* Let $\Sigma = \{a, b, c\}$ and let $L = \{a^n b^n c^n \mid n > 0\} \subset \Sigma^*$ be a string language. Then $L = \pi(fr(S))$, where $S$ contains pictures like the following.

$$p = \begin{array}{|cccc|cccc|}
\hline
X & X & X & 1 & Y & Y & Y & 2c \\
X & X & 1 & X & Y & Y & 2 & Yc \\
X & 1 & X & X & Y & 2 & Y & Yc \\
1 & X & X & X & 2 & Y & Y & Yc \\
\hline
a & a & a & a & b & b & b & b \\
\hline
\end{array}$$

The set $\Theta$ for $S$ is given by all different subpictures of size $2 \times 2$ of the bordered version of $p$. Intuitively, to recognize the word $w = a^n b^n c^n$, we "force" $w$ to be in the frontier of a picture $p$ of size $(2n, n+1)$ that can be represented as the concatenation of two squares of side $n$ with one more row on top of them. For simplicity we used a local alphabet that contains also compositions with symbols in $\Sigma$, then $\pi$ just "extract" symbols in $\Sigma$: thus $w = \pi(fr(p))$. Similar techniques applies to string language $L = \{a^n b^n c^n d^n\}$ (take three squares of side $n$).

*Example 2.* Let $L = \{ww \,|\, w \in \Sigma^*\}$. Then $L = \pi(fr(S))$, where $S$ contains pictures like the following $p$ that corresponds to the string $ww$ with $w = a_1 a_2 a_3 a_4 a_5 a_6$.

$$
p = \begin{array}{|cccccc|}
\hline
0 & 0 & 0 & 0 & 0 & a_6 6 \\
0 & 0 & 0 & 0 & 5 & a_5 6 \\
0 & 0 & 0 & 4 & 45 & a_4 6 \\
0 & 0 & 3 & 34 & 35 & a_3 6 \\
0 & 2 & 23 & 24 & 25 & a_2 6 \\
1 & 12 & 13 & 14 & 15 & a_1 6 \\
\hline
a_1 & a_2 & a_3 & a_4 & a_5 & a_6 \\
\hline
\end{array}
$$

Again we need pictures of size $(n+1) \times n$ (that is a square on top of a row). Essentially, we let $w$ "reflect" on the reverse diagonal of the square to carry the information from the bottom to the right border. (Again assume that $\pi$ "extracts" the elements of $\Sigma$ from symbols of $\Gamma$).

Similar technique applies for the string language: $L = \{www \,|\, w \in \Sigma^*\}$.

Further interesting examples are the following languages.

$L = \{a^{n^2} | n > 1\}$ $\qquad\qquad\qquad$ $L = \{a^{2^n} | n > 1\}$
$L = \{a^p \,|\, p \text{ prime }\}$ $\qquad\qquad\quad$ $L = \{a^f \,|\, f \text{ not prime }\}$
$L = \{w\tilde{w} \,|\, w \in \Sigma^*\}$ (palindromes) $\quad$ $L = \{w^{|w|} | \text{ and } |w| \geq 2\}$
$L = \{w \,|\, |w|_a = |w|_b = |w|_c\}$ (same number of $a$'s, $b$'s and $c$'s)

For lack of space we cannot give all the corresponding local picture languages. The techniques for the constructions are always the same (despite more sophisticated): we fold in some way the input string to be the frontier of a rectangle. Then, using a local alphabet, we fill the rectangle by drawing geometric pattern that is a compositions of triangles, square diagonals, square regions that effectively "count" or "reflect" symbols. Using an appropriate number of different symbols for the local alphabet, one can make all such geometric drawings, i.e. the pictures, to be local. Regarding pictures for words of prime length, recall that any possible factor of a number $n$ is odd and $\leq \sqrt{n}$ and therefore if we sum up all such possible factors , we get a number $\leq \frac{n}{2}$. Then to test that the length $n$ of the input word is a prime number we consider a square of side $\frac{n}{2}$ and there we have "space to draw" sequences of triangles to test all possible factors of $n$.

## 4.2 Closure Properties

In this section we state some closure properties of the family of Bgrid-CS languages under the usual operations on strings and sets. Unfortunately, for lack of space, we cannot give the proofs that are rather technical. The techniques used have the same flavor as the ones in [5] to prove closure properties for recognizable picture languages (projections of local picture languages). Nevertheless, to apply them here, it is necessary to somehow "manipulate" the pictures.

**Theorem 3.** *The family of Bgrid-CS languages is closed under concatenation and Kleene's star.*

**Theorem 4.** *The family of Bgrid-CS languages is closed under Boolean union and intersection.*

To give a rough idea of the proof techniques, let $L_1$ and $L_2$ be two Bgrid-CS languages and let $S_1$ and $S_2$ be the corresponding local picture languages. Suppose, we want to show that the concatenation $L = L_1 \cdot L_2$ is also a Bgrid-CS language. We construct a local picture language $S_1'$ such that for each $p \in S_1$ there is $p' \in S_1'$ with same number of rows as $p$ such that the last non-border row of $p'$ is equal to $fr(p)$. Then we can define a local picture language $S$ corresponding to $L$, by defining a particular picture concatenation between $S_1'$ and $S_2$ (to define all possible concatenations of string pairs we "adapt" the number of rows of the corresponding pictures by introducing an extra symbol).

## 5    Final Discussions and Open Problems

We defined the family of bounded-grids context sensitive languages as a more natural "next step" after context-free ones in the hierarchy of languages characterized by projections of frontiers of local sets of structures. Observe that, starting from the examples listed in Sect. 4.1 together with many others one can construct explicitly, and using the closure properties given in Sect. 4.2, we can show that quite a large subfamily of context-sensitive languages are projections of frontiers of bounded-grids local languages (bounded-grids context sensitive languages). This leads to the following open problem.

*Problem 1.* Does the family of Bgrid-CS languages coincide with CS languages?

Interestingly, the same open problem was already formulated by R. Book in [1] by asking whether an LBA with a quadratic time bounding function can recognize all the context-sensitive languages. Recall that, by a result due to Gladkij, there are context-sensitive languages that have no linear bounded derivations ([7]).

A big open problem for the context-sensitive languages concerns the existence of a deterministic machine for them (recall that CS languages are closed under complement). We did not have here much space to discuss the determinism of local sets of structures. We just mention that determinism involve somehow a direction in the computation therefore we need to refer to the local machine defined in Sect. 3. In the case of bounded-grids it is convenient to choose the direction of counter-diagonals from the bottom-right to the top-left corners as in the proof of Theorem 2. Then, one can define a set $\Theta$ of elementary structures to be *deterministic* when, for each triple $x_1, x_2, x_3 \in \Gamma \cup \{\#\}$ there is at most one element $\begin{array}{|c|c|} \hline y & x_3 \\ \hline x_1 & x_2 \\ \hline \end{array}$ in $\Theta$. The definition becomes more clear if one thinks to line-structures and deterministic finite automata. It could be very interesting to investigate whether there exists some determinization process (may be like the subset construction for finite automata) to be applied to this bounded-grids local sets. A positive answer would let Bgrid-CS to be closed also under complement. We remark that, give a positive answer to both open problems 1 and 2 is equivalent to prove that DSPACE(n)=NSPACE(n). We leave open the following more specific problem.

*Problem 2.* Is the family of Bgrid-CS languages closed under complement?

Finally we turn back to our characterization for Chomsky's hierarchy and we consider also languages generated by unrestricted grammars.

*Problem 3.* Is there a kind of structure such that every recursive language can be represented as projection of the frontier of a local set of such structures?

# References

1. R.V. Book. Time-bounded grammars and their languages. *Journal of Computer System Science.* Vol 5, pp. 397–429, 1971.
2. G. Buntrok and F. Otto Growing Context-Sensitive Languages and Church-Rosser Languages . *Information and Computation.* Vol 141, pp. 1–36, 1998.
3. S. Eilenberg. *Automata, Languages and Machines.* Vol. A, Academic Press, 1974.
4. M.J. Fischer Two characterizations of the context-sensitive languages. Prc. 10th Annual IEEE Symp. on Switching and Automata Theory, pp. 157–165, 1969.
5. D. Giammarresi and A. Restivo. Two-dimensional finite state recognizability. *Fundamenta Informaticae.* vol. 25, no. 3,4 , pp. 399–422, 1996.
6. D. Giammarresi, A. Restivo. "Two-dimensional languages". In *Handbook of Formal Languages* , G. Rosenberg *et al.,* Eds, Vol. III, pp. 215 – 268. Springer Verlag, 1997.
7. A.W. Gladkij. "On the complexity of derivations in context-sesitive grammars". In *Algebri i Logika Sem.*, vol. 3, pp. 29–44, 1964.
8. J.E. Hopcroft, and J.D. Ullman. *Introduction to Automata Theory, Languages and Computation.* Addison-Wesley, Reading, MA, 1979.
9. M. Latteux and D. Simplot. Context-sensitive string languages and recognizable picture languages. *Information and Computation,* Vol. 138, 2, pp. 160–169, 1997.
10. M. Latteux, D. Simplot and A. Terlutte. Iterated length-preserving rational trasductions. Proc. *MFCS'98*, LNCS Vol. 1450, pp. 286–295. Springer-Verlag, 1998.
11. R. McNaughton. An insertion into the Chomsky Hierarchy? *Jewels are Forever, Contributions on Theoretical Computer Science in Honor of Arto Salomaa,* J. Karhumaki et al., Eds), pp. 204–212. Springer-Verlag, 1999.
12. J. Mezei and J.B. Wright. Algebraic automata and context-free sets. *Information and Computation,* Vol. 11, pp. 3–29, 1967.

# About Duval's Conjecture

T. Harju and D. Nowotka

Turku Centre for Computer Science (TUCS)
Department of Mathematics, University of Turku, Finland

**Abstract.** A word is called unbordered if it has no proper prefix which is also a suffix of that word. Let $\mu(w)$ denote the length of the longest unbordered factor of a word $w$. Let a word where the longest unbordered prefix equal to $\mu(w)$ be called Duval extension. A Duval extension is called trivial, if its longest unbordered factor is of the length of the period of that Duval extension. In 1982 it was shown by Duval that every Duval extension $w$ longer than $3\mu(w) - 4$ is trivial. We improve that bound to $5\mu(w)/2 - 1$ in this paper, and with that, move closer to the bound $2\mu(w)$ conjectured by Duval. Our proof also contains a natural application of the Critical Factorization Theorem.

## 1 Introduction

The periodicity and the borderedness of words are two basic subjects of interest in the study (e.g., Chapter 8 in [10]) and application (e.g., [8]) of combinatorics on words. We investigate the relation between periodicity and the length of borders in this paper. In particular, we improve a result by Duval [4] which relates the length of the longest unbordered factor of a word to the period of that word.

Let us consider an arbitrary finite word $w$ of length $n$. The period of $w$, denoted by $\partial(w)$, is the smallest positive integer $p$ such that the $i$th letter equals the $(i+p)$th letter for all $1 \leq i \leq n-p$. Let $\mu(w)$ denote the length of the longest unbordered factor of $w$, where a word $v$ is bordered, if it has a proper prefix $u$, which is neither empty nor $v$ itself, such that $u$ is also a suffix of $v$. Assume $w$ has an unbordered prefix $u$ of maximum length $\mu(w)$, then $w$ is called *Duval extension* of $u$, and $w$ is called trivial Duval extension if $u$ is of length $\partial(w)$.

In 1979 Ehrenfeucht and Silberger started a line of research [5,1,4] exploring the relation between the length of the longest unbordered factor, $\mu(w)$, and the period, $\partial(w)$. In 1982 these efforts culminated in Duval's result: If $n \geq 4\mu(w) - 6$ then $\partial(w) = \mu(w)$. However, it is believed that $\partial(w) = \mu(w)$ holds for $n \geq 3\mu(w)$ which follows if Duval's conjecture [4] holds true.

*Conjecture 1 (Duval's conjecture).* If $n \geq 2\mu(w)$ and $w$ is a Duval extension, then $\partial(w) = \mu(w)$.

After that, no progress was recorded, to the best of our knowledge, for 20 years. However, the topic remained popular; see for example Problem 8.2.13 on page 308 in Chapter 8 of [10]. Only recently the theme was independently picked up again by Mignosi and Zamboni [11] and us [7]. However, these papers investigate not

Z. Ésik and Z. Fülöp (Eds.): DLT 2003, LNCS 2710, pp. 316–324, 2003.

Duval's conjecture but rather its opposite, that is: which words admit only trivial Duval extensions? It is shown in [11] that unbordered Sturmian words allow only trivial Duval extensions, in other words, if an unbordered Sturmian word of length $\mu(w)$ is a prefix of $w$, then $\partial(w) = \mu(w)$. That result was improved in [7] by showing that Lyndon words allow only trivial Duval extensions and the fact that every unbordered Sturmian word is a Lyndon word.

We show in this paper that, if $w$ is a Duval extension and $n \geq 5\mu(w)/2$ then $\partial(w) = \mu(w)$. This is a first improvement of Duval's result which is given implicitly in [4], namely that, if $w$ is a Duval extension and $n \geq 3\mu(w) - 3$ then $\partial(w) = \mu(w)$. Duval's conjecture, which is actually believed to hold even in the slightly stronger version where $n \geq 2\mu(w) - 1$ is required, may lead to an ultimate bound for the length of words that have $\partial(w) \neq \mu(w)$. Note, that the bound $n \geq 2\mu(w) - 1$ is sharp by the following example. Let $w = a^i b a^j b b a^j b a^i$ with $1 \leq i < j$. Then $\mu(w) = i + j + 3$ and $\partial(w) = \mu(w) + (j - i) = 2j + 3$ and $|w| = 2\mu(w) - 2 = 2(i + j + 2)$.

Section 4 presents the proof of our main result, Theorem 2, which uses the notations from Sect. 2 and preliminary results from Sect. 3. We conclude with Sect. 5.

## 2   Notations

In this section we define the notations of this paper. We refer to [9,10] for more basic and general definitions.

We consider a finite alphabet $A$ of letters. Let $A^*$ denote the monoid of all finite words over $A$ including the empty word, denoted by $\varepsilon$. A nonempty word $u$ is called a *border* of a word $w$, if $w = uv = v'u$ for some suitable words $v$ and $v'$. We call $w$ *bordered* if it has a border that is shorter than $w$, otherwise $w$ is called *unbordered*. Note, that every bordered word $w$ has a minimum border $u$ such that $w = uvu$, where clearly $u$ is unbordered. Let $w = w_{(1)}w_{(2)} \cdots w_{(n)}$ where $w_{(i)}$ is a letter, for every $1 \leq i \leq n$. Then we denote the length $n$ of $w$ by $|w|$. An integer $1 \leq p \leq n$ is a *period* of $w$, if $w_{(i)} = w_{(i+p)}$ for all $1 \leq i \leq n - p$. The smallest period of $w$ is called the *minimum period* of $w$. Let $w = uv$. Then $u$ is called a *prefix* of $w$, denoted by $u \leq w$, and $v$ is called a *suffix* of $w$, denoted by $v \preccurlyeq w$. Let $u, v \neq \varepsilon$. Then we say that $u$ *overlaps* $v$ *from the left* or *from the right*, if there is a word $w$ such that $|w| < |uv|$, and $u \leq w$ and $v \preccurlyeq w$, or $v \leq w$ and $u \preccurlyeq w$, respectively. We say that $u$ *overlaps* $v$, if either $v$ is a factor of $u$, or $u$ overlaps $v$ from the left or right.

Let $w$ be a nonempty word of length $n$. We call $wu$ a *Duval extension* of $w$, if every factor of $wu$ longer than $n$ is bordered. A Duval extension $wu$ of $w$ is called *trivial*, if there exists a positive integer $j$ such that $u \leq w^j$, that is, the minimum period of $wu$ is $n$.

*Example 1.* Let $w = abaabb$ and $u = aaba$. Then

$$wu = abaabbaaba$$

is a nontrivial Duval extension of $w$ where $\partial(wu) = 7$ and $\mu(wu) = |w| = 6$. Actually, $wu$ is the longest possible nontrivial Duval extension of $w$. The word

$$w = aababb$$

is not a prefix of any nontrivial Duval extension $wu$ where $\mu(wu) = |w| = 6$.

Let an integer $p$ with $1 \leq p < |w|$ be called *position* or *point* in $w$. Intuitively, a position $p$ denotes the place between $w_{(p)}$ and $w_{(p+1)}$ in $w$. A word $u \neq \varepsilon$ is called a *repetition word* at position $p$ if $w = xy$ with $|x| = p$ and there exist $x'$ and $y'$ such that $u \preccurlyeq x'x$ and $u \leq yy'$. For a point $p$ in $w$, let

$$\partial(w, p) = \min\{|u| \mid u \text{ is a repetition word at } p\}$$

denote the *local period* at point $p$ in $w$. Note, that the repetition word of length $\partial(w, p)$ at point $p$ is unbordered, and we have $\partial(w, p) \leq \partial(w)$. A factorization $w = uv$, with $u, v \neq \varepsilon$ and $|u| = p$, is called *critical* if $\partial(w, p) = \partial(w)$, and, if this holds, then $p$ is called *critical point*.

# 3    Preliminary Results

We state some auxiliary and well-known results in this section which will be used to prove our main contribution, Theorem 2, in Sect. 4.

**Lemma 1.** *Let $zf = gzh$ where $f, g \neq \varepsilon$. Let $az'$ be the maximum unbordered prefix of $az$. If $az$ does not occur in $zf$, then $agz'$ is unbordered.*

*Proof.* Assume $agz'$ is bordered, and let be $y$ its shortest border. In particular, $y$ is unbordered. If $|z'| \geq |y|$ then $y$ is a border of $az'$ which is a contradiction. If $|az'| = |y|$ or $|az| < |y|$ then $az$ occurs in $zf$ which is again a contradiction. If $|az'| < |y| \leq |az|$ then $az'$ is not maximum since $y$ is unbordered; a contradiction. □

**Lemma 2.** *Let $w$ be an unbordered word and $u \leq w$ and $v \preccurlyeq w$. Then $uw$ and $wv$ are unbordered.*

*Proof.* Obvious. □

The critical factorization theorem (CFT) was discovered by Césari and Vincent [2] and developed into its current form by Duval [3]. We refer to [6] for a short proof of the CFT.

**Theorem 1 (CFT).** *Every word $w$, with $|w| \geq 2$, has at least one critical factorization $w = uv$, with $u, v \neq \varepsilon$ and $|u| < \partial(w)$, i.e., $\partial(w, |u|) = \partial(w)$.*

**Lemma 3.** *Let $w = uv$ be unbordered and $|u|$ be a critical position of $w$. Then $u$ and $v$ do not overlap.*

*Proof.* Note, that $\partial(w, |u|) = \partial(w) = |w|$ since $w$ is unbordered. Let $|u| \le |v|$ without restriction of generality. Assume that $u$ and $v$ overlap. If $u = u's$ and $v = sv'$, then $\partial(w, |u|) \le |s| < |w|$. If $u = su'$ and $v = v's$, then $w$ is bordered with $s$. If $v = sut$ then $\partial(w, |u|) \le |su| < |w|$.     $\square$

**Lemma 4.** *Let $uv$ be unbordered and $|u|$ be a critical position of $uv$. Then either $vxu$ is unbordered or has a minimum border $g$ such that $|g| \ge |uv|$, for any word $x$.*

*Proof.* Follows directly from Lemma 3.     $\square$

The following Lemmas 5, 6 and 7 and Corollary 1 are given in [4]. Let $a_0, a_1 \in A$, with $a_0 \ne a_1$, and $t_0 \in A^*$. Let the sequences $(a_i)$, $(s_i)$, $(s'_i)$, $(s''_i)$, and $(t_i)$, for $i \ge 1$, be defined by

- $a_i = a_{i \pmod 2}$, that is, $a_i = a_0$ or $a_i = a_1$ if $i$ is even or odd, respectively,
- $s_i$ such that $a_i s_i$ is the shortest border of $a_i t_{i-1}$,
- $s'_i$ such that $a_{i+1} s'_i$ is the longest unbordered prefix of $a_{i+1} s_i$,
- $s''_i$ such that $s'_i s''_i = s_i$,
- $t_i$ such that $t_i s''_i = t_{i-1}$.

For any parameters of the above definition, we have the following.

**Lemma 5.** *For any $a_0$, $a_1$, and $t_0$ there exists an $m \ge 0$ such that*

$$|s_1| < |s_2| < \cdots < |s_m| = |t_{m-1}| \le \cdots \le |t_1| \le |t_0|$$

*and $s_m = t_{m-1}$.*

**Lemma 6.** *Let $z \le t_0$ such that $a_0 z$ and $a_1 z$ do not occur in $t_0$. Let $a_0 z_0$ and $a_1 z_1$ be the longest unbordered prefixes of $a_0 z$ and $a_1 z$, respectively. Then*

1. *if $m = 1$ then $a_0 t_0$ is unbordered,*
2. *if $m > 1$ is odd, then $a_1 s_m$ is unbordered and $|t_0| \le |s_m| + |z_0|$,*
3. *if $m > 1$ is even, then $a_0 s_m$ is unbordered and $|t_0| \le |s_m| + |z_1|$.*

**Lemma 7.** *Let $v$ be an unbordered factor of $w$ of length $\mu(w)$. If $v$ occurs twice in $w$, then $\mu(w) = \partial(w)$.*

**Corollary 1.** *Let $wu$ be a Duval extension of $w$. If $w$ occurs twice in $wu$, then $wu$ is a trivial Duval extension.*

## 4    An Improved Bound for Duval Extensions

We present the main result of this paper. Note that the proof of Theorem 2 rests to a great extend on the application of the CFT by applying Lemma 4.

**Theorem 2.** *Let $wu$ be a nontrivial Duval extension of $w$. Then $|u| < 3|w|/2$.*

*Proof.* Recall that every factor of $wu$ which is longer than $|w|$ is bordered since $wu$ is a Duval extension.

Let $z$ be the longest suffix of $w$ that occurs twice in $zu$. If $z = \varepsilon$ then $a \preccurlyeq w$ and $u = b^j$, where $a, b \in A$ and $a \neq b$ and $j \geq 1$, but now $|u| < |w|$ since $ab^j$ is unbordered. So, $z \neq \varepsilon$. We have that $z \neq w$ since $wu$ is otherwise trivial by Corollary 1. Let $a, b \in A$ such that

$$w = w'az \qquad \text{and} \qquad u = u'bzr$$

and $z$ occurs in $zr$ only once, that is, $bz$ matches the rightmost occurrence of $z$ in $u$. Note, that $bz$ does not overlap $az$ from the right, and therefore $u'$ exists, by Lemma 2. Naturally, $a \neq b$ by the maximality of $z$, and $w' \neq \varepsilon$, otherwise $azu'bz \leq wu$ has either no border or $w$ is bordered or $az$ occurs in $zu$; a contradiction in any case. Let $a_0 = a$ and $a_1 = b$ and $t_0 = zr$, and let $z_0$ and $z_1$ and the sequences $(a_i)$, $(s_i)$, $(s_i')$, $(s_i'')$, $(t_i)$, and the integer $m$ be defined as in Lemma 6. Let

$$t_0 = s_m t' \ .$$

Consider $azu'bz_0$. We have that $az$ and $azu'bz_0$ are both prefixes of $a_0 zu$, and $bz_0$ is a suffix of $azu'bz_0$ and $az$ does not occur in $zu'bz_0$. From Lemma 1 it follows that $azu'bz_0$ is unbordered, and hence,

$$|azu'bz_0| \leq |w| \tag{1}$$

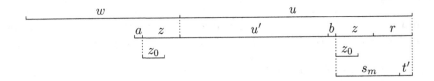

**Case:** Suppose that $m$ is even. Then $as_m \ (= a_m s_m)$ is unbordered and $|t_0| \leq |s_m| + |z_1|$ by Lemma 6. From (1) it follows that $|z_1| \leq |z| \leq |w| - 2$. If $|s_m| \leq |z_0|$, then $|azu| \leq |w| + |z_1|$, and hence, $|u| \leq |w|$, since we have

$$|azu| = |azu'bz_0| - |z_0| + |t_0| \leq |azu'bz_0| - |z_0| + |s_m| + |z_1| \ .$$

Suppose then that $|s_m| > |z_0|$. We have that $as_m$ is unbordered, and since $az_0$ is the longest unbordered prefix of $az$, we have $az \leq as_m$, and hence, $|z| \leq |s_m|$. Now, the word $azu'as_m$ is unbordered or otherwise its shortest border is longer than $az$, since no prefix of $az$ is a suffix of $as_m$, and $az$ occurs in $u$; a contradiction. So, $|azu'as_m| \leq |w|$ and $|u| < |w|$, since $|z_1| \leq |z|$.

**Case:** Suppose that $m$ is odd. Then $bs_m \ (= a_m s_m)$ is unbordered and $|t_0| \leq |s_m| + |z_0|$ (see Lemma 6). If $s_m = \varepsilon$. Then $|t_0| \leq |z_0|$ and $t_0 = z_0$, since

$z_0 \leq t_0$, and hence, $|azu| \leq |w|$, by (1). So, assume $s_m \neq \varepsilon$. If $|s_m| < |z|$, then $|u| < |w|$ since

$$|u| = |azu'bz_0| - |bz_0| + |bt_0| - |az|$$

and

$$|azu'bz_0| \leq |w| \quad \text{and} \quad |t_0| \leq |s_m| + |z_0| \;.$$

So, assume $|s_m| \geq |z|$. Since $|bs_m| \geq 2$ there exists a critical point $p$ in $bs_m$ such that $bs_m = v_0 v_1$, where $|v_0| = p$, by the CFT.

Consider $wu'bz$ which is bordered and must have a shortest border longer than $z$, otherwise $w$ is bordered since we have $z \preccurlyeq w$. So, $bz$ occurs in $w$. Note, that $|az_0| \leq \partial(az)$ and that $bz$ occurs left from $az$ in $w$. If $az$ and $bz$ do not overlap in $w$ then $|az_0| \leq |az| \leq |w|/2$. Also, if $bz$ overlaps $az$ from the left, then $|az_0| \leq \partial(az) \leq |w|/2$. It follows that

$$\text{if} \quad |u| < |w| + |z_0| \quad \text{then} \quad |u| < 3|w|/2 \;. \tag{2}$$

Let

$$w = w_0 bz w_1$$

where $bz$ occurs only once in $w_0 bz$, that is, we consider the leftmost occurrence of $bz$ in $w$. Consider the factor

$$f = bz w_1 u' b s_m \;. \tag{3}$$

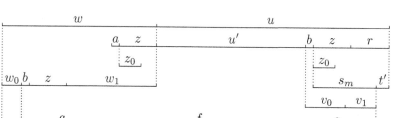

If $f$ is unbordered then $|f| \leq |w|$, and hence, $|u| \leq |w| - 2$. Assume $f$ is bordered, and let $g$ be its shortest border. If $g \leq bz$ then $bs_m$ is bordered which is a contradiction. If $|bz| < |g| \leq |s_m|$ then $bz$ occurs in $zr$; a contradiction as well. Hence, $|bs_m| \leq |g|$. Note, that also $|g| < |bzw_1|$ since $az$ does not occur in $u$. So, we have

$$w = w_0' b s_m w_2 = w_0' v_0 v_1 w_2$$

where $w_0 \leq w_0'$ and $w_2 \neq \varepsilon$. Consider

$$f_0 = v_1 w_2 u' v_0 \;.$$

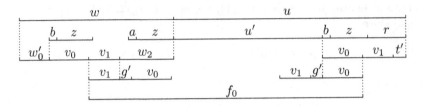

If $f_0$ is unbordered, then $|f_0| \leq |w|$ and

$$|u| = |v_1 w_2 u' v_0| - |v_0| + |bt_0| - |v_1 w_2| \leq |w| + |z_0| - |w_2|$$

since $|bt_0| \leq |bs_m| + |z_0| = |v_0 v_1| + |z_0|$, and we have $|u| < 3|w|/2$ by (2). Assume that $f_0$ is bordered. Then its shortest border $g_0$ is longer than $|v_0 v_1|$ by Lemma 4. Let

$$g_0 = v_1 g' v_0 .$$

If $|g_0| > |v_1 w_2|$. Then $|v_1 w_2| < |az|$ otherwise $az$ occurs in $u$; a contradiction. But, now $v_1$ occurs in $z$, and hence, overlaps with $v_0$ since $bz \leq v_0 v_1$ and $w_2 \neq \varepsilon$. This contradicts Lemma 3.

Assume that $|g_0| \leq |v_1 w_2|$, and let $w_2'$ be such that $v_0 w_2' \preccurlyeq w_2$. Consider

$$f_1 = v_0 w_2' u' v_0 v_1 .$$

If $f_1$ is unbordered, then $|f_1| \leq |w|$ and $|u| < |f_1| + |t'|$. We have $|u| < |w| + |z_0|$, and hence, $|u| < 3|w|/2$ by (2). Assume now that $g_1$ is the shortest border of $f_1$. We have $|g_1| \geq |v_0 v_1|$ by Lemma 4, and therefore $v_0 v_1 \preccurlyeq g_1$. If $g_1 \leq v_0 w_2'$ then $v_0 v_1$ has two nonoverlapping occurrences in $w$ and is therefore at most half as long as $w$. We have then

$$|u| \leq |a z u' b z_0| - |az| - |b z_0| + |v_0 v_1| + |t'|$$

and from $|a z u' b z_0| \leq |w|$ and $|v_0 v_1| \leq |w|/2$ and $|t'| \leq |z_0|$ and (2) it follows that $|u| \leq 3|w|/2 - 3$. So, assume $v_0 w_2' \leq g_1$. We have $|az| > |v_0 w_2'|$ since $az$ does not occur in $u$. Certainly, $v_0 v_1$ occurs in $u'$. Indeed, it does not overlap with $az$ from the right by Lemma 2 since $bz \leq v_0 v_1$. Let

$$x v_0 v_1 \leq u' .$$

Consider

$$f' = b z w_1 x v_0 v_1 .$$

If $f'$ is unbordered than we have two nonoverlapping occurrences of $v_0 v_1$ in a factor that is at most as long as $|w|$ and $|u| < 3|w|/2 - 2$. Suppose, $f'$ is bordered, then its shortest border is longer than $v_0 v_1$ since otherwise $g$ is not the shortest border of $f$; see (3). In fact, the shortest border of $f'$ is $g$ since otherwise we have again two nonoverlapping occurrences of $v_0 v_1$ in a factor that is at most as long as $|w|$ and $|u| < 3|w|/2 - 2$. Moreover, $|azx| > |gv_0v_1|$ otherwise $az$ occurs in $u$ (in the border of $f$). Actually, $|x| \geq |gv_0v_1|$ since $bz \leq g$ and $bz$ does not overlap $w$ from the right by Lemma 2. Let now $x'bz \leq xbz$ where $bz$ occurs only once in $x'bz$. Then the shortest border of $wx'bz$ is at least as long as $|w_0bz|$ and at most as long as $|x'bz|$. So, $|w_0'| \leq |x|$. We have

$$w = w_0'v_0v_1g'v_0w_2' \tag{4}$$

and

$$u = xv_0v_1u''v_0w_2'xbt_0 . \tag{5}$$

Moreover, $v_1g' \preccurlyeq xv_0v_1u''v_0w_2'x$ by the border $g_0$ of $f_0$. Note, that

$$|g'| \leq |u''v_0w_2'x| \tag{6}$$

otherwise $|g'| \geq |v_0v_1u''v_0w_2'x|$ since $v_0$ and $v_1$ do not overlap by Lemma 4, but then $v_0v_1$ has two nonoverlapping occurrences in $w$ and $|v_0v_1| \leq |w|/2 - 2$, where the constant 2 comes from the fact that $v_0$ has at least four nonoverlapping occurrences in $w$, and $|u| \leq 3|w|/2 - 3$. Now,

$$|azxv_0v_1u''v_0w_2'xbz_0| \leq |w|$$

by (1) and (5), and

$$|azxu''xbz_0| \leq |w_0'g'|$$

by (4) and $|w_0'| \leq |x|$, and

$$|azbz_0| \leq |v_0w_2'|$$

by (6), and

$$|az| < |v_0w_2'|$$

and hence, $az$ occurs in $u$; a contradiction.    □

## 5    Conclusions

We have lowered the bound of nontrivial Duval extentions from $3\mu(w) - 4$ in [4] to $5\mu(w)/2 - 1$ which brings us closer to the improved Duval's conjecture of $2\mu(w) - 2$. It should be noted that our result rests to a great part on the critical factorization theorem, which is new application in this context, and might help in finding a new approach to eventually solving the slightly improved version of Duval's conjecture by finding a sharp bound for the length of words which contain no unbordered factor of the length of of their period or longer.

324    T. Harju and D. Nowotka

# References

1. R. Assous and M. Pouzet. Une caractérisation des mots périodiques. *Discrete Math.*, 25(1):1–5, 1979.
2. Y. Césari and M. Vincent. Une caractérisation des mots périodiques. *C. R. Acad. Sci. Paris Sér. A*, 286:1175–1177, 1978.
3. J.-P. Duval. Périodes et répétitions des mots de monoïde libre. *Theoret. Comput. Sci.*, 9(1):17–26, 1979.
4. J.-P. Duval. Relationship between the period of a finite word and the length of its unbordered segments. *Discrete Math.*, 40(1):31–44, 1982.
5. A. Ehrenfeucht and D. M. Silberger. Periodicity and unbordered segments of words. *Discrete Math.*, 26(2):101–109, 1979.
6. T. Harju and D. Nowotka. Density of critical factorizations. *Theor. Inform. Appl.*, 36(3):315–327, 2002.
7. T. Harju and D. Nowotka. Duval's conjecture and Lyndon words. submitted to Discrete Math., 2002.
8. D. E. Knuth, J. H. Morris, and V. R. Pratt. Fast pattern matching in strings. *SIAM J. Comput.*, 6(2):323–350, 1977.
9. M. Lothaire. *Combinatorics on Words*, volume 17 of *Encyclopedia of Mathematics*. Addison-Wesley, Reading, Massachusetts, 1983.
10. M. Lothaire. *Algebraic Combinatorics on Words*, volume 90 of *Encyclopedia of Mathematics and its Applications*. Cambridge University Press, Cambridge, United Kingdom, 2002.
11. F. Mignosi and L. Q. Zamboni. A note on a conjecture of Duval and Sturmian words. *Theor. Inform. Appl.*, 36(1):1–3, 2002.

# Computation with Absolutely No Space Overhead

Lane A. Hemaspaandra[1,*], Proshanto Mukherji[1], and Till Tantau[2,**]

[1] Department of Computer Science
University of Rochester
Rochester, NY 14627, USA
{lane,mukherji}@cs.rochester.edu
[2] Fakultät IV – Elektrotechnik und Informatik
Technische Universität Berlin
Berlin, Germany
tantau@cs.tu-berlin.de

**Abstract.** We study Turing machines that are allowed absolutely no space overhead. The only work space the machines have, beyond the fixed amount of memory implicit in their finite-state control, is that which they can create by cannibalizing the input bits' own space. This model more closely reflects the fixed-sized memory of real computers than does the standard complexity-theoretic model of linear space. Though some context-sensitive languages cannot be accepted by such machines, we show that subclasses of the context-free languages can even be accepted in polynomial time with absolutely no space overhead.

**Keywords:** space overhead, space reuse, overhead-free computation, context-sensitive languages, context-free languages, linear space, deterministic linear languages, metalinear languages

## 1 Introduction

While recursion theory studies which problems can be solved *in principle* on a computer, complexity theory focuses on which problems can be solved *realistically*. Since it depends on context what resources are deemed "realistic," different resource bounds on various models have been studied. For example, deterministic linear space can be seen as a formalization of the limited memory of computers. Unfortunately, the standard complexity-theoretic formalizations may be too "rough" in realistic contexts, as most have hidden constants tucked away inside their definitions. Polynomial-time algorithms with a time bound of $n^{100}$ and linear-space algorithms that need one megabyte of extra memory per input bit will typically be unhelpful from a practical point of view.

* Supported in part by grants NSF-CCR-9322513 and NSF-INT-9815095 / DAAD-315-PPP-gü-ab.
** Work done in part while visiting the University of Rochester, supported by a TU Berlin Erwin-Stephan-Prize grant.

Z. Ésik and Z. Fülöp (Eds.): DLT 2003, LNCS 2710, pp. 325–336, 2003.
© Springer-Verlag Berlin Heidelberg 2003

In this paper we study a model that we believe more realistically captures the fixed-sized memory of computers. We use (non)deterministic one-tape Turing machines that may both read and write their tape. Crucially, we require that the machine may write only on tape cells that are nonempty at the beginning of the computation, when the tape is initialized with the input string (with an unwritable left endmarker ⊢ to its immediate left and an unwritable right endmarker ⊣ to its immediate right). The head may neither move left of the left endmarker nor right of the right endmarker. All (and only) words over the input alphabet, which is typically {0, 1}, are allowed as input strings. Also crucially, we require that the machine may write only symbols drawn from the *input* alphabet.

Thus, in our model of overhead-free computation the input initially completely fills the machine's memory and no auxiliary space is available. The machine can create work space by "cannibalizing" the space occupied by its input. However, the price of doing so is that the overwritten parts will potentially be lost (unless stored elsewhere via overwriting other parts of the input or unless stored in the machine's finite-state control). Note that the machine is not allowed to cheat by using an enriched tape alphabet. Allowing such cheating would transform our model into one accepting exactly the (non)deterministic linear-space languages.

Although deterministic overhead-free computation is a natural model, its nondeterministic counterpart might appear to be of only theoretical interest. After all, nondeterministic computations are hardly "realistic" even if they are overhead-free. However, nondeterministic computations are useful in understanding the inherent limitations of overhead-free computation. An example of such a limitation is the fact that some context-sensitive languages cannot be accepted by overhead-free machines—*not even by nondeterministic ones*.

The class of languages accepted by deterministic overhead-free machines will be denoted DOF, and its nondeterministic counterpart will be denoted NOF. Although these classes "realistically" limit the *space* resources, the underlying machines can still potentially run exponentially long before they decide whether to accept. We will also study which languages can be accepted *efficiently* by overhead-free machines, that is, in polynomial time. Let $\text{DOF}_{\text{poly}}$ denote the class of those languages in DOF that are accepted by overhead-free machines running in polynomial time, and let $\text{NOF}_{\text{poly}}$ denote the class of languages in NOF that are accepted by overhead-free machines running in polynomial time.

Previous work on machines with limited alphabet size mostly concerned the *limitations* of such machines. For example, machines somewhat similar to overhead-free machines have been studied in a note by Feldman and Owings [5], namely linear bounded automata with bounded alphabet size. The work of Feldman and Owings shows that DOF is a proper subset of DCSL, the class of all deterministic context-sensitive languages. The work of Seiferas [21] shows that NOF is a proper subset of CSL, the class of all context-sensitive languages.

In this paper we are mostly concerned with the *power* of overhead-free machines. We show that all deterministic linear languages [18] are contained even in the most restrictive of our four classes, namely $\text{DOF}_{\text{poly}}$. An even larger class of context-free languages, the metalinear languages [2,20], is contained in NOF$_{poly}$. In both cases we give an explicit algorithm. As additional indicators of the power of overhead-free computation, we point out that DOF$_{poly}$ contains non-context-free sets and that DOF even contains PSPACE-complete sets.

To anyone who might think, "your model, which allows no extra space, is unnatural compared with the standard, natural notion of linear space," we would reply that actually the standard models of the field such as linear space are somewhat unnatural in a way that jumps off the page to students, but that professors have grown so used to that we rarely think about the unnaturalness (though we know well why we tolerate it). In the standard models, given an input the machine magically gets larger and larger amounts of extra space, and students often point out that they have yet to see machines whose amount of extra memory grows based on the input. The model is unnatural—yet we study things within it for all the standard, reasonable reasons that we learned long ago and know, love, and teach. Our point is not that standard models or classes are bad, but rather that ours is a natural model.

In addition, we point out that our model has strong roots in the literature. There is a large body of work, dating back decades and active through the present day, on *in situ* or "in-place" algorithms. Though that work is, loosely speaking, interested in computing functions (transformations) with almost no overhead rather than accepting languages with almost no overhead, it suggests that allowing essentially no additional space is a natural notion. As just as few examples of the literature on in-place algorithms, we mention [3,4,6,17,7,1].

This paper is organized as follows. In Sect. 2, we review basic concepts and define the classes DOF, NOF, DOF$_{poly}$, and NOF$_{poly}$. In Sect. 3, we demonstrate the power of overhead-free computation by giving explicit algorithms for accepting the above-mentioned subclasses of the context-free languages with absolutely no overhead. In Sect. 4, we study the class of languages that can be accepted in an almost-overhead-free way, namely via using just one extra tape symbol. In Sect. 5, we discuss the limitations of overhead-free computation.

Due to space limitations, this conference version omits various proofs. We refer the reader to the technical report version [10], and the in-preparation expanded, revised version of that, for omitted proofs and discussion.

## 2   Definitions and Review of Basic Concepts

In this section we first review basic concepts that will be needed in later sections. We then define the four models of overhead-freeness studied in this paper.

Given two alphabets $\Sigma$ and $\Gamma$, a *substitution* is a mapping $s\colon \Sigma \to \mathcal{P}(\Gamma^*)$ that assigns a language $s(a)$ to every symbol $a \in \Sigma$. A substitution is extended to words by $s(a_1 \cdots a_n) := \{w_1 \cdots w_n \mid w_i \in s(a_i)\}$ and to languages by $s(L) := \bigcup_{w \in L} s(w)$. A *homomorphism* is a mapping $h\colon \Sigma \to \Gamma^*$. A homomorphism is *isometric* if all words in the range of $h$ have the same length. A homomorphism is extended to words by $h(a_1 \cdots a_n) := h(a_1) \cdots h(a_n)$ and to languages by $h(L) := \{h(w) \mid w \in L\}$. Let $h^{-1}(L) := \{w \mid h(w) \in L\}$ denote the *inverse image* of $L$.

Let DLINSPACE denote $\bigcup_{k>0}$ DSPACE$[kn]$ and let NLINSPACE denote $\bigcup_{k>0}$ NSPACE$[kn]$. DCFL denotes the class of all deterministic context-free languages [13]. CFL denotes the class of all context-free languages. CSL denotes the class of all context-sensitive languages. It is well-known that CSL = NLINSPACE. The class DCSL (the "deterministic context-sensitive languages") is by definition DLINSPACE. It is not hard to see that DLINSPACE and NLINSPACE contain the languages accepted by deterministic, respectively nondeterministic, one-tape Turing machines that write on only the cells occupied by the input—but in the model (not ours) in which machines *are* allowed to write arbitrary symbols of a possibly large tape alphabet.

We next review different classes of *linear* languages. These have been studied extensively in the literature [2] and have applications in probabilistic finite automata theory [19]. They are defined in terms of the following special types of context-free grammars.

**Definition 2.1 ([20]).** *A grammar is* linear *if it is a context-free grammar in which the right-hand sides of all productions contain at most one nonterminal.*

**Definition 2.2 ([20]).** *A context-free grammar $G = (N, T, S, P)$ is said to be a $k$-linear grammar if it has the form of a linear grammar plus one additional rule of the form $S \to S_1 S_2 \cdots S_k$, where none of the $S_i$ may appear on the right-hand side of any other rule and $S$ may not appear in any other rule at all.*

**Definition 2.3 ([18,20]).** *A linear grammar $G = (N, T, S, P)$ is* deterministic linear *if it has the following two properties. First, all right-hand sides containing nonterminals have this nonterminal at the second position. Second, for every $X \in N$ and $a \in T$ there is at most one production with $X$ on the left-hand side whose right-hand side starts with $a$.*

A language is called *linear* (*$k$-linear, deterministic linear*) if it is generated by a grammar that is linear (*$k$-linear, deterministic linear*). It is called *metalinear* if it is $k$-linear for some $k$.

The above standard definition of deterministic linear grammars is the one given by Nasu and Honda in 1969 [18]. It is rather restrictive, which will make our proofs more transparent. The full power of deterministic linear languages can in fact be better appreciated by looking at the far more flexible-seeming definition—which includes a much broader range of grammars—given by Ibarra, Jiang, and Ravikumar in 1988 [14]. Crucially, Holzer and Lange [11] proved that the latter definition in fact yields exactly the same class of languages as the Nasu–Honda definition, except for the pathological case of the language containing exactly the empty string. (Since that pathological case is in fact in DOF$_{\text{poly}}$, all results of this paper hold under either definition.)

We now define four classes modeling overhead-free computation. To do this rigorously we must tackle one technical issue: The notion of overhead-freeness is sensible only if languages "carry around" their underlying alphabet. Normally, the difference between, say, the language $A = \{1^p \mid p \text{ is prime}\}$ taken over the alphabet $\{1\}$ and the same language $A$ taken over the alphabet $\{0,1\}$ is irrelevant in complexity theory, since we can enlarge the tape alphabet appropriately. In contrast, for overhead-freeness it makes a difference whether the input alphabet is unary or binary, since a unary input alphabet rids us of the possibility of interestingly writing *anything* onto the tape, see Theorem 5.1. Thus, from a formal point of view we consider our complexity classes to contain tuples $(L, \Sigma)$ consisting of a language $L$ and an alphabet $\Sigma$ such that $L \subseteq \Sigma^*$.

For the following definition, recall that we called a machine *overhead-free* if it writes on only those cells that were initially filled with the input and if it writes only symbols drawn from the input alphabet.

**Definition 2.4.** *A pair $(L, \Sigma)$ is in the class* DOF *if $L$ is accepted by a deterministic overhead-free machine with input alphabet $\Sigma$. A pair $(L, \Sigma)$ is in* $\text{DOF}_{\text{poly}}$ *if $L$ is accepted by a deterministic overhead-free, polynomial-time machine with input alphabet $\Sigma$. The counterparts to* DOF *and* $\text{DOF}_{\text{poly}}$ *defined in terms of nondeterministic machines are denoted* NOF *and* $\text{NOF}_{\text{poly}}$.

As differing input alphabets are mainly a technical subtlety, we will in the following speak just of $L$ when the $\Sigma$ is clear from context.

# 3 The Power of Overhead-Free Computation

In this section we demonstrate the power of overhead-free computation. We first give an explicit example of a non-context-free set that nonetheless is in the smallest of our classes, namely $\text{DOF}_{\text{poly}}$. Then we show that DOF contains a PSPACE-complete set. The final part of this section is taken up by a sequence of theorems that establish containment in $\text{DOF}_{\text{poly}}$, or at least $\text{NOF}_{\text{poly}}$, for larger and larger subclasses of CFL.

As an introductory example of overhead-free computation, we mention that the regular languages are clearly (via machines that move their heads steadily to the right and never write at all) even in $\text{DOF}_{\text{linear}}$.

**Theorem 3.1.** *All regular languages are in* $\text{DOF}_{\text{poly}}$.

As a second introductory example, the language $A := \{0^n 1^n 0^n \mid n \geq 1\}$ is in $\text{DOF}_{\text{poly}}$. Since this set is not context-free, $\text{DOF}_{\text{poly}}$ contains non-context-free sets.

**Theorem 3.2.** *There is a set in* $\text{DOF}_{\text{poly}}$ *that is not context-free.*

Our next aim is to show that DOF contains a PSPACE-complete set. To prove this, we first establish that given any language $L \in \text{DLINSPACE}$ we can find a closely related language $L' \subseteq \{0, 1\}^*$ that is accepted by an overhead-free machine.

**Lemma 3.3.** *Let $L \in \text{DLINSPACE}$ with $L \subseteq \Sigma^*$. Then there exists an isometric homomorphism $h \colon \Sigma \to \{0, 1\}^*$ such that for $L' := h(L) \subseteq \{0, 1\}^*$ we have $L' \in$ DOF. Similarly, if $L \in \text{NLINSPACE}$ there exists an isometric homomorphism $h$ such that $L' := h(L) \in$ NOF.*

Since DLINSPACE and NLINSPACE are clearly closed under inverse isometric homomorphisms and since their closure under $\leq_{\mathrm{m}}^{\log}$-reductions is PSPACE, the following corollaries hold.

**Corollary 3.4.** *The closure of* DOF *under inverse isometric homomorphisms is exactly* DLINSPACE. *Likewise, the closure of* NOF *under inverse isometric homomorphisms is exactly* NLINSPACE.

**Corollary 3.5.** *The class* DOF *contains a* $\leq_{\mathrm{m}}^{\log}$-*complete set for* PSPACE.

The fact that powerful sets can reduce to DOF does not say that those sets are in DOF themselves. In fact, since DOF is obviously a subset of DLINSPACE, by the Space Hierarchy Theorem some PSPACE languages are not in DOF. So, what pre-existing classes *are* are computable via overhead-free computation?

We next prove containment in $\mathrm{DOF_{poly}}$ or $\mathrm{NOF_{poly}}$ of ever larger subclasses of CFL. It was noted earlier that all regular sets are clearly in $\mathrm{DOF_{poly}}$. Our first aim now is to prove that all deterministic linear languages (see Sect. 2 for a detailed definition) are in $\mathrm{DOF_{poly}}$. We state a useful lemma on the effect of a *constant* amount of additional space at the beginning or end of the input on overhead-free computation. Roughly put, it has no effect.

**Lemma 3.6.** *Let* $C \in \{\mathrm{DOF}, \mathrm{NOF}, \mathrm{DOF_{poly}}, \mathrm{NOF_{poly}}\}$, *let* $L$ *be a language over the alphabet* $\Sigma$, *and let* $u, v \in \Sigma^*$. *Then* $\{uzv \mid z \in L\} \in C$ *iff* $L \in C$.

*Proof.* Let $L \in C$ via $M$. To show $\{uzv \mid z \in L\} \in C$, on input $w$ we first check whether $w = uzv$ for some word $z \in \Sigma^*$. If so, we simulate $M$ on input $z$. During this simulation we pretend to see a simulated left endmarker whenever we are actually $|u|$ many symbols from the real left end, and to see a simulated right endmarker whenever we are $|v|$ many symbols from the real right end. We do so by, between each simulated step, moving up to $|u|$ extra squares left and up to $|v|$ extra squares right to detect whether we are this close to the real ends.

For the only-if part, let $\{uzv \mid z \in L\} \in C$ via machine $M$. Then $L \in C$ via the machine $M'$ that, essentially, simulates $M$ but when over the "$u$" or "$v$" parts (which do not exist in its own input) uses its finite-state control to keep track of its head location, and that globally keeps in its finite-state control the content currently on those hypothetical cells. By "using its finite-state control to keep track of" we refer to the standard fact that given a machine with state set $F$ we can build a machine with state set $F' := F \times \{0,1\}^k$, thus in effect adding a constant number of usable extra "memory" bits to our finite control. $\square$

**Theorem 3.7.** *All deterministic linear languages belong to* $\mathrm{DOF_{poly}}$.

Our next aim is to show that all metalinear languages can be accepted by nondeterministic overhead-free machines. To prove this, we show the following stronger theorem first.

**Theorem 3.8.** *Let* $A \subseteq \Delta^*$ *be a regular language and let* $s \colon \Delta \to \mathcal{P}(\Sigma^*)$ *be a substitution such that for all* $d \in \Delta$ *the language* $s(d)$ *is linear. Then* $s(A) \in \mathrm{NOF_{poly}}$.

*Proof.* By definition we have $z \in s(A)$ iff there exists a word $y = y_1 \cdots y_k \in A$ and words $z_1, \ldots, z_k \in \Sigma^*$ such that $z = z_1 \cdots z_k$ and $z_i \in s(y_i)$ for all $i \in \{1, \ldots, k\}$. Let $D$ be a deterministic finite automaton that accepts $A$. In the following we first show $s(A) \in \mathrm{NOF}$ via some machine $M$. Later on we will add some safe-guards to ensure that $M$ runs in polynomial time. This will prove $s(A) \in \mathrm{NOF}_{\mathrm{poly}}$.

We first give a rough sketch of $M$'s behavior. On input $z$ it starts a main loop. In this main loop the machine guesses a word $y_1 \cdots y_k \in A$, that is, for every word in $A$ there is a nondeterministic path on which this word is guessed. Next, for each symbol $y_i$ the machine guesses a word $z_i \in s(y_i)$ in a subloop and compares this word with an appropriate part of the input. At the end, if the machine has verified that the input $z$ can be decomposed in the form $z = z_1 \cdots z_k$ with $z_i \in s(y_i)$ for all $i \in \{1, \ldots, k\}$, it accepts. What remains to be shown is how the machine avoids producing any overhead.

We may assume that there exist two distinct symbols $0, 1 \in \Sigma$. We show $\{1z \mid z \in L\} \in \mathrm{NOF}_{\mathrm{poly}}$. Let $w = 1z = 1c_1 \cdots c_n$ be given as input. For $d \in \Delta$, let $G_d = (N_d, \Sigma, S_d, P_d)$ denote a linear grammar that generates $s(d)$.

In the main loop the machine nondeterministically guesses a word $y \in A$ by guessing an accepting computation of the automaton $D$. However, this word $y$ is not written down anywhere. Rather, the machine stores just the current symbol $d := y_i$ and the current state of the automaton $D$ in its finite-state control (see the related comments in the proof of Lemma 3.6). When the $i$th symbol is guessed, the machine will already have verified that $z$ starts with $z_1 \cdots z_{i-1}$, with $z_j \in s(y_j)$ for $j \in \{1, \ldots, i-1\}$. At this point the content of the tape will be $0^\ell 1 c_{\ell+1} \cdots c_n$, where $\ell = |z_1 \cdots z_{i-1}|$. In other words, the $z_1 \cdots z_{i-1}$ will have been replaced by 0's followed by a stop marker. We will call this the "land" stop marker as it marks the end of the "land" $0^\ell 1$ before the rough "sea" $c_{\ell+1} \cdots c_n$.

The tricky part is verifying that after the land stop marker comes a $z_i$ for which there exists a derivation $S_d = l_0 X_0 r_0 \Rightarrow_{G_d} l_1 X_1 r_1 \Rightarrow_{G_d} l_2 X_2 r_2 \Rightarrow_{G_d} \cdots \Rightarrow_{G_d} l_t X_t r_t \Rightarrow_{G_d} z_i$.

Starting from the land stop marker we move right until we nondeterministically guess that we have reached the point where the last rule $X_t \to a$ with $a \in T^*$ was employed. We check whether we really find $a$ at that position. If so, we replace $a$ by $10^{|a|-2}1$ (if $|a| < 2$ we defer this replacement until we have gathered enough space in the following steps). The tape's content will now be $0^\ell 1 c_{\ell+1} \cdots c_{\ell+|l_t|} 10^{|a|-2} 1 c_{\ell+|l_t|+|a|+1} \cdots c_n$ and the head will be inside the "island" of 0's in the middle. Next, the machine nondeterministically guesses which rule $X_{t-1} \to y X_t y'$ was employed in the last step of the derivation of $z_i$. It then checks whether it finds $y$ immediately left of the left island stop marker and $y'$ immediately right of the right island stop marker. If so, it pushes the left island stop marker $|y|$ many cells to the left and the right island stop marker $|y'|$ many cells to the right.

Note that the machine will *not notice* if it inadvertently pushes the left island stop marker over the land stop marker—after all, 0 and 1 are perfectly legitimate input symbols. However, the machine will notice such a mistake later on, when it has eaten away a complete derivation of $z_i$. At this point, if all went well, the land stop marker must be directly adjacent to the left island stop marker. The machine must hence check whether the tape left of the island (whose boundary the machine knows) looks like this: 0*1. If this is the case, the land stop marker can be pushed to the right end of the island and the next stage can be entered.

When the machine has guessed a complete word $y \in A$, it can check whether the land stop marker has hit the right end. If so, it accepts.

It remains to show how we can ensure that $M$ runs in polynomial time. There are two places where $M$ might "spend too much time." First, in a derivation of a word $z_i$ rules of the form $X \to Y$ with $X, Y \in N_d$ might be applied over and over again, that is, a nonterminal might repeatedly be replaced by a nonterminal without any terminals being read. Since every row of more than $|N_d|$ many such replacements must necessarily contain the same nonterminal twice, no derivation of a word $z_i$ *needs* such long rows of replacement. Thus, we keep a counter of how often we applied rules of the form $X \to Y$ in a row. We reject if this counter exceeds the cardinality of $N_d$. Since this cardinality is fixed, we can store this counter in our finite control.

The second place where $M$ might loop is a long row of $z_i$'s that are all the empty word. In this case $M$ constantly switches the stored internal state and the output $d$ of the automaton $D$ that accepts $A$, but always guesses that the empty word is derived from $S_d$. Similarly to first case, if we make more internal switches in a row than there are states in $D$, we must visit the same state twice. Thus we do not really *need* more empty derivations in a row than there are states in $D$. We keep a counter of the number of times in a row the empty string was substituted for $z_i$. If this counter exceeds the number of states in $D$, we reject. This counter can also be kept in our finite control.    □

**Corollary 3.9.** *All metalinear languages belong to* $\mathrm{NOF}_{\mathrm{poly}}$.

*Proof.* Every $k$-linear language $L$ can be written as $L = s(A)$, where $A$ contains just one word $a_1 \cdots a_k$ consisting of $k$ different symbols, and $s$ is a substitution that assigns a linear language $L_i$ to each $a_i$.    □

We now present a final example of a class of context-free languages that is contained in $\mathrm{NOF}_{\mathrm{poly}}$. It contains all context-free languages whose underlying alphabet is not fully "used." By this we mean that some symbol in the alphabet is not contained in any word of the language. Note that this class is incomparable to the class of metalinear languages.

**Theorem 3.10.** *Let* $L \subseteq \Sigma^*$ *be a context-free language. If there exists a symbol in $\Sigma$ that is not contained in any word of $L$, then $L$ is in* $\mathrm{NOF}_{\mathrm{poly}}$.

In the next section, where we define and study almost-overhead-free computation, we will revisit this theorem, via looking at the equivalent statement that every context-free language can be accepted by almost-overhead-free computation (see Theorem 4.2).

# 4  Almost-Overhead-Free Computation

In the previous section we saw that many languages can be accepted (at least nondeterministically) in an overhead-free manner. In the next section we will see that some languages in NLINSPACE cannot be accepted in an overhead-free way. One natural question to ask is: What about "middle ground"—for example, computations that are similar to overhead-free computation, but slightly more forgiving regarding overhead? One could look at various middle grounds, but perhaps particularly natural is one very close to overhead-free computation: Consider languages over $k$-letter alphabets that can be accepted by machines having all the restrictions of overhead-free machines except that they are allowed a $(k+1)$-ary richness to their tape alphabets. One might naturally call this "almost overhead-free computing."

Though one must be careful in the proof, it is not too hard to show that the following claim holds.

**Theorem 4.1.** *Let $L$ be a set over a $k$-token alphabet such that for some $\alpha < 1 - \log_{k+1} k$ it holds that $L$ can be accepted by a nondeterministic Turing machine (in the model in which the input is on a separate, one-head, nonwritable tape, and the single, writable worktape has one head and is over the $k$-token alphabet) using space $\alpha n$. Then there is an almost-overhead-free machine (that is, overhead-free in our standard sense except it may use a $(k+1)$-token alphabet) accepting $L$.*

In effect, this says that the logistics and limitations of almost-overhead-free machines do not prevent us from fully exploiting the compression offered by the richer tape alphabet.

Almost-overhead-free machines have the power to accept all context-free languages.

**Theorem 4.2.** *Let $L$ be a context-free language over the alphabet $\Sigma$ and let $\bullet \notin \Sigma$. Then $(L, \Sigma \cup \{\bullet\})$ is in $\mathrm{NOF}_{\mathrm{poly}}$.*

# 5  Limitations of Overhead-Free Computation

The previous two sections demonstrated that several interesting languages can be accepted by (almost-)overhead-free machines. In this section we discuss what cannot be done using overhead-free machines.

We begin this section with an observation that shows that overhead-free machines on unary alphabets are just as powerless (or powerful, depending on your point of view) as finite automata. Theorems 5.2 and 5.3 then show that there are (non)deterministic context-sensitive languague that cannot be accepted by (non)deterministic overhead-free machines. Both results are based on diagonalization techniques. In the rest of the section we discuss whether certain *natural* sets can be accepted in an overhead-free fashion.

**Theorem 5.1.** *Let $L$ be a tally set. Then the following are equivalent: (a) $L$ is regular, (b) $L \in \mathrm{DOF}_{\mathrm{poly}}$, (c) $L \in \mathrm{DOF}$, (d) $L \in \mathrm{NOF}_{\mathrm{poly}}$, and (e) $L \in \mathrm{NOF}$.*

**Theorem 5.2.** DOF $\subsetneq$ DLINSPACE.

*Proof.* This follows immediately from Corollary 2 of a paper by Feldman and Owings [5]. They show that for every constant $m$ deterministic linear-bounded automata with alphabet size at most $m$ cannot accept all deterministic context-sensitive languages.     □

**Theorem 5.3.** NOF $\subsetneq$ NLINSPACE.

*Proof.* Seiferas [21] has shown that for every $m$ there exists a language in NLINSPACE that cannot be accepted by any nondeterministic off-line Turing machine that uses only $m$ different symbols on its tape. Since any overhead-free machine can be simulated by a linear space off-line machine that first copies its input to its work tape, we get the claim.     □

An alternative proof of Theorem 5.3 can be based on combining Corollary 1 of the paper of Feldman and Owings [5] with the Immerman-Szelepcsényi theorem [15,22]. The corollary states that for each $m$ there is a language whose complement is context-sensitive and cannot be accepted by a nondeterministic linear-bounded automaton whose alphabet size is bounded by $m$. This is an example of the often encountered effect that the Immerman-Szelepcsényi technique can be used to simplify nondeterministic space hierarchy proofs (see [15,8]).

Theorems 5.2 and 5.3 show that overhead-free computation is less powerful than linear-space computation *in principle*. A next step would be to prove that certain simple, natural context-sensitive languages cannot be accepted in an overhead-free fashion. Our candidate for a language that is not in NOF is $L_1 := \{ww \mid w \in \{0,1\}^*\}$. Our candidate for a language that is not in DOF is $L_2 := \{uu^{-1}vv^{-1} \mid u, v \in \{0,1\}^*\}$. Since $L_2$ is metalinear, proving $L_2 \notin$ DOF would be especially satisfying as it would also separate nondeterministic and deterministic overhead-free computation.

Interestingly, the languages $L_1$ and $L_2$, though we name them candidate non-NOF (respectively non-DOF) languages, are in "2-head-DOF$_{\text{poly}}$," the analog of DOF$_{\text{poly}}$ in which our overhead-free machine has two heads. For $L_1$ this is very easily seen. For $L_2$ the algorithm works as follows: In a big loop, head 1 iterates over all symbols of the input, while head 2 stays at the left endmarker. The body of the main loop consists of two checks, in which the machine tries to verify whether the subwords before and after head 1 are palindromes. For the first check, head 2 is moved right and head 1 is moved left until head 1 hits the left endmarker. Then head 1 is moved to the right endmarker. For the second check, head 1 is moved left and head 2 is moved right until head 2 hits the right endmarker. Then head 2 is moved back to the left endmarker once more. If at any stage of either check the symbols under the heads differ, the checks will be said to "fail," but they are completed nevertheless. If the subwords before and after head 1 are palindromes, the checks will not "fail" and the machine can accept its input. In any case, the heads will have resumed their previous positions at the end of each iteration of the loop.

These observations raise the question of how powerful extra heads make our model. First, by a classic result of Hartmanis [9], even $\mathcal{O}(1)$-head *finite automata* taken collectively yield the power of logarithmic-space Turing computation. Thus, at least in that different context, additional heads are well-known to be a valuable resource. However, using the same argument as in Theorem 5.3, Seiferas' results [21] can be used to show that for every $m$ there exists a context-sensitive language that is not in $m$-head-NOF.

# 6   Conclusion

In this paper we introduced a computational model, namely overhead-free computation, in which Turing machines are allowed to manipulate their input, but may not use any symbols other than the input alphabet. For the case of a binary input alphabet this models what a computer can compute in its fixed-sized noncontrol memory, if the input initially fills the whole noncontrol memory. Building on this model we defined the four complexity classes DOF, NOF, $DOF_{poly}$, and $NOF_{poly}$ and studied how these classes relate to standard formal-language and complexity classes. The most "realistic" of the four classes is $DOF_{poly}$, which contains the languages that can be accepted efficiently (that is, in polynomial time) with absolutely no space overhead. We showed by means of an explicit overhead-free algorithm that an important class of languages, namely the class of deterministic linear languages, is a subset of $DOF_{poly}$. For the larger class of metalinear languages we proved containment in $NOF_{poly}$.

Our results suggest that $CFL \subseteq NOF$ might hold. Our research gave neither proof nor disproof of this inclusion. We recommend further research to characterize exactly those context-free grammars that generate languages in NOF. The ultimate goal of this line of research would be to prove $CFL \subseteq DOF_{poly}$, which would improve on the well-known $CFL \subseteq P$ result due to Cocke, Younger, and Kasami [16,23] with respect to the space used (though not necessarily time), or to prove the lack of that and other inclusions.

It is known [12, Sect. 11.3] that some context-free languages inherently need logarithmic space on deterministic off-line Turing machines and that some context-free languages inherently need linear space on deterministic on-line Turing machines. However, these results do not imply that these languages are not in NOF, as both in the on-line and off-line models of [12] the input is read-only. In contrast, in our model the input *may* be overwritten, albeit at the cost of losing the overwritten input bits. In fact, the very set used in the logspace-overhead space lower bound of [12] can be shown to belong to $DOF_{poly}$.

**Acknowledgments.** We thank Edith Hemaspaandra, Mitsunori Ogihara, and Jonathan Shaw for helpful discussions. We are very grateful to the anonymous DLT 2003 referees for many helpful comments, most particularly for generously and insightfully suggesting studying almost-overhead-free computation and that it should contain the power of context-free languages.

# References

1. A. Amir, G. Landau, and D. Sokol. Inplace run-length 2D compressed search. *Theoretical Computer Science*, 290(3):1361–1383, 2003.
2. N. Chomsky and M. Schützenberger. The algebraic theory of context-free languages. In P. Braffort and D. Hirschberg, editors, *Computer Programming and Formal Systems*, pages 118–161. North Holland, Amsterdam, 1963.
3. E. Dijkstra. Smoothsort, an alternative for sorting in situ. *Science of Computer Programming*, 1(3):223–233, 1982.
4. E. Dijkstra and A. van Gastern. An introduction to three algorithms for sorting in situ. *Information Processing Letters*, 15(3):129–134, 1982.
5. E. Feldman and J. Owings, Jr. A class of universal linear bounded automata. *Information Sciences*, 6:187–190, 1973.
6. K. Fishkin. Performing in-place affine transformations in constant space. In *Proceedings of Graphics Interface '92*, pages 106–114, 1992.
7. V. Geffert, J. Katajainen, and T. Pasanen. Asymptotically efficient in-place merging. *Theoretical Computer Science*, 237(1–2):159–181, 2000.
8. J. Geske. Nondeterminism, bi-immunity and almost-everywhere complexity. *IEICE Trans. on Communications, Electronics, Information, and Systems*, E76, 1993.
9. J. Hartmanis. On non-determinancy in simple computing devices. *Acta Informatica*, 1:336–344, 1972.
10. L. Hemaspaandra, P. Mukherji, and T. Tantau. Computation with absolutely no space overhead. Technical Report TR-779, Department of Computer Science, University of Rochester, Rochester, NY, May 2002.
11. M. Holzer and K. Lange. On the complexities of linear LL(1) and LR(1) grammars. In *Proc. of the 9th Conference on Fundamentals of Computation Theory*, volume 710 of *Lecture Notes in Computer Science*, pages 299–308. Springer-Verlag, 2003.
12. J. Hopcroft and J. Ullman. *Formal Languages and their Relation to Automata*. Addison-Wesley, 1969.
13. J. Hopcroft and J. Ullman. *Introduction to Automata Theory, Languages, and Computation*. Addison-Wesley, 1979.
14. O. Ibarra, T. Jiang, and B. Ravikumar. Some subclasses of context-free languages in $NC^1$. *Information Processing Letters*, 29(3):111–117, 1988.
15. N. Immerman. Nondeterministic space is closed under complementation. *SIAM Journal on Computing*, 17(5):935–938, 1988.
16. T. Kasami. An efficient recognition and syntax algorithm for context-free languages. Scientific Report AFCRL-65-758, Air Force Cambridge Research Lab., Bedford, Mass., 1965.
17. J. Katajainen and T. Pasanen. In-place sorting with fewer moves. *Information Processing Letters*, 70:31–37, 1999.
18. M. Nasu and N. Honda. Mappings induced by PGSM-mappings and some recursively unsolvable problems of finite probabilistic automata. *Information and Control*, 15(3):250–273, 1969.
19. A. Paz. *Introduction to Probabilistic Automata*. Academic Press, New York, 1971.
20. A. Salomaa. *Formal Languages*. Academic Press, 1973.
21. J. Seiferas. Relating refined space complexity classes. *Journal of Computer and System Sciences*, 14:100–129, 1977.
22. R. Szelepcsényi. The method of forced enumeration for nondeterministic automata. *Acta Informatica*, 26(3):279–284, 1988.
23. D. Younger. Recognition and parsing of context-free languages in time $n^3$. *Information and Control*, 10(2):189–208, 1967.

# Deleting String Rewriting Systems Preserve Regularity

Dieter Hofbauer[1] and Johannes Waldmann[2]

[1] Fachbereich Mathematik/Informatik, Universität Kassel
D-34109 Kassel, Germany
`dieter@theory.informatik.uni-kassel.de`
[2] Fakultät für Mathematik und Informatik, Universität Leipzig
D-04109 Leipzig, Germany
`joe@informatik.uni-leipzig.de`

**Abstract.** A string rewriting system $R$ is called *deleting* if there exists a partial ordering on its alphabet such that each letter in the right hand side of a rule is less than some letter in the corresponding left hand side. We show that the rewrite relation $R^*$ induced by $R$ can be represented as the composition of a finite substitution (into an extended alphabet), a rewrite relation of an inverse context-free system (over the extended alphabet), and a restriction (to the original alphabet). Here, a system is called *inverse context-free* if $|r| \leq 1$ for each rule $\ell \to r$. The decomposition result directly implies that deleting systems preserve regularity, and that inverse deleting systems preserve context-freeness. The latter result was already obtained by Hibbard [Hib74].

## 1 Introduction

In order to analyze the global behaviour of a string rewriting system $R$ on a set of strings $L$, we are interested in the set $R^*(L)$ of descendants of $L$ modulo the rewrite relation induced by $R$. It is particularly convenient if the set of descendants is in a class of languages with nice decidability and closure properties, e.g., if it is regular or context-free. A system $R$ is said to *preserve* regularity (context-freeness) if $R^*(L)$ is regular (context-free) whenever $L$ is.

Consider a few generic examples. A rewriting rule $\ell \to r$ is *context-free* (*inverse context-free*) if $|\ell| \leq 1$ ($|r| \leq 1$ resp.). A rule is *monadic* if it is inverse context-free and length-reducing, i.e, if $|\ell| > |r| \leq 1$, see [BJW82,BO93]. A rewriting system is (inverse) context-free or monadic if all its rules are of the respective form. Context-free systems preserve context-freeness. On the other hand, inverse context-free systems preserve regularity. Another well-known result is the regularity preservation of prefix rewriting [Büc64]. Typically, the latter results are proved via constructions of finite automata.

There has been quite some work on generalizations, e.g., finding classes of systems that are not context-free but still produce only context-free languages. For example, Ginsburg and Greibach [GG66] prove that "terminal grammars" (where each right hand side of a production contains at least one letter that

Z. Ésik and Z. Fülöp (Eds.): DLT 2003, LNCS 2710, pp. 337–348, 2003.

does not occur in any left hand side) preserve context-freeness. Hibbard [Hib74] relaxes this restriction (non-occurence), by introducing an ordering on the alphabet. Systems that respect this ordering are called *context-limited*.

As the example of context-free rewriting systems shows, it is often the case that a system $R$ preserves context-freeness, and at the same time the inverse system $R^-$ preserves regularity. In the present paper, we prove another instance of this phenomenon. We define *deleting* string rewriting systems (i.e., inverse context-limited grammars in case terminal and non-terminal symbols are distinguished), and prove that they preserve regularity.

The main theorem is a decomposition result: The rewrite relation induced by a deleting system can be represented as the composition of a finite substitution (into an extended alphabet), a rewrite relation of an inverse context-free system (over the extended alphabet), and a restriction (back to the original alphabet). This decomposition will be obtained by a sequence of transformations of the original system. Each transformation step may produce a larger system over a larger alphabet, so the crucial point is to find a terminating transformation strategy. As immediate corollaries, we obtain a slightly generalized version of Hibbard's result [Hib74] (inverse deleting systems preserve context-freeness), as well as our dual result (deleting systems preserve regularity).

The paper is organized as follows. We first define deleting systems as rewriting systems respecting a particular termination ordering; this restriction turns out to be strong enough to imply linearly bounded derivational complexity for deleting systems. The main theorem and corollaries are stated and proved in Section 4. Many well-known regularity preservation results are direct consequences, as shown in Section 5. Finally, we discuss some variants and possible extensions.

## 2   Preliminaries

We will mostly stick to standard notations for strings and string rewriting, as in [BO93], for instance. A string rewriting system (SRS) over an alphabet $\Sigma$ is a relation $R \subseteq \Sigma^* \times \Sigma^*$, and the rewrite relation induced by $R$ is $\to_R = \{(x\ell y, xry) \mid x, y \in \Sigma^*, (\ell, r) \in R\}$. Unless indicated otherwise, all rewriting systems are finite. Pairs $(\ell, r)$ from $R$ are frequently referred to as rules $\ell \to r$. By lhs($R$) and rhs($R$) we denote the sets of left (resp. right) hand sides of $R$. The reflexive and transitive closure of $\to_R$ is $\to_R^*$, often abbreviated as $R^*$.

If $\rho$ is a relation on a set $A$ then $\rho(b) = \{a \in A \mid (b, a) \in \rho\}$ for $b \in A$ and $\rho(B) = \bigcup_{b \in B} \rho(b)$ for $B \subseteq A$, so the set of descendants of a language $L \subseteq \Sigma^*$ modulo $R$ is $R^*(L)$. The inverse of $\rho$ is $\rho^{-1} = \{(b, a) \mid (a, b) \in \rho\}$, and often we simply write $\rho^-$ instead of $\rho^{-1}$.

For a relation $\rho \subseteq \Sigma^* \times \Sigma^*$ and a set $\Delta \subseteq \Sigma$, let $\rho|_\Delta$ denote $\rho \cap \Delta^* \times \Delta^*$. Note the difference between $R^*|_\Delta$ and $(R|_\Delta)^*$ for a SRS $R$. For $R = \{a \to b, b \to c\}$ over $\Sigma = \{a, b, c\}$ and $\Delta = \{a, c\}$, e.g., we have $(a, c) \in R^*|_\Delta$, but $(a, c) \notin (R|_\Delta)^*$.

Let $\Delta(x)$ (resp. $\Delta_{\mathrm{mset}}(x)$) denote the set (resp. multiset) of letters from $\Delta \subseteq \Sigma$ occurring in a string $x \in \Sigma^*$, i.e., $\Delta_{(\mathrm{mset})}(a_1 \ldots a_n) = \{a_i \in \Delta \mid 1 \le i \le n\}$. For $L \subseteq \Sigma^*$, let $\Delta(L) = \bigcup_{x \in L} \Delta(x)$. The number of occurrences of a letter $a$

in $x$ is $|x|_a$, i.e., $|\epsilon|_a = 0$, $|ax|_a = 1 + |x|_a$, and $|bx|_a = |x|_a$ for $b \neq a$. Define $|x|_\Delta = \sum_{a \in \Delta} |x|_a$ for $\Delta \subseteq \Sigma$, thus the length of $x$ is $|x| = |x|_\Sigma$.

Given an ordering $>$ on a set $A$, let $>_{set}$ denote the extension of $>$ to finite subsets of $A$, defined by

$$B >_{set} C \quad \text{iff} \quad B \neq C \text{ and } \forall c \in C \; \exists b \in B : b > c.$$

(Note that instead of requiring $B \neq C$ we could equivalently exclude the case $B \neq \emptyset$.) Further, let $>_{mset}$ denote the extension of $>$ to finite multisets over $A$,

$$B >_{mset} C \quad \text{iff} \quad B \neq C \text{ and } \forall c \in C \setminus B \; \exists b \in B \setminus C : b > c,$$

where $\setminus$ denotes multiset difference.

*Remark 1.* If $B_0 \supseteq B_1 >_{set} B_2 \supseteq B_3$, then $B_0 >_{set} B_3$ for subsets $B_i$ of $A$.

*Remark 2.* Let $\max B_i$ denote the set of maximal elements of $B_i \subseteq A$ modulo $>$. Then $B_1 >_{set} B_2$ if and only if $\max B_1 >_{set} \max B_2$.

*Remark 3.* For a total ordering $>$, the above definitions are equivalent to $B >_{set} C$ iff $\exists b \in B \; \forall c \in C : b > c$ and to $B >_{mset} C$ iff $\exists b \in B \setminus C \; \forall c \in C \setminus B : b > c$.

*Remark 4.* We can identify subsets of $A$ with mappings $B : A \to \{0, 1\}$ and multisets over $A$ with mappings $B : A \to \mathbb{N}$, as usual. For a multiset $B$, define the set $set(B)$ by $set(B)(a) = 0$ iff $B(a) = 0$ for $a \in A$. Then $set(B) = B' >_{set} C' = set(C)$ implies $B >_{mset} C$ for multisets $B$ and $C$.

*Remark 5.* The ordering $>_{set}$ is well-founded on finite subsets of $A$ if and only if $>$ is well-founded on $A$ if and only if $>_{mset}$ is well-founded on finite multisets over $A$. This is a consequence of a result from [DM79] and the previous remark.

## 3    Deleting String Rewriting Systems

A *precedence* on an alphabet $\Sigma$ is an irreflexive (partial) ordering $>$ on $\Sigma$. We extend such a precedence to an ordering on $\Sigma^*$ by $x > y$ iff $\Sigma(x) >_{set} \Sigma(y)$ for $x, y \in \Sigma^*$.

**Definition 1.** *A string rewriting system $R$ over an alphabet $\Sigma$ is $>$-deleting for a precedence $>$ on $\Sigma$ if $R \subseteq >$, and the system $R$ is deleting if it is $>$-deleting for some precedence $>$.*

*Remark 6.* The extension from $>$ to $>_{set}$ is monotone in the sense that $> \subseteq >'$ implies $>_{set} \subseteq >'_{set}$ for orderings $>$ and $>'$. As a consequence, if a rewriting system is $>$-deleting then it is $>'$-deleting for any ordering $>' \supseteq >$. In particular, if a system is deleting then it is $>$-deleting for some total ordering $>$.

*Example 1.* The rewriting system $R = \{ba \to cb, bd \to d, cd \to de, d \to \epsilon\}$ is $>$-deleting for the precedence $a > b > d$, $a > c > e$, $c > d$. For instance, we have $R^*(ba^*d) \cap NF(R) = c^*b \cup c^*e^*$, where $NF(R)$ denotes the set of $R$-normal forms.

*Example 2.* A system $R$ over $\Sigma$ is said to be $\Delta$-*deleting* for $\Delta \subseteq \Sigma$ if $|\ell|_\Delta >$ $|r|_\Delta = 0$ for each rule $\ell \to r$ in $R$. This appears as a special case of the above definition: Choose $a > b$ for $a \in \Delta$ and $b \in \Sigma \setminus \Delta$. (Notation: For singleton sets we write $a$-*deleting* instead of $\{a\}$-deleting in the sequel.) For instance, consider a grammar with terminal alphabet $\Delta$ and non-terminal alphabet $\Sigma \setminus \Delta$. This grammar is the inverse of a $\Delta$-deleting rewriting system if each right hand side of a rule contains at least one terminal symbol. Ginsburg and Greibach [GG66] have proved that a grammar of this kind generates always a context-free language.

Deleting string rewriting systems are terminating, and their derivational complexity is linearly bounded, i.e., there is a number $c$ such that every $R$-derivation $x = x_0 \to_R x_1 \to_R \cdots$ with $x_i \in \Sigma^*$ has length at most $c \cdot |x|$. The constant $c$, however, can be exponential in $|\Sigma|$.

**Proposition 1.** *Deleting string rewriting systems have linearly bounded derivational complexity.*

*Proof.* By Remark 4, if $\ell > r$ then $\Sigma_{\mathrm{mset}}(\ell) >_{\mathrm{mset}} \Sigma_{\mathrm{mset}}(r)$, therefore $x_i \to_R x_{i+1}$ implies $\Sigma_{\mathrm{mset}}(x_i) >_{\mathrm{mset}} \Sigma_{\mathrm{mset}}(x_{i+1})$. Note that $|\Sigma_{\mathrm{mset}}(x_{i+1}) \setminus \Sigma_{\mathrm{mset}}(x_i)| \le m$ where $m$ is the maximal length of a right hand side of a rule from $R$. Let $\mathrm{ht}(a)$ denote the height of a letter $a \in \Sigma$ (see [Fra86], e.g.); for finite $\Sigma$ this is the maximal length of a $>$-chain of letters starting with $a$ (so minimal elements have height 0). By a remark in [DM79] (p. 468) we know that the length of any $R$-derivation starting with string $a_1 \ldots a_n$ ($a_i \in \Sigma$) is bounded by $\sum_{i=0}^{n}(m+1)^{\mathrm{ht}(a_i)}$. Therefore, we can choose $c = (m+1)^h$ with $h = \max_{a \in \Sigma} \mathrm{ht}(a)$.    $\square$

## 4   Main Theorem

**Theorem 1.** *Let $R$ be a deleting string rewriting system over $\Sigma$. Then there are an extended alphabet $\Sigma' \supseteq \Sigma$, a finite substitution $S \subseteq \Sigma^* \times \Sigma'^*$, and an inverse context-free string rewriting system $M$ over $\Sigma'$ such that*

$$R^* = (S \circ M^*)|_\Sigma.$$

**Corollary 1.** *Every deleting string rewriting system preserves regularity.*

*Proof.* If $R$ is a deleting SRS, then $R^*(L) = (S \circ M^*)|_\Sigma(L) = M^*(S(L)) \cap \Sigma^*$ for $L \subseteq \Sigma^*$. Now, the class of regular languages is closed under finite substitution, inverse context-free rewriting, and restriction (i.e., intersection with $\Sigma^*$). Note that closure under inverse context-free rewriting follows from closure under monadic rewriting, since in typical proofs, length reduction is not used.    $\square$

**Corollary 2.** *Every inverse deleting string rewriting system preserves context-freeness.*

*Proof.* Similarly, if $R$ is deleting then $R^{-*}(L) = (R^*)^-(L) = ((S \circ M^*)|_\Sigma)^-(L) = S^-(M^{-*}(L))$ for $L \subseteq \Sigma^*$. And the class of context-free languages is closed under context-free rewriting, inverse finite substitution, and restriction.    $\square$

The decomposition in Theorem 1 can be obtained by a sequence of transformations, *splitting* (Lemma 5) and *elimination* (Lemma 2). We first give some technical lemmas that prove the correctness of these transformation steps. Then we describe a transformation strategy and prove it terminating.

## 4.1   Extending the Alphabet

In the following, we will often extend an alphabet $\Delta$ by introducing 'fresh' intermediary letters $a_i \notin \Delta$ that are extensions of existing letters $a \in \Delta$. Some of these fresh letters will again be extended to $a_{ij}, a_{ijk}, \ldots$, so we will now fix a basic alphabet $\Sigma$, and introduce an infinite alphabet $\Sigma' \supset \Sigma$ that contains, once and for all, every fresh letter that will ever be needed. This alphabet is $\Sigma' = \Sigma \times \mathbb{N}^*$, so a letter $x' \in \Sigma'$ has the form $x' = (x, [n_1, \ldots, n_k])$ with $x \in \Sigma$ and $n_j \in \mathbb{N}$. We call $x$ the *base* of $x'$, written base($x'$), and we will identify $x'$ with its base $x$ if $k = 0$ (i.e., if the sequence is empty). For $x'$ as above, let $x'_j$ denote the letter $(x, [n_1, \ldots, n_k, j])$. So whenever we write "let $a \in \Delta$, and let $a_1$ be a fresh letter not in $\Delta$", it is understood that $\Delta$ is a finite subset of $\Sigma'$, and $a_1$ does in fact denote a letter $a_i \in \Sigma'$ with the property that neither $a_i$ nor any of its extension is in $\Delta$ (such an index $i$ exists by finiteness of $\Delta$).

A partial ordering $>$ on $\Sigma$ is extended to a partial ordering on $\Sigma'$ by $(x, \bar{m}) > (y, \bar{n})$ iff $x > y$, or $x = y$ and $\bar{m}$ is a proper prefix of $\bar{n}$. Note that if $<$ (sic) (defined as $>^{-1}$) is well-founded on $\Sigma$ (e.g., if $\Sigma$ is finite), then $<$ is well-founded on $\Sigma'$, i.e., there is no infinite ascending chain of letters from $\Sigma'$. (Observe that $<$ on $\Sigma'$ is the lexicographic combination of $<$ on $\Sigma$ and the well-founded prefix ordering on $\mathbb{N}^*$.) On the other hand, $>$ is not well-founded on $\Sigma'$ since there is the infinite chain $a > a_1 > a_{11} > \cdots$.

## 4.2   Separation

Given a string rewriting system $R$, we want to find systems $R_1$ and $R_2$ such that $R^* = R_1^* \circ R_2^*$ and $R_1$ is of a particularly simple form (a finite substitution). The idea is to replace rules $\{x \to y, y \to z\} \subseteq R$ by $\{y \to z\} \subseteq R_1$ and $\{x \to y, x \to z\} \subseteq R_2$. We need additional restrictions to ensure correctness of this transformation.

For a SRS $R$ over $\Sigma$ and a letter $a \in \Sigma$, define the SRSs

$$R_a = \{\ell \to r \mid (\ell \to r) \in R, \ell = a\},$$
$$\overline{R_a} = \{\ell \to r' \mid (\ell \to r) \in R \setminus R_a, r' \in R_a^*(r)\}.$$

Note that $\overline{R_a}$ might be infinite (if $a$ occurs in $r$). We will require finiteness only later on.

**Lemma 1 (Separation).** *For $R$ and $a$ as above we have $R^* = R_a^* \circ \overline{R_a}^*$.*

*Proof.* See the full version [HW03].                                      □

Now, if the letter $a$ does not occur in any right hand side of $R_a$, then $R_a^*$ is in fact a finite substitution, and $\overline{R_a}$ is a finite string rewriting system.

**Corollary 3.** *If $a \notin \mathrm{rhs}(R_a)$, then $R_a^* = \phi_a$ where $\phi_a$ is the finite substitution that maps letter $a$ to $\{a\} \cup \{r \mid (a \to r) \in R_a\}$, and each letter $b \neq a$ to $\{b\}$.* □

**Corollary 4.** *If $a \notin \mathrm{rhs}(R_a)$, then $\overline{R_a}$ is a finite SRS.* □

We will apply the separation lemma to eliminate auxiliary letters.

**Lemma 2 (Elimination).** *Let $R$ be a SRS over $\Sigma$, and assume that $a \in \Sigma$ never occurs in a left hand side $\ell$ of $R$ except in case $|\ell| = 1$. Then $R^*|_\Delta = (\overline{R_a}|_\Delta)^*|_\Delta$, where $\Delta = \Sigma \setminus \{a\}$.*

*Proof.* By Lemma 1, $R^* = R_a^* \circ \overline{R_a}^*$. Let $(u, w) \in R^*|_\Delta$. There exists $v$ such that $(u, v) \in R_a^*$ and $(v, w) \in \overline{R_a}^*$. Since $a \notin \Sigma(u)$, no rule of $R_a$ can be applied, so we have $u = v$. By construction of $\overline{R_a}$, it follows from the precondition that $a$ does not occur in $\mathrm{lhs}(\overline{R_a})$. That is, if a letter $a$ is ever produced, it cannot be removed by $\overline{R_a}$ rules. But since $a \notin \Sigma(w)$, we cannot apply at all a rule from $\overline{R_a}$ that would produce an $a$. This proves $R^*|_\Delta \subseteq (\overline{R_a}|_\Delta)^*|_\Delta$. For $(\overline{R_a}|_\Delta)^*|_\Delta \subseteq R^*|_\Delta$, observe that $(\overline{R_a}|_\Delta)^*|_\Delta \subseteq \overline{R_a}^*|_\Delta \subseteq R^*|_\Delta$. □

We ensure that it is enough to consider systems where in each rule, left hand side letters are distinct.

**Lemma 3.** *For each deleting SRS $R$ over $\Sigma$, there is a deleting SRS $R'$ over some alphabet $\Sigma' \supseteq \Sigma$ such that $R^* = R'^*|_\Sigma$, and for each rule $(\ell \to r) \in R'$, all letters in $\ell$ are distinct (i.e., there are no $a \in \Sigma$, $\ell_1, \ell_2, \ell_3 \in \Sigma^*$ such that $\ell = \ell_1 a \ell_2 a \ell_3$).*

*Proof.* This is proved by iterating the following construction that removes duplicate letter occurences. Assume that $R$ over $\Sigma$ contains a rule $\ell = \ell_1 a \ell_2 a \ell_3 \to r$ with $a \in \Sigma$ and $\ell_1, \ell_2, \ell_3 \in \Sigma^*$, i.e., the letter $a$ occurs at least twice in $\ell$. We will use a fresh letter $a_1$ (see the remark in Section 4.1) and define $R' = R \setminus \{\ell \to r\} \cup \{a \to a_1, \ell_1 a \ell_2 a_1 \ell_3 \to r\}$. This system is deleting for the extended ordering. We apply Lemma 2 to the system $R'^-$ and the letter $a_1$. Note that $\overline{(R'^-)}_{a_1}|_\Sigma = R^-|_\Sigma$, thus $R'^{-*}|_\Sigma = ((R'^-)_{a_1}|_\Sigma)^*|_\Sigma = (R^-|_\Sigma)^*|_\Sigma = R^{-*}$. By reversing the direction again, we get $R^* = R'^*|_\Sigma$. □

### 4.3    Splitting

Now we describe how to introduce auxiliary letters.

**Lemma 4 (Chaining).** *Let $R$ be a SRS over $\Sigma$, and let $(\ell_1 \ell_2 \to r_1 r_2) \in R$. For a letter $a \notin \Sigma$, the system $R'$ is obtained from $R$ by replacing this one rule by the two rules $\{\ell_1 \to r_1 a, a \ell_2 \to r_2\}$. Then $R^* = R'^*|_\Sigma$.*
    *If $R$ is $>$-deleting and if $>$ can be extended to a precedence on $\Sigma \cup \{a\}$ such that $\ell_1 > a > r_2$, then $R'$ is deleting as well.*

*Proof.* See the full version [HW03]. □

By symmetry, the statement holds true for $\{\ell_2 \rightarrow ar_2, \ell_1 a \rightarrow r_1\}$ as well.

Next we show how to simulate one rule that is not inverse context-free by splitting it into one context-free rule and two inverse context-free rules.

**Lemma 5 (Splitting).** *Let $R$ be a SRS over $\Sigma$, and let $(x_1ax_2 \rightarrow y_1zy_2) \in R$ with $a \in \Sigma$, $x_i, y_i, z \in \Sigma^*$. Let $a_1, a_2$ be two fresh letters not in $\Sigma$. The system $R'$ over $\Sigma'$ is obtained from $R$ by replacing this one rule by the set of rules*

$$\{a \rightarrow a_1za_2, \ x_1a_1 \rightarrow y_1, \ a_2x_2 \rightarrow y_2\}.$$

*Then $R^* = R'^*|_\Sigma$. If $R$ is $>$-deleting with $a > d$ for each $d \in \Sigma(y_1zy_2)$, then $R'$ is deleting as well.*

*Proof.* Note that letters $a_1, a_2$ do not occur elsewhere in $R'$. Apply chaining w.r.t. letter $a_1$ (Lemma 4) to replace $x_1ax_2 \rightarrow y_1zy_2$ by $\{ax_2 \rightarrow a_1zy_2, x_1a_1 \rightarrow y_1\}$, and chaining w.r.t. letter $a_2$ to replace $ax_2 \rightarrow a_1zy_2$ by $\{a \rightarrow a_1za_2, a_2x_2 \rightarrow y_2\}$.

For each $d \in \Sigma(y_1zy_2)$ we have $a > a_i > d$ by construction of the extended letters. This proves that the newly built rules are deleting as well.    □

## 4.4 Transformation

To prove Theorem 1, we may assume by Lemma 3 that for each left hand side $\ell$ of $R$, all letters in $\ell$ are distinct. Further, by Remark 6 we may assume that the precedence $>$ is total on $\Sigma$.

We construct a sequence $(S_i, M_i, N_i)_{i \in \mathbb{N}}$ such that for all $i$, $S_i$ is a finite substitution, $M_i$ is an inverse context-free rewriting system (i.e., right hand sides have length $\leq 1$) and $N_i$ consists of rules that are not inverse context-free (i.e., right hand sides have length $\geq 2$). For all $i$, we will ensure that

$$R^* = \left((S_1 \circ \ldots \circ S_i) \circ (M_i \cup N_i)^*\right)\big|_\Sigma. \tag{1}$$

We start with $S_0 = $ identity, $M_0 = $ the inverse context-free rules of $R$, and $N_0 = R \setminus M_0$. We will arrive at some $i$ with $N_i = \emptyset$. Then, $S = S_1 \circ \cdots \circ S_i$ and $M = M_i$ have the properties that the theorem asserts. For each $i$, we obtain $(S_{i+1}, M_{i+1}, N_{i+1})$ from $(S_i, M_i, N_i)$ with $N_i \neq \emptyset$ by the following algorithm:

- Pick a rule $(\ell \rightarrow r) \in N_i$ and a maximal letter $a$ in $\ell$, that is, a factorization $\ell = \ell_1 a \ell_2$, such that $\max \Sigma(\ell)$ is maximal among $\Sigma(\text{lhs}(N_i))$ modulo $>$. We will show that the choice of $a$ in $\ell$ is unique. (The choice of $\ell \rightarrow r$ may be not, however, but this does not matter.) Call $\ell \rightarrow r$ the *pivot rule*, and $a$ the *pivot letter*.
- Since $\ell \rightarrow r$ is not inverse context-free, its right hand side $r$ has the form $b_1zb_2$ for letters $b_1, b_2$ and a string $z$. By Lemma 5, obtain $N_i'$ from $N_i$ by replacing rule $\ell \rightarrow r$ with the rules $\{a \rightarrow a_1za_2, \ell_1a_1 \rightarrow b_1, a_2\ell_2 \rightarrow b_2\}$.
- By Lemma 1, compute $R_i' = M_i \cup N_i'$. By Corollary 3, $(R_i')_a$ is associated with a substitution, which we choose as $S_{i+1}$. Let $M_{i+1}$ (resp. $N_{i+1}$) be the sets of inverse context-free (resp. not inverse context-free) rules of $(R_i')_a$.

By construction, we have *correctness*, that is, Equation (1) holds for each $i$. We are now going to prove *termination*.

*Claim.* For each $i$, the pivot letter $a$ does not occur in $\mathrm{rhs}(N_i)$.

*Proof.* Assume there is some $(\ell' \to r') \in N_i$ with $a \in \Sigma(r')$. Since $N_i$ is deleting, we have $\Sigma(\ell') >_{\mathrm{set}} \Sigma(r')$. This implies that there is some letter $b \in \Sigma(\ell')$ with $b > a$, contradicting the choice of $a$ as a maximal element among $\Sigma(\mathrm{lhs}(N_i))$.  □

So the substitution $S_{i+1} = (R_i')_a$ is in fact only applied to $\mathrm{rhs}(M_i)$, but never to $\mathrm{rhs}(N_i)$.

*Claim.* For each $i$ and for each $(\ell \to r) \in M_i \cup N_i$, all positions in $\ell$ have pairwise distinct bases. Therefore, the choice of the pivot letter in the pivot rule is unique.

*Proof.* For $i = 0$, this is true by assumption. New left hand sides in $M_{i+1} \cup N_{i+1}$ are of the form $\ell_1 a_1$ resp. $a_2 \ell_2$. The claim holds by induction since $\mathrm{base}(a_i) = \mathrm{base}(a)$.

Since $>$ is total on $\Sigma$ by assumption, its extension to $\Sigma'$ is total on each $\Sigma(\ell)$ by construction, so there is exactly one maximum letter position in each $\ell$.  □

Denote by $\mathrm{base}\, N_i$ the multiset $\{\mathrm{base}\max \Sigma(\ell) \mid (\ell \to r) \in N_i\}$. Recall that the multiset extension of $<$ is $<_{\mathrm{mset}}$. Note that $<_{\mathrm{mset}}$ is *not* the same as $(>_{\mathrm{mset}})^{-1}$.

*Claim.* For each $i$, we have $\mathrm{base}\, N_i <_{\mathrm{mset}} \mathrm{base}\, N_{i+1}$.

*Proof.* Let $(\ell_1 a \ell_2 \to b_1 z b_2) \in N_i$ be the pivot rule in step $i$. This rule is not in $N_{i+1}$, since the rule is split and removed.

Consider a rule $(\ell' \to r') \in N_{i+1} \setminus N_i$. It is a substitution instance of some $(\ell \to r) \in M_i$. In this case, we know $r = a$, since rules in $M_i$ are inverse context-free, and the substitution $S_i$ replaces only $a$ (by $a_1 z a_2$). So we have $r = a$, $r' = a_1 z a_2$, and $\ell = \ell'$ (since we do not substitute on the left). Since $a \in \Sigma(r)$, and $\ell \to r$ is deleting, there must be some letter $d \in \Sigma(\ell)$ with $d > a$. This implies $a = \max \Sigma(\ell_1 a \ell_2) < \max \Sigma(\ell) = \max \Sigma(\ell')$.  □

By construction, $<$ is well-founded on $\Sigma'$ (there is no infinite *ascending* chain $x_0 < x_1 < \cdots$). Therefore, also the multiset extension of $<$ is well-founded. This implies that there is no infinite chain $\mathrm{base}\, N_0 <_{\mathrm{mset}} \mathrm{base}\, N_1 <_{\mathrm{mset}} \cdots$, so the algorithm must stop. When it stops, we have $N_i = \emptyset$, establishing the theorem.

*Example 3.* For the rewriting system from Example 1, the algorithm produces the following sequence.

| $i$ | pivot rule | $S_i$ | $M_i$ | $N_i$ |
|---|---|---|---|---|
| 0 |  | $\emptyset$ | $\{bd \to d, d \to \epsilon\}$ | $\{ba \to cb, cd \to de\}$ |
| 1 | $ba \to cb$ | $\{a \to a_1 a_2\}$ | $bd \to d, d \to \epsilon,$ $ba_1 \to c, a_2 \to b$ | $\{cd \to de\}$ |
| 2 | $cd \to de$ | $\{c \to c_1 c_2\}$ | $bd \to d, d \to \epsilon, ba_1 \to c,$ $a_2 \to b, c_1 \to d, c_2 d \to e$ | $\{ba_1 \to c_1 c_2\}$ |
| 3 | $ba_1 \to c_1 c_2$ | $\{a_1 \to a_{1,1} a_{1,2}\}$ | $bd \to d, d \to \epsilon, ba_1 \to c,$ $a_2 \to b, c_1 \to d, c_2 d \to e,$ $ba_{1,1} \to c_1, a_{1,2} \to c_2$ | $\emptyset$ |

Here, the sequence $(\text{base } N_i)$ is $\{a,c\} <_{\text{mset}} \{c\} <_{\text{mset}} \{a\} <_{\text{mset}} \emptyset$.

An implementation of the above algorithm is accessible via a CGI-interface at `http://theo1.informatik.uni-leipzig.de/~joe/bounded/`

## 5 Applications

Many of the well-known results concerning classes of regularity preserving rewriting systems can be obtained as corollaries. Here, we sketch a few applications of this kind. We show that regularity is preserved under prefix rewriting, under transductions, and under monadic rewriting. The simplicity of our approach is substantiated by the fact that also combinations of these rewriting mechanisms can easily be handled.

In order to define prefix and suffix rewriting systems, consider a version of Post's *canonical systems* [Pos43] (with single premise rules). We need a set $X$ of *variables*, disjoint to the underlying alphabet $\Sigma$. Here, a *substitution* is a mapping $\sigma : X \to \Sigma^*$, extended to $X \cup \Sigma$ by $\sigma(a) = a$ for $a \in \Sigma$, and extended to $(X \cup \Sigma)^*$ as a morphism. A canonical system is a set of rules $C \subseteq (X \cup \Sigma)^* \times (X \cup \Sigma)^*$, inducing a rewrite relation $\to_C$ on $\Sigma^*$ by $x \to_C y$ iff $x = \sigma(\ell)$ and $y = \sigma(r)$ for some rule $\ell \to r$ in $C$ and some substitution $\sigma$.

A *prefix rewriting system* is a canonical system with rules of the form $\ell\xi \to r\xi$ with $\xi \in X$ and $\ell, r \in \Sigma^*$. Symmetrically, a *suffix rewriting system* allows rules of the form $\xi\ell \to \xi r$, and a *mixed prefix/suffix system* allows both forms. A *(standard) rewriting system* has rules of the form $\xi_1\ell\xi_2 \to \xi_1 r\xi_2$ with $\xi_i \in X$ and $\ell, r \in \Sigma^*$; as before, such a rule is written as $\ell \to r$.

**Prefix Rewriting.** For a given prefix rewriting system $P$ we define a (standard) rewriting system $P_\nabla$ over $\Sigma \cup \{\nabla\}$ (assuming $\nabla \notin \Sigma$) by

$$P_\nabla = \{\nabla\ell \to r \mid (\ell\xi \to r\xi) \in P\}.$$

Note that $P_\nabla$ is $\nabla$-deleting. Now, as is easily seen, for any language $L \subseteq \Sigma^*$ we have

$$\nabla^* \cdot P^*(L) = P_\nabla^*(\nabla^* \cdot L),$$

thus $P^*(L) = \pi_\nabla(P_\nabla^*(\nabla^* \cdot L))$, where $\pi_\nabla : (\Sigma \cup \{\nabla\})^* \to \Sigma^*$ is the projection morphism induced by $\pi_\nabla : \nabla \mapsto \epsilon, a \mapsto a$ ($a \neq \nabla$). As a consequence, regularity of $L$ implies regularity of $P^*(L)$ by Corollary 1, a theorem due to Büchi [Büc64].

**Mixed Prefix and Suffix Rewriting.** Let $M$ be a mixed prefix/suffix rewriting system over $\Sigma$. As a direct generalization of the previous construction, define the $\nabla$-deleting rewriting system

$$M_\nabla = \{\nabla\ell \to r \mid (\ell\xi \to r\xi) \in M\} \cup \{\ell\nabla \to r \mid (\xi\ell \to \xi r) \in M\}.$$

Then, for $L \subseteq \Sigma^*$,

$$\nabla^* \cdot M^*(L) \cdot \nabla^* = M_\nabla^*(\nabla^* \cdot L \cdot \nabla^*),$$

thus $M^*(L) = \pi_\nabla(M_\nabla^*(\nabla^* \cdot L \cdot \nabla^*))$. Again, $M^*(L)$ is regular if $L$ is regular. This is a particular case of a result by Büchi and Hosken [BH70] and Kratko [Kra65], cf. [Büc89].

**Transducers.** A *(rational) transducer* (see [Ber79,Yu98], e.g.) with *states* $Q$ (a finite set), *input alphabet* $\Sigma$ and *output alphabet* $\Gamma$ is a string rewriting system $T \subseteq Q \cdot \Sigma^* \times \Gamma^* \cdot Q$. The *transduction* induced by $T$ with *initial states* $I \subseteq Q$ and *final states* $F \subseteq Q$ is the mapping $T[I, F] : \Sigma^* \to 2^{\Gamma^*}$ with $T[I, F](x) = \{y \in \Gamma^* \mid \exists i \in I, f \in F : ix \to_T^* yf\}$. This mapping is extended to languages $L \subseteq \Sigma^*$ by $T[I, F](L) = \bigcup_{x \in L} T[I, F](x)$. Observe that

$$T[I, F](L) = (T^*(I \cdot L) \cap \Gamma^* \cdot Q)/F.$$

We will demonstrate that regularity of $L$ implies regularity of $T[I, F](L)$, again using Corollary 1. For this purpose, define the rewriting system

$$T_\nabla = \{\ell\nabla \to r \mid (\ell \to r) \in T\}$$

over $\Sigma \cup \Gamma \cup Q \cup \{\nabla\}$, again a $\nabla$-deleting system. Then

$$T^*(I \cdot L) \cap \Gamma^* \cdot Q = T_\nabla^*(I \cdot \pi_\nabla^{-1}(L)) \cap \Gamma^* \cdot Q,$$

thus $T[I, F](L)$ is regular if $L$ is regular.

**Monadic Rewriting.** For a monadic system $N$, define a $\triangle$-deleting system

$$N_\triangle = \{h_\triangle(x) \to \epsilon \mid (x \to \epsilon) \in N\} \cup \{h_\triangle(x)a \to b \mid (xa \to b) \in N,\ a, b \in \Sigma\}$$

over $\Sigma \cup \{\triangle\}$ ($\triangle \notin \Sigma$), where $h_\triangle : \Sigma^* \to (\Sigma \cup \{\triangle\})^*$ is the morphism induced by $h_\triangle : a \mapsto a\triangle$. For $L \subseteq \Sigma^*$ we get

$$N^*(L) = \pi_\triangle(N_\triangle^*(h_\triangle(L))),$$

where the projection $\pi_\triangle$ is defined analogously to $\pi_\nabla$. Thus regularity of $L$ implies regularity of $N^*(L)$, see Book, Jantzen and Wrathall [BJW82].

**Mixed Prefix, Suffix, and Monadic rewriting.** Let $R = M \cup N$ be the union of a mixed prefix/suffix system $M$ and a monadic system $N$ over $\Sigma$. In this case, we need two different symbols $\nabla, \triangle \notin \Sigma$. Combining the above constructions, let $R_\diamond = M_\diamond \cup N_\triangle$ with

$$M_\diamond = \{\nabla h_\triangle(\ell) \to h_\triangle(r) \mid (\ell\xi \to r\xi) \in M\} \cup$$
$$\{h_\triangle(\ell)\nabla \to h_\triangle(r) \mid (\xi\ell \to \xi r) \in M\}.$$

Let $\pi_\diamond$ be the projection $\pi_\diamond : \nabla \mapsto \epsilon, \triangle \mapsto \epsilon, a \mapsto a$ ($a \in \Sigma$). Then, for $L \subseteq \Sigma^*$,

$$R^*(L) = \pi_\diamond(R_\diamond^*(\nabla^* \cdot h_\triangle(L) \cdot \nabla^*)).$$

Since $R_\diamond$ is $>$-deleting for the precedence $\nabla > \triangle > a$ ($a \in \Sigma$), once again regularity of $L$ implies regularity of $R^*(L)$.

## 6    Discussion

Here, we discuss some conceivable variants of Definition 1, and show why they have to be abandoned.

First observe that length-preserving rewriting systems (where $|\ell| = |r|$ for each rule $\ell \to r$) do not necessarily preserve regularity. The generic example for this case is the 'bubble sort' system $B = \{ba \to ab\}$. Here we have $w \to_B^* a^{|w|_a} b^{|w|_b}$ for $w \in \{a, b\}^*$, thus $B^*((ab)^*) \cap a^* b^* = \{a^n b^n \mid n \geq 0\}$, a language that is not regular.

Neither do length-reducing rewriting systems (where $|\ell| > |r|$ for each rule $\ell \to r$) preserve regularity: Modify bubble sort by introducing 'food' that is consumed in each bubble step. We have to ensure that food (the letter $c$) can move to any place, and we feed at the right end of the string. For the resulting system $B_1 = \{bac \to ab, acc \to ca, bcc \to cb\}$ we have $B_1^*((ab)^* c^*) \cap a^* b^* = \{a^n b^n \mid n \geq 0\}$. This system is extracted from an example by Otto [Ott98], where it is shown that even additional constraints such as confluence are not sufficient to make the property of preserving regularity decidable.

We have seen in Example 2 that all $\Delta$-deleting systems are deleting. Here, weakening the requirement $|\ell|_\Delta > |r|_\Delta = 0$ into $|\ell|_\Delta > |r|_\Delta$ for rules $\ell \to r$, we can no longer guarantee regularity preservation, as the system $B_1$ reveals (take $\Delta = \{c\}$).

Finally, how about using the multiset extension $>_{\text{mset}}$ of a precedence $>$ instead of the extension $>_{\text{set}}$? A first variant would be to require $\Sigma_{\text{mset}}(\ell) >_{\text{mset}} \Sigma_{\text{mset}}(r)$ for each rule $\ell \to r$, which is again rendered impossible by system $B_1$. As a second variant, we call a system $>$-multiset-deleting if $\Sigma(\ell) >_{\text{mset}} \Sigma(r)$ for each rule $\ell \to r$, where sets of letters are used, but compared as multisets. In order to verify that also this definition doesn't serve our purpose, consider the system $B_2 = \{bac \to ab, ac\bar{c} \to ca, a\bar{c}c \to \bar{c}a, bc\bar{c} \to cb, b\bar{c}c \to \bar{c}b\}$, a variant of $B_1$. This system is not deleting in the sense of Definition 1 (the second and third rule would require $\bar{c} > c > \bar{c}$), but it is $>$-multiset-deleting, no matter which precedence is chosen. Nevertheless, $B_2$ does not preserve regularity since $B_2^*((ab)^* \{c, \bar{c}\}^*) \cap a^* b^* = \{a^n b^n \mid n \geq 0\}$.

## 7    Conclusion

We have shown that deleting systems can be reduced (by two rational transductions) to inverse context-free systems. This implies regularity preservation in one direction (our new result), and context-freeness preservation in the other (Hibbard's result [Hib74]). We have also shown that many well-known classes of regularity preserving string rewriting systems can be reduced (again, by rational transductions) to deleting systems. This indicates that our result is not just a coincidence, but rather deleting systems and regularity preservation seem to be intrinsically related.

Our work contributes to linking rewriting theory, in particular termination orderings, and formal language theory. We aim at extending this approach to include term rewriting as well. Here, the question of regularity preservation, and the related tree automata constructions, constitute an active field of research, see [Tis00] for a survey. In light of the duality mentioned above, it should be worthwhile to make applicable classical results on context-free tree languages.

**Acknowledgements.** This research was supported in part by the National Aeronautics and Space Administration (NASA) while the authors were visiting scientists at the Institute for Computer Applications in Science and Engineering (ICASE), NASA Langley Research Center (LaRC), Hampton, VA, in September 2002. Particular thanks to Alfons Geser for the invitation and hospitality.

# References

[Ber79]   Jean Berstel. *Transductions and Context-Free Languages.* Teubner, Stuttgart, 1979.

[BH70]    J. Richard Büchi and William H. Hosken. Canonical systems which produce periodic sets. *Math. Syst. Theory*, 4:81–90, 1970.

[BJW82]   Ronald V. Book, Matthias Jantzen, and Celia Wrathall. Monadic Thue systems. *Theoret. Comput. Sci.*, 19:231–251, 1982.

[BO93]    Ronald V. Book and Friedrich Otto. *String-Rewriting Systems.* Texts and Monographs in Computer Science. Springer-Verlag, New York, 1993.

[Büc64]   J. Richard Büchi. Regular canonical systems. *Arch. Math. Logik und Grundlagenforschung*, 6:91–111, 1964.

[Büc89]   J. Richard Büchi. *Finite Automata, Their Algebras and Grammars – Towards a Theory of Formal Expressions.* D. Siefkes (Ed.). Springer-Verlag, New York, 1989.

[DM79]    Nachum Dershowitz and Zohar Manna. Proving termination with multiset orderings. *Commun. ACM*, 22(8):465–476, 1979.

[Fra86]   Roland Fraïssé. *Theory of Relations*, volume 118 of *Studies in Logic and the Foundations of Mathematics*. North-Holland, Amsterdam, 1986.

[GG66]    Seymour Ginsburg and Sheila A. Greibach. Mappings which preserve context sensitive languages. *Inform. and Control*, 9(6):563–582, 1966.

[Hib74]   Thomas N. Hibbard. Context-limited grammars. *J. ACM*, 21(3):446–453, 1974.

[HW03]    Dieter Hofbauer and Johannes Waldmann. Deleting string rewriting systems preserve regularity. Mathem. Schriften 07/03, Univ. Kassel, Germany, 2003.

[Kra65]   M. I. Kratko. A class of Post calculi. *Soviet Math. Doklady*, 6(6):1544–1545, 1965.

[Ott98]   Friedrich Otto. Some undecidability results concerning the property of preserving regularity. *Theoret. Comput. Sci.*, 207:43–72, 1998.

[Pos43]   Emil L. Post. Formal reductions of the general combinatorial decision problem. *Amer. J. Math.*, 65:197–215, 1943.

[Tis00]   Sophie Tison. Tree automata and term rewrite systems. In Leo Bachmair (Ed.), *Proc. 11th Int. Conf. Rewriting Techniques and Applications RTA-00*, *Lect. Notes Comp. Sci.* Vol. 1833, pp. 27–30. Springer-Verlag, 2000.

[Yu98]    Sheng Yu. Regular languages. In G. Rozenberg and A. Salomaa (Eds.), *Handbook of Formal Languages*, Vol. 1, pp. 41–110. Springer-Verlag, 1998.

# On Deterministic Finite Automata and Syntactic Monoid Size, Continued

Markus Holzer and Barbara König

Institut für Informatik, Technische Universität München
Boltzmannstraße 3, D-85748 Garching bei München, Germany
{holzer,koenigb}@informatik.tu-muenchen.de

**Abstract.** We continue our investigation on the relationship between regular languages and syntactic monoid size. In this paper we confirm the conjecture on two generator transformation semigroups. We show that for every prime $n \geq 7$ there exist natural numbers $k$ and $\ell$ with $n = k + \ell$ such that the semigroup $U_{k,\ell}$ is maximal w.r.t. its size among all (transformation) semigroups which can be generated with two generators. This significantly tightens the bound on the syntactic monoid size of languages accepted by $n$-state deterministic finite automata with binary input alphabet. As a by-product of our investigations we are able to determine the maximal size among all semigroups generated by two transformations, where one is a permutation with a single cycle and the other is a non-bijective mapping.

## 1 Introduction

Finite automata are used in several applications and implementations in software engineering, programming languages and other practical areas in computer science. They are one of the first and most intensely investigated computational models. Since regular languages have many representations in the world of finite automata it is natural to investigate the succinctness of their different representations. Recently, the size of the syntactic monoid as a natural measure of descriptive complexity for regular languages was proposed in [3] and studied in detail. Recall, that the syntactic monoid of a language $L$ is the smallest monoid recognizing the language under consideration. It is uniquely defined up to isomorphism and is induced by the syntactic congruence $\sim_L$ defined over $\Sigma^*$ by $v_1 \sim_L v_2$ if and only if for every $u, w \in \Sigma^*$ we have $uv_1w \in L \iff uv_2w \in L$. The syntactic monoid of $L$ is the quotient monoid $M(L) = \Sigma^* / \sim_L$.

In particular, the size of transformation monoids of $n$-state (minimal) deterministic finite automata was investigated in [3]. In most cases tight upper bounds on the syntactic monoid size were obtained. It was proven that an $n$-state deterministic finite automaton with singleton input alphabet (input alphabet with at least three letters, respectively) induces a linear ($n^n$, respectively) size syntactic monoid. In the case of two letter input alphabet, a lower bound of $n^n - \binom{n}{\ell}\ell!n^k - \binom{n}{\ell}k^k\ell^\ell$, for some natural numbers $k$ and $\ell$ close to $\frac{n}{2}$, and a trivial

Z. Ésik and Z. Fülöp (Eds.): DLT 2003, LNCS 2710, pp. 349–360, 2003.
© Springer-Verlag Berlin Heidelberg 2003

non-matching upper bound of $n^n - n! + g(n)$, where $g(n)$ denotes Landau's function [5,6,7], which gives the maximal order of all permutations in $S_n$, for the size of the syntactic monoid of a language accepted by an $n$-state deterministic finite automaton was given. This induces a family of deterministic finite automata such that the fraction of the size of the induced syntactic monoid and $n^n$ tends to 1 as $n$ goes to infinity, and is the starting point of our investigations.

In this paper we tighten the bound on the syntactic monoid size on two generators, confirming the conjecture, that for every prime $n \geq 7$ there exist natural numbers $k$ and $\ell$ with $n = k + \ell$ such that the semigroup $U_{k,\ell}$ as introduced in [3] is maximal w.r.t. its size among all (transformation) semigroups which can be generated with two generators. Since $U_{k,\ell}$, for suitable $k$ and $\ell$ is a syntactic monoid, this sharpens the above given bound for syntactic monoids induced by $n$-state deterministic finite automata with binary input alphabet. In order to show that there is no larger subsemigroup of $T_n$ with two generators, we investigate all possible combinations of generators. In principle the following situations for generators appear:

1. Two permutations,
2. a permutation with one cycle and a non-bijective transformation,
3. a permutation with two or more cycles and a non-bijective transformation – the semigroup $U_{k,\ell}$ is of this type, and
4. two non-bijective transformations.

In the forthcoming we will show that for a large enough $n$ the maximal subsemigroup is of type (3) and that whenever $n$ is *prime* the semigroup is isomorphic to some $U_{k,\ell}$. The entire argument relies on a series of lemmata covering the above mentioned cases, where the second case plays a major role. In fact, as a by-product we are able to determine the maximal size among all semigroups generated by two transformations, where one transformation is a permutation with a single cycle and the other is a non-bijective mapping. In order to achieve our goal we use diverse techniques from algebra, analysis, and even computer verified results for a finite number of cases.

The paper is organized as follows. In the next section we introduce the necessary notations. Then in Sect. 3 we start our investigations with the case where one is a permutation with a single cycle and the other is a non-bijective mapping. Next, two permutations and two non-bijective mappings are considered. Section 5 deals with the most complicated case, where the permutation contains two or more cycles, and Sect. 6 is devoted to the main result of this paper, on the size maximality of the semigroup under consideration. Finally, we summarize our results and state some open problems.

## 2    Definitions

We assume the reader to be familiar with the basic notions of formal language theory and semigroup theory, as contained in [4] and [9]. In this paper we are dealing with regular languages and their syntactic monoids. A *semigroup* is a non-empty set $S$ equipped with an associative binary operation, i.e., $(\alpha\beta)\gamma = \alpha(\beta\gamma)$ for all $\alpha, \beta, \gamma \in S$. The semigroup $S$ is called a *monoid* if it contains an identity element *id*. If $E$ is a set, then we denote by $T(E)$ the monoid of functions from $E$ into $E$ together with the composition of functions. We read composition from left to right, i.e., first $\alpha$, then $\beta$. Because of this convention, it is natural to write the argument $i$ of a function to the left: $(i)\alpha\beta = ((i)\alpha)\beta$. The image of a function $\alpha$ in $T(E)$ is defined as $img(\alpha) = \{ (i)\alpha \mid i \in E \}$ and the kernel of $\alpha$ is the equivalence relation $\equiv$, which is induced by $i \equiv j$ if and only if $(i)\alpha = (j)\alpha$. In particular, if $E = \{1, \ldots, n\}$, we simply write $T_n$ for the monoid $T(E)$. The monoid of all permutations over $n$ elements is denoted by $S_n$ and trivially is a sub-monoid of $T_n$.

The semigroup we are interested in is defined below and was introduced in [3] in order to study the relation between $n$-state deterministic finite automata with binary input alphabet and the size of syntactic monoids.

**Definition 1.** *Let $n \geq 2$ such that $n = k + \ell$ for some natural numbers $k$ and $\ell$. Furthermore, let $\alpha = (1\,2\,\ldots\,k)(k+1\,k+2\,\ldots\,n)$ be a permutation of $S_n$ consisting of two cycles. We define the semigroup $U_{k,\ell}$ as a subset of $T_n$ as follows: A transformation $\gamma$ is an element of $U_{k,\ell}$ if and only if*

1. *there exists a natural number $m \in \mathbb{N}$ such that $\gamma = \alpha^m$ or*
2. *the transformation $\gamma$ satisfies that*
   a) *there exist $i \in \{1, \ldots, k\}$ and $j \in \{k+1, \ldots, n\}$ such that $(i)\gamma = (j)\gamma$ and*
   b) *there exists $h \in \{k+1, \ldots, n\}$ such that $h \notin img(\gamma)$.*

Observe, that it is always better to choose the element $h$ which is missing in the image of $\gamma$ from the larger cycle of $\alpha$ since this yields a larger semigroup $U_{k,\ell}$. Therefore we can safely assume that $k \leq \ell$.

In [3] it was shown that if $\gcd\{k, \ell\} = 1$, then the semigroup $U_{k,\ell}$ can be generated by two generators only. Moreover, in this case, $U_{k,\ell}$ is the syntactic monoid of a language accepted by an $n$-state deterministic finite automaton, where $n = k + \ell$.

Finally, we need some additional notation. If $A$ is an arbitrary non-empty subset of a semigroup $S$, then the family of subsemigroups of $S$ containing $A$ is non-empty, since $S$ itself is one such semigroup; hence the intersection of the family is a subsemigroup of $S$ containing $A$. We denote it by $\langle A \rangle$. It is characterized within the set of subsemigroups of $S$ by the properties: (1) $A \subseteq \langle A \rangle$ and (2) if $U$ is a subsemigroup of $S$ containing $A$, then $\langle A \rangle \subseteq U$. The semigroup $\langle A \rangle$ consists of all elements of $S$ that can be expressed as finite products of elements in $A$. If $\langle A \rangle = S$, then we say that $A$ is a set of generators for $S$. If $A = \{\alpha, \beta\}$ we simply write $\langle A \rangle$ as $\langle \alpha, \beta \rangle$.

## 3 Semigroup Size – The Single Cycle Case

In this section we consider the case where one generator is a permutation containing a single cycle and the other is a non-bijective transformation. This situation is of particular interest, since it allows us to completely characterize this case and moreover it is very helpful in the sequel when dealing with two permutations or two non-bijective transformations.

The outline of this section is as follows: First we define a subset of $T_n$ by some easy properties – as in the case of the $U_{k,\ell}$ semigroup, verify that it is a semigroup and that it is generated by two generators. The subset of $T_n$ we are interested in, is defined as follows:

**Definition 2.** Let $n \geq 2$ and $1 \leq d < n$. Furthermore, let $\alpha = (1\,2\,3\,\ldots\,n)$ be a permutation of $S_n$ consisting of one cycle. We define $V_n^d$ as a subset of $T_n$ as follows: A transformation $\gamma$ is an element of $V_n^d$ if and only if

1. there exists a natural number $m \in \mathbb{N}$ such that $\gamma = \alpha^m$ or
2. there exists an $i \in \{1, \ldots, n\}$ such that $(i)\gamma = (i +_n d)\gamma$, where $+_n$ denotes the addition modulo $n$.

The intuition behind choosing this specific semigroup $V_n^d$ is the following: Without loss of generality we can assume that $\alpha = (1\,2\,3\,\ldots\,n)$. By choosing a non-bijective transformation $\beta$ which maps two elements $1 \leq i < j \leq n$ onto the same image one can infer that every transformation $\gamma$ generated by $\alpha$ and $\beta$ is either a multiple of $\alpha$ or maps two elements of distance $d := j - i$ to the same value. Next we show that $V_n^d$ is indeed a semigroup and that $V_n^d$ is isomorphic to $V_n^{d'}$ if $\gcd\{n, d\} = \gcd\{n, d'\}$. Therefore, it will be sufficient to consider only divisors of $n$ in the following. We omit the proof of the following lemma.

**Lemma 1.** The set $V_n^d$ is closed under composition and is therefore a (transformation) semigroup. Moreover, $V_n^d$ is isomorphic to $V_n^{d'}$ whenever $d = \gcd\{n, d'\}$.

Before we can prove that $V_n^d$ can be generated by two elements of $T_n$ we need a result, which constitutes how to find a complete basis for the symmetric group $S_n$. The result given below was shown in [8].

**Theorem 1.** Given a non-identical element $\alpha$ in $S_n$, then there exists $\beta$ such both generate the symmetric group $S_n$, provided that it is not the case that $n = 4$ and $\alpha$ is one of the three permutations $(1\,2)(3\,4)$, $(1\,3)(2\,4)$, and $(1\,4)(2\,3)$.

Now we are ready for the proof that two elements are enough to generate all of the semigroup $V_n^d$. Due to the lack of space we omit the proof of the following theorem, which is heavily based on Theorem 1.

**Theorem 2.** Let $n \geq 2$ and $1 \leq d < n$. The semigroup $V_n^d$ can be generated by two elements of $T_n$, where one element is the permutation $\alpha = (1\,2\,3\,\ldots\,n)$ and the other is an element $\beta$ of kernel size $n - 1$.[1]

In order to determine the size of $V_n^d$, the following theorem, relating size and number of colourings of a particular graph, is very useful in the sequel.

---

[1] Observe, that there is an $n$-state minimal deterministic finite automaton $A$ with binary input alphabet the transition monoid of which equals $V_n^d$. Hence, $V_n^d$ is the syntactic monoid of $L(A)$. Since this statement can be easily seen, we omit its proof.

**Theorem 3.** *Let $n \geq 2$ and $1 \leq d < n$ with $d \mid n$. Denote the undirected graph consisting of $d$ circles, each of length $\frac{n}{d}$, by $G$. Then*

$$|V_n^d| = n + N,$$

*where $N = n^n - \left((n-1)^{\frac{n}{d}} + (-1)^{\frac{n}{d}}(n-1)\right)^d$ is the number of invalid colourings of $G$ with $n$ colours.*

*Proof.* The subsemigroup $V_n^d$ can be obtained from $T_n$ by removing all transformations not satisfying the second part of Definition 2 and by adding the $n$ multiples of $\alpha$ afterwards. The number of the former transformations can be determined as follows: Assume that a graph $G$ has nodes $V = \{1, \dots, n\}$ where a circle $C_k$ consists of nodes $\{k, k+d, \dots, k+id, \dots, k+n-d\}$, for $1 \leq k \leq d$. Then one can easily verify that the colourings of $G$ are exactly the transformations which do not satisfy the second part of Definition 2. The number of colourings of a graph $G$ with $k$ colours is described by its chromatic polynomial, see, e.g. [10]. Since the chromatic polynomial of a circle $C_n$ with $n$ nodes is $(k-1)^n + (-1)^n(k-1)$ and the chromatic polynomial of a graph consisting of disconnected components is the product of the chromatic polynomials of its components, the desired result follows.     □

Now we are ready to prove some asymptotics on the size of $V_n^d$ for some particular values of $d$, which are determined first.

**Theorem 4.** *The size of $V_n^d$ is maximal whenever*

$$d = \max(\{1\} \cup \{d' \mid d' \text{ divides } n \text{ and } \tfrac{n}{d'} \text{ is odd}\}).$$

*Let $V_n$ denote the semigroup $V_n^d$ of maximal size. Then*

$$\lim_{n \to \infty} \frac{|V_n|}{n^n} = 1 - \frac{1}{e},$$

*where $e$ is the base of the natural logarithm.*

*Proof.* The maximality of $V_n^d$ w.r.t. its size is seen as follows. We first define two real-valued functions

$$u_{n,k}^{even}(x) = \left((n-1)^{\frac{k}{x}} + (n-1)\right)^x \quad \text{and} \quad u_{n,k}^{odd}(x) = \left((n-1)^{\frac{k}{x}} - (n-1)\right)^x.$$

The additional index $k$ is present for later use – see Lemma 5. For now we assume that $k = n$.

We have $|V_n^d| = n^n + n - u_{n,n}^{even}(d)$ whenever $\frac{n}{d}$ is even and $|V_n^d| = n^n + n - u_{n,n}^{odd}(d)$ whenever $\frac{n}{d}$ is odd. Obviously $u_{n,k}^{odd} < u_{n,k}^{even}$. First we show that $u_{n,k}^{even}$ is strictly monotone by taking the first derivation of $\ln u_{n,k}^{even}(x)$. We obtain

$$\frac{d}{dx} \ln u_{n,k}^{even}(x) = \ln\left((n-1)^{\frac{k}{x}} + (n-1)\right) + x \frac{(n-1)^{\frac{k}{x}} \ln(n-1)\left(-\frac{k}{x^2}\right)}{(n-1)^{\frac{k}{x}} + (n-1)}$$

$$> \ln\left((n-1)^{\frac{k}{x}}\right) - \frac{k}{x} \frac{(n-1)^{\frac{k}{x}} \ln(n-1)}{(n-1)^{\frac{k}{x}}}$$

$$= \frac{k}{x} \ln(n-1) - \frac{k}{x} \ln(n-1) = 0$$

Analogously one can show that $u_{n,k}^{odd}$ is strictly antitone.

So if there exist divisors $d'$ such that $\frac{n}{d'}$ is odd, the semigroup $V_n^d$ is maximal w.r.t. its size whenever we choose the largest such $d'$. Otherwise there are only divisors $d'$ such that $\frac{n}{d'}$ is even and we choose the smallest of these divisors which is 1.

Next consider the semigroup $V_n = V_n^d$, for some $1 \leq d < n$. From our previous investigations one can infer that the following inequalities hold:

$$n^n + n - (n-1)^n - (n-1) \leq n^n + n - \left((n-1)^{\frac{n}{d}} + (-1)^{\frac{n}{d}}(n-1)\right)^d$$
$$\leq n^n + n - \left((n-1)^3 - (n-1)\right)^{\frac{n}{3}}.$$

The second half of the inequality follows since the size of $V_n^d$ is maximal whenever $\frac{n}{d}$ is odd and $1 \leq d < n$ is maximal. This is achieved ideally whenever $\frac{n}{d} = 3$. The rest follows with the monotonicity and antitonicity of the functions $u_{n,n}^{even}$ and $u_{n,n}^{odd}$, respectively.

We now determine the limits of the lower and upper bounds. There we find that

$$\lim_{n \to \infty} \frac{n^n + n - (n-1)^n - (n-1)}{n^n} = \lim_{n \to \infty} \left(1 + \frac{1}{n^n} - \left(\frac{n-1}{n}\right)^n\right)$$
$$= 1 - \lim_{n \to \infty} \left(\frac{n-1}{n}\right)^{n-1} \cdot \lim_{n \to \infty} \frac{n-1}{n}$$
$$= 1 - \frac{1}{e},$$

since $\lim_{n \to \infty} (1 + \frac{1}{n})^n = e$, and the limit of the upper bound tends also to $1 - \frac{1}{e}$ by similar reasons as above. Hence $\lim_{n \to \infty} \frac{|V_n|}{n^n} = 1 - \frac{1}{e}$.    □

From the asymptotic behaviour of the semigroups $V_n$ and $U_{k,\ell}$ we immediately infer the following theorem.

**Theorem 5.** *There exists a natural number $N$ such that for every $n \geq N$, there exist $k$ and $\ell$ with $n = k + \ell$ such that $|V_n| < |U_{k,\ell}|$.*

*Proof.* The existence of a natural number $N$ satisfying the requirements given above follows from Theorem 4 and a result from [3], which state that

$$\lim_{n \to \infty} \frac{|V_n|}{n^n} = 1 - \frac{1}{e} \quad \text{and} \quad \lim_{n \to \infty} \frac{|U_{k(n),\ell(n)}|}{n^n} = 1,$$

for suitable $k(n)$ and $\ell(n)$.    □

The following lemma shows that whenever we have a permutation consisting of a single cycle and a non-bijective transformation, we obtain at most as many elements as contained in $V_n$.

**Lemma 2 (A cycle and a non-bijective transformation).** *If $\alpha \in S_n$ such that $\alpha$ consists of a single cycle and $\beta \in T_n \backslash S_n$, then $|\langle \alpha, \beta \rangle| \leq |V_n|$.*

*Proof.* Since the permutation $\alpha$ consists of a single cycle, there is a permutation $\pi$ such that $\pi\alpha\pi^{-1} = (1\,2\,3\,\ldots\,n)$. We set $\alpha' = \pi\alpha\pi^{-1}$ and $\beta' = \pi\beta\pi^{-1}$. Because $\pi$ is a bijection, we can infer that $|\langle\alpha,\beta\rangle| = |\langle\alpha',\beta'\rangle|$. There are two elements $i < j$ such that $(i)\beta' = (j)\beta'$. We define $d = j - i$. It can be easily seen that $\alpha'$ and $\beta'$ generate at most the transformations specified in Definition 2. Therefore we conclude that $|\langle\alpha',\beta'\rangle| \le |V_n|$.    $\square$

Observe, that because of Theorem 5, Lemma 2 implies that there exists a natural number $N$ such that for every $n \ge N$ there exist $k$ and $\ell$ with $n = k + \ell$ such that $|\langle\alpha,\beta\rangle| < |U_{k,\ell}|$, for every $\alpha \in S_n$ such that $\alpha$ consists of a single cycle and $\beta \in T_n \backslash S_n$.

## 4    Semigroup Size – Two Permutations or Non-bijective Mappings

In this section we show that two permutations or two non-bijective transformation are inferior in size to an $U_{k,\ell}$ semigroup, for large enough $n = k + \ell$. Here it turns out, that the semigroup $V_n$ is very helpful in both cases. If we take two permutations as generators, then we can at most obtain the symmetric group $S_n$.

**Lemma 3 (Two permutations).** *Let $n \ge 2$. If $\alpha, \beta \in S_n$, then $|\langle\alpha,\beta\rangle| < |V_n|$.*

*Sketch of Proof.* Obviously, for permutations $\alpha$ and $\beta$ we have $|\langle\alpha,\beta\rangle| \le n!$. In order to prove the stated inequality it suffices to show that $n! < |V_n^1|$. The details are left to the reader.    $\square$

Next we consider the case of two non-bijective transformations.

**Lemma 4 (Two non-bijective transformations).** *Let $n \ge 2$. If both $\alpha$ and $\beta$ in $T_n \setminus S_n$, then $|\langle\alpha,\beta\rangle| < |V_n|$.*

*Proof.* Since $\alpha$ and $\beta$ are both non-bijective, there are indices $j_1 < k_1$ and $j_2 < k_2$ such that $(j_1)\alpha = (k_1)\alpha$ and $(j_2)\beta = (k_2)\beta$. In this case we can construct a permutation $\pi$ such that $(i_1)\pi = j_1$, $(i_1 +_n 1)\pi = k_1$ for some index $i_1$ and $(i_2)\pi = j_2$, $(i_2 +_n 1)\pi = k_2$ for some index $i_2$. If $j_1 = j_2$, then it is the case that $i_1 = i_2$, similarly if $j_1 = k_2$, then $i_1 = i_2 +_n 1$, etc. This means that all transformations generated by $\pi\alpha\pi^{-1}$ and $\pi\beta\pi^{-1}$ satisfy the second part of Definition 2 for $d = 1$. According to Definition 2 the set $\langle\pi\alpha\pi^{-1}, \pi\beta\pi^{-1}\rangle$, and therefore also $\langle\alpha,\beta\rangle$ which is isomorphic, have less elements than $V_n^1$, since at least the permutations are missing. Thus, the stated claim follows.    $\square$

## 5    Semigroup Size – Two and More Cycles

Finally we consider the case where one of the generators is a permutation $\alpha$ consisting of two or more cycles and the other is a non-bijective transformation. In this case we distinguish two sub-cases, according to whether the non-bijective transformation $\beta$ merges elements from the same or different cycles of $\alpha$. We start our investigation with the case where there are $i$ and $j$ such that $(i)\beta = (j)\beta$ and both are located within the same cycle of $\alpha$.

**Lemma 5 (An arbitrary permutation and a non-bijective mapping merging elements from the same cycle).** *There exists a natural number $N$ such that for every $n \geq N$ the following holds: Let $\alpha, \beta \in T_n$ be transformations where $\alpha$ is a permutation. Furthermore let $\beta$ be a non-bijective transformation such that $(i)\beta = (j)\beta$ and both $i$ and $j$ are located in the same cycle of $\alpha$. Then there exist $k$ and $\ell$ with $n = k + \ell$ such that $|\langle \alpha, \beta \rangle| < |U_{k,\ell}|$.*

*Proof.* We assume that $i$ and $j$ are located in the same cycle of length $m$ with distance $d$ w.r.t. their location within the cycle. We can assume that $d$ divides $m$, otherwise we can find an isomorphic semigroup where this is the case, following the ideas of the proof of Lemma 1.

With a similar argument as in the proof of Theorem 3 we can deduce that the semigroup generated by $\alpha$ and $\beta$ contains at most some permutations and the invalid colourings of a graph $G$, where $G$ consists of $d$ circles of length $\frac{m}{d}$ and $n - m$ isolated nodes. The number of valid colourings of such a graph equals

$$((n-1)^{\frac{m}{d}} + (-1)^{\frac{m}{d}}(n-1))^d n^{n-m}.$$

Therefore we conclude $|\langle \alpha, \beta \rangle| \leq n^n + n! - \left((n-1)^{\frac{m}{d}} + (-1)^{\frac{m}{d}}(n-1)\right)^d n^{n-m}$. Similar reasoning as in the proof of Theorem 4 shows that

$$n^n + n! - \left((n-1)^{\frac{m}{d}} + (-1)^{\frac{m}{d}}(n-1)\right)^d n^{n-m}$$
$$\leq n^n + n! - n^n \left(\frac{(n-1)(n-2)}{n^2}\right)^{\frac{n}{3}}$$

and

$$\lim_{n \to \infty} \frac{n^n + n! - n^n \left(\frac{(n-1)(n-2)}{n^2}\right)^{\frac{n}{3}}}{n^n} = 1 - \frac{1}{e}.$$

Hence, a similar asymptotic argument as in the proof of Theorem 5 shows that there is a natural number $N$ such for every $n \geq N$ the size of the semigroups on $n$ elements under consideration is strictly less than the size of $U_{k,\ell}$, for suitable $k$ and $\ell$ with $n = k + \ell$. □

Finally, we consider the case where the non-bijective transformation $\beta$ merges elements from different cycles of the permutation $\alpha$. In the remainder of this section we assume $n = k + \ell$ to be a prime number. The reasons for this assumption is that $k$ and $\ell$ are always coprime, which guarantees that $U_{k,\ell}$ can be generated by two generators only.

**Lemma 6 (A permutation with two or more cycles and a non-bijective mapping merging elements from different cycles).** *Let $n$ be a prime number and let $\alpha, \beta \in T_n$ be transformations where $\alpha$ is a permutation consisting of $m \geq 2$ cycles. Furthermore let $\beta$ be a non-bijective transformation such that $(i)\beta = (j)\beta$ and $i$ and $j$ are located in different cycles of $\alpha$. Then there exist $k$ and $\ell$ with $n = k + \ell$ such that $|\langle \alpha, \beta \rangle| \leq |U_{k,\ell}|$.*

*Proof.* We define $U := \langle \alpha, \beta \rangle$ and show that $|U| \le |U'|$, where $U'$ is generated by a two-cycle permutation $\alpha'$ and a non-bijective mapping $\beta'$ that merges elements of different cycles, as described below in detail.

Now assume that the $m$ cycles in $\alpha$ have lengths $k_1, \dots, k_m$, i.e., $n = \sum_{i=1}^{m} k_i$. Furthermore the sets of elements of the $m$ cycles are denotes by $C_1, \dots, C_m$ and $|C_i| = k_i$. Without loss of generality we may assume that $\beta$ merges elements of the first two cycles $C_1$ and $C_2$. We now consider the following two cases according to which element is missing in the image of $\beta$:

1. There is an element $h$ which is not contained in the image of $\beta$ and moreover, $h$ is not located in the first two cycles of $\alpha$. So let us assume that it is located in the third cycle $C_3$. Let $\alpha'$ be a permutation with two cycles, where the elements of the first cycle are $C_1' = C_2 \cup \bigcup_{i=4}^{m} C_i$ and the elements of the second cycle are $C_2' = C_1 \cup C_3$. In the cycles these elements can be arranged in an arbitrary way. We now set $k = k_2 + \sum_{i=4}^{m} k_i$ and $\ell = k_1 + k_3$ . Since $n = k + \ell$ and $n$ is prime, it follows that $\gcd\{k, \ell\} = 1$. Similar to the construction for the $U_{k,\ell}$ one can now find a transformation $\beta'$ such that $\alpha'$ and $\beta'$ generate a semigroup $U'$ isomorphic to $U_{k,\ell}$. That means, the elements of $U'$ are exactly the multiples of $\alpha'$ and all transformations $\gamma$ which satisfy $(i)\gamma = (j)\gamma$, for $i \in C_1'$ and $j \in C_2'$, and where at least one element of $C_2'$ is missing in the image of $\gamma$.

   Now let us compare the sizes of $U$ and $U'$. First consider only the non-bijective transformations of $U'$. This includes at least all non-bijective transformations generated by $\alpha$ and $\beta$, since the first cycle of $\alpha'$ includes $C_2$ and the second cycle of $\alpha'$ includes $C_1$ and $C_3$. So for any non-bijective $\gamma$ generated by $\alpha$ and $\beta$ there are indices $i \in C_1$, $j \in C_2$, $h \in C_3$ such that $(i)\gamma = (j)\gamma$ and $h \notin img(\gamma)$. This implies that $\gamma$ can be generated by $\alpha'$ and $\beta'$ as well. However, $U$ may contain more permutations than $U'$. In the worst case, if $\gcd\{k_i, k_j\} = 1$ for all pairs of cycle lengths with $i \ne j$, then $U$ contains $\prod_{i=1}^{m} k_i$ permutations, whereas $U'$ contains only $k\ell$ permutations, which might be less. We show that this shortcoming is already compensated by the number of transformations with image size $n - 1$.

   The semigroup $U$ contains $k_1 k_2 k_3 (n - 1)!$ mappings with image size $n - 1$. We first choose the two elements which are in the same kernel equivalence class, for which there are $k_1 k_2$ possibilities, then we choose the element of the image that is missing, for which there are $k_3$ possibilities, and finally we distribute the $n-1$ elements of the image onto the kernel equivalence classes. In the same way we can show that there are $k\ell^2 (n-1)!$ transformations with image size $n - 1$ in $U'$. Now define $k' = \sum_{i=4}^{m} k_i$ and observe, that $k'$ might be equal to 0. Then we conclude that

$$
\begin{aligned}
k\ell^2 - k_1 k_2 k_3 &= (k_2 + k')(k_1 + k_3)^2 - k_1 k_2 k_3 \\
&= (k_2 + k')(k_1^2 + 2k_1 k_3 + k_3^2) - k_1 k_2 k_3 \\
&= k_1^2 k_2 + k_1 k_2 k_3 + k_2 k_3^2 + k' k_1^2 + 2k' k_1 k_3 + k' k_3^2 \\
&\ge k_1 + k_2 + k_3 + k' = n.
\end{aligned}
$$

Therefore $U'$ contains at least $n!$ more transformations of image size $n-1$ than $U$. This makes up for the missing permutations, since there are at most $n!$ of them.

2. The missing element $h$ of the image of $\beta$ is located in one of the first two cycles. Then an analogous construction as in (1) shows how to construct suitable $\alpha'$ and $\beta'$ such that $|U| \leq |\langle \alpha', \beta' \rangle|$. Due to the lack of space the details are left to the reader.

This completes our proof and shows that $|\langle \alpha, \beta \rangle| \leq |U_{k,\ell}|$, because in both cases semigroup $U'$ is isomorphic to some $U_{k,\ell}$, for appropriate $k$ and $\ell$.    □

# 6    On the Maximality of $U_{k,\ell}$ Semigroups

Now we are ready to prove the main theorem of this paper, namely that the size maximal semigroup has $|U_{k,\ell}|$ elements, for some $k$ and $\ell$, whenever $n = k + \ell$ is a prime greater or equal than 7. Observe, that the following theorem strengthens Lemma 5.

**Theorem 6.** *Let $n \geq 7$ be a prime number. Then the semigroup $U_{k,\ell}$, for some $k$ and $\ell$ with $n = k + \ell$, is maximal w.r.t. its size among all semigroups which can be generated with two generators.*

*Proof.* Since all other cases have already been treated in the Lemmata 3, 4, and 6, it is left to show that $U_{k,\ell}$ has more elements than the semigroup $V$, where $V$ is generated by $\alpha$ and $\beta$ and latter mapping merges elements located in the same cycle of $\alpha$. Note that $k$ and $\ell$ are trivially coprime whenever $n = k + \ell$ is a prime.

We have shown in Lemma 5 that

$$|V| \leq n^n + n! - n^n \left( \frac{n(n-1)(n-2)}{n^3} \right)^{\frac{n}{3}} = n^n + n! - (n(n-1)(n-2))^{\frac{n}{3}}.$$

Furthermore from [3] it follows that

$$|U_{k,\ell}| \geq n^n - \binom{n}{\ell} \ell! n^k - \binom{n}{\ell} k^k \ell^\ell.$$

We use Stirling's approximation in the version

$$\sqrt{2\pi n} \left( \frac{n}{e} \right)^n < n! < \sqrt{2\pi n} \left( \frac{n}{e} \right)^n e^{\frac{1}{12}}$$

given in [1,11]. In this way we obtain an upper bound for $|V|$ and a lower bound for $|U_{k,\ell}|$, see the proof in [3], as follows:

$$|V| \leq n^n + \sqrt{2\pi n} \left( \frac{n}{e} \right)^n e^{\frac{1}{12}} - (n(n-1)(n-2))^{\frac{n}{3}}$$

and

$$|U_{k,\ell}| \geq n^n - \left( \sqrt{2} \left( \frac{2}{e} \right)^{\frac{n}{2}} e^{\frac{1}{12}} + \sqrt{8} \frac{1}{\sqrt{n}} e^{\frac{1}{12}} \right) n^n.$$

The upper bound for $|V|$ is smaller than the lower bound for $|U_{k,\ell}|$ whenever

$$\sqrt{2} \left( \frac{2}{e} \right)^{\frac{n}{2}} e^{\frac{1}{12}} + \sqrt{8} \frac{1}{\sqrt{n}} e^{\frac{1}{12}} < \underbrace{\left( \frac{(n-1)(n-2)}{n^2} \right)^{\frac{n}{3}}}_{A(n)} - \underbrace{\sqrt{2\pi n} \left( \frac{1}{e} \right)^n e^{\frac{1}{12}}}_{B(n)}.$$

The function $A(n)$ is monotone and converges to $\frac{1}{e} \approx 0.3678794412$ while the function $B(n)$ is antitone and converges to 0. For $n \geq 20$ we have $A(n) > 0.358$ and $B(n) < 10^{-7}$, and therefore $A(n) - B(n) > 0.35 =: c$. We set $c_1 = 0.01$ and $c_2 = 0.34$ and solve the equations

$$\sqrt{2} \left( \frac{2}{e} \right)^{\frac{n}{2}} e^{\frac{1}{12}} < c_1 \quad \text{and} \quad \sqrt{8} \frac{1}{\sqrt{n}} e^{\frac{1}{12}} < c_2.$$

These equations are satisfied if

$$n > 2 \frac{\log \left( c_1 \frac{1}{\sqrt{2}} e^{-\frac{1}{12}} \right)}{\log \frac{2}{e}} \approx 32.81753852 \quad \text{and} \quad n > \left( \frac{\sqrt{8}}{c_2} e^{\frac{1}{12}} \right)^2 \approx 81.75504594,$$

i.e., whenever $n \geq 82$.

The remaining cases for $7 \leq n \leq 81$ have been checked with the help of the Groups, Algorithms and Programming (GAP) system for computational discrete algebra. To this end we have verified that $|V| \leq |U_{k,\ell}|$, for some $k$ and $\ell$, where the upper bound for $|V|$ from Lemma 5 and the exact value of $|U_{k,\ell}|$ was used.[2] It turned out that $|V_n|$ is maximal w.r.t. size for all $V$ semigroups.    □

## 7    Conclusions

We have confirmed the conjecture in [3] on the size of two generator semigroups. In the end, we have shown that for prime $n$, such that $n \geq 7$, the semigroup generated by two generators with maximal size can be characterized in a very nice and accurate way. The cases $2 \leq n \leq 6$ are not treated in this paper, but we

---

[2] The formula given below did not appear in [3] and gives the exact size of the $U_{k,\ell}$ semigroup: Let $k, \ell \in \mathbb{N}$ such that $\gcd\{k, \ell\} = 1$. The semigroup $U_{k,\ell}$ contains exactly

$$|U_{k,\ell}| = k\ell + \sum_{i=1}^{n} \left( \binom{n}{i} - \binom{n-\ell}{i-\ell} \right) \left( \left\{ {n \atop i} \right\} - \sum_{r=1}^{i} \left\{ {k \atop r} \right\} \left\{ {\ell \atop i-r} \right\} \right) i!$$

elements, where $n = k + \ell$. Here $\left\{ {n \atop i} \right\}$ stands for the Stirling numbers of the second kind and denotes the number of possibilities to partition an $n$-element set into $i$ non-empty subsets.

were able to show that in all these cases the semigroup $V_n$ contains a maximal number of elements. Here $2 \leq n \leq 5$ were done by brute force search using the GAP system and $n = 6$ by additional quite involved considerations, which we have to omit to due the lack of space. Moreover, we have completely classified the case when one generator is a permutation consisting of a single cycle.

Nevertheless, some questions remain unanswered. First of all, what about the case when $n \geq 7$ is not a prime number. We conjecture, that Theorem 6 also holds in this case, but we have no proof yet. Also, the question how to choose $k$ and $\ell$ properly remains unanswered. In order to maximize the size of $U_{k,\ell}$ one has to minimize the number of valid colourings – see [3] – which is minimal if $k$ and $\ell$ are close to $\frac{n}{2}$. This clashes with the observation that the cycle $\alpha$ from which an element in the image of $\beta$ is missing should be as large as possible. Nevertheless, to maximize the size of $U_{k,\ell}$ we conjecture that for large enough $n$ both $k$ and $\ell$ are as close to $\frac{n}{2}$ as the condition that $k$ and $\ell$ should be coprime allows. Again a proof of this statement is still missing. In order to understand the very nature of the question much better, a step towards its solution would be to show that the sequence $|U_{k,\ell}|$ for fixed $n = k + \ell$ and varying $k$ is unimodal.

**Acknowledgments.** Thanks to Rob Johnson, and Paul Pollack, and John Robertson for their help in managing the essential step in the proof of Lemma 1.

# References

1. W. Feller. Stirling's formula. In *An Introduction to Probability Theory and Its Applications*, volume 1, chapter 2.9, pages 50–53. Wiley, 3rd edition, 1968.
2. G.H. Hardy and E.M. Wright. *An Introduction to the Theory of Numbers*. Clarendon, 5th edition, 1979.
3. M. Holzer and B. König. On deterministic finite automata and syntactic monoid size. In M. Ito and M. Toyama, editors, *Preproceedings of the 6th International Conference on Developments in Language Theory*, pages 229–240, Kyoto, Japan, September 2002. Kyoto Sangyo University. To appear in LNCS.
4. J.M. Howie. *An Introduction to Semigroup Theory*, volume 7 of *L. M. S. Monographs*. Academic Press, 1976.
5. E. Landau. Über die Maximalordnung der Permutationen gegebenen Grades. *Archiv der Mathematik und Physik*, 3:92–103, 1903.
6. J.-L. Nicolas. Sur l'ordre maximum d'un élément dans le groupe $s_n$ des permutations. *Acta Arithmetica*, 14:315–332, 1968.
7. J.-L. Nicolas. Ordre maximum d'un élément du groupe de permutations et highly composite numbers. *Bulletin of the Mathematical Society France*, 97:129–191, 1969.
8. S. Piccard. *Sur les bases du groupe symétrique et les couples de substitutions qui engendrent un groupe régulier*. Librairie Vuibert, Paris, 1946.
9. J.-E. Pin. *Varieties of formal languages*. North Oxford, 1986.
10. R.C. Read. An introduction to chromatic polynomials. *Journal of Combinatorial Theory*, 4:52–71, 1968.
11. H. Robbins. A remark of Stirling's formula. *American Mathematical Monthly*, 62:26–29, 1955.

# Flip-Pushdown Automata: Nondeterminism is Better than Determinism

Markus Holzer[1] and Martin Kutrib[2]

[1] Institut für Informatik, Technische Universität München
Boltzmannstraße 3, D-85748 Garching bei München, Germany
holzer@informatik.tu-muenchen.de
[2] Institut für Informatik, Universität Giessen
Arndtstraße 2, D-35392 Giessen, Germany
kutrib@informatik.uni-giessen.de

**Abstract.** Flip-pushdown automata are pushdown automata with the additional ability to flip or reverse its pushdown. We investigate deterministic and nondeterministic flip-pushdown automata accepting by final state or empty pushdown. In particular, for nondeterministic flip-pushdown automata both acceptance criterion are equally powerful, while for determinism, acceptance by empty pushdown is strictly weaker. This nicely fits into the well-known results on ordinary pushdown automata. Moreover, we consider hierarchies of flip-pushdown automata w.r.t. the number of pushdown reversals. There we show that nondeterminism is better than determinism. Moreover, since there are languages which can be recognized by a deterministic flip-pushdown automaton with $k + 1$ pushdown reversals but which cannot be recognized by a $k$-flip-pushdown (deterministic or nondeterministic) as shown in [9] we are able to complete our investigations with incomparability results on different levels of the hierarchies under consideration.

## 1 Introduction

A pushdown automaton is a one-way finite automaton with a separate pushdown store, that is a last-in first-out (LIFO) storage structure, which is manipulated by pushing and popping. Probably, such machines are best known for capturing the family of context-free languages, which was independently established by Chomsky [2] and Evey [4]. Pushdown automata have been extended in various ways. Examples of extensions are variants of stacks [6,8], queues or dequeues, while restrictions are for instance counters or one-turn pushdowns [7]. The results obtained for these classes of machines hold for a large variety of formal language classes, when appropriately abstracted. This led to the rich theory of abstract families of automata, which is the equivalent of abstract families of languages theory. For the general treatment of machines and languages we refer to Ginsburg [5].

In this paper, we consider a recently introduced extension of pushdown automata, so-called flip-pushdown automata [11]. Basically, a flip-pushdown automaton is an ordinary pushdown automaton with the additional ability to flip

Z. Ésik and Z. Fülöp (Eds.): DLT 2003, LNCS 2710, pp. 361–372, 2003.
© Springer-Verlag Berlin Heidelberg 2003

its pushdown during the computation. This allows the machine to push and pop at both ends of the pushdown. Hence, a flip-pushdown is a form of a dequeue storage structure, and thus becomes equally powerful to Turing machines, since a dequeue automaton can simulate two pushdowns. On the other hand, if the number of pushdown flips or pushdown reversals is zero, obviously the family of context-free languages is characterized. Thus it remains to investigate the number of pushdown reversals as a natural computational resource.

By Sarkar [11] it was shown for nondeterministic computations that if the number of pushdown flips is bounded by a constant, then a nonempty hierarchy of language classes is introduced, and it was conjectured that the hierarchy is strict. In [9] it is shown that $k+1$ pushdown reversals are better than $k$. To this end, a technique is developed to decrease the number of pushdown reversals, which simply speaking shows that flipping the pushdown is equivalent to reverse part of the remaining input for nondeterministic automata. An immediate consequence is that every flip-pushdown language accepted by a flip-pushdown with a constant number of pushdown reversals obeys a semi-linear Parikh mapping. It turned out, that the family of nondeterministic flip-pushdown languages share similar closure and non-closure properties as the family of context-free languages like, e.g., closure under intersection with regular sets, or the non-closure under complementation. Nevertheless, there are some interesting differences as, e.g., the non-closure under concatenation and Kleene star.

Here mainly we investigate several variants of deterministic flip-pushdown automata, nondeterministic flip-pushdown automata, and their relationships among each other. In particular, we distinguish flip-pushdown automata where the number of pushdown reversals has to be exactly $k$ or has to be at most $k$. For nondeterministic automata this distinction is shown to make no difference. This is not true for deterministic automata that accept by final state. Interestingly, it also makes no difference if deterministic automata are considered that accept by empty pushdown. The two modes of acceptance are the second considered distinction. Our main contribution are results yielding, for a fixed $k$, strict inclusions in between the different types of deterministic flip-pushdown automata and, furthermore, between deterministic and nondeterministic automata. By an adaption of the hierarchy result in [9], all considered classes are separated for $k$ and $k+1$. The question, how the number of pushdown reversals relates to determinism/nondeterminism, at most $k$/exactly $k$ reversals, or acceptance by final state/empty pushdown, is answered by deriving incomparability for all classes which are not related by a strict inclusion.

The paper is organized as follows: The next section contains preliminaries and basics on flip-pushdown automata. Then Sect. 3 is devoted to the separation of deterministic and nondeterministic flip-pushdown automata. The next section deals with the comparison between flip-pushdown automata making exactly $k$ and at most $k$ pushdown reversals. In Sect. 5 we consider the relationships caused by the acceptance modes empty pushdown and final state. The penultimate Sect. 6 is devoted to the separation of the flip-pushdown hierarchies for all considered classes. Finally, we summarize our results and pose a few open questions in Sect. 7.

## 2    Preliminaries

We denote the powerset of a set $S$ by $2^S$. The empty word is denoted by $\lambda$, the reversal of a word $w$ by $w^R$, and for the length of $w$ we write $|w|$. For the number of occurrences of a symbol $a$ in $w$ we use the notation $|w|_a$. In the following we consider pushdown automata with the ability to flip their pushdowns. These machines were recently introduced by Sarkar [11] and are defined as follows:

**Definition 1.** A nondeterministic flip-pushdown automaton (NFPDA) *is a system* $A = (Q, \Sigma, \Gamma, \delta, \Delta, q_0, Z_0, F)$, *where* $Q$ *is a finite set of states,* $\Sigma$ *is the finite input alphabet,* $\Gamma$ *is a finite pushdown alphabet,* $\delta$ *is a mapping from* $Q \times (\Sigma \cup \{\lambda\}) \times \Gamma$ *to finite subsets of* $Q \times \Gamma^*$ *called the transition function,* $\Delta$ *is a mapping from* $Q$ *to* $2^Q$, $q_0 \in Q$ *is the initial state,* $Z_0 \in \Gamma$ *is a particular pushdown symbol, called the bottom-of-pushdown symbol, which initially appears on the pushdown store, and* $F \subseteq Q$ *is the set of final states.*

A *configuration* or *instantaneous description* of a flip-pushdown automaton is a triple $(q, w, \gamma)$, where $q$ is a state in $Q$, $w$ a string of input symbols, and $\gamma$ is a string of pushdown symbols. A flip-pushdown automaton $A$ is said to be in configuration $(q, w, \gamma)$ if $A$ is in state $q$ with $w$ as remaining input, and $\gamma$ on the pushdown store, the rightmost symbol of $\gamma$ being the top symbol on the pushdown. If $p, q$ are in $Q$, $a$ is in $\Sigma \cup \{\lambda\}$, $w$ in $\Sigma^*$, $\gamma$ and $\beta$ in $\Gamma^*$, and $Z$ is in $\Gamma$, then we write $(q, aw, \gamma Z) \vdash_A (p, w, \gamma\beta)$, if the pair $(p, \beta)$ is in $\delta(q, a, Z)$, for "ordinary" pushdown transitions and $(q, aw, Z_0\gamma) \vdash_A (p, aw, Z_0\gamma^R)$, if $p$ is in $\Delta(q)$, for pushdown-flip or pushdown-reversal transitions. Whenever there is a choice between an ordinary pushdown transition or a pushdown reversal, the automaton nondeterministically chooses the next move. Observe, that we do not want the flip-pushdown automaton to move the bottom-of-pushdown symbol when the pushdown is flipped. As usual, the reflexive transitive closure of $\vdash_A$ is denoted by $\vdash_A^*$. The subscript $A$ will be dropped from $\vdash_A$ and $\vdash_A^*$ whenever the meaning remains clear.

A deterministic flip-pushdown automaton (DFPDA) is a flip-pushdown automaton for which there is at most one choice of action for any possible configuration. In particular, there must never be a choice of using an input symbol or of using $\lambda$ input. Formally, a flip-pushdown automaton $A = (Q, \Sigma, \Gamma, \delta, \Delta, q_0, Z_0, F)$ is *deterministic* if the following conditions are satisfied: (1) $\delta(q, a, Z)$ contains at most one element, for all $a$ in $\Sigma \cup \{\lambda\}$, $q$ in $Q$, and $Z$ in $\Gamma$. (2) If $\delta(q, \lambda, Z)$ is not empty, then $\delta(q, a, Z)$ is empty, for all $a$ in $\Sigma$, $q$ in $Q$, and $Z$ in $\Gamma$. (3) Set $\Delta(q)$ contains at most one element, for all $q$ in $Q$. (4) If $\Delta(q)$ is not empty, then $\delta(q, a, Z)$ is empty, for all $a$ in $\Sigma \cup \{\lambda\}$, $q$ in $Q$, and $Z$ in $\Gamma$.

Let $k \geq 0$. For a (deterministic) flip-pushdown automaton $A$ we define $T_{\leq k}(A)$, the language *accepted by final state and at most $k$ pushdown reversals,* to be

$$T_{\leq k}(A) = \{\, w \in \Sigma^* \mid (q_0, w, Z_0) \vdash_A^* (q, \lambda, \gamma) \text{ with at most } k$$
$$\text{pushdown reversals, for some } \gamma \in \Gamma^* \text{ and } q \in F \,\}.$$

The family of languages accepted by nondeterministic (deterministic, resp.) flip-pushdown automata by final state making at most $k$ pushdown reversals is denoted by $\mathcal{L}(\text{NFPDA}_{\leq k})$ ($\mathcal{L}(\text{DFPDA}_{\leq k})$, resp.). Furthermore, let $\mathcal{L}(\text{NFPDA}_{\leq fin}) = \bigcup_{k=0}^{\infty}(\text{NFPDA}_{\leq k})$ and similarly $\mathcal{L}(\text{DFPDA}_{\leq fin})$.

An essential technique for flip-pushdown automata is the so-called "flip-pushdown input-reversal" technique, which has been developed and proved in [9]. It allows to simulate flipping the pushdown by reversing the (remaining) input, and reads as follows.

**Theorem 2.** *Let $k \geq 0$. Then language $L$ is accepted by a nondeterministic flip-pushdown automaton $A_1 = (Q, \Sigma, \Gamma, \delta, \Delta, q_0, Z_0, F)$ by final state with at most $k + 1$ pushdown reversals, i.e., $L = T_{\leq(k+1)}(A_1)$, if and only if language*

$$L_R = \{\, wv^R \mid (q_0, w, Z_0) \vdash_{A_1}^* (q_1, \lambda, Z_0\gamma) \text{ with at most } k \text{ reversals, } q_2 \in \Delta(q_1),$$
$$\text{and } (q_2, v, Z_0\gamma^R) \vdash_{A_1}^* (q_3, \lambda, q_4) \text{ without any reversal, } q_4 \in F \,\}$$

*is accepted by a nondeterministic flip-pushdown automaton $A_2$ by final state with at most $k$ pushdown reversals, i.e., $L_R = T_{\leq k}(A_2)$.*

## 3   Determinism versus Nondeterminism

We show that nondeterminism is better than determinism. In particular, the family of languages accepted by deterministic flip-pushdown automata with at most $k$ pushdown reversals is a strict subset of the family of languages accepted by nondeterministic flip-pushdown automata with at most $k$ pushdown reversals. To this end, we need the closure of deterministic flip-pushdown languages under intersection with regular sets and under the prefix operation. The first property is straightforward by simulating a deterministic finite automaton in the control of the flip-pushdown automaton at the same time. For the second property let $w = a_1 a_2 \cdots a_n$ with $a_i \in \Sigma$, for $1 \leq i \leq n$, be some word over $\Sigma$. The *set of prefixes* of $w$ is defined to be $\{\lambda, a_1, a_1 a_2, \ldots, a_1 \cdots a_n\}$. For a language $L$ in $\Sigma^*$ and a natural number $i \geq 1$ define the following set of prefixes

$$P_i(L) = \{\, w \in L \mid \text{exactly } i \text{ prefixes of } w \text{ are belonging to } L \,\}.$$

The following theorem shows that deterministic flip-pushdown languages are closed under the $P_i$ operation. Since the proof of the following theorem is an adaption from the deterministic context-free case, we omit the proof.

**Theorem 3.** *Let $i \geq 1$ and $k \geq 0$. If $L \in \mathcal{L}(\text{DFPDA}_{\leq k})$, then $P_i(L) \in \mathcal{L}(\text{DFPDA}_{\leq k})$.*     □

The following corollary is an immediate consequence of the above given theorem.

**Corollary 4.** *Let $i \geq 1$. If $L \in \mathcal{L}(\text{DFPDA}_{\leq fin})$, then $P_i(L) \in \mathcal{L}(\text{DFPDA}_{\leq fin})$.*     □

Now we are ready to separate determinism from nondeterminism for flip-pushdown languages.

**Theorem 5.** *Let $k \geq 0$. Then $\mathcal{L}(\text{DFPDA}_{\leq k}) \subset \mathcal{L}(\text{NFPDA}_{\leq k})$.*

*Proof.* The inclusion immediately follows for structural reasons. For the strictness we argue as follows. For $k = 0$ the theorem states the well-known result for deterministic context-free languages. So let $k \geq 1$ and $L = \{ ww \mid w \in \{a,b\}^+ \}$. Since $L$ belongs to $\mathcal{L}(\text{NFPDA}_{\leq 1})$, it belongs to $\mathcal{L}(\text{NFPDA}_{\leq k})$ as well. In the following we show that $L \notin \mathcal{L}(\text{DFPDA}_{\leq fin})$. This immediately implies that $L \notin \mathcal{L}(\text{DFPDA}_{\leq k})$, for any $k$.

Assume to the contrary, that $L \in \mathcal{L}(\text{DFPDA}_{\leq k})$, for some $k$. Then by Theorem 3 we obtain that $P_2(L) \in \mathcal{L}(\text{DFPDA}_{\leq k})$, and since this language family is closed under intersection with regular sets, also the language $L' = P_2(L) \cap ba^*ba^*ba^*$ is a deterministic $k$-flip language. Thus, language $L'$ belongs to $\mathcal{L}(\text{NFPDA}_{\leq fin})$, which is closed under rational $a$-transduction as shown in [9]. Since $L'$ contains exactly the words of the form $ba^nba^na^mba^nba^na^m$, this implies that the language $L'' = \{ a^nb^{n+m}c^nd^{n+m} \mid m,n \geq 1 \}$ belongs also to $\mathcal{L}(\text{NFPDA}_{\leq fin})$. In order to obtain a contradiction, it remains to show that this is not the case. To this end, we use Theorem 2 and a generalization of Ogden's Lemma, which is due to Bader and Moura [1].

Assume to the contrary, that language $L'' = \{ a^nb^{n+m}c^nd^{n+m} \mid m,n \geq 1 \}$ belongs to $\mathcal{L}(\text{NFPDA}_{\leq k})$ for some $k$. Then we apply $k$ times the flip-pushdown input-reversal Theorem 2 to $L$ obtaining a context-free language. Since we do the input reversal from right-to-left, the block of $d$'s remains adjacent in all words. Hence a word $w$ in the context-free language reads as $w = ud^{n+m}v$, where $|uv|_a = |uv|_c = n$ and $|uv|_b = n + m$. Then it is an easy exercise to show that this language cannot be context-free using the generalized version of Ogden's lemma. This contradicts our assumption on $L''$, and thus, it does not belong to $\mathcal{L}(\text{NFPDA}_{\leq k})$, for any $k \geq 0$. This shows that $L'' \notin \mathcal{L}(\text{NFPDA}_{\leq fin})$. Hence, we can conclude that $L$ does not belong to $\mathcal{L}(\text{DFPDA}_{\leq k})$, for any $k$.  □

**Corollary 6.** $\mathcal{L}(\text{DFPDA}_{\leq fin}) \subset \mathcal{L}(\text{NFPDA}_{\leq fin})$.  □

## 4  Exactly $k$ Flips versus At Most $k$ Flips

Let $k \geq 0$. For a flip-pushdown automaton $A = (Q, \Sigma, \Gamma, \delta, \Delta, q_0, Z_0, F)$ we define $T_{=k}(A)$, the language *accepted by final state and exactly $k$ pushdown reversals*, to be

$$T_{=k}(A) = \{ w \in \Sigma^* \mid (q_0, w, Z_0) \vdash_A^* (q, \lambda, \gamma) \text{ with exactly } k$$
$$\text{pushdown reversals, for some } \gamma \in \Gamma^* \text{ and } q \in F \}.$$

Similarly, we use the notations $\mathcal{L}(\text{DFPDA}_{=k})$, $\mathcal{L}(\text{DFPDA}_{=fin})$, etc. The next result shows that for nondeterministic computations the classes for exactly $k$ flips and at most $k$ flips coincide. It turns out that this is not true for deterministic computations, but the result shows also that exactly $k$ flips are included in at most $k$ flips for deterministic computations. This does not follow for structural reasons since the automaton with at most $k$ flips may accept some input with strictly less than $k$ flips. By definition these inputs do not belong to the accepted language of the automaton with exactly $k$ flips.

The idea for both directions of the proof relies on the simple fact that a flip-pushdown automaton can count the number of reversals performed during its computation in its finite control. Moreover, the construction must ensure that it is possible to increase the number of pushdown flips without changing the accepted language.

**Theorem 7.** Let $k \geq 0$. Then (1) $\mathcal{L}(\text{NFPDA}_{=k}) = \mathcal{L}(\text{NFPDA}_{\leq k})$ and (2) $\mathcal{L}(\text{DFPDA}_{=k}) \subseteq \mathcal{L}(\text{DFPDA}_{\leq k})$.

*Proof.* Let $A_1 = (Q, \Sigma, \Gamma, \delta, \Delta, q_0, Z_0, F)$ be a (deterministic) flip-pushdown automaton. Then we define

$$A_2 = (Q \times \{0, 1, \ldots, k\}, \Sigma, \Gamma, \delta', \Delta', (q_0, 0), Z_0, \{(q, k) \mid q \in F\}),$$

where $\delta'$ and $\Delta'$ are defined as follows: For all $p, q \in Q$, $a \in \Sigma \cup \{\lambda\}$, and $Z \in \Gamma$ let

1. $((p, i), \gamma) \in \delta'((q, i), a, Z)$, if $(p, \gamma) \in \delta(q, a, Z)$, for $0 \leq i \leq k$,
2. $\Delta'((q, i)) = \{(p, i+1) \mid p \in \Delta(q)\}$, for $0 \leq i < k$,
3. $\Delta'((q, k)) = \{(p, k) \mid p \in \Delta(q)\}$.

By construction one sees, that whenever $A_1$ is a deterministic flip-pushdown automaton, then $A_2$ is a deterministic machine, too.

Observe, that each state in $A_2$ is a tuple, where in the second component the number of pushdown reversals performed so far is stored. The transitions from (1) cause $A_2$ to simulate the original flip-pushdown $A_1$. If a pushdown flip is performed, the state's second component is increased by one, which is seen in (2). Since only the states of the form $(q, k)$, for $q \in F$, are accepting states, the automaton $A_2$ must have done at least $k$ pushdown reversals in order to reach a state of this form.

Assume that $w \in T_{=k}(A_1)$. Then $(q_0, w, Z_0) \vdash^*_{A_1} (q, \lambda, \gamma)$ for some $q \in F$ and $\gamma \in \Gamma^*$ with exactly $k$ reversals. Thus, $((q_0, 0), w, Z_0) \vdash^*_{A_2} ((q, k), \lambda, \gamma)$. Therefore, $w \in T_{\leq k}(A_2)$, since the number of reversals is at most $k$ now and no state of the form $(q, i)$ with $0 \leq i < k$ is an accepting one. By similar reasoning, if $w \in T_{\leq k}(A_2)$, then $w \in T_{=k}(A_1)$.

Conversely we argue as follows. Let $A_2 = (Q, \Sigma, \Gamma, \delta, \Delta, q_0, Z_0, F)$ be a flip pushdown automaton. Then we define

$$A_1 = ((Q \cup \{q_0'\}) \times \{0, 1, \ldots, k\}, \Sigma, \Gamma, \delta', \Delta', (q_0', 0), Z_0, \{(q, k) \mid q \in F\}),$$

where $\delta'$ and $\Delta'$ are defined as follows: For all $p, q \in Q$, $a \in \Sigma \cup \{\lambda\}$, and $Z \in \Gamma$ let

1. $((q_0, i), Z_0) \in \delta((q'_0, i), \lambda, Z_0)$, for $0 \leq i \leq k$,
2. $((p, i), \gamma) \in \delta'((q, i), a, Z)$, if $(p, \gamma) \in \delta(q, a, Z)$,
3. $\Delta'((q'_0, i)) = \{(q'_0, i + 1)\}$, for $0 \leq i < k$,
4. $\Delta'((q, i)) = \{(p, i + 1) \mid p \in \Delta(q)\}$, for $0 \leq i < k$.

The transitions from (1) cause $A_1$ to enter the initial configuration of $A_2$, while those transitions in (2) simulate the original flip-pushdown $A_2$. If a pushdown flip is performed, the state's second component is increased by one, which is seen in (4). In order to obtain exactly $k$ flips, additional pushdown reversals can be performed using (3) at the very beginning of the computation.

Assume that $w \in T_{\leq k}(A_2)$. Then $(q_0, w, Z_0) \vdash^*_{A_2} (q, \lambda, \gamma)$ for some $q \in F$ and $\gamma \in \Gamma^*$ with $\ell$ reversals, for some $0 \leq \ell \leq k$. Thus,

$$((q'_0, 0), w, Z_0) \vdash^{k-\ell}_{A_1} ((q'_0, k - \ell), w, Z_0) \vdash_{A_1} ((q_0, k - \ell), w, Z_0) \vdash^*_{A_1} ((q, k), \lambda, \gamma),$$

where the first $k - \ell$ steps are pushdown flips only. Therefore, $w \in T_{=k}(A_1)$, since the number of reversals is exactly $k$ now. By similar reasoning, if $w \in T_{=k}(A_1)$, then $w \in T_{\leq k}(A_2)$. □

Due to the equality of at most $k$ and exactly $k$ pushdown reversals for nondeterministic flip pushdown automata we simplify the notation to $\mathcal{L}(\text{NFPDA}_k)$, etc., in the sequel. The next step is to separate the families $\mathcal{L}(\text{DFPDA}_{=k})$ and $\mathcal{L}(\text{DFPDA}_{\leq k})$ for $k \geq 1$. Trivially, they coincide for $k = 0$.

**Theorem 8.** *Let $k \geq 1$. Then $\mathcal{L}(\text{DFPDA}_{=k}) \subset \mathcal{L}(\text{DFPDA}_{\leq k})$.*

*Proof.* The inclusion has been shown in the previous theorem. The strictness is seen as follows. Define $L = L' \cup L''$, where

$$L' = \{ \#w\# \mid w \in \{a, b\}^* \} \quad \text{and} \quad L'' = \{ \#w_0\#w_1\$w_1\# \mid w_0, w_1 \in \{a, b\}^* \}.$$

Obviously, language $L$ is accepted by some deterministic flip-pushdown automaton making at most one flip. So, language $L$ is accepted by some $\text{DFPDA}_{\leq k}$, for any $k \geq 1$.

Assume to the contrary that language $L$ is accepted by some deterministic flip-pushdown automaton $A = (Q, \Sigma, \Gamma, \delta, \Delta, q_0, Z_0, F)$ with exactly $k$ pushdown reversals. Consider an arbitrary word $\#w_0\#$ that belongs to $L'$ and, hence, also to $L$. There is an accepting computation of $A$ which reads as $(q_0, \#w_0\#, Z_0) \vdash^*_A (q, \lambda, \gamma)$, where $q \in F$ and $\gamma \in \Gamma^*$. Since for every $w_1 \in \{a, b\}^*$ we find that $v = \#w_0\#w_1\$w_1\#$ belongs to $L$, there must be also an accepting computation on $v$. Since $A$ is deterministic and, thus, has already performed $k$ pushdown reversals on the prefix $\#w_0\#$ of $v$, we can construct an ordinary deterministic pushdown automaton accepting the language $\tilde{L} = \{ w\$w\# \mid w \in \{a, b\}^* \}$. The deterministic pushdown automaton accepting $\tilde{L}$ initially creates the pushdown content $\gamma$ with a series of $\lambda$-moves, changes to state $q$, simulates $A$ step-by-step, and accepts if $A$ does. Since $\tilde{L}$ is not even context free, we obtain a contradiction to our assumption. So $L \notin \mathcal{L}(\text{DFPDA}_{=k})$, and the assertion follows. □

**Corollary 9.** $\mathcal{L}(\text{DFPDA}_{=fin}) \subset \mathcal{L}(\text{DFPDA}_{\leq fin})$. $\qquad\qquad\qquad\qquad$ □

*Proof.* In the proof of the previous theorem we obtained for the witness language $L \notin \mathcal{L}(\text{DFPDA}_{=k})$. Since $k$ was chosen arbitrarily, it follows that $L \notin \mathcal{L}(\text{DFPDA}_{=fin})$. $\qquad\qquad\qquad\qquad\qquad\qquad\qquad\qquad\qquad\qquad\qquad$ □

Thus, for $k \geq 1$, we have $\mathcal{L}(\text{DFPDA}_{=k}) \subset \mathcal{L}(\text{DFPDA}_{\leq k}) \subset \mathcal{L}(\text{NFPDA}_k)$ and $\mathcal{L}(\text{DFPDA}_{=fin}) \subset \mathcal{L}(\text{DFPDA}_{\leq fin}) \subset \mathcal{L}(\text{NFPDA}_{fin})$.

## 5   Empty Pushdown versus Final State

Now we are going to consider flip-pushdown automata with a different mode of acceptance. For ordinary deterministic pushdown automata it is well known that acceptance by empty pushdown yields strictly weaker devices than acceptance by final state. In the following we distinguish for both modes also acceptance with at most and with exactly $k$ pushdown reversals, respectively.

Let $k \geq 0$. For a flip-pushdown automaton $A = (Q, \Sigma, \Gamma, \delta, \Delta, q_0, Z_0, F)$ we define $N_{\leq k}(A)$, the language *accepted by empty pushdown and at most $k$ pushdown reversals*, to be

$$N_{\leq k}(A) = \{\, w \in \Sigma^* \mid (q_0, w, Z_0) \vdash_A^* (q, \lambda, \lambda) \text{ with at most } k$$
$$\text{pushdown reversals, for some } q \in Q \,\}.$$

As before we define $N_{=k}(A)$, and use subscript $N$ in the class notation to denote acceptance by empty pushdown store. Thus, our nomenclature reads as $\mathcal{L}_N(\text{DFPDA}_{\leq k})$, $\mathcal{L}_N(\text{DFPDA}_{=k})$, etc.

The first result in this section concerns nondeterministic automata. It generalizes the theorem on ordinary pushdown automata, that languages accepted by nondeterministic flip-pushdown automata by final state are exactly those languages accepted by nondeterministic flip-pushdown automata by empty pushdown. We state the theorem without proof since the proof is a simple adaption of the proof for ordinary pushdown automata.

**Theorem 10.** *Let $k \geq 0$. Then (1) $\mathcal{L}_N(\text{NFPDA}_{\leq k}) = \mathcal{L}(\text{NFPDA}_{\leq k})$ and (2) $\mathcal{L}_N(\text{NFPDA}_{=k}) = \mathcal{L}(\text{NFPDA}_{=k})$.* $\qquad\qquad\qquad$ □

In particular, the above given theorem together with Theorem 7 shows that $\mathcal{L}_N(\text{NFPDA}_{\leq k}) = \mathcal{L}(\text{NFPDA}_{\leq k}) = \mathcal{L}(\text{NFPDA}_{=k}) = \mathcal{L}_N(\text{NFPDA}_{=k})$ and moreover $\mathcal{L}_N(\text{NFPDA}_{fin}) = \mathcal{L}(\text{NFPDA}_{fin})$. So we have the robust situation that for the acceptance power of flip-pushdown automata it makes neither a difference whether the mode is by empty pushdown or by final state nor whether at most $k$ or exactly $k$ flips are considered. For deterministic classes it does make a difference, but interestingly, we can show that in case of acceptance by empty pushdown the families with at most $k$ and exactly $k$ pushdown reversals coincide. In case of acceptance by final state the classes have already been separated.

**Theorem 11.** *Let $k \geq 0$. Then $\mathcal{L}_N(\text{DFPDA}_{\leq k}) = \mathcal{L}_N(\text{DFPDA}_{=k})$.*

*Proof.* In order to show the inclusion $\mathcal{L}_N(\text{DFPDA}_{\leq k}) \subseteq \mathcal{L}_N(\text{DFPDA}_{=k})$ let $A_1 = (Q, \Sigma, \Gamma, \delta, \Delta, q_0, Z_0, \emptyset)$ be a deterministic flip-pushdown automaton. Basically, the idea is to simulate $A_1$ until it empties its pushdown and then to perform the missing number of flips. To this end, the simulating automaton $A_2$ has to count the number of simulated flips in its finite control, and has to provide a new bottom-of-pushdown symbol $Z_0'$. The latter is necessary to continue the computation when $A_1$ has emptied its pushdown. Since now the old bottom-of-pushdown symbol $Z_0$ is object of pushdown reversals, special attention has to be paid. Every time $A_2$ wishes to simulate a pushdown reversal of $A_1$, it has to push $Z_0$, to perform the reversal, and to pop $Z_0$. Due to the lack of space we omit a detailed construction.

It remains to show the inclusion $\mathcal{L}_N(\text{DFPDA}_{=k}) \subseteq \mathcal{L}_N(\text{DFPDA}_{\leq k})$. The construction is similar to the corresponding one given in the proof of Theorem 7. The main difference is, that now the simulating automaton is prevented from popping the bottom-of-pushdown symbol whenever the number of performed pushdown reversals is not equal to $k$. □

**Corollary 12.** $\mathcal{L}_N(\text{DFPDA}_{\leq fin}) = \mathcal{L}_N(\text{DFPDA}_{=fin})$. □

Due to the equalities, again, in the sequel we can simplify the notation to $\mathcal{L}_N(\text{DFPDA}_k)$ and $\mathcal{L}_N(\text{DFPDA}_{fin})$. Now the question for the relationships between $\mathcal{L}_N(\text{DFPDA}_k)$ and the other classes under consideration arises. The next result nicely extends the chain of strict inclusions known so far.

**Theorem 13.** *Let* $k \geq 0$. *Then* $\mathcal{L}_N(\text{DFPDA}_k) \subset \mathcal{L}(\text{DFPDA}_{=k})$.

*Proof.* Due to Theorem 11 it suffices to show $\mathcal{L}_N(\text{DFPDA}_{=k}) \subset \mathcal{L}(\text{DFPDA}_{=k})$. Both, for inclusion and for its strictness we can adapt the proofs for ordinary deterministic pushdown automata. For inclusion a new bottom-of-pushdown symbol is provided in order to detect when the simulated automaton empties its pushdown. If and only if this occurs, the simulating automaton enters an accepting state. Finally, strictness of the inclusion it is easily seen because a language $L \in \mathcal{L}_N(\text{DFPDA}_{=k})$ must have the prefix property. That is, no word in $L$ is a proper prefix of another word in $L$. Since there exist deterministic context-free languages that do not have the prefix property, there exist languages which do not belong to $\mathcal{L}_N(\text{DFPDA}_{=k})$, but which do belong to $\mathcal{L}(\text{DFPDA}_{=k})$, for all $k \geq 0$. □

**Corollary 14.** $\mathcal{L}_N(\text{DFPDA}_{fin}) \subset \mathcal{L}(\text{DFPDA}_{=fin})$. □

So far we have shown the inclusions $\mathcal{L}_N(\text{DFPDA}_k) \subset \mathcal{L}(\text{DFPDA}_{=k}) \subset \mathcal{L}(\text{DFPDA}_{\leq k}) \subset \mathcal{L}(\text{NFPDA}_k)$, for all $k \geq 1$, and correspondingly for "fin."

## 6    Flip-Pushdown Hierarchies

In this section we consider hierarchies on flip-pushdown automata induced by the number of pushdown reversals. Recently, it was shown shown in [9] that for nondeterministic flip-pushdown automata $k + 1$ pushdown reversals are better than $k$. The proof of this result is based on the "flip-pushdown input-reversal" technique. The witness language

$$L_{k+1} = \{\, \#w_1\$w_1\#w_2\$w_2\# \ldots \#w_{k+1}\$w_{k+1}\# \mid w_i \in \{a, b\}^* \text{ for } 1 \leq i \leq k+1 \,\}$$

for the hierarchy has shown to belong to $\mathcal{L}(\text{DFPDA}_{=(k+1)})$ but not to the family $\mathcal{L}(\text{NFPDA}_k)$. Actually, the language is also accepted by some deterministic flip-pushdown automaton by empty pushdown making $k + 1$ pushdown reversals.

**Theorem 15.** *Let $k \geq 0$. Then $\mathcal{L}_N(\text{DFPDA}_{k+1}) \setminus \mathcal{L}(\text{NFPDA}_k) \neq \emptyset$.* □

So we have four hierarchies with separated levels and, furthermore, separated families on every level. In order to explore the relationships between any two families somewhere in the inclusion structure (cf. Fig. 1), we have to answer the question how the number of pushdown reversals relates to determinism/nondeterminism, at most $k$/exactly $k$ reversals, or acceptance by final state/empty pushdown. In principle, these answers have already been given. Either two families are related by a strict inclusion or they are incomparable. One direction of the following theorems is an immediate consequence of the proof of the hierarchy Theorem 15.

**Theorem 16.** *Let $\ell > k \geq 0$. (1) The family $\mathcal{L}(\text{NFPDA}_k)$ is incomparable with each of the language families $\mathcal{L}(\text{DFPDA}_{\leq \ell})$, $\mathcal{L}(\text{DFPDA}_{=\ell})$, and $\mathcal{L}_N(\text{DFPDA}_\ell)$. (2) The family $\mathcal{L}(\text{DFPDA}_{=k})$ is incomparable with the family $\mathcal{L}_N(\text{DFPDA}_\ell)$.*

*Proof.* (1) By the proof of Theorem 5 the language $\{\, ww \mid w \in \{a, b\}^+ \,\}$ belongs to $\mathcal{L}(\text{NFPDA}_1)$ but does not belong to $\mathcal{L}(\text{DFPDA}_{\leq fin})$. This proves the assertion for $k \geq 1$. The case $k = 0$ is seen by the language $\{\, ww^R \mid w \in \{a, b\}^+ \,\}$ which belongs to $\mathcal{L}(\text{NFPDA}_0)$ but does not belong to $\mathcal{L}(\text{DFPDA}_{\leq fin})$. The latter can be shown by the same argumentation used to show that the language $\{\, ww \mid w \in \{a, b\}^+ \,\}$ is not a member of $\mathcal{L}(\text{DFPDA}_{\leq fin})$. (2) Finally, as witness we may take any deterministic context-free language $L$ that does not have the prefix property. Trivially, language $L$ belongs to the language family $\mathcal{L}(\text{DFPDA}_{=0})$, but $L$ is not in $\mathcal{L}_N(\text{DFPDA}_{fin})$. □

The next result holds for $k \geq 1$ only, since trivially the family $\mathcal{L}(\text{DFPDA}_{\leq 0})$ equals $\mathcal{L}(\text{DFPDA}_{=0})$.

**Theorem 17.** *Let $\ell > k \geq 1$. The family $\mathcal{L}(\text{DFPDA}_{\leq k})$ is incomparable with each of the families $\mathcal{L}(\text{DFPDA}_{=\ell})$ and $\mathcal{L}_N(\text{DFPDA}_\ell)$.*

*Proof.* Language $L = \{\, \#w\# \mid w \in \{a, b\}^* \,\} \cup \{\, \#w_0\#w_1\$w_1\# \mid w_0, w_1 \in \{a, b\}^* \,\}$ belongs to $\mathcal{L}(\text{DFPDA}_{\leq 1})$ but does not belong to $\mathcal{L}(\text{DFPDA}_{=fin})$, because of Theorem 8. □

## 7   Conclusions

We have investigated (deterministic) flip-pushdown automata with a constant number of pushdown reversals, which were recently introduced by Sarkar [11].

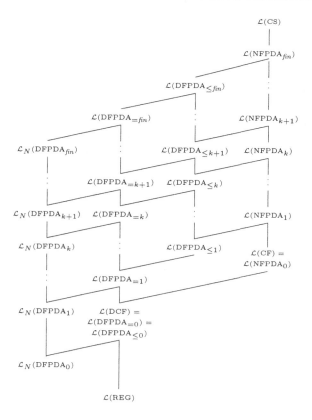

**Fig. 1.** Inclusion structure

The main interest was on deterministic computations. We distinguished deterministic automata, that accept with at most $k$ pushdown reversals or with exactly $k$ pushdown reversals, that accept by final state or by empty pushdown. For all models we showed a strict hierarchy induced by the number of pushdown reversals. Furthermore, we have separated deterministic from nondeterministic automata and have considered the relationships between the distinguished deterministic classes. The major technique is the "flip-pushdown input-reversal" technique developed in [9].

A natural property for further distinctions is whether ordinary $\lambda$-moves are allowed or not. Here we assume that a pushdown reversal never consumes an input symbol. A well-known language which separates deterministic context-free languages without $\lambda$-moves from deterministic context-free languages with $\lambda$-moves is $\{a^m b^n x c^m \mid m, n \geq 1\} \cup \{a^m b^n y c^n \mid m, n \geq 1\}$ as shown in [3]. Obviously, this language is accepted by some deterministic flip-pushdown automaton without $\lambda$-moves by empty pushdown making one pushdown reversal. Nevertheless, we can generalize this language to $L = \{a^\ell b^m c^n x d^\ell \mid \ell, m, n \geq 1\} \cup \{a^\ell b^m c^n y d^m \mid \ell, m, n \geq 1\} \cup \{a^\ell b^m c^n z d^n \mid \ell, m, n \geq 1\}$. It is easy to see that $L$ is accepted by an ordinary deterministic pushdown automaton by empty pushdown or by final state that makes some $\lambda$-moves. On the other hand, language $L$ cannot be accepted by any deterministic flip-pushdown automaton not allowed to make $\lambda$-moves.

Conversely, the language $L_k$ of the hierarchy Theorem 15 is accepted by some deterministic flip-pushdown automaton without $\lambda$-moves by empty pushdown making $k$ pushdown reversals. But $L_k$ is not accepted by any deterministic flip-pushdown automaton making strictly less than $k$ pushdown reversals, even if $\lambda$-moves are allowed. This shows incomparability between each two of the families.

Nevertheless, several questions for flip-pushdown languages remain unanswered. We mention a few of them: (1) What is the power of $\lambda$-moves for nondeterministic flip-pushdown automata? Can they be removed without affecting the computational capacity? (2) What are the closure properties of the deterministic families? (3) Which properties are decidable? (4) What are the relationships between these language families and other well-known formal language classes? Especially, the latter question is of some interest, because not even the relationship between the family of flip-pushdown languages and some Lindenmayer families like, e.g. E0L or ET0L languages is known. For more on Lindenmayer languages we refer to Rozenberg and Salomaa [10]. We conjecture incomparability, but have no proof yet. Obviously, $\{a^n b^n c^n \mid n \geq 0\}$ is an E0L language which does not belong to $\mathcal{L}(\text{NFPDA}_{fin})$ [9], but for the other way around we need a language with a semi-linear Parikh mapping which is not an ET0L language.

# References

1. Ch. Bader and A. Moura. A generalization of Ogden's lemma. *Journal of the ACM*, 29(2):404–407, 1982.
2. N. Chomsky. *Handbook of Mathematic Psychology*, volume 2, chapter Formal Properties of Grammars, pages 323–418. Wiley & Sons, New York, 1962.
3. S. N. Coole. Deterministic pushdown store machines and real-time computations. *Journal of the ACM*, 18:306–328, 1971.
4. R. J. Evey. *The Theory and Applications of Pushdown Store Machines*. Ph.D thesis, Harvard University, Massachusetts, May 1963.
5. S. Ginsburg. *Algebraic and Automata-Theoretic Properties of Formal Languages*. North-Holland, Amsterdam, 1975.
6. S. Ginsburg, S. A. Greibach, and M. A. Harrison. One-way stack automata. *Journal of the ACM*, 14(2):389–418, April 1967.
7. S. Ginsburg and E. H. Spanier. Finite-turn pushdown automata. *SIAM Journal on Computing*, 4(3):429–453, 1966.
8. S. A. Greibach. An infinite hierarchy of context-free languages. *Journal of the ACM*, 16(1):91–106, January 1969.
9. M. Holzer and M. Kutrib. Flip-pushdown automata: $k + 1$ pushdown reversals are better than $k$. In *Proceedings of the 30th International Colloquium on Automata, Languages, and Programming*, LNCS, Eindhoven, Netherlands, June–July 2003. Springer. To appear.
10. G. Rozenberg and A. Salomaa. *The Mathematical Theory of L Systems*, volume 90 of *Pure and Applied Mathematics*. Academic Press, 1980.
11. P. Sarkar. Pushdown automaton with the ability to flip its stack. Report TR01-081, Electronic Colloquium on Computational Complexity (ECCC), November 2001.

# Deciding the Sequentiality of a Finitely Ambiguous Max-Plus Automaton

Ines Klimann, Sylvain Lombardy, Jean Mairesse, and Christophe Prieur

LIAFA, CNRS (UMR 7089), Université Paris 7
2, place Jussieu, 75251 Paris Cedex 5, France
{klimann,lombardy,mairesse,prieur}@liafa.jussieu.fr

**Abstract.** Finite automata with weights in the max-plus semiring are considered. The main result is: it is decidable in an effective way whether a series that is recognized by a finitely ambiguous max-plus automaton is sequential. A collection of examples is given to illustrate the hierarchy of max-plus series with respect to ambiguity.

## 1 Introduction

A *max-plus automaton* is a finite automaton with multiplicities in the max-plus semiring $R_{max} = (R \cup \{-\infty\}, max, +)$. Roughly speaking, it is an automaton with two tapes: an input tape labeled by a finite alphabet $\Sigma$, and an output tape weighted in $R_{max}$. The weight of a word in $\Sigma^*$ is the maximum over all accepting paths of the sum of the weights along the paths.

Max-plus automata, and their min-plus counterparts, are studied under various names in the literature: distance automata, finance automata, cost automata. They have also appeared in various contexts: to study logical problems in formal language theory (star height, finite power property) [13,23], to model the dynamic of some Discrete Event Systems [10,12], or in the context of automatic speech recognition [18].

Two automata are *equivalent* if they recognize the same series, *i.e.* if they have the same input/output behavior. The problem of equivalence of two max-plus automata is undecidable [15]. The same problem for finitely ambiguous max-plus automata is decidable [14,25].

The *sequentiality* problem is defined as follows: given a max-plus automaton, is there an equivalent max-plus automaton which is deterministic in input (= sequential). Sequentiality is decidable for unambiguous max-plus automata [18]. In the present paper, we prove that sequentiality is decidable for finitely ambiguous max-plus automata. To the best of our knowledge, it is not known if the finite ambiguity of a max-plus series (defined via an infinitely ambiguous automaton) is a decidable problem. In particular, the status of the sequentiality problem is still open for a general max-plus automaton (even if the multiplicities are restricted to be in $Z_{max}$, $N_{max}$, or $Z_{max}^-$). To be complete, it is necessary to mention that in [18, §3.5], it is claimed that any max-plus automaton admits an effectively computable equivalent unambiguous one. If that was true, it

Z. Ésik and Z. Fülöp (Eds.): DLT 2003, LNCS 2710, pp. 373–385, 2003.
© Springer-Verlag Berlin Heidelberg 2003

would imply the decidability of the sequentiality for general max-plus automata. However, the statement is erroneous and counter-examples are provided in §3[1].

The sequentiality problem can be asked for automata over any semiring K. For transducers, *i.e.* when K is the set of rational subsets of a free monoid (with union and concatenation as the two laws), the problem is completely solved in the functional case (when the output is a language of cardinal at most one) [3,8, 7]. For a general transducer, the problem is wide open. Observe that the semiring $\{a^{\geqslant n}, n \in \mathbb{N}\} = \{a^n a^*, n \in \mathbb{N}\}$ is isomorphic to $\mathbb{N}_{\min}$: $a^n a^* + a^m a^* = a^{\min(n,m)} a^*$ and $a^n a^* \cdot a^m a^* = a^{n+m} a^*$. Similarly, the semiring $\{a^{\leqslant n}, n \in \mathbb{N}\}$ is isomorphic to $\mathbb{N}_{\max}$. Hence automata over $\mathbb{N}_{\max}$ translate into transducers, but *not* functional (or finite valued) ones. Also the translation does not work for automata over $\mathbb{R}_{\max}$. Hence, the vast literature on transducers is of limited use in our context.

In the present paper, we work with $\mathbb{R}_{\max}$. Decidability and complexity should be interpreted under the assumption that two real numbers can be added or compared in constant time.

## 2   Preliminaries

The free monoid over a finite set (alphabet) $\Sigma$ is denoted by $\Sigma^*$ and the empty word is denoted by $\varepsilon$. The structure $\mathbb{R}_{\max} = (\mathbb{R} \cup \{-\infty\}, \max, +)$ is a semiring, which is called the *max-plus semiring*. It is convenient to use the notations $\oplus = \max$ and $\otimes = +$. The neutral elements of $\oplus$ and $\otimes$ are denoted respectively by $0 = -\infty$ and $1 = 0$. The subsemirings $\mathbb{N}_{\max}, \mathbb{Z}_{\max}, \dots$, are defined in a natural way. The *min-plus semiring* $\mathbb{R}_{\min}$ is obtained by replacing max by min and $-\infty$ by $+\infty$ in the definition of $\mathbb{R}_{\max}$. The results of this paper can be easily adapted to the min-plus setting. Observe that the subsemiring $\mathbb{B} = (\{0, 1\}, \oplus, \otimes)$ is isomorphic to the Boolean semiring. For matrices $A, B$ of appropriate sizes with entries in $\mathbb{R}_{\max}$, we set $(A \oplus B)_{ij} = A_{ij} \oplus B_{ij}$, $(A \otimes B)_{ij} = \bigoplus_k A_{ik} \otimes B_{kj}$, and for $a \in \mathbb{R}_{\max}$, $(a \otimes A)_{ij} = a \otimes A_{ij}$. We usually omit the $\otimes$ sign, writing for instance $AB$ instead of $A \otimes B$.

Consider the set $\mathbb{R}_{\max}\langle\langle \Sigma^* \rangle\rangle$ of *(formal power) series (over $\Sigma^*$ with coefficients in $\mathbb{R}_{\max}$*), that is the set of maps from $\Sigma^*$ into $\mathbb{R}_{\max}$. The *support* of a series $S$ is the set $\operatorname{Supp} S = \{u \in \Sigma^* \mid \langle S, u \rangle \neq 0\}$. It is convenient to use the notation $S = \bigoplus_{u \in \Sigma^*} \langle S, u \rangle u = \bigoplus_{u \in \operatorname{Supp} S} \langle S, u \rangle u$. Equipped with the addition ($\oplus$) and the Cauchy product ($\otimes$), the set $\mathbb{R}_{\max}\langle\langle \Sigma^* \rangle\rangle$ forms a semiring. The image of $\lambda \in \mathbb{R}_{\max}$ by the canonical injection into $\mathbb{R}_{\max}\langle\langle \Sigma^* \rangle\rangle$ is still denoted by $\lambda$. In particular, the neutral elements of $\mathbb{R}_{\max}\langle\langle \Sigma^* \rangle\rangle$ are 0 and 1.

Let $Q$ and $\Sigma$ be two finite sets. A *max-plus automaton of set of states (dimension) $Q$* over the alphabet $\Sigma$, is a triple $\mathcal{A} = (\alpha, \mu, \beta)$, where $\alpha \in \mathbb{R}_{\max}^{1 \times Q}$, $\beta \in \mathbb{R}_{\max}^{Q \times 1}$, and where $\mu : \Sigma^* \to \mathbb{R}_{\max}^{Q \times Q}$ is a morphism of monoids. The morphism $\mu$ is uniquely determined by the family of matrices $\{\mu(a), a \in \Sigma\}$, and for $w = a_1 \cdots a_n$, we have $\mu(w) = \mu(a_1) \otimes \cdots \otimes \mu(a_n)$. The series *recognized by* $\mathcal{A}$ is by definition $S(\mathcal{A}) = \bigoplus_{u \in \Sigma^*} (\alpha \mu(u) \beta) u$. This is just a specialization to the max-plus semiring of the classical notion of an automaton with multiplicities

---

[1] The version of [18] available on the author's website has been correctly modified.

over a semiring [4,9,17]. By the Kleene-Schützenberger Theorem [21], the set of series recognized by a max-plus automaton is equal to the set of rational series over $R_{max}$. We denote it by Rat.

A state $i \in Q$ is *initial*, resp. *final*, if $\alpha_i \neq 0$, resp. $\beta_i \neq 0$. As usual a max-plus automaton is represented graphically by a labeled weighted digraph with ingoing and outgoing arcs for initial and final states, see Fig. 4. The terminology of graph theory is used accordingly (*e.g.* (simple) path or circuit of an automaton, union of automata, ... ). A path which is both starting with an ingoing arc and ending with an outgoing arc is called a *successful path*. The *label of a path* is the concatenation of the labels of the successive arcs, the *weight of a path* is the product ($\otimes$) of the weights of the successive arcs (including the ingoing and the outgoing arc, need it be). We use the following notations for paths in an automaton $\mathcal{A} = (\alpha, \mu, \beta)$:

$$p \to q, \quad \to p \to q, \quad p \to q \to, \quad p \xrightarrow{u|x} q, \quad \left[ p \xrightarrow{u|x} q \right]_{\mathcal{A}}, \text{ if } \mu(u)_{pq} = x \,.$$

An automaton is *trim* if any state belongs to a successful path.

A *heap* or *Tetris model* [24], consists of a finite set of slots $\mathcal{R}$, and a finite set of rectangular pieces $\Sigma$. Each piece $a \in \Sigma$ is of height 1 and occupies a determined subset $\mathcal{R}(a)$ of the slots. To a word $u = u_1 \cdots u_k \in \Sigma^*$ is associated the *heap* obtained by piling up in order the pieces $u_1, \dots, u_k$, starting with an horizontal ground and according to the Tetris game mechanism (pieces are subject to gravity and fall down vertically until they meet either a previously piled up piece or the ground). Consider the morphism generated by the matrices $M(a) \in R_{max}^{\mathcal{R} \times \mathcal{R}}, a \in \Sigma$, defined by

$$M(a)_{ij} = \begin{cases} 1 & \text{if } i, j \in \mathcal{R}(a), \\ 0 & \text{if } i = j \notin \mathcal{R}(a), \\ -\infty & \text{otherwise} \,. \end{cases}$$

Let $x(u)_i$ be the height of the heap $u$ on slot $i \in \mathcal{R}$. We have ([5,11,12]): $x(u)_i = 1M(u)\delta_i$, where $1 = (1, \dots, 1) \in R_{max}^{1 \times \mathcal{R}}$ and $\delta_i \in R_{max}^{\mathcal{R} \times 1}$ is defined by $(\delta_i)_j = 1$ if $j = i$ and 0 otherwise. In other words, $x(\cdot)_i : \Sigma^* \to R_{max}$ is recognized by the max-plus automaton $(1, M, \delta_i)$. We call $(1, M, \delta)$, $\delta = \bigoplus_{i \in I} \delta_i, I \subseteq \mathcal{R}$, a *heap automaton* (associated with the heap model). Among max-plus automata, heap automata are particularly convenient and playful, due to the underlying geometric interpretation. Here, they are used as a source of examples and counter-examples, *e.g.* Figs. 2, 3, and 5.

Consider a finite family of max-plus automata $(\mathcal{A}_i)_{i \in I}$ with respective dimensions $(Q_i)_{i \in I}$. Set $\mathcal{A}_i = (\alpha^{i}, \mu^{i}, \beta^{i})$. The corresponding *product automaton* is $\mathcal{P} = \mathcal{P}(\mathcal{A}_i)_{i \in I} = (A, M, B)$ defined as follows. It is an automaton with multiplicities in the product semiring $R_{max}^I$, its dimension is $Q = \prod_{i \in I} Q_i$, and

$$\forall p, q \in Q, \quad A_p = (\alpha_{p_i}^{i})_{i \in I}, \quad \forall a \in \Sigma, \quad M(a)_{p,q} = (\mu^{i}(a)_{p_i, q_i})_{i \in I}, \quad B_p = (\beta_{p_i}^{i})_{i \in I} \,.$$

Clearly, the automaton $\mathcal{P}$ satisfies: $\forall u \in \Sigma^*$, $\bigoplus_{i \in I} \langle \mathcal{P}, u \rangle_i = \langle \bigoplus_{i \in I} S(\mathcal{A}_i), u \rangle = \bigoplus_{i \in I} \alpha^{i} \mu^{i}(u) \beta^{i}$. The *tensor product automaton* of $(\mathcal{A}_i)_{i \in I}$, denoted by $\odot_{i \in I} \mathcal{A}_i$, is defined in the same way as $\mathcal{P}(\mathcal{A}_i)_i$ except that a multiplicity $(x_i)_i \in \mathrm{R}_{\max}^I$ is replaced by the multiplicity $\bigotimes_i x_i \in \mathrm{R}_{\max}$.

## 2.1    Ambiguity and Sequentiality

Consider a max-plus automaton $\mathcal{A} = (\alpha, \mu, \beta)$ over $\Sigma$. The automaton is *sequential* if there is a unique initial state and if for all $i$, and for all $a \in \Sigma$, there is at most one $j$ such that $\mu(a)_{ij} \neq 0$. In the case of a Boolean automaton, we also say *deterministic* for sequential. The automaton $\mathcal{A}$ is *unambiguous* if for any word $u \in \Sigma^*$, there is at most one successful path of label $u$. The automaton is *finitely ambiguous* if there exists some $k \in \mathrm{N}$ such that for any word $u \in \Sigma^*$, there are at most $k$ successful paths of label $u$. The minimal such $k$ is called the *degree of ambiguity* of the automaton. Clearly, 'sequential' implies 'unambiguous' which implies 'finitely ambiguous'. The automaton is *infinitely ambiguous* if it is not finitely ambiguous.

Consider a series $S \in \mathrm{Rat}$. The series is *sequential* (resp. *unambiguous, finitely ambiguous*) if there exists a sequential (resp. unambiguous, finitely ambiguous) max-plus automaton recognizing it. The *degree of ambiguity* of a finitely ambiguous series is the minimal degree of ambiguity of an automaton recognizing it. The set of sequential, unambiguous, and finitely ambiguous series are denoted respectively by Seq, NAmb, and FAmb. Define $\mathrm{FSeq} = \{S \mid \exists k, \exists S_1, \ldots, S_k \in \mathrm{Seq}, \ S = S_1 \oplus \cdots \oplus S_k\}$.

Consider a total order on $\Sigma$ and the corresponding lexicographic order on $\Sigma^*$. Given a series $S \neq 0$, define the *normalized series* $\varphi(S)$ by $\varphi(S) = \bigoplus_{u \in \Sigma^*} (\langle S, u \rangle - \langle S, u_0 \rangle)u$, where $u_0$ is the smallest word of Supp $S$. The *(left) quotient* of a series $S$ by a word $w$ is the series $w^{-1}S$ defined by $w^{-1}S = \bigoplus_{u \in \Sigma^*} \langle S, wu \rangle u$.

A series $S$ is rational if and only if the semi-module of series $\langle w^{-1}S, w \in \Sigma^* \rangle$ is finitely generated, *i.e.* if there exists $S_1, \ldots S_k$, such that: $\forall w \in \Sigma^*, \exists \lambda_1, \ldots, \lambda_k \in \mathrm{R}_{\max}$, $w^{-1}S = \bigoplus_i \lambda_i S_i$. A series $S$ is sequential if and only if the set of series $\{\varphi(w^{-1}S), w \in \Sigma^*\}$ is finite.

**Proposition 1.** *A trim automaton $\mathcal{A}$ of dimension $Q$ is infinitely ambiguous if and only if there exist $p, q \in Q, p \neq q$, and $v \in \Sigma^*$, such that $p \xrightarrow{v} p$, $p \xrightarrow{v} q$, $q \xrightarrow{v} q$. This can be checked in polynomial time.*

For a proof, see [27] and the references therein. Observe that the (in)finite ambiguity is independent of the underlying semiring. Next result is due to Mohri [18] and is an adaptation of a classical result of Choffrut on functional transducers, see [3,7,8] (for the decidability) and [2,26] (for the polynomial complexity).

**Theorem 2.** *Let $\mathcal{A}$ be an unambiguous max-plus automaton. There exists a polynomial time algorithm to decide whether $S(\mathcal{A})$ is a sequential series.*

If $\mathcal{A}$ is unambiguous and $S(\mathcal{A})$ is sequential, a sequential automaton recognizing the series can be effectively constructed from $\mathcal{A}$ using an adaptation of the subset construction of Boolean automata [1,6,18].

## 3 Hierarchy of Series

The examples in this section illustrate the classes of series on which we work.

$$\text{Seq} \subsetneq \underset{(\S3.1)}{\text{NAmb} \cap \text{FSeq}} \subsetneq \underset{(\S3.2)}{\overset{\text{FSeq}}{}} \subsetneq \underset{(\S3.3)}{\text{NAmb}} \subsetneq \underset{(\S3.4)}{\text{FAmb}} \subsetneq \underset{(\S3.5)}{\text{Rat}} \subsetneq \underset{(\S3.6)}{\text{Series}}$$

### 3.1  A Series in $\overline{\text{Seq}} \cap \text{NAmb} \cap \text{FSeq}$

An example over a one-letter alphabet is provided in Fig. 1. The recognized
series is

$$\langle S, a^n \rangle = \begin{cases} 0 & \text{if } n \text{ is odd,} \\ n & \text{if } n \text{ is even.} \end{cases}$$

In fact, any max-plus rational series over a one-letter alphabet is unambiguous and a sum of sequential series [16,19].

### 3.2  A Series in $\text{FSeq} \cap \overline{\text{NAmb}}$

The series $\langle S, u \rangle = |u|_a \oplus |u|_b$ over the alphabet $\{a, b\}$ is a sum of two sequential
series: the heap automaton of Fig. 2 recognizes this series. On the other hand,
one can prove that $S$ is ambiguous.

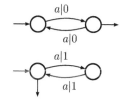

**Fig. 1.** $\overline{\text{Seq}} \cap \text{NAmb} \cap \text{FSeq}$

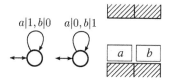

**Fig. 2.** $\text{FSeq} \cap \overline{\text{NAmb}}$

### 3.3 Series in NAmb ∩ $\overline{\text{FSeq}}$

Here are two examples of such series (without justifications).

The first example is the series $S$ given by the heap automaton of Fig. 3a, or equivalently by the automaton of Fig. 3b. The second example is the series given by the automaton of Fig. 4.

The series recognized by this automaton is: $\langle S, a^{m_1} b^{n_1} \cdots a^{m_p} b^{n_p} \rangle = \sum_{m_i \text{even}} m_i,$

where $m_1 \in \mathbb{N}$, $m_{k+1} \in \mathbb{N} - \{0\}$ and $n_k \in \mathbb{N} - \{0\}$ for $1 \leqslant k \leqslant p-1$ and $n_p \in \mathbb{N}$. This series is clearly unambiguous.

### 3.4 Series in $\overline{\text{NAmb}} \cap \overline{\text{FSeq}} \cap \text{FAmb}$

Consider the heap automaton given in Fig. 5a. The corresponding series is at most two-ambiguous since it is also recognized by the two-ambiguous automaton of Fig. 5b. It cannot be unambiguous: on $\{a, b\}^*$, it coincides with the series of Fig. 2 which is in $\overline{\text{NAmb}}$. It cannot be a finite sum of sequential series: on $\{b, c\}^*$, it coincides with the series of Fig. 3 which is in $\overline{\text{FSeq}}$.

Besides, Weber has given examples of series which are $k$-ambiguous and not $(k-1)$-ambiguous [25, Theorem 4.2].

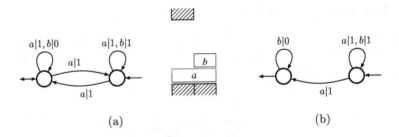

(a)                    (b)

**Fig. 3.** NAmb ∩ $\overline{\text{FSeq}}$

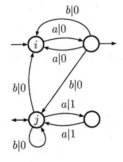

**Fig. 4.** NAmb ∩ $\overline{\text{FSeq}}$

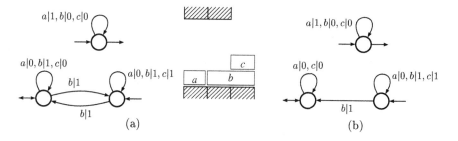

**Fig. 5.** $\overline{\mathrm{NAmb}} \cap \overline{\mathrm{FSeq}} \cap \mathrm{FAmb}$

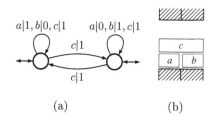

**Fig. 6.** $\overline{\mathrm{FAmb}} \cap \mathrm{Rat}$

## 3.5    Series in $\overline{\mathrm{FAmb}} \cap \mathbf{Rat}$

Consider the series $S$ recognized by the automaton of Fig. 6. Assume that $S$ is finitely ambiguous. Using the result of Corollary 8 below, $S$ is recognized by a finite union of unambiguous automata of the same support, say $\mathcal{A}_1, \dots \mathcal{A}_k$. Denote by $n$ the maximal size of an automaton $\mathcal{A}_i$.

Using words of the form $(a^{n+\lambda_1 n!} b^{n+\mu_1 n!} c) \cdots (a^{n+\lambda_k n!} b^{n+\mu_k n!} c)$, we can prove that for any $u \in \{01, 10\}^k$, there exists $i$ such that $\mathcal{A}_i$ contains a path of the form

$$\pi(u):$$

where the $j$-th $a$-loop has a mean weight $u_{2j+1}$ and the $j$-th $b$-loop has a mean weight $u_{2j+2}$, for any $j$. But an automaton $\mathcal{A}_i$ cannot have paths $\pi(u)$ and $\pi(u')$ for $u \neq u'$ since it is unambiguous. So there is a contradiction.

## 3.6    Rational and Non-rational Series ($\overline{\mathrm{Rat}}$)

A max-plus series is non-rational as soon as its support is a non-rational language. Here, we present a less trivial example of non-rational max-plus series.

Clearly, we have[2]

$$R_{\max}NAmb = R_{\min}NAmb = NAmb.$$

On the other hand, it is easy to find $S \in R_{\min}FSeq \cap R_{\min}\overline{NAmb}$ such that $S \notin R_{\max}Rat$. Consider for instance the series $S = \min(|w|_a, |w|_b)$ (recognized by the automaton of Fig. 2 seen as a min-plus automaton).

Let us prove that $S$ does not belong to $R_{\max}Rat$. If it does: let $S_1, \ldots S_n$ be a minimal generating family of $\langle u^{-1}(S), u \in \Sigma^* \rangle$ (see §2.1), we have: $\forall u \in \Sigma^*, \exists \lambda_1^{(u)}, \ldots \lambda_n^{(u)}, u^{-1}(S) = \bigoplus_i \lambda_i^{(u)} \otimes S_i$. The restrictions of the quotients of $S$ to $b^*$ are bounded, hence so are the restrictions of the $S_i$'s. Let $k_i$ be such that: $\langle S_i, b^{k_i} \rangle = \max_k \langle S_i, b^k \rangle$. It follows that for any word $u$: $\max_k \langle u^{-1}S, b^k \rangle = \max_{k_i} \langle u^{-1}S, b^{k_i} \rangle$. Consider $k > \max_i k_i$. Then arises a contradiction:
$$\max_l \langle (a^k)^{-1}S, b^l \rangle = \langle (a^k)^{-1}S, b^k \rangle = k > \max_{k_i} \langle (a^k)^{-1}S, b^{k_i} \rangle.$$

## 4 From Finitely Ambiguous to Union of Unambiguous

Weber [25] has proved that a finitely ambiguous $N_{\max}$-automaton can be turned into a union of unambiguous ones. We present a completely different and simpler proof that holds in $R_{\max}$.

In this section, we work on the structure of the automaton. This is the reason why we consider Boolean automata.

Let $\mathcal{A}$ be an automaton. The *past* of a state $p$ is the set of words that label a path from some initial state to $p$. The *future* of $p$ is the set of words that label a path from $p$ to some final state. We write:

$$\mathsf{Past}_{\mathcal{A}}(p) = \{w \in \Sigma^* \mid (\alpha\mu(w))_p = 1\}, \quad \mathsf{Fut}_{\mathcal{A}}(p) = \{w \in \Sigma^* \mid (\mu(w)\beta)_p = 1\}.$$

Let $\mathcal{A} = (\alpha, \mu : \Sigma^* \to B^{Q \times Q}, \beta)$ be an automaton. Let us recall the usual determinization procedure *via* the subset construction. Let $R$ be the least subset of $B^{1 \times Q}$ inductively defined by: $\alpha \in R, \quad X \in R \Rightarrow \forall a \in \Sigma, X\mu(a) \in R$.

Let $\mathcal{D} = \mathcal{D}(\mathcal{A}) = (J, \nu : \Sigma^* \to B^{R \times R}, U)$ be the *determinized automaton* of $\mathcal{A}$ defined by:

$$J = \{\alpha\}, \qquad U = \{P \in R \mid P\beta = 1\}, \qquad \nu(a)_{P,P'} = 1 \iff P' = P\mu(a).$$

**Lemma 3.** *i) Let $\mathcal{A}$ be an automaton and $\mathcal{D}$ its determinized automaton. Then, $\forall P$ state of $\mathcal{D}$,* $\mathsf{Past}_{\mathcal{D}}(P) \subseteq \bigcap_{p \in P} \mathsf{Past}_{\mathcal{A}}(p),$ *and* $\mathsf{Fut}_{\mathcal{D}}(P) = \bigcup_{p \in P} \mathsf{Fut}_{\mathcal{A}}(p).$
*ii) Let $\mathcal{A}$ and $\mathcal{B}$ be two automata and $\mathcal{A} \odot \mathcal{B}$ their tensor product (c.f. §2.1), then, for all $(p, q)$ state of $\mathcal{A} \odot \mathcal{B}$,*

$$\mathsf{Past}_{\mathcal{A} \odot \mathcal{B}}(p, q) = \mathsf{Past}_{\mathcal{A}}(p) \cap \mathsf{Past}_{\mathcal{B}}(q), \qquad \mathsf{Fut}_{\mathcal{A} \odot \mathcal{B}}(p, q) = \mathsf{Fut}_{\mathcal{A}}(p) \cap \mathsf{Fut}_{\mathcal{B}}(q).$$

---

[2] In this paragraph, it is necessary to distinguish between $R_{\min}$ and $R_{\max}$: for $R = $ Rat or NAmb, we use the respective notations $R_{\min}R$, $R_{\max}R$. If $S \in R_{\max}\langle\langle \Sigma^* \rangle\rangle$, we identify $S$ with $\tilde{S} \in R_{\min}\langle\langle \Sigma^* \rangle\rangle$ such that $\langle \tilde{S}, w \rangle = \langle S, w \rangle$ if $w \in \mathrm{Supp}\, S$ and $\langle \tilde{S}, w \rangle = +\infty$ if $\langle S, w \rangle = -\infty$.

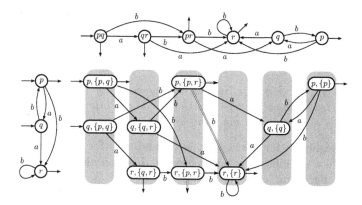

**Fig. 7.** $\mathcal{A}$ (left), its determinized automaton (top) and its Schützenberger covering

The constructions and results given in Propositions 4 and 6 are inspired by Schützenberger [22]. They have been explicitly stated by Sakarovitch in [20].

Let $\mathcal{A}$ be an automaton and $\mathcal{D}$ its determinized automaton. The product $\mathcal{S} = \mathcal{A} \odot \mathcal{D}$ is called the *Schützenberger covering* of $\mathcal{A}$.

**Proposition 4.**
*i) The states of $\mathcal{S}$ are exactly the pairs $(p, P)$, where $P$ is a state of $\mathcal{D}$ and $p \in P$. We call a set $\{(p, P) \mid p \in P\}$ of states of $\mathcal{S}$ a* column *(in gray in Fig. 7).*
*ii) The canonical surjection $\psi$ from the transitions of $\mathcal{S}$ onto the transitions of $\mathcal{A}$ induces a one-to-one mapping between the successful paths of $\mathcal{S}$ and $\mathcal{A}$.*
*iii) Let $P$ be a state of $\mathcal{D}$.*
*Then, for every $p$ in $P$, $\mathsf{Past}_{\mathcal{S}}(p, P) = \mathsf{Past}_{\mathcal{D}}(P)$,     $\mathsf{Fut}_{\mathcal{S}}(p, P) = \mathsf{Fut}_{\mathcal{A}}(p)$.*
*Thus, all the states of a given column have the same past.*

**Definition 5.** *In $\mathcal{S}$, different transitions with the same label, the same destination and whose origins belong to the same column are* competing. *Likewise, different final states of the same column are* competing. *A* competing set *is a maximal set of competing transitions or competing final states.*

Let $\mathcal{U}$ be the automaton obtained from $\mathcal{S}$ by removing all transitions except one in every competing set and by turning all final states of a column, except one, into non-final states. The choice of the transition (or the final state) to keep in a competing set is arbitrary.

**Proposition 6.** *Let $\mathcal{S}$ and $\mathcal{U}$ be automata defined as above. Then,*
*i) $\mathsf{Past}_{\mathcal{U}}(p, P) = \mathsf{Past}_{\mathcal{S}}(p, P)$.*
*ii) Futures of states in a column of $\mathcal{U}$ are disjoint and*

$$\bigcup_{p \in P} \mathsf{Fut}_{\mathcal{U}}(p, P) = \bigcup_{p \in P} \mathsf{Fut}_{\mathcal{S}}(p, P).$$

*Consequently, the automaton $\mathcal{U}$ is an unambiguous automaton equivalent to $\mathcal{A}$.*

For instance, the covering of Fig. 7 has two competing sets (drawn with double lines); the first one contains two transitions with label $b$ that arrive in $(r, \{r\})$ the second one contains states $(p, \{p, r\})$ and $(r, \{p, r\})$ which are both final. The above selection principle gives rise to four possible automata.

**Proposition 7.** *Let $S$ be the Schützenberger covering of a finitely ambiguous automaton. Then, competing transitions of $S$ do not belong to any circuits of $S$. Thus a path of $S$ contains at most one transition of each competing set.*

As a consequence, for every path in $S$ (and thus for every path in $\mathcal{A}$), one can compute an unambiguous automaton $\mathcal{U}$ that contains this path. Consider the following algorithm.

As they do not belong to any circuit, competing sets of $S$ are partially ordered. Let $C$ be the set of maximal competing sets of $S$ (there is no path from any element of $C$ to another competing set).
– Let $S_1$ and $S_2$ be two copies of $S$. For every competing set $X$ in $C$, let $x$ be an element of $X$;
– if $x$ is a transition, remove every transition of $X \backslash \{x\}$ in $S_1$ and remove $x$ in $S_2$;
– if $x$ is a final state, make every state of $X \backslash \{x\}$ in $S_1$ and $x$ in $S_2$ non-final.
– Apply inductively this algorithm to $S_1$ and $S_2$.

The result is a finite set of unambiguous automata, all recognizing the language of $\mathcal{A}$, and containing all the paths of $\mathcal{A}$. Notice that the cardinality of this set may be larger than the degree of ambiguity.

If $\mathcal{A}$ is an automaton with multiplicities, since there is a canonical mapping from the transitions (resp. initial states, resp. final states) of the Schützenberger covering $S$ of $\mathcal{A}$ onto the transitions (resp. initial states, resp. final states) of $\mathcal{A}$, one can decorate every transition (resp. initial state, resp. final state) of $S$ with the corresponding multiplicity in $\mathcal{A}$.

Obviously, since there is a one-to-one mapping between the paths of $\mathcal{A}$ and those of $S$, the series realized by $S$ is equal to the one realized by $\mathcal{A}$.

**Corollary 8.** *A finitely ambiguous max-plus automaton can be effectively turned into an equivalent finite union of unambiguous max-plus automata that all have the same support.*

## 5    The Decidability Result

In this section, we show that if a series is realized by a finite union of unambiguous automata having the same support and satisfying a certain property denoted by **(P)** and stated below, then it can be realized by one unambiguous automaton. If **(P)** is not satisfied, then the series is not sequential. Associated with the results of the previous section, this enables to prove Theorem 11.

**Definition 9.** *Let $\mathcal{P}$ be the product automaton of $(\mathcal{A}_i)_{i \in I}$, a finite family of trim max-plus automata. Let $\theta$ be a simple circuit of $\mathcal{P}$, whose weight is $(x^{\iota^i})_{i \in I}$. The set of victorious coordinates of $\theta$, denoted by $Vict(\theta)$, is the set of coordinates on which the weight of $\theta$ is maximal, i.e. : $Vict(\theta) = \{i \in I | x^{\iota^i} = \max_{j \in I}\{x^{\iota^j}\}\}$.*

This definition and the notation can be naturally extended to a path or a strongly connected subgraph of $\mathcal{P}$, as the intersection of the victorious coordinates of all the simple circuits of, respectively, the path or the subgraph.

Let us define the 'dominance' property (**P**):

*For each successful path $\pi$ of the product automaton $\mathcal{P}$, the set of victorious coordinates of $\pi$ is not empty.*

Obviously, the number of simple circuit is finite. Hence (**P**) is a decidable property.

Let $\left(\mathcal{A}_i = (\alpha^{L^i} \in \mathrm{R}_{max}^{Q_i}, \mu^{L^i} : \Sigma^* \to \mathrm{R}_{max}^{Q_i \times Q_i}, \beta^{L^i} \in \mathrm{R}_{max}^{Q_i})\right)_{i \in I}$ be a finite family of unambiguous trim automata, all with the same support, and let $\mathcal{P}$ be the product automaton with set of states $Q = \Pi_{i \in I} Q_i$. We assume that $\mathcal{P}$ satisfies the dominance property (**P**).

Let $N = |Q|$ and $M = \max(\max_{i,a,p,q} \mu^{L^i}(a)_{p,q}, \max_{i,p} \beta_p^{L^i}) - \min(\min_{i,a,p,q} \mu^{L^i}(a)_{p,q}, \min_{i,p} \beta_p^{L^i})$.

We use the following notations as shortcuts. For $\mathbf{x} = (x_i)_{i \in I} \in \mathrm{R}_{max}^I$, set $\check{\mathbf{x}} = \min_{i \in I}\{x_i | x_i \neq -\infty\}$ and $\underline{\mathbf{x}} = \mathbf{x} - (\check{\mathbf{x}}, \ldots, \check{\mathbf{x}})$.

Set $I = \{1, \ldots, n\}$. The automaton $\mathcal{U}$ is now defined. The states of $\mathcal{U}$ belong to $\mathrm{R}_{max}^n \times Q$.

*Initial states.* Given a tuple $(q^{L^1}, \ldots, q^{L^n})$ such that $q^{L^i}$ is an initial state of $\mathcal{A}_i$, if we set $\mathbf{a} = (\alpha_{q^{L^1}}^{L^1}, \ldots, \alpha_{q^{L^n}}^{L^n})$, then $(\underline{\mathbf{a}}, q^{L^1}, \ldots, q^{L^n})$ is an initial state of $\mathcal{U}$ and the weight of the ingoing arc is $\check{\mathbf{a}}$.

*States and transitions.* If $\mathbf{p} = (z_1, \ldots, z_n, p^{L^1}, \ldots, p^{L^n})$ is a state of $\mathcal{U}$, then for each transition in $\mathcal{P}$ of type: $(p^{L^1}, \ldots, p^{L^n}) \xrightarrow{a|\mathbf{x}} (q^{L^1}, \ldots, q^{L^n})$ such that $x_i \neq -\infty$ for all $i$, there is a transition in $\mathcal{U}$ leaving $\mathbf{p}$, labeled by the letter $a$ and that we now describe. Let $\mathbf{y} = (z_1 + x_1, \ldots, z_n + x_n)$. Let $V$ be the set of victorious coordinates of the maximal strongly connected subgraph of $(q^{L^1}, \ldots, q^{L^n})$ in $\mathcal{P}$. Since $\mathcal{P}$ satisfies (**P**), the set $V$ is non-empty. Let $j \in V$ be such that $y_j = \min_{k \in V}\{y_k | y_k \neq -\infty\}$, and let $\mathbf{y}' \in \mathrm{R}_{max}^n$ be defined by:

$$\forall i, \quad y_i' = \begin{cases} -\infty & \text{if } y_i < y_j - NM, \\ y_i & \text{otherwise.} \end{cases}$$

Now $(\underline{\mathbf{y}'}, q^{L^1}, \ldots, q^{L^n})$ is a state of $\mathcal{U}$ and we have the following transition:

$$\left[(z_1, \ldots, z_n, p^{L^1}, \ldots, p^{L^n}) \xrightarrow{a|\check{\mathbf{y}}'} (\underline{\mathbf{y}'}, q^{L^1}, \ldots, q^{L^n})\right]_{\mathcal{U}}.$$

*Final states.* If $\mathbf{q} = (z_1, \ldots, z_n, q^{L^1}, \ldots, q^{L^n})$ is a state of $\mathcal{U}$, and if $q^{L^i}$ is a final state of $\mathcal{A}_i$ for all $i$, then $\mathbf{q}$ is a final state of $\mathcal{U}$ and the weight of the outgoing arc is $\max_{i \in I}\{z_i + \beta_{q^{L^i}}^{L^i}\}$.

**Proposition 10.** *Consider a finite family $(\mathcal{A}_i)_{i \in I}$ of trim and unambiguous max-plus automata having the same support. Let $\mathcal{P}$ be the corresponding product automaton. If $\mathcal{P}$ satisfies the property (**P**), then the automaton $\mathcal{U}$ defined above is finite, unambiguous and realizes the series $\bigoplus_{i \in I} S(\mathcal{A}_i)$.*

*If $\mathcal{P}$ does not satisfy the property (**P**), then the series $\bigoplus_{i \in I} S(\mathcal{A}_i)$ is not sequential.*

Note that if (**P**) is not satisfied, we may have either $\bigoplus_{i \in I} S(\mathcal{A}_i) \in \overline{\text{Seq}} \cap$ NAmb or $\bigoplus_{i \in I} S(\mathcal{A}_i) \in \overline{\text{NAmb}}$. We can now state the main result of this paper.

**Theorem 11.** *One can decide in an effective way, whether the series recognized by a finitely ambiguous max-plus automaton is sequential.*

More precisely, turn first the finitely ambiguous automaton into an equivalent finite union of unambiguous automata, all having the same support (Corollary 8). Then check the property (**P**) on the new family of automata. If (**P**) is satisfied, build the unambiguous automaton $\mathcal{U}$ (Proposition 10), then decide the sequentiality of $\mathcal{U}$ (Theorem 2).

# References

1. C. Allauzen and M. Mohri. Efficient algorithms for testing the twins property. *Journal of Automata, Languages and Combinatorics*, 2003. (To appear).
2. M.-P. Béal, O. Carton, C. Prieur, and J. Sakarovitch. Squaring transducers: An efficient procedure for deciding functionality and sequentiality. *Theor. Comput. Sci.*, 292:45–63, 2003.
3. J. Berstel. *Transductions and context-free languages*. B. G. Teubner, 1979.
4. J. Berstel and C. Reutenauer. *Rational Series and their Languages*. Springer Verlag, 1988.
5. M. Brilman and J.M. Vincent. Dynamics of synchronized parallel systems. *Stochastic Models*, 13(3):605–619, 1997.
6. A.L. Buchsbaum, R. Giancarlo, and J.R. Westbrook. On the determinization of weighted finite automata. *SIAM J. Comput.*, 30(5):1502–1531, 2000.
7. C. Choffrut. Une caractérisation des fonctions séquentielles et des fonctions sous-séquentielles en tant que relations rationnelles. *Theor. Comput. Sci.*, 5:325–337, 1977.
8. C. Choffrut. *Contribution à l'étude de quelques familles remarquables de fonctions rationnelles*. Thèse d'état, Univ. Paris VII, 1978.
9. S. Eilenberg. *Automata, languages and machines*, vol. A. Academic Press, 1974.
10. S. Gaubert. Performance evaluation of (max,+) automata. *IEEE Trans. Aut. Cont.*, 40(12):2014–2025, 1995.
11. S. Gaubert and J. Mairesse. Task resource models and (max,+) automata. In J. Gunawardena, editor, *Idempotency*, volume 11, pages 133–144. Cambridge University Press, 1998.
12. S. Gaubert and J. Mairesse. Modeling and analysis of timed Petri nets using heaps of pieces. *IEEE Trans. Aut. Cont.*, 44(4):683–698, 1999.
13. K. Hashigushi. Algorithms for determining relative star height and star height. *Inf. Comput.*, 78(2):124–169, 1988.

14. K. Hashigushi, K. Ishiguro, and S. Jimbo. Decidability of the equivalence problem for finitely ambiguous finance automata. *Int. J. Algebra Comput.*, 12(3):445–461, 2002.

15. D. Krob. The equality problem for rational series with multiplicities in the tropical semiring is undecidable. *Int. J. Algebra Comput.*, 4(3):405–425, 1994.

16. D. Krob and A. Bonnier-Rigny. A complete system of identities for one-letter rational expressions with multiplicities in the tropical semiring. *Theor. Comput. Sci.*, 134:27–50, 1994.

17. W. Kuich and A. Salomaa. *Semirings, Automata, Languages*, volume 5 of *EATCS*. Springer-Verlag, 1986.

18. M. Mohri. Finite-state transducers in language and speech processing. *Comput. Linguist.*, 23(2):269–311, 1997.

19. P. Moller. *Théorie algébrique des systèmes à événements discrets*. PhD thesis, École des Mines de Paris, 1988.

20. J. Sakarovitch. A construction in finite automata that has remained hidden. *Theor. Comput. Sci.*, 204:205–231, 1998.

21. M.-P. Schützenberger. On the definition of a family of automata. *Information and Control*, 4(2–3):245–270, 1961.

22. M.-P. Schützenberger. Sur les relations rationnelles entre monoïdes libres. *Theor. Comput. Sci.*, 3:243–259, 1976.

23. I. Simon. Recognizable sets with multiplicities in the tropical semiring. In *Mathematical Foundations of Computer Science, Proc. 13th Symp.*, number 324 in LNCS, pages 107–120, 1988.

24. G.X. Viennot. Heaps of pieces, I: Basic definitions and combinatorial lemmas. In Labelle and Leroux, editors, *Combinatoire Énumérative*, number 1234 in Lect. Notes in Math., pages 321–350. Springer, 1986.

25. A. Weber. Finite-valued distance automata. *Theor. Comput. Sci.*, 134:225–251, 1994.

26. A. Weber and R. Klemm. Economy of description for single-valued transducers. *Information and Computation*, 118(2):327–340, 1995.

27. A. Weber and H. Seidl. On the degree of ambiguity of finite automata. *Theor. Comput. Sci.*, 88(2):325–349, 1991.

# Minimizing Finite Automata Is Computationally Hard

Andreas Malcher

Institut für Informatik, Johann Wolfgang Goethe Universität
D-60054 Frankfurt am Main, Germany
malcher@psc.informatik.uni-frankfurt.de

**Abstract.** It is known that deterministic finite automata (DFAs) can be algorithmically minimized, i.e., a DFA $M$ can be converted to an equivalent DFA $M'$ which has a minimal number of states. The minimization can be done efficiently [6]. On the other hand, it is known that unambiguous finite automata (UFAs) and nondeterministic finite automata (NFAs) can be algorithmically minimized too, but their minimization problems turn out to be NP-complete and PSPACE-complete, respectively [8]. In this paper, the time complexity of the minimization problem for two restricted types of finite automata is investigated. These automata are nearly deterministic, since they only allow a small amount of nondeterminism to be used. The main result is that the minimization problems for these models are computationally hard, namely NP-complete. Hence, even the slightest extension of the deterministic model towards a nondeterministic one, e.g., allowing at most one nondeterministic move in every accepting computation or allowing two initial states instead of one, results in computationally intractable minimization problems.

## 1 Introduction

Finite automata are a well-investigated concept in theoretical computer science with a wide range of applications such as lexical analysis, pattern matching, or protocol specification in distributed systems. Due to time and space constraints it is often very useful to provide minimal or at least succinct descriptions of such automata. Deterministic finite automata (DFAs) and their corresponding language class, the set of regular languages, possess many nice properties such as, for example, closure under many language operations and many decidable questions. In addition, most of the decidability questions for DFAs, such as membership, emptiness, or equivalence, are efficiently solvable (cf. Sect. 5.2 in [15]). Furthermore, in [6] a minimization algorithm for DFAs is provided working in time $O(n \log n)$, where $n$ denotes the number of states of the given DFA.

It is known that both nondeterministic finite automata (NFAs) and DFAs accept the set of regular languages, but NFAs can achieve exponential savings in size when compared to DFAs [13]. Unfortunately, certain decidability questions, which are solvable in polynomial time for DFAs, are computationally hard for NFAs such as equivalence, inclusion, or universality [14,15]. Furthermore,

Z. Ésik and Z. Fülöp (Eds.): DLT 2003, LNCS 2710, pp. 386–397, 2003.

the minimization of NFAs is proven to be PSPACE-complete in [8]. In the latter paper, it is additionally shown that unambiguous finite automata (UFAs) have an NP-complete minimization problem.

Therefore, we can summarize that determinism permits efficient solutions whereas the use of nondeterminism often makes solutions computationally intractable. Thus, one might ask what amount of nondeterminism is necessary to make things computationally hard, or, in other words, what amount of nondeterminism may be allowed so that efficiency is preserved.

Measures of nondeterminism in finite automata were first considered in [12] and [2] where the relation between the amount of nondeterminism of an NFA and the succinctness of its description is studied. Here, we look at computational complexity aspects of NFAs with a fixed finite amount of nondeterminism. In particular, these NFAs are restricted such that within every accepting computation at most a fixed number of nondeterministic moves is allowed to be chosen. It is easily observed that certain decidability questions then become solvable in polynomial time in contrast to arbitrary NFAs. However, the minimization problem for such NFAs is proven to be NP-complete.

We further investigate a model where the nondeterminism used is not only restricted to a fixed finite number of nondeterministic moves, but is additionally cut down such that only the first move is allowed to be a nondeterministic one. Hence we come to DFAs with multiple initial states (MDFAs) which were introduced in [5] and recently studied in [11] and [3]. The authors of the latter paper examine the minimization problem for MDFAs and prove its PSPACE-completeness. Their proof is a reduction from Finite State Automata Intersection [4] which states that it is PSPACE-complete to answer the question whether there is a string $x \in \Sigma^*$ accepted by each $A_i$, where DFAs $A_1, A_2, \ldots, A_n$ are given. As is remarked in [4], the problem becomes solvable in polynomial time when the number of DFAs is fixed. We would like to point out that the number of initial states is not part of the instance of the minimization problem for MDFAs discussed in [3]. Thus, one might ask whether minimization of MDFAs with a fixed number of initial states is possible in polynomial time. We will show in Sect. 3 that the minimization problem of such MDFAs is NP-complete even if only two initial states are given. In analogy to NFAs with fixed finite branching, certain decidability questions can be shown to be efficiently solvable.

## 2    Preliminaries and Definitions

Let $\Sigma^*$ denote the set of all strings over the finite alphabet $\Sigma$, $\epsilon$ the empty string, and $\Sigma^+ = \Sigma^* \setminus \{\epsilon\}$. By $|w|$ we denote the length of a string $w$ and by $|S|$ the cardinality of a set $S$. We assume that the reader is familiar with the common notions of formal language theory as presented in [7] as well as with the common notions of computational complexity theory that can be found in [4]. Let $L$ be a regular set; then $size(L)$ denotes the number of states of the minimal DFA accepting $L$. We say that two finite automata are equivalent if both accept the same language. The size of an automaton $M$, denoted by $|M|$, is defined to be the number of states. A state of a finite automaton will be called *trap state* when no accepting state can be obtained from that state on every input.

Concerning the definitions of NFAs with finite branching and MDFAs we follow the notations introduced in [2] and [11].

A nondeterministic finite automaton over $\Sigma$ is a tuple $M = (Q, \Sigma, \delta, q_0, F)$, with $Q$ a finite set of states, $q_0 \in Q$ the initial state, $F \subseteq Q$ the set of accepting states, and $\delta$ a function from $Q \times \Sigma$ to $2^Q$. A move of $M$ is a triple $\mu = (p, a, q) \in Q \times \Sigma \times Q$ with $q \in \delta(p, a)$. A computation for $w = w_1 w_2 \ldots w_n \in \Sigma^*$ is a sequence of moves $\mu_1 \mu_2 \ldots \mu_n$ where $\mu_i = (q_{i-1}, w_i, q_i)$ with $1 \leq i \leq n$. It is an accepting computation if $q_n \in F$. The language accepted by $M$ is $T(M) = \{w \in \Sigma^* \mid \delta(q_0, w) \cap F \neq \emptyset\}$. $M$ is an (incomplete) deterministic finite automaton if $|\delta(q, a)| \leq 1$ for all pairs $(q, a)$. The branching $\beta_M(\mu)$ of a move $\mu = (q, a, p)$ is defined to be $\beta_M(\mu) = |\delta(q, a)|$. The branching is extended to computations $\mu_1 \mu_2 \ldots \mu_n$, $n \geq 0$, by setting $\beta_M(\mu_1 \mu_2 \ldots \mu_n) = \beta_M(\mu_1) \cdot \beta_M(\mu_2) \cdot \ldots \cdot \beta_M(\mu_n)$. For each word $w \in T(M)$, let $\beta_M(w) = \min \beta_M(\mu_1 \mu_2 \ldots \mu_n)$ where $\mu_1 \mu_2 \ldots \mu_n$ ranges over all accepting computations of $M$ with input $w$. The branching $\beta_M$ of the automaton $M$ is $\beta_M = \sup \{\beta_M(w) \mid w \in T(M)\}$. The set of all NFAs with branching $\beta = k$ is defined as $\text{NFA}(\beta = k) = \{M \mid M \text{ is NFA and } \beta_M = k\}$.

A DFA with multiple initial states (MDFA) is a tuple $M = (Q, \Sigma, \delta, Q_0, F)$ and $M$ is identical to a DFA except that there is a set of initial states $Q_0$. The language accepted by an MDFA $M$ is $T(M) = \{w \in \Sigma^* \mid \delta(Q_0, w) \cap F \neq \emptyset\}$. An MDFA with $k = |Q_0|$ initial states is denoted by $k$-MDFA.

Let $\mathcal{A}, \mathcal{B}$ be two classes of finite automata. Following the notation of [8], we say that $\mathcal{A} \longrightarrow \mathcal{B}$ denotes the problem of converting a type-$\mathcal{A}$ finite automaton to a minimal type-$\mathcal{B}$ finite automaton. Formally:

PROBLEM $\mathcal{A} \longrightarrow \mathcal{B}$

INSTANCE A type-$\mathcal{A}$ finite automaton $M$ and an integer $l$.

QUESTION Is there an $l$-state type-$\mathcal{B}$ finite automaton $M'$ such that
$$T(M') = T(M)?$$

## 3   Minimizing MDFAs Is Computationally Hard

In this section we are going to show that the minimization problem for $k$-MDFAs is NP-complete. Throughout this section, $k \geq 2$ denotes a constant integer.

**Theorem 1.** $k$-MDFA $\longrightarrow k$-MDFA *is* NP-*complete.*

*Proof.* The problem is in NP, since a $k$-MDFA $M'$ with $|M'| \leq l$ can be determined nondeterministically and the equality $T(M) = T(M')$ can be tested in polynomial time as is shown below. At first $M$ and $M'$ are converted to DFAs in the following manner. Let $M = (Q, \Sigma, \delta, \{q_0^1, q_0^2, \ldots, q_0^k\}, F)$, $M_1 = (Q, \Sigma, \delta, q_0^1, F)$, $M_2 = (Q, \Sigma, \delta, q_0^2, F), \ldots, M_k = (Q, \Sigma, \delta, q_0^k, F)$. Then $T(M_1) \cup T(M_2) \cup \ldots \cup T(M_k) = T(M)$ and we construct a DFA $\hat{M}$ as the Cartesian product of $M_1, M_2, \ldots, M_k$ accepting $T(M_1) \cup \ldots \cup T(M_k)$ in the usual way. A DFA $\hat{M}'$ can be constructed from $M'$ analogously. The time complexity of the inequivalence problem of two DFAs is in NLOGSPACE $\subseteq$ P [9]. Hence $T(\hat{M}) = T(\hat{M}')$ can be tested in polynomial time.

The NP-hardness of the problem will be shown by reduction from the Minimum Inferred DFA problem. In [8] the NP-hardness of the Minimum Inferred DFA problem is used to prove that the Minimum Union Generation problem is NP-complete.

PROBLEM  Minimum inferred DFA [1]

INSTANCE  Finite alphabet $\Sigma$, two finite subsets $S, T \subset \Sigma^*$, integer $l$.

QUESTION  Is there an $l$-state DFA that accepts a language $L$ such that
$$S \subseteq L \text{ and } T \subseteq \Sigma^* \setminus L?$$
Such an $l$-state DFA will be called *consistent* with $S$ and $T$.

The essential idea of the reduction in [8] is to design a language $L_5$ depending on $S$ and $T$ such that $L_5$ can be decomposed into the union of two DFAs with certain size bounds if and only if there is a DFA consistent with $S$ and $T$ satisfying a certain size bound. The difficult part is the "only-if"-portion. To this end, the notions of a "tail" and a "waist" are introduced, i.e., a sequence of states connected with #-edges only and ending at some state with no outgoing edges or with at least one outgoing edge, respectively. With the help of these two elements in $L_5$, it is possible to show that exactly one DFA contains a tail or a waist, respectively. Then it is not difficult to construct a DFA consistent with $S$ and $T$.

We now want to adopt the basic ideas of the above construction. We have to show that a modified language $L_5$ is accepted by a $k$-MDFA if and only if there is a DFA consistent with $S$ and $T$ satisfying a certain size bound. To apply the above result, our goal here is to decompose the $k$-MDFA into two sub-DFAs whose union is the language $L_5$ from [8]. To this end, the beginning of all words in $L_5$ is suitably modified. This enables us to show that there are two initial states such that from each of these states either all words ending in the tail or all words ending in the waist are accepted. This fact finally leads to the desired decomposition.

We follow the notations given in [8]. W.l.o.g. we may assume that $S \cap T = \emptyset$. Let #, \$ and £ be symbols not in $\Sigma$. Let $\Sigma' = \Sigma \cup \{\#, \$, £\}$, $m = 1 + size(\overline{T} \cap \overline{S})$, and $t = \max(k, m)$.

$$L_1 = \overline{T},$$
$$L_2 = \overline{T} \cap \overline{S},$$
$$L_3 = \{\$, £\} \#^t L_2 \#^m (£ \#^t L_2 \#^m)^*,$$
$$L_4 = \$ \#^t L_1 \#^m,$$
$$L_5 = L_3 \cup L_4.$$

Following [8], it is easy to show the following lemma:

**Lemma 1.** *Let $L$ be regular and $M'$ a DFA consistent with $S$ and $T$.*

1. $size(\$ \#^t L \#^m) = size((\$ \#^t L \#^m)^+) = t + m + 1 + size(L)$
2. $size(L_3) = t + m + 1 + size(L_2)$
3. $\$ \#^t L_1 \#^m = \$ \#^t (L_2 \cup T(M')) \#^m$

*Proof.* The claims 1. and 3. can be shown similarly to the Claims 4.1. and 4.2. in [8]. Claim 2. can be shown similarly to the first claim.                                    □

We now present the reduction. Let $M_1 = (Q_1, \Sigma', \delta_1, q_0^1, F_1)$ and $M_2 = (Q_2, \Sigma', \delta_2, q_0^2, F_2)$ be two minimal DFAs such that $T(M_1) = L_3$ and $T(M_2) = L_4$. W.l.o.g. we may assume that $Q_1 \cap Q_2 = \emptyset$. We choose $k - 2$ additional states $\{q_0^3, \ldots, q_0^k\}$ not in $Q_1 \cup Q_2$. Then we can construct a $k$-MDFA $M = (Q_1 \cup Q_2 \cup \{q_0^3, \ldots, q_0^k\}, \Sigma', \delta, \{q_0^1, q_0^2, \ldots, q_0^k\}, F_1 \cup F_2)$. For $\sigma \in \Sigma'$ we define $\delta(q, \sigma) = \delta_1(q, \sigma)$ if $q \in Q_1$, $\delta(q, \sigma) = \delta_2(q, \sigma)$ if $q \in Q_2$, and $\delta(q_0^i, \sigma) = \delta(q_0^1, \sigma)$ for $i \in \{3, \ldots, k\}$. Then $T(M) = L_5$. The instance $S, T, l$ has been transformed to the instance $M, 3m + 2t + k$. Let $m' = \sum_{w \in S \cup T} |w| + l$ be the size of the instance of the Minimum Inferred DFA problem, then it is easily seen that $M$ can be constructed from $S, T, l$ in time bounded by a polynomial in $m'$. We next show the correctness of the reduction.

Claim: There is an $l$-state DFA consistent with $S$ and $T$ if and only if $T(M) = L_5$ is accepted by a $k$-MDFA $M'$ having at most $3m + 2t + k$ states.

"$\Rightarrow$":

Let $M''$ be a DFA consistent with $S$ and $T$ and $|M''| \leq l$. Let $M_1$ and $M_2$ be the minimal DFAs with $T(M_1) = L_3$ and $T(M_2) = \$\#^t T(M'')\#^m$. Then we have $|M_1| = t + m + 1 + size(L_2) = t + 2m + 1 - l$, $|M_2| \leq t + m + l + 1$ and therefore $|M_1| + |M_2| \leq 3m + 2t + 2$. Considering the two symbols $\$, £$ we can show analogously to [8] that $T(M_1) \cup T(M_2) = L_5$. Now we choose $k - 2$ additional initial states $\{q_0^3, \ldots, q_0^k\} \not\subseteq Q_1 \cup Q_2$ and construct a $k$-MDFA $M' = (Q_1 \cup Q_2 \cup \{q_0^3, \ldots, q_0^k\}, \Sigma', \delta, \{q_0^1, q_0^2, \ldots, q_0^k\}, F_1 \cup F_2)$ in the above-mentioned manner. We thus obtain a $k$-MDFA such that $|M'| \leq 3m + 2t + k$ and $T(M') = L_5$.

"$\Leftarrow$": (sketch)

Let $M = (Q, \Sigma', \delta, \{q_0^1, q_0^2, \ldots, q_0^k\}, F)$ be a $k$-MDFA such that $T(M) = L_5$ and $|M| \leq 3m + 2t + k$. We may assume that $M$ is minimal. We have to construct an $l$-state DFA $M'$ consistent with $S$ and $T$. At first we show that $M$ can be modified such that $M$ has the form depicted in Fig. 1. Then we prove that $M$ can be easily decomposed into two DFA $M_1$ and $M_2$ such that $|M_1| + |M_2| \leq 3m$ and $T(M_1) \cup T(M_2) = L_1 \#^m \cup (L_2 \#^m)^+$. This situation is exactly the situation of the "if"-part in Claim 4.3 of [8]. Hence we can conclude that an $l$-state DFA $M'$ consistent with $S$ and $T$ can be constructed.

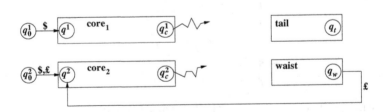

**Fig. 1.** The modified $k$-MDFA $M$

1. W.l.o.g. $S \neq \emptyset$. If $S = \emptyset$, then any DFA accepting the empty set is a DFA consistent with $S$ and $T$. Hence there is a one-state DFA accepting the empty set, and there is in particular an $l$-state DFA $M'$ consistent with $S$ and $T$.

2. Let $w = \$w_1$ with $w_1 \in \#^t S \#^m$ and $w' = w'_1 w'_2$ with $w'_1, w'_2 \in \pounds \#^t L_2 \#^m$ be two words in $L_5$. Then there are initial states $q_0^i$ and $q_0^j$ such that $\delta(q_0^i, w) \in F$ and $\delta(q_0^j, w') \in F$. We remark that $q_0^i$ and $q_0^j$ may be identical.

3. According to [8] a waist is defined as a sequence of states $q_1, q_2, \ldots, q_m$ such that $\delta(q_i, \#) = q_{i+1}$ for all $i \in \{1, 2, \ldots, m-1\}$ and $q_m$ is an accepting state and has an outgoing $\pounds$-edge. A tail is defined as a sequence of states $q_1, q_2, \ldots, q_m$ such that $\delta(q_i, \#) = q_{i+1}$ for all $i \in \{1, 2, \ldots, m-1\}$ and $q_m$ is an accepting state and has no outgoing edges. A core is defined as a sequence of states $q_1, q_2, \ldots, q_{t+1}$ such that $\delta(q_i, \#) = q_{i+1}$ for all $i \in \{1, 2, \ldots, t\}$ and $q_{t+1}$ is non-accepting and has outgoing edges, but no outgoing $\pounds$-edge. It can be shown that $M$ contains exactly one waist, one tail and two distinct cores.

   W.l.o.g. we may assume that $w$ will be accepted from $q_0^i$ passing through $\text{core}_1$ and the tail and $w'$ will be accepted from $q_0^j$ passing through $\text{core}_2$ and the waist. Let $q_t = \delta(q_0^i, w)$ and $q_w = \delta(q_0^j, w'_1)$ denote the last states in the tail and the waist. Let $q^1 = \delta(q_0^i, \$)$ and $q^2 = \delta(q_0^j, \pounds)$ denote the first states of $\text{core}_1$ and $\text{core}_2$. By $q_c^1 = \delta(q_0^i, \$\#^t)$ and $q_c^2 = \delta(q_0^j, \pounds\#^t)$ we denote the last states of $\text{core}_1$ and $\text{core}_2$. Since $w$ is accepted passing through $\text{core}_1$, we can conclude that $q^2 = \delta(q_w, \pounds)$ is the starting state of the loop.

4. It can be shown that all initial states have no incoming edges.

5. It can be shown that $\delta(q^1, \#^t S \#^m) \subseteq F$ and $\delta(q^2, \#^t S \#^m) \cap F = \emptyset$.

6. It can be shown that $\delta(q^2, \#^t L_2 \#^m (\pounds \#^t L_2 \#^m)^*) \subseteq F$.

7. $M$ can be modified to the form depicted in Fig. 1 on page 390. (The initial states $q_0^3, \ldots, q_0^k$ are not included.)

   At first we remove all edges from initial states to any other states. We choose two different initial states $q_0^1$ and $q_0^2$ and then insert the following edges: $q_0^1 \xrightarrow{\$} q^1$, $q_0^2 \xrightarrow{\$, \pounds} q^2$, and $q_0^i \xrightarrow{\$, \pounds} q^2$ for $i \in \{3, \ldots, k\}$. We observe that due to 4., 5., and 6. the modified automaton still recognizes $L_5$. In particular, $L_3$ is accepted from $q_0^2$ and all words in $\$\#^t S \#^m$ are accepted only from $q_0^1$.

8. We now look at the two DFAs obtained when considering only one initial state in $M$. We define the set of reachable states as follows: $\mathcal{E}(q_0^1) = \{q \in Q \mid \exists x, x' \in (\Sigma')^* : \delta(q_0^1, x) = q \wedge \delta(q, x') \in F\}$. $\mathcal{E}(q_0^2)$ is defined analogously. We first claim that there is no edge from $p \in \mathcal{E}(q_0^2)$ to a state $q$ from which $q_t$ can be obtained. Assume by way of contradiction that there are $p \in \mathcal{E}(q_0^2)$, $q \in Q$, $s \in \Sigma'$, and $u \in (\Sigma')^*$ such that $\delta(p, s) = q$ and $\delta(q, u) = q_t \in F$. Since $p \in \mathcal{E}(q_0^2)$, there are strings $x, x' \in (\Sigma')^*$ such that $\delta(q_0^2, x) = p$ and $\delta(p, x') \in F$. Due to 7., we may assume that $x$ starts with $\pounds$. We then know that $\delta(q_0^2, xsu) = q_t \in F$, but $\delta(q_0^2, xsuxsu) \notin F$, because $q_t$ has no outgoing edges. Moreover, $\delta(q_0^1, xsuxsu) \notin F$, since $q_0^1$ has no outgoing $\pounds$-edge. Hence $xsuxsu \notin L_5$ which is a contradiction, because $xsu \in L_3$ and therefore $xsuxsu \in L_3 \subset L_5$.

   Furthermore, we observe that all edges from states in $\mathcal{E}(q_0^1)$ to states in $\mathcal{E}(q_0^2)$ can be removed. If we have such an edge, all words passing this edge will be accepted in the waist and therefore are in $L_3$. Due to 6. and 7., these words can already be accepted from $q_0^2$. So, removing such edges does not affect the accepted language. We observe that this modification yields $\mathcal{E}(q_0^1) \cap \mathcal{E}(q_0^2) = \emptyset$.

9. Since the sets of reachable states are distinct, we obtain two DFAs $M_1' = (Q_1', \Sigma', \delta_1', q_0^1, F_1')$ and $M_2' = (Q_2', \Sigma', \delta_2', q_0^2, F_2')$ after having minimized the DFAs $(\mathcal{E}(q_0^1), \Sigma', \delta, q_0^1, F)$ and $(\mathcal{E}(q_0^2), \Sigma', \delta, q_0^2, F)$. Due to 5. and 7., we know that $L_4 \supseteq T(M_1') \supseteq \$\#^t S\#^m$ and $T(M_2') = L_3$. Furthermore, $|M_1'| + |M_2'| \leq 3m + 2t + 2$, since $Q_1' \cap Q_2' = \emptyset$.

10. Starting from $M_1'$ we define another DFA $M_1$ by removing $q_0^1$ and the first $t$ states of $\text{core}_1$. We define $q_c^1$ as new initial state and observe that $L_1\#^m \supseteq T(M_1) \supseteq S\#^m$. Starting from $M_2'$ we define another DFA $M_2$ by removing $q_0^2$ and the first $t$ states of $\text{core}_2$. We define $q_c^2$ as new initial state. The $\pounds$-edge from $q_w$ to $q^2$ is replaced by the following edges: if $\delta_2'(q_c^2, \sigma) = q$ for $\sigma \in \Sigma$, we add a $\sigma$-edge from $q_w$ to $q$. It is easy to see that $T(M_2) = (L_2\#^m)^+$. Hence we have $T(M_1) \cup T(M_2) = L_1\#^m \cup (L_2\#^m)^+$. Moreover, $|M_1| + |M_2| \leq 3m$.

11. We have $|M_2| = size((L_2\#^m)^+) = m + size(\overline{T} \cap \overline{S}) = 2m - l$ and therefore $|M_1| \leq 3m - |M_2| = 3m - 2m + l = m + l$. Removing the tail in $M_1$ yields an $l$-state DFA $M'$ consistent with $S$ and $T$.     $\square$

**Corollary 1.** *Let $k, k' \geq 2$ be two constant numbers. Then DFA $\longrightarrow$ $k$-MDFA and $k$-MDFA $\longrightarrow$ $k'$-MDFA are NP-complete.*

The following theorem is a simple observation of the fact that $k$-MDFAs can be efficiently converted to DFAs whose size is bounded by a polynomial in $k$, and that the below-mentioned decidability questions are efficiently solvable for DFAs.

**Theorem 2.** *Let $M$ be a $k$-MDFA and $M'$ be a $k'$-MDFA. Then the following problems are solvable in polynomial time. Is $T(M) = T(M')$? Is $T(M) \subseteq T(M')$? Is $T(M) \subset T(M')$? Is $T(M) = \Sigma^*$?*

# 4     Minimizing NFAs with Fixed Finite Branching Is Computationally Hard

In this section we are going to show that the minimization problem for NFAs with branching $\beta = k$ (NFA($\beta = k$)) is NP-complete for $k \geq 3$.

**Lemma 2.** *Let $M$ be an NFA and $k \geq 2$ be a constant integer. Then the question whether $M$ has branching $k$ can be solved in polynomial time.*

*Proof.* Let $i$ be an integer. We consider the language

$$T_i(M) = \{w \in \Sigma^* \mid \exists \text{ accepting computation } \pi \text{ of } M \text{ of } w \text{ with } \beta(\pi) \leq i\}.$$

In [2] it is shown that a DFA $M_i$ accepting $T_i(M)$ can be effectively constructed. It can be observed that the construction can be done in time polynomially bounded in $|M|$ and that the resulting DFA has size $O(|M|^k)$.

Since $T_k(M) \subseteq T(M)$, we have: $T(M) \setminus T_k(M) = \emptyset \Leftrightarrow \beta_M \leq k$. Since $M_k$ is a DFA, we can simply construct a DFA $M_k'$ accepting the complement $\Sigma^* \setminus T_k(M)$.

$$\beta_M \leq k \Leftrightarrow T(M) \setminus T_k(M) = \emptyset \Leftrightarrow T(M) \cap \overline{T_k(M)} = \emptyset \Leftrightarrow T(M) \cap T(M_k') = \emptyset$$

Since $M$ is an NFA and $M_k'$ is a DFA, we can construct, in polynomial time, an NFA $\hat{M}$ of size $O(|M| \cdot |M|^k)$ as the Cartesian product of $M$ and $M_k'$ accepting $T(M) \cap T(M_k')$. The non-emptiness of $T(\hat{M})$ can be tested in NLOGSPACE $\subseteq$ P [10]. If $T(\hat{M}) \neq \emptyset$, then $\beta_M > k$. If $T(\hat{M}) = \emptyset$, then we know that $\beta_M \leq k$. To find out whether $\beta_M = k$, we construct $T_{k-1}(M)$ if $k - 1 \geq 1$. This can be done in polynomial time as well as the test for inequivalence of $T_{k-1}(M)$ and $T_k(M)$. If both sets are inequivalent, then $\beta_M = k$; otherwise $\beta_M < k$.    □

**Theorem 3.** *NFA $(\beta = k) \longrightarrow$ NFA $(\beta = k)$ is NP-complete for $k \geq 3$.*

*Proof.* We first show that the problem is in NP. To this end we determine non-deterministically an NFA $M'$ with $|M'| \leq l$. Due to Lemma 2, we can test whether $M'$ has branching $k$ in polynomial time. We next convert $M$ and $M'$ to $k$-MDFAs $\hat{M}$ and $\hat{M}'$ with at most $k|M| + 1$ and $k|M'| + 1$ states applying the construction presented in [11]. The equality of $T(\hat{M})$ and $T(\hat{M}')$ can then be tested in polynomial time analogous to the considerations of Theorem 1. Hence the above problem is in NP.

The NP-hardness of the problem will be shown by reduction from the Minimum Inferred DFA problem similar to the proof for MDFAs. In detail, we want to transform an NFA with fixed finite branching $k$ into a 2-MDFA accepting $L_5$. Due to Theorem 1, we then can construct an appropriate DFA consistent with $S$ and $T$. An obvious, but essential observation for NFAs with finite branching is that a move with a branching $\beta > 1$ cannot be located within a loop, since otherwise the branching of the NFA would be infinite. Thus, we modify the language $L_5$ by adding loops at the beginning of all words. Therefore, we can show that a given NFA with fixed finite branching $k$ has exactly one nondeterministic move with branching $k$. Furthermore, this move has to be the first move. It is then easy to convert this NFA to a 2-MDFA accepting $L_5$.

Let $m = l + size(\overline{T} \cap \overline{S})$ and $n = 5m + 1$. In addition to the previous definitions we define:

$$L_3' = \{\$, \pounds\} \#^m (\#^{m+1})^* L_2 \#^m (\pounds \#^m (\#^{m+1})^* L_2 \#^m)^*,$$
$$L_4' = \$ \#^m (\#^{m+1})^* L_1 \#^m,$$
$$L^i = \{(\$ \#^{in^k - 1})^+\} \ (1 \leq i \leq k - 2),$$
$$L_5' = L^1 \cup L^2 \cup \ldots \cup L^{k-2},$$
$$L_6' = L_3' \cup L_4' \cup L_5'.$$

**Lemma 3.** *Let $L$ be regular and $M'$ a DFA consistent with $S$ and $T$.*

*1. $size(\$ \#^m (\#^{m+1})^* L \#^m) = size((\$ \#^m (\#^{m+1})^* L \#^m)^+) = 2m + 1 + size(L)$*

2. $size(L_3') = 2m + 1 + size(L_2)$
3. $\$\#^m(\#^{m+1})^*L_1\#^m = \$\#^m(\#^{m+1})^*(L_2 \cup T(M'))\#^m$
4. $size(L^i) = in^k + 1$
5. $size(\{\$\#^m(\#^{m+1})^*\} \cup L_5') \geq n^k + 2n^k + \ldots + (k-2)n^k + (k-2)n^k + 1 + (m+1)$

*Proof.* The claims 1., 2., and 3. can be shown analogously to those of Lemma 1. Claim 4. is obvious. The proof of 5. is not difficult, but lengthy and is omitted here.  □

We now present the reduction. Let $M_1 = (Q_1, \Sigma', \delta_1, q_0^1, F_1)$ and $M_2 = (Q_2, \Sigma', \delta_2, q_0^2, F_2)$ be two minimal DFAs such that $T(M_1) = L_3'$ and $T(M_2) = L_4'$. Furthermore, let $M_i = (Q_i, \Sigma', \delta_i, q_0^i, F_i)$, $3 \leq i \leq k$ be $k - 2$ minimal DFAs accepting $L^1, L^2, \ldots, L^{k-2}$. W.l.o.g. we may assume that $Q_1$, $Q_2$, $\ldots$, and $Q_k$ are pairwise distinct. We observe that for $3 \leq i \leq k$ the states $q_0^i$ have no incoming edges and only one outgoing edge to a non-trap state, namely a $-edge. Moreover, $q_0^1$ has no incoming edges and only two outgoing edges to non-trap states, namely a $-edge and a £-edge. We remove $q_0^1$ from $M_1$ and $q_0^i$ from $M_i$ for $3 \leq i \leq k$ and construct an NFA $M = ((Q_1 \setminus \{q_0^1\}) \cup Q_2 \cup (Q_3 \setminus \{q_0^3\}) \cup \ldots \cup (Q_k \setminus \{q_0^k\}), \Sigma', \delta, q_0^2, F_1 \cup F_2 \cup \ldots \cup F_k)$. For $\sigma \in \Sigma'$ and $1 \leq i \leq k$ we define $\delta(q, \sigma) = \delta_i(q, \sigma)$ if $q \in Q_i$. Furthermore, $\delta(q_0^2, \$) = \delta_1(q_0^1, \$)$, $\delta(q_0^2, £) = \delta_1(q_0^1, £)$, and $\delta(q_0^2, \$) = \delta_i(q_0^i, \$)$ for $3 \leq i \leq k$. Then $T(M) = L_6'$ and $M$ is an NFA with branching $k$.

The instance $S, T, l$ has been transformed to $M, 5m + 1 + \sum_{i=1}^{k-2} in^k$. Let $m' = \sum_{w \in S \cup T} |w| + l$ be the size of the instance of the Minimum Inferred DFA problem, then $M$ can be constructed from $S, T, l$ in time bounded by a polynomial in $m'$. We next show the correctness of the reduction.

Claim: There is an $l$-state DFA consistent with $S$ and $T$ if and only if $T(M) = L_6'$ is accepted by an NFA $M'$ with branching $\beta_M = k$ that has at most $5m + 1 + \sum_{i=1}^{k-2} in^k$ states.

"$\Rightarrow$":

Let $M''$ be a DFA consistent with $S$ and $T$ and $|M''| \leq l$. Let $M_1$ and $M_2$ be the minimal DFAs with $T(M_1) = L_3'$, $T(M_2) = \$\#^m(\#^{m+1})^*T(M'')\#^m$. Furthermore, $M_3, \ldots, M_k$ are minimal DFAs accepting $L^1, \ldots, L^{k-2}$. Analogous to the proof of Theorem 1 and the above considerations we can construct an NFA $M'$ with $\beta_{M'} = k$ such that $T(M') = L_6'$ and $|M'| \leq 5m + 1 + \sum_{i=1}^{k-2} in^k$.

"$\Leftarrow$": (sketch)

Let $M = (Q, \Sigma', \delta, q_0, F)$ be an NFA with branching $\beta_M = k$ such that $T(M) = L_6'$ and $|M| \leq 5m + 1 + \sum_{i=1}^{k-2} in^k$. We may assume that $M$ is minimal. We have to construct an $l$-state DFA $M'$ consistent with $S$ and $T$. In consequence of the definition of $L_6'$, we can show that the nondeterministic moves of $M$ have to start in $q_0$. Then, $M$ can be converted to a 2-MDFA $M''$ such that $|M''| \leq 3m + 2t + 2$ (setting $t = m$) and $T(M'') = L_5$. Due to the proof of Theorem 1, we then can conclude that an $l$-state DFA $M'$ consistent with $S$ and $T$ can be constructed.

1. W.l.o.g. $S \neq \emptyset$. Let $w = \$\#^m w_1 \#^m$ with $w_1 \in S$ and $w' = w_1' w_2'$ with $w_1', w_2' \in £\#^m L_2 \#^m$ be two words in $L_6'$.

2. A loop-core is defined as a sequence of states $q_1, q_2, \ldots, q_m, q_{m+1}$ such that $\delta(q_i, \#) = q_{i+1}$ for all $i \in \{1, 2, \ldots, m\}$ and $q_{m+1}$ is non-accepting, has outgoing edges, in particular a #-edge to $q_1$, but no outgoing £-edge.
   A \$-#-loop of length $jn^k$ with $1 \leq j \leq k - 2$ is defined as a sequence of states $q_1, q_2, \ldots, q_{jn^k}$ such that $\delta(q_i, \#) = q_{i+1}$ for all $i \in \{1, 2, \ldots, jn^k - 1\}$ and $q_{jn^k}$ is accepting and has an outgoing \$-edge to $q_1$.
   It can be shown that $M$ contains exactly one waist, one tail, two distinct loop-cores, and $k - 2$ \$-#-loops of length $n^k, 2n^k, \ldots, (k - 2)n^k$.
   W.l.o.g. we may assume that $w$ will be accepted passing through loop-core$_1$ and the tail and $w'$ will be accepted passing through loop-core$_2$ and the waist. Let $q_t \in \delta(q_0, w)$ and $q_w \in \delta(q_0, w_1')$ denote the last states in the tail and the waist. By $q^1$ and $q^2$ we denote the states obtained after having read \$ and £ when $M$ passes through the accepting computations of $w$ and $w_1'$. Since $w$ is accepted passing through loop-core$_1$, we can conclude that $\{q^2\} = \delta(q_w, \text{£})$ is the starting state of the loop in $L_3'$.
3. All computations starting in $q^2 \in \delta(q_0, \text{£})$ and leading to an accepting state, thus computations of words in $\text{£}\#^m(\#^{m+1})^* L_2 \#^m (\text{£}\#^m(\#^{m+1})^* L_2 \#^m)^*$, have branching 1. This is obvious, since even one move with a branching greater than one would imply that $M$ contains accepting computations with infinite branching due to the £-edge from $q_w$ to $q^2$.
4. The loop-cores and the \$-#-loops contain no moves with branching greater than one, since due to their loops there would be computations with infinite branching.
5. All computations starting in $\delta(q_0, \$)$ and leading to an accepting state, thus computations of words in $\$\#^m(\#^{m+1})^* S'\#^m$ with $S \subseteq S' \subseteq L_1, L_5'$, and $\$\#^m(\#^{m+1})^* L_2 \#^m (\text{£}\#^m(\#^{m+1})^* L_2 \#^m)^*$, have branching 1.
   Due to 3. and 4. the moves with branching greater than one have to be located either in the states before entering the loop-core and the \$-#-loops, or in the states recognizing $S'\#^m$.
   First of all, we assume that all moves with branching greater than one start before entering the loop-core and the \$-#-loops. Then we can shift the branching to $q_0$: we remove any outgoing \$-edges from $q_0$ and insert $k - 2$ \$-edges to the first states of the \$-#-loops and two \$-edges to loop-core$_1$ and loop-core$_2$. It follows that the modified automaton still recognizes $L_6'$, but there is at least one unnecessary state $q \in \delta(q_0, \$)$. Hence $M$ was not minimal which is a contradiction.
   We now assume that there is at least one move with branching 2 within the states recognizing $S'\#^m$. Then $L = \$\#^m(\#^{m+1})^* \cup L_5'$ must be recognized by an NFA with a branching of at most $\lfloor \frac{k}{2} \rfloor$. Due to Lemma 3 we know that a DFA for $L$ needs at least $n^k + 2n^k + \ldots + (k - 2)n^k + (k - 2)n^k + 1 + (m + 1)$ states. Analogous to the (here omitted) considerations in 2., one can show that every NFA accepting $L$ with finite branching contains $k - 2$ different \$-#-loops of length $n^k, 2n^k, \ldots, (k - 2)n^k$, a loop-core of length $m + 1$ and an initial state. In comparison with the minimal DFA, an NFA with finite branching can therefore achieve savings in size only through nondeterministic moves that start in states which are not part of a loop. Subtracting the

loop-states from $n^k + 2n^k + \ldots + (k-2)n^k + (k-2)n^k + 1 + (m+1)$, there remain $(k-2)n^k + 1$ states. In [2] it is shown that the best possible reduction of states that an NFA with branching $i$ can achieve in comparison with the corresponding minimal DFA is at most the $i$-th root of the size of the DFA. Hence an NFA accepting $L$ with branching $\lfloor \frac{k}{2} \rfloor$ has at least $n^k + 2n^k + \ldots + (k-2)n^k + (m+1) + ((k-2)n^k + 1)^{1/\lfloor \frac{k}{2} \rfloor}$ states. Since $((k-2)n^k + 1)^{1/\lfloor \frac{k}{2} \rfloor} \geq n^2$, we have that $|M| \geq \sum_{i=1}^{k-2} in^k + n^2$ which is a contradiction to $|M| \leq 5m + 1 + \sum_{i=1}^{k-2} in^k = n + \sum_{i=1}^{k-2} in^k$.

It follows that $|\delta(q_0, \$)| > 1$. From $q_0$ we then have a \$-edge to $q^1$ and the first states of the $k-2$ \$-#-loops. Furthermore, we can assume to have a \$-edge to $q^2$. If there is no such edge, we can insert one without affecting the accepted language. We next remove the $k-2$ \$-#-loops and reduce the two loop-cores to cores by removing their #-loops. We then have an NFA with branching 2 with $3m + 2t + 1$ states ($t = m$) accepting $L_5$. Now, we remove the \$-edge from $q_0$ to $q^1$ and we insert an additional state $q_0'$ which has an outgoing \$-edge to $q^1$. Thus, we have a 2-MDFA with $3m + 2t + 2$ states accepting $L_5$. Due to Theorem 1 we can construct an $l$-state DFA $M'$ consistent with $S$ and $T$.    □

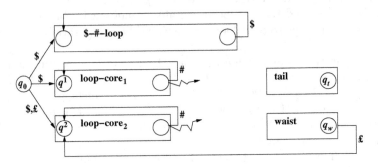

**Fig. 2.** The NFA($\beta = 3$) $M$ accepting $L_6'$.

**Corollary 2.** *Let $k \geq 2$ and $k' \geq 3$ be constant integers. Then the problems* $DFA \longrightarrow NFA(\beta = k')$ *and* $NFA(\beta = k) \longrightarrow NFA(\beta = k')$ *are* NP-*complete.*

**Theorem 4.** *The following problems, which are* PSPACE-*complete when arbitrary NFAs are considered, are solvable in polynomial time.*

1. *Given two NFAs $M, M'$ with $\beta_M = k$ and $\beta_{M'} = k'$. Is $T(M) = T(M')$? Is $T(M) \subseteq T(M')$? Is $T(M) \subset T(M')$? Is $T(M) = \Sigma^*$?*
2. *Given any NFA $M$ and an NFA $M'$ with $\beta_{M'} = k$. Is $T(M) \subseteq T(M')$?*

*Proof.* The proof of 1. is similar to the proof of Theorem 2. To prove 2., we observe that $T(M) \subseteq T(M') \Leftrightarrow T(M) \cap \overline{T(M')} = \emptyset$. The claim can then be shown using a similar argumentation as in Lemma 2.    □

# 5   Conclusions

We have shown that the minimization of finite automata equipped with a very small and fixed amount of nondeterminism is computationally hard. Hence the question arises whether there are extensions of the deterministic model at all that preserve polynomial time minimization algorithms. Two candidates result from our considerations. At first, the computational complexity of the problem $NFA(\beta = k) \longrightarrow NFA(\beta = 2)$ remains open. Moreover, the two constructions in Theorem 1 and Theorem 3 present finite automata which are not unambiguous. It is currently unknown whether unambiguous $k$-MDFAs or unambiguous NFAs with branching $k$ provide efficient minimization algorithms.

# References

1. Gold, E.M.: Complexity of automaton identification from given data. Information and Control **37:3** (1978) 302–320
2. Goldstine, J., Kintala, C.M.R., Wotschke, D.: On measuring nondeterminism in regular languages. Information and Computation **86:2** (1990) 179–194
3. Holzer, M., Salomaa, K., Yu, S.: On the state complexity of $k$-entry deterministic finite automata. Journal of Automata, Languages and Combinatorics **6:4** (2001) 453–466
4. Garey, M.R., Johnson, D.S.: Computers and Intractability. W.H. Freeman and Co., San Francisco (1979)
5. Gill, A., Kou, L.: Multiple-entry finite automata. Journal of Computer and System Sciences **9** (1974) 1–19
6. Hopcroft, J.E.: An $n \log n$ algorithm for minimizing states in a finite automaton. In: Kohavi, Z. (ed.): Theory of machines and computations. Academic Press, New York (1971) 189–196
7. Hopcroft, J.E., Ullman, J.D.: Introduction to Automata Theory, Languages and Computation. Addison-Wesley, Reading MA (1979)
8. Jiang, T., Ravikumar, B.: Minimal NFA problems are hard. SIAM Journal on Computing **22:6** (1993) 1117–1141
9. Jones, N.D., Lien, E.Y., Laaser, W.T.: New problems complete for nondeterministic log space. Mathematical Systems Theory **10:1** (1976) 1–17
10. Jones, N.D.: Space-bounded reducibility among combinatorial problems. Journal of Computer and System Sciences **11:1** (1975) 68–85
11. Kappes, M.: Descriptional complexity of deterministic finite automata with multiple initial states. Journal of Automata, Languages and Combinatorics **5:3** (2000) 269–278
12. Kintala, C.M.R., Wotschke, D.: Amounts of nondeterminism in finite automata. Acta Informatica **13:2** (1980) 199–204
13. Meyer, A.R., Fischer, M.J.: Economy of descriptions by automata, grammars, and formal systems. IEEE Symposium on Foundations of Computer Science (1971) 188–191
14. Stockmeyer, L., Meyer, A.R.: Word problems requiring exponential time: preliminary report. Fifth Annual ACM Symposium on Theory of Computing (1973) 1–9
15. Yu, S.: Regular languages. In: Rozenberg, G., Salomaa, A. (eds.): Handbook of Formal Languages Vol. 1. Springer-Verlag, Berlin Heidelberg (1997) 41–110

# Boolean Grammars

Alexander Okhotin

School of Computing, Queen's University, Kingston, Ontario, Canada K7L3N6
okhotin@cs.queensu.ca

**Abstract.** As a direct continuation of the earlier research on conjunctive grammars – context-free grammars equipped with intersection – this paper introduces a new class of formal grammars, which allow the use of all set-theoretic operations as an integral part of the formalism of rules. Rigorous semantics for such grammars is defined by language equations in a way that allows to generalize some techniques from the theory of context-free grammars, including Chomsky normal form, Cocke–Kasami–Younger recognition algorithm and some limited extension of the notion of a parse tree, which together allow to conjecture the practical applicability of the new concept.

## 1 Introduction

The observation that the generative power of context-free grammars is insufficient for denoting the languages that appear in practice is probably as old as the notion of a context-free grammar itself. Numerous candidates to fill the gap between too weak context-free and too powerful context-sensitive grammars were suggested in course of many years of efforts, and the search for new formalisms with good properties continues.

Most of the attempts of this sort naturally start from a context-free backbone and extend it with some features that are lacking in context-free grammars. Conjunctive grammars [4] were introduced in 2000 as an extension of context-free grammars with an explicit intersection operation. While context-free rules are of the form $A \to \alpha$, the rules in conjunctive grammars allow the use of conjunction: $A \to \alpha_1 \& \ldots \& \alpha_n$ $(n \geqslant 1)$. The semantics of these grammars is defined using derivation [4], but a characterization by least solutions of systems of language equations with union, intersection and concatenation is also known [5]. Several parsing algorithms for conjunctive grammars with worst-case cubic time performance were developed and implemented in a parser generator [7], which confirms the practical applicability of the concept.

The question is: could this formalism be extended a little further, making it even more expressive but still computationally feasible? This paper attempts to give a positive answer to this question by introducing a new class of grammars that additionally allows the use of negation. However, even the question of defining the semantics of these grammars turns out to be complicated, as a derivation-based approach could hardly work due to negation. Using language equations [1,3] as a starting point reveals complications of a different kind, for

Z. Ésik and Z. Fülöp (Eds.): DLT 2003, LNCS 2710, pp. 398–410, 2003.

systems with arbitrary set-theoretic operations can have multiple incomparable solutions or no solutions at all [8], the set of systems that have exactly one solution is in the second level of arithmetical hierarchy [8], and the class of languages defined by such unique solutions coincides with the class of recursive sets [8].

So, in order to obtain a practically applicable device, one has to consider more restrictive semantics than the semantics of unique solution. Two such semantics are proposed in Sect. 2; one of them is backward compatible with conjunctive and context-free grammars and allows to define parse trees of strings. Then, in Sect. 3, a normal form for Boolean grammars that naturally extends binary normal form for conjunctive [4] and Chomsky normal form for context-free grammars is proposed, and it is shown that every grammar compliant to either of the given semantics can be effectively transformed to this normal form. Since every grammar in the normal form is compliant to both semantics of grammars, this implies the computational equivalence of the two semantics.

Section 4 develops a cubic-time recognition algorithm for Boolean grammars that is a natural generalization of the algorithm for conjunctive grammars [4], which in turns extends the Cocke–Kasami–Younger algorithm for context-free grammars. Section 5 proves some basic properties of the language family defined by Boolean grammars, in particular its inclusion in the family of deterministic context-free languages.

## 2 Equations and Grammars

### 2.1 Language Equations

**Definition 1 (Language formula).** *Let $\Sigma$ be a finite nonempty alphabet and let $X = (X_1, \ldots X_n)$ $(n \geqslant 1)$ be a vector of language variables. Language formulae over the alphabet $\Sigma$ in variables $X$ are defined inductively as follows: (i) the empty string $\epsilon$ is a formula; (ii) any symbol from $\Sigma$ is a formula; (iii) any variable from $X$ is a formula; (iv) if $\varphi$ and $\psi$ are formulae, then $(\varphi\psi)$, $(\varphi\&\psi)$, $(\varphi \vee \psi)$ and $(\neg\varphi)$ are formulae.*

As in logic formulae, we shall omit the parentheses whenever possible, using the following default precedence of operators: the concatenation has the highest precendence and is followed by the logical connectives arranged in their usual order $\neg$, $\&$ and $\vee$. If needed, this default precedence will be overridden with parentheses. For instance, $XY \vee \neg aX\&aY$ means the same as $(X \cdot Y) \vee ((\neg(a \cdot X))\&(a \cdot Y))$. Note that all the mentioned binary logical operations, as well as concatenation, are associative, and therefore there is no need to disambiguate formulae like $XYZ$ or $X \vee Y \vee Z$ with extra parentheses.

We have defined the syntax of formulae; let us now define their semantics by interpreting the connectives with operations on languages, associating a language function with every formula:

**Definition 2 (Value of a formula).** *Let $\varphi$ be a formula over an alphabet $\Sigma$ in variables $X = (X_1, \ldots, X_n)$. Let $L = (L_1, \ldots, L_n)$ be a vector of languages* *over $\Sigma$. The value of the formula $\varphi$ on the vector of languages L, denoted as $\varphi(L)$, is defined inductively on the structure of $\varphi$: $\epsilon(L) = \{\epsilon\}$, $a(L) = \{a\}$ for every $a \in \Sigma$, $X_i(L) = L_i$ for every $i$ $(1 \leqslant i \leqslant n)$, $\psi\xi(L) = \psi(L) \cdot \xi(L)$, $(\psi \vee \xi)(L) = \psi(L) \cup \xi(L)$, $(\psi\&\xi)(L) = \psi(L) \cap \xi(L)$ and $(\neg\psi)(L) = \Sigma^* \setminus \psi(L)$.*

*The value of a vector of formulae $\varphi = (\varphi_1, \ldots, \varphi_l)$ on a vector of languages $L = (L_1, \ldots, L_n)$ is the vector of languages $\varphi(L) = (\varphi_1(L), \ldots, \varphi_l(L))$.*

**Definition 3 (System of equations).** *Let $\Sigma$ be an alphabet. Let $n \geqslant 1$. Let $X = (X_1, \ldots, X_n)$ be a set of language variables. Let $\varphi = (\varphi_1, \ldots, \varphi_n)$ be a vector of formulae in variables $X$ over the alphabet $\Sigma$. Then*

$$\begin{cases} X_1 = \varphi_1(X_1, \ldots, X_n) \\ \quad\vdots \\ X_n = \varphi_n(X_1, \ldots, X_n) \end{cases} \tag{1}$$

*is called a resolved system of equations over $\Sigma$ in variables $X$. (1) can also be denoted in the vector form as $X = \varphi(X)$.*

**Definition 4 (Solution of a resolved system).** *A vector of languages $L = (L_1, \ldots, L_n)$ is said to be a solution of a system (1), if for every $i$ $(1 \leqslant i \leqslant n)$ it holds that $L_i = \varphi_i(L_1, \ldots, L_n)$. In the vector form, this is denoted as $L = \varphi(L)$.*

Let us introduce some simple language-theoretic terminology that will be used in the following. For a pair of languages $L_1, L_2 \subseteq \Sigma^*$ and another language $M \subseteq \Sigma^*$, we say that $L_1$ and $L_2$ are equal modulo $M$ if $L_1 \cap M = L_2 \cap M$; this is denoted $L_1 = L_2 \pmod{M}$. The relation of equality modulo $M$ is easily extended to vectors of languages. A vector of languages $L$ is a solution of a system $X = \varphi(X)$ modulo $M$, if the vectors $L$ and $\varphi(L)$ are equal modulo $M$.

For every string $w \in \Sigma^*$, $y \in \Sigma^*$ is a *substring* of $w$ if $w = xyz$ for some $x, z \in \Sigma^*$; $y$ is a *proper substring* of $w$ if additionally $y \neq w$. A language $L$ is *closed under substring*, if all substrings of every $w \in L$ are in $L$.

*Example 1.* The following system of equations over the alphabet $\Sigma = \{a, b\}$

$$\begin{array}{ll} X_1 = \neg X_2 X_3 \& \neg X_3 X_2 \& X_4 & X_3 = (a \vee b)X_3(a \vee b) \vee b \\ X_2 = (a \vee b)X_2(a \vee b) \vee a & X_4 = (aa \vee ab \vee ba \vee bb)X_4 \vee \epsilon \end{array} \tag{2}$$

has unique solution $(\{ww \mid w \in \{a,b\}^*\}, \{xay \mid x,y \in \{a,b\}^*\}, \{xby \mid x,y \in \{a,b\}^*\}, \{u \mid u \in \{a,b\}^{2n}, n \geqslant 0\})$.

It we use the first variable as a kind of "start symbol", then the grammar given in Example 1 can be said to denote the language $\{ww \mid w \in \{a,b\}^*\}$. This abstract language is often given as an example that captures the notion of "reduplication", which is viewed as an essential property of natural languages. The language $\{ww \mid w \in \{a,b\}^*\}$ is co-context-free, but not context-free; additionally, it is known not to be representable as a finite intersection of context-free languages. Although conjunctive grammars can denote a very similar language $\{wcw \mid w \in \{a,b\}^*\}$ [4], it is not known whether $\{ww \mid w \in \{a,b\}^*\}$ is a conjunctive language (or, equivalently, whether it can be denoted using a system of language equations containing concatenation, union and intersection).

*Example 2 ([3]).* The equation $X = a(\neg(\neg(\neg X)^2)^2)^2$, where where $\varphi^2$ abbreviates $\varphi \cdot \varphi$, has unique solution $L = \{a^n \mid 2^{3k} \leqslant n < 2^{3k+2}$ for some $k \geqslant 0\}$, which is a nonregular unary language.

## 2.2  Semantics of Unique Solution in the Strong Sense

In [8] it was proved that the class of languages defined as a component of the unique solution of a system of language equations as in Definition 3 is the class of recursive sets. Although uniqueness of solution is a very natural requirement to impose on a system, the mentioned result implies that it cannot be used for practical purposes. Let us impose an additional restriction upon the systems of equations that have unique solution: we require that the system has a unique solution *modulo every language closed under substring*. For any system satisfying this requirement there exists a simple method of computing its unique solution modulo every given finite language, which implies a straightforward algorithm for deciding the membership of strings in the components of this unique solution.

**Lemma 1.** *Let $X = \overline{\varphi}(X)$ be a system of equations that has unique solution modulo every language closed under substring. Let $M$ be an arbitrary finite language closed under substring, let $u \in \Sigma^*$ be a string not in $M$, such that all of its proper substrings are in $M$, and let $L = (L_1, \ldots, L_n)$ be the unique solution of the system modulo $M$. Then there exists a unique Boolean vector $\overline{b} = (b_1, \ldots, b_n) \in \mathbb{B}^n$, such that the vector $L_{\overline{b}} = (L'_1, \ldots, L'_n)$, where $L'_i = L_i$ if $b_i = 0$ or $L'_i = L_i \cup \{u\}$ if $b_i = 1$, is a solution of the system modulo $M \cup \{u\}$.*

Therefore, the unique solution of the system modulo $M \cup \{u\}$ can be determined by trying all $2^n$ Boolean vectors and checking the appropriate $L_{\overline{b}}$ for being a solution of the system modulo this language. It should be noted that one cannot effectively decide whether a system complies to this semantics.

**Theorem 1.** *The set of systems compliant to the semantics of unique solution in the strong sense is co-RE-complete.*

## 2.3  Semantics of Naturally Reachable Solution

Semantics of context-free grammars is the semantics of the least fixed point, which well corresponds to the notion of derivability; semantics of conjunctive grammars [4,5] continues with this tradition. Things change once the non-monotonous operation of negation is introduced: here the possible absense of the least solution of a system well corresponds to the quite predictable difficulties with developing a set of derivation rules equivalent to any denotational semantics, such as the semantics of the unique solution of the system. However, in order to maintain the connection with conjunctive and context-free grammars, it would be highly desirable to find a way to choose, if possible, the most natural among multiple solutions of some system, which would be the least solution in the monotonous case and something else in the general case. For that purpose we slightly modify the iterative approach of Lemma 1 to define *the semantics of naturally reachable solution*, which has some traits of operational semantics:

**Definition 5 (Naturally reachable solution).** *Let $X = \overline{\varphi}(X)$ be a system of equations. A solution $L = (L_1, \dots, L_n)$ of the system is called a naturally reachable solution if for every finite modulus $M$ closed under substring and for every string $u \notin M$ (such that all proper substrings of $u$ are in $M$) every sequence of vectors of the form*

$$L^{(0)}, L^{(1)}, \dots, L^{(i)}, \dots \tag{3}$$

*(where $L^{(0)} = (L_1 \cap M, \dots, L_n \cap M)$ and every next vector $L^{(i+1)} \neq L^{(i)}$ in the sequence is obtained from the previous vector $L^{(i)}$ by substituting some $j$-th component with $\varphi_j(L^{(i)}) \cap (M \cup \{u\}))$ converges to $(L_1 \cap (M \cup \{u\}), \dots, L_n \cap (M \cup \{u\}))$ in finitely many steps regardless of the choice of components.*

Note that all vectors forming the sequence (3) are equal modulo $M$ (because the initial term is a solution modulo $M$), and therefore the derivation is confined to transforming Boolean vectors of the membership of $u$ in the components, similarly to Lemma 1. It is possible to compute the naturally reachable solution modulo every finite language by just following Definition 5 and carrying out the derivation (3), repeating this inductively on the cardinality of the modulus. While this semantics does not have the same theoretical justifications as the semantics of the unique solution does, the semantics of naturally reachable solution can be shown to define the least fixed point of any system which does not make use of negation, and thus ensures "backward compatibility" with conjunctive and context-free grammars. The same undecidability result as in Theorem 1 can be proved for this semantics as well:

**Theorem 2.** *The set of systems compliant to the semantics of naturally reachable solution is co-RE-complete.*

It should be noted that if a system complies to both semantics, then it defines the same vector under both semantics, because the solution is unique by the first semantics, and therefore the second semantics cannot define any other solution. Therefore they do not contradict each other, but the classes of systems compliant to the two semantics can be shown to be incomparable. For instance, the system $\{X = X, Y = \neg Y \& \neg X\}$ has unique solution $(\Sigma^*, \emptyset)$ modulo every language, but does not comply to the semantics of naturally reachable solution. On the other hand, the system $\{X = \neg Y, Y = Y\}$ has naturally reachable solution $(\Sigma^*, \emptyset)$, but it has many other solutions as well and therefore is not compliant to the semantics of unique solution in the strong sense. Later on we shall prove that these two semantics nevertheless define a single class of languages.

## 2.4    Boolean Grammars

Now let us represent the systems of language equations using the notation of formal grammars.

**Definition 6.** *A Boolean grammar is a quadruple $G = (\Sigma, N, P, S)$, where $\Sigma$ and $N$ are disjoint finite nonempty sets of terminal and nonterminal symbols respectively; $P$ is the set of rules, each of the form*

$$A \rightarrow \alpha_1 \& \ldots \& \alpha_m \& \neg \beta_1 \& \ldots \& \neg \beta_n \quad (m + n \geqslant 1, \; \alpha_i, \beta_i \in (\Sigma \cup N)^*) \qquad (4)$$

*Objects of the form $A \rightarrow \alpha_i$ and $A \rightarrow \neg \beta_j$ are called conjuncts, positive and negative respectively; $S \in N$ is the start symbol of the grammar.*

*The right hand side of every rule is a formula, and a grammar is interpreted as a system of equations over $\Sigma$ in variables $N$ of the form*

$$A = \bigvee_{A \rightarrow \varphi \in P} \varphi \quad (\text{for all } A \in N) \qquad (5)$$

*The vector of languages $\overline{L}$ generated by a grammar is then defined using either the semantics of unique solution in the strong sense or the semantics of naturally reachable solution (see Sects. 2.2 and 2.3 respectively).*

*For every formula $\varphi$, denote the language of the formula $L_G(\varphi) = \varphi(\overline{L})$. Denote the language generated by the grammar as $L(G) = L_G(S)$.*

Every conjunctive grammar is a Boolean grammar, in which every rule (4) contains only positive conjuncts, i.e., $m \geqslant 1$ and $n = 0$. In particular, every context-free grammar is a Boolean grammar, in which every rule (4) contains a single positive conjunct ($m = 1$, $n = 0$).

Consider the system of equations from Example 1, and let us use its general idea to produce a Boolean grammar that will generate the language $\{ww \mid w \in \{a, b\}^*\}$ under both semantics defined in this paper:

*Example 3.* Let $G = (\{a, b\}, \{S, A, B, C, X\}, P, S)$ be a Boolean grammar, where $P$ consists of the following rules:

$$
\begin{array}{lll}
S \rightarrow \neg AB \& \neg BA \& C & B \rightarrow XBX & C \rightarrow \epsilon \\
A \rightarrow XAX & B \rightarrow b & X \rightarrow a \\
A \rightarrow a & C \rightarrow XXC & X \rightarrow b
\end{array}
\qquad (6)
$$

## 2.5    Parse Trees

A serious drawback of semantics of unique solution in the strong sense is that there is no general way to represent the membership in the components of this solution in the form of a tree. Consider the following grammar comprised of two nonterminals $\{S, A\}$ and two rules, $S \rightarrow S$ and $A \rightarrow \neg A \& \neg S$. $(\Sigma^*, \emptyset)$ is the unique solution of this system modulo every language, and therefore the system is compliant to this semantics and $L(G) = \Sigma^*$. However, it seems impossible to denote in the form of a parse tree *why* some particular string $w$ is in $L(G)$.

On the other hand, the semantics of naturally reachable solution is based on a certain type of derivation, and this derivation could be to some extent represented as a tree. Let $G = (\Sigma, N, P, S)$ be a Boolean grammar compliant to this semantics, and suppose without loss of generality that every rule in $P$ contains at least one positive conjunct (every grammar can be converted to this form by adding a new nonterminal that generates $\Sigma^*$ and referring to it in every rule). Let $r = |N|$. A parse tree of a string $w \in \Sigma^*$ from $A \in N$ is an acyclic directed graph with shared leaves that has a terminal leaf for every symbol in $w$. Define it inductively on the length of $w$.

Let $L^{(0)}, \dots, L^{(z)}$ be a sequence of vectors satisfying Definition 5 for the string $w$ and a modulus $M$. For all $p$ $(0 \leqslant p \leqslant z)$, let $L^{(p)} = (L_1^{(p)}, \dots, L_r^{(p)})$. Let us construct the sequence $\{(t_1^{(p)}, \dots, t_r^{(p)})\}_{p=0}^z$ of vectors of trees corresponding to the sequence of vectors of languages. For the initial term of the sequence, define all $t_i^{(0)}$ to be empty. For every next $p$-th term, if the string $u$ is added to some $k$-th component, then there should exist a rule

$$A_k \to \alpha_1 \& \dots \& \alpha_m \& \neg\beta_1 \& \dots \& \neg\beta_n \tag{7}$$

such that $w \in \alpha_i(L^{(p-1)})$ for all $i$ and $w \notin \beta_j(L^{(p-1)})$ for all $j$. In our construction, we completely ignore the negative conjuncts and make use of the positive ones. We want to construct a tree with a root labeled with (7), which will have $|\alpha_1| + \dots + |\alpha_m|$ descendants corresponding to all symbols from these positive conjuncts. For each $\alpha_i = s_1 \dots s_l$ we have to construct $l$ descendants. There exists a factorization $u = v_1 \dots v_l$, such that $v_j \in s_j(L^{(p-1)})$ for all $j$. For $s_j \in \Sigma$, we simply add a leaf labeled $s_j$. For $s_j \in N$, if the corresponding $v_j$ is shorter than $u$, then, by the induction hypothesis, we know how to construct a derivation tree of $v_j$ from $s_j$, and hence we use it as a subtree. If $v_j$ is of the same length as $u$, then $v_j = u$, and therefore $u \in L_j^{(p-1)}$. Then we have this subtree as $t_j^{(p-1)}$, and also connect it to the new root. The subtrees collected for all positive conjuncts have identical set of terminal leaves – those corresponding to the symbols from $u$. So we identify the corresponding leaves in these subtrees, connect them to a single root and place the newly constructed tree in $t_k^{(p)}$.

Note that this technique is a direct generalization of the tree construction method for conjunctive grammars, which in turn generalizes the context-free case. For grammars that use negation as the principal logical connective, such as the one from Example 3, this method does not have much sense, because the resulting tree contains no meaningful information. But if negation is used sparingly, such trees can contain enough "positive" information on the syntactic structure of the string according to the grammar.

## 3   Normal Forms

### 3.1   Epsilon Conjuncts

Given a Boolean grammar that generates a vector of languages $L = (L_1, \dots, L_n)$ under one of the mentioned semantics, we aim to construct a Boolean grammar that generates $L' = (L_1 \setminus \{\epsilon\}, \dots, L_n \setminus \{\epsilon\})$ under the same semantics. As in the cases of context-free and conjunctive grammars, this is being done by removing the so-called *positive epsilon conjuncts* of the form $A \to \epsilon$; in our case we shall also add *negative* epsilon conjuncts of the form $A \to \neg\epsilon$.

Given a Boolean grammar $G = (\Sigma, N, P, S)$ and a solution $L_\epsilon = \langle L_A \rangle_{A \in N}$ of the corresponding system modulo $\{\epsilon\}$, let $Nullable \subseteq N$ denote the nonterminals, for which $\epsilon \in L_A$. For any string $\alpha \in (\Sigma \cup N)^*$, define $\rho(\alpha)$ to be the set of all nonempty strings $\alpha' = s_1 \dots s_k$ ($s_i \in \Sigma \cup N$), such that $\alpha = \nu_0 s_1 \nu_1 s_2 \dots \nu_{k-1} s_k \nu_k$ for some $\nu_0, \dots, \nu_k \in Nullable^*$.

**Lemma 2.** *Let $G = (\Sigma, N, P, S)$ be a Boolean grammar. Let $L = (L_1, \dots, L_n)$ be an arbitrary vector of languages that satisfies modulo $\{\epsilon\}$ the system corresponding to $G$. Let $L' = (L_1 \setminus \{\epsilon\}, \dots, L_n \setminus \{\epsilon\})$. Let $\alpha \in (\Sigma \cup N)^*$. Let $w \in \Sigma^+$. Then $w \in \alpha(L)$ holds if and only if $w \in \alpha'(L')$ for some $\alpha' \in \rho(\alpha)$.*

Then construct a grammar $G' = (\Sigma, N, P', S)$, such that for every rule

$$A \to \alpha_1 \& \dots \& \alpha_m \& \neg\beta_1 \& \dots \& \neg\beta_n, \tag{8}$$

from $P$, where $\rho(\alpha_i) = \{\mu'_{i1}, \dots, \mu'_{ik_i}\}$ ($k_i \geqslant 0$; for all $i$) and $\rho(\beta_j) = \{\nu'_{j1}, \dots, \nu'_{jl_j}\}$ ($l_i \geqslant 0$; for all $i$), the set $P'$ contains the rule

$$A \to \mu_{1t_1} \& \dots \& \mu_{mt_m} \& \neg\nu_{11} \& \dots \& \neg\nu_{1l_1} \& \dots \& \neg\nu_{n1} \& \dots \& \neg\nu_{1l_n} \& \neg\epsilon \tag{9}$$

for every vector of numbers $(t_1, \dots, t_m)$ ($1 \leqslant t_i \leqslant k_i$ for all $i$).

**Lemma 3.** *Let $G = (\Sigma, N, P, S)$ be a Boolean grammar. Let $L_\epsilon$ be a solution modulo $\{\epsilon\}$ of the system corresponding to $G$. Let the Boolean grammar $G'$ be constructed out of $G$ and $L_\epsilon$ by the method above. Let $X = \varphi(X)$ and $X = \varphi'(X)$ be systems of language equations corresponding to $G$ and $G'$ respectively.*

*Let $L = (L_1, \dots, L_n)$ (where $L_i \subseteq \Sigma$ and $n = |N|$) be a vector of languages that equals $L_\epsilon$ modulo $\{\epsilon\}$. Let $L' = (L_1 \setminus \{\epsilon\}, \dots, L_n \setminus \{\epsilon\})$. Then $\varphi(L) = \varphi'(L') \setminus \{\epsilon\}$.*

**Lemma 4.** *Under the conditions of Lemma 3, for every $M$ closed under substring, $L$ is a solution of the first system iff $L'$ is a solution of the second system.*

**Theorem 3.** *For every Boolean grammar $G = (\Sigma, N, P, S)$ compliant to the semantics of unique solution in the strong sense there exists and can be effectively constructed a Boolean grammar $G'$ compliant to the semantics of unique solution in the strong sense, such that $L(G') = L(G) \setminus \{\epsilon\}$.*

**Theorem 4.** *For every Boolean grammar $G = (\Sigma, N, P, S)$ compliant to the semantics of naturally reachable solution there exists and can be effectively constructed a Boolean grammar $G'$ compliant to the semantics of naturally reachable solution, such that $L(G') = L(G) \setminus \{\epsilon\}$.*

## 3.2   Unit Conjuncts

Conjuncts of the form $A \to B$ and $A \to \neg B$ are called *positive* and *negative unit conjuncts* respectively. They will be collectively referred to as *unit conjuncts*, and our next challenge will be to devise an algorithm to get rid of them.

Let $G = (\Sigma, N, P, S)$ be a Boolean grammar compliant to one of the two semantics, such that $\epsilon \notin L_G(A)$ for every $A \in N$. Let $M \subseteq \Sigma^*$ be a finite language closed under substring and let $w \notin M$ be a string, such that all of its proper substrings are in $M$. Let $L = (L_1, \dots, L_n)$ be a solution of the system modulo $M$ and let $L' = (L_1 \cup L'_1, \dots, L_n \cup L'_n)$ be a vector, such that $L'_i \subseteq \{w\}$. Note that the membership of $w$ in $\alpha(L')$ depends on $L$ alone if $\alpha \notin N$ and on $\{L'_i\}$ alone if $\alpha \in N$. Each time a solution modulo $M \cup \{w\}$ is obtained out of $L$ by choosing $\{L'_i\}$ (as in Lemma 1 or in Definition 5), the values of non-unit conjuncts are fixed (as $L$ does not change), while the values of unit conjunct do not depend on $L$ at all.

Let $R \subseteq (\Sigma \cup N)^* \setminus N$ be a finite set of strings that contains a string $\gamma \notin N$ if and only if there is a conjunct $A \to \gamma$ or $A \to \neg\gamma$ in the grammar. A fixed solution $L$ modulo $M$ defines a certain *assignment to conjuncts* – a mapping $f_{L,w} : R \to \mathbb{B}$, such that $f_{L,w}(\alpha) = 1$ iff $w \in \alpha(L)$.

The idea of our construction is to precompute the processing of unit conjuncts for every possible assignment to non-unit conjuncts. For every assignment to conjuncts $f : R \to \mathbb{B}$ (there are $2^{|R|}$ such assignments), let us trace the method of Lemma 1 or Definition 5 for this assignment of conjuncts $f$. If the method fails, then, taking into account that the grammar is assumed to be compliant to the chosen semantics, this means that this situation is artificial and could never happen on a real modulus $M$, string $w$ and a solution $L$ modulo $M$. If the method terminates and produces a certain set $N' \subseteq N$ of nonterminals that evaluate to true, then for every nonterminal $A \in N'$ construct a rule $A \to \alpha_1 \& \dots \& \alpha_k \& \neg\beta_1 \& \dots \& \neg\beta_l$, which lists *all* strings from $R$, where $f(\alpha_1) = \dots = f(\alpha_k) = 1$ and $f(\beta_1) = \dots = f(\beta_l) = 0$.

**Lemma 5.** *Let $G = (\Sigma, N, P, S)$ comply to one of the two semantics and let $\epsilon \notin L_G(A)$ for all $A \in N$. Let $M$ be a finite modulus closed under substring and let all proper substrings of $w \notin M$ be in $M$. Let $L$ be the solution modulo $M$ according to the semantics. Then the membership of $w$ in $L_G(A)$ depends entirely on $f_{L,w}$ constructed as above.*

**Theorem 5.** *Let $G = (\Sigma, N, P, S)$ be a Boolean grammar that generates an $\epsilon$-free vector of languages $L$ under the semantics of unique solution in the strong sense **or** under the semantics of naturally reachable solution. Then there exists a Boolean grammar $G' = (\Sigma, N, P', S)$ that has no unit conjuncts, such that $G'$ generates $L$ under **both** semantics.*

**Corollary 1.** *The classes of languages defined by Boolean grammars under the semantics of unique solution in the strong sense and under the semantics of naturally reachable solution coincide.*

### 3.3   The Binary Normal Form

**Definition 7.** *A Boolean grammar* $G = (\Sigma, N, P, S)$ *is said to be in the binary normal form if every rule in* $P$ *is of the form*

$$A \to B_1 C_1 \& \ldots \& B_m C_m \& \neg D_1 E_1 \& \ldots \& \neg D_n E_n \& \neg \epsilon \quad (m + n \geqslant 0) \tag{10a}$$

$$A \to a \tag{10b}$$

$$S \to \epsilon \quad \text{(only if } S \text{ does not appear in right parts of rules)} \tag{10c}$$

Exactly as in the case of context-free and conjunctive grammars [4], the transformation of a grammar to the binary normal form can be carried out by first removing the epsilon conjuncts, then eliminating the unit conjuncts, cutting the bodies of the "long" conjuncts by adding extra nonterminals, and, if the original grammar generated the empty string, by adding a new start symbol with the rule (10c).

**Theorem 6.** *For any Boolean grammar* $G = (\Sigma, N, P, S)$ *that generates some language* $L$ *under some of the two given semantics, there exists and can be effectively constructed a Boolean grammar* $G' = (\Sigma, N', P', S')$ *in the binary normal form that generates this language* $L$ *under both semantics.*

Note that the effectiveness of the transformation does not include checking whether the original grammar actually complies to the semantics.

### 3.4   Linear Normal Form

**Definition 8.** *A Boolean grammar* $G = (\Sigma, N, P, S)$ *is said to be linear if every rule in* $P$ *is of the form*

$$A \to u_1 B_1 v_1 \& \ldots \& u_m B_m v_m \& \neg x_1 C_1 y_1 \& \ldots \& \neg x_n C_n y_n, \tag{11}$$

*where* $u_i, v_i, x_j y_j \in \Sigma^*$ *and* $m + n \geqslant 1$. $G$ *is said to be in the linear normal form if every rule in* $P$ *is of the form*

$$\begin{aligned} A \to b B_1 \& \ldots b B_m \& C_1 c \& \ldots \& C_n c \& \neg b D_1 \& \ldots \& \neg b D_k \& \\ \& \neg E_1 c \& \ldots \& \neg E_l c \ (m, n, k, l \geqslant 0; m + n + k + l \geqslant 1), \end{aligned} \tag{12a}$$

$$A \to a \tag{12b}$$

$$S \to \epsilon \quad \text{(only if } S \text{ does not appear in right parts of rules)} \tag{12c}$$

**Theorem 7.** *For any linear Boolean grammar* $G = (\Sigma, N, P, S)$ *that generates some language* $L$ *under some of the two given semantics, there exists and can be effectively constructed a linear Boolean grammar* $G' = (\Sigma, N', P', S')$ *in the linear normal form that generates this language* $L$ *under both semantics.*

Theorem 7 can be proved very much like Theorem 6, if we note that the transformations given in Sects. 3.1 and 3.2 preserve the linearity of a grammar.

This result implies an interesting property of linear Boolean grammars: using an effective transformation almost identical to the one applicable to linear conjunctive grammars [6], one can convert any given Boolean grammar in the linear normal form to an equivalent *trellis automaton* [2], which is a very simple square-time and linear-space computing device. The construction is omitted due to space considerations. Since trellis automata are computationally equivalent to linear conjunctive grammars [6], we obtain the following:

**Theorem 8.** *The family of languages generated by linear Boolean grammars (under the semantics of unique solution in the strong sense or under semantics of naturally reachable solution) coincides with the family generated by linear conjunctive grammars and the family accepted by trellis automata.*

## 4    Recognition and Parsing

The transformation of a Boolean grammar to the binary normal form given by Theorem 6 allows us to devise a cubic time recognition algorithm for every language generated by a Boolean grammar.

This algorithm is an extension of a similar algorithm for conjunctive grammars [4], which in turn generalizes the well-known Cocke–Kasami–Younger algorithm for context-free grammars in Chomsky normal form. The idea is to compute the sets of nonterminals deriving the proper substrings of the input string, starting from the shorter ones and continuing with the longer ones.

**Definition 9.** *Let $G = (\Sigma, N, P, S)$ be a Boolean grammar in the binary normal form. Let $w = a_1 \ldots a_n \in \Sigma^+$ $(n \geqslant 1)$ be the input string. Let $1 \leqslant i \leqslant j \leqslant n$. Define $T_{ij} = \{A \mid A \in N, \ a_i \ldots a_j \in L_G(A)\}$.*

The following algorithm computes all $T_{ij}$ starting from $T_{11}, \ldots, T_{nn}$ and ending with $T_{1n}$:

**Algorithm 1** *Let $G = (\Sigma, N, P, S)$ be a Boolean grammar in the binary normal form. For each $R \subseteq N \times N$ denote*

$$
\begin{aligned}
f(R) = \{A \mid \ &A \in N, \text{ there exists a rule} \\
&A \rightarrow B_1 C_1 \& \ldots \& B_m C_m \& \neg D_1 E_1 \& \ldots \& \neg D_n E_n \& \neg \epsilon \in P, \\
&\text{such that for any } p \ (1 \leqslant p \leqslant m) \ (B_p, C_p) \in R \\
&\text{and for any } q \ (1 \leqslant q \leqslant n) \ (D_q, E_q) \notin R\}
\end{aligned}
\tag{13}
$$

*Let $w = a_1 \ldots a_n \in \Sigma^*$ $(n \geqslant 1)$ be the input string. For all $i, j$, such that $1 \leqslant i \leqslant j \leqslant n$, compute $T_{ij}$.*

> for $i = 1$ to $n$
>> $T_{ii} = \{A \mid A \rightarrow a_i \in P \text{ or } A \rightarrow \neg D_1 E_1 \& \ldots \& \neg D_n E_n \& \neg \epsilon \in P\}$
>
> for $k = 1$ to $n - 1$
>> for $i = 1$ to $n - k$
>>> {

```
    let j = k + i
    let R = ∅  ( R ⊆ N × N )
    for l = i to j − 1
        R = R ∪ T_il × T_{l+1,j}
    T_ij = f(R)
}
```

This algorithm can be extended to construct a parse tree of the input string (in the sense of Sect. 2.5) in the same way as its prototypes.

## 5  General Properties

It is easy to prove that the language family denoted by Boolean grammars is closed under all set-theoretic operations, concatenation, reversal and star. The following result on the scope of Boolean grammars holds:

**Theorem 9.** *Every language generated by a Boolean grammar is deterministic context-sensitive, i.e., is in* **DSPACE**$(n)$.

The proof is by construction of DLBA and is omitted due to space considerations.

This result also has direct implications on conjunctive grammars. It was mentioned in [4] that every language generated by a conjunctive grammar is in **NSPACE**$(n)$. Now we know that a presumably stronger claim of being in **DSPACE**$(n)$ is true. The relation between these families of grammars is as follows:

$$\mathcal{L}(Conjunctive) \subseteq \mathcal{L}(Boolean) \subseteq \mathcal{L}(DetCS) \subseteq \mathcal{L}(CS) \tag{14}$$

The strictness of each of these three inclusions is left as an open problem.

It was conjectured in [4] that every unary conjunctive language is regular, which would imply proper inclusion in the context-sensitive languages. If this conjecture were proved, it would mean that the first inclusion in (14) is proper, because Boolean grammars can generate some nonregular unary languages (which can be inferred from Example 2, taken from [3]). However, no results on unary conjunctive languages have been obtained so far. Separating Boolean grammars from deterministic and nondeterministic context-sensitive languages is a different problem. It is very likely that this inclusion is proper, because the converse would mean that **P** = **PSPACE**.

Turning to applications, generalizing some of the parsing algorithms for conjunctive grammars [7] for the case of Boolean grammars would be a good argument for the practical value of the newly introduced concept.

## References

1. J. Autebert, J. Berstel and L. Boasson, "Context-Free Languages and Pushdown Automata", *Handbook of Formal Languages*, Vol. 1, 111–174, Springer-Verlag, Berlin, 1997.

2. K. Culik II, J. Gruska, A. Salomaa, "Systolic trellis automata", I and II, *Internat. Journal of Computer Mathematics*, 15 (1984), 195–212, and 16 (1984), 3–22.

3. E.L. Leiss, *Language equations*, Springer-Verlag, New York, 1999.

4. A. Okhotin, "Conjunctive grammars", *Journal of Automata, Languages and Combinatorics*, 6:4 (2001), 519–535.

5. A. Okhotin, "Conjunctive grammars and systems of language equations", *Programming and Computer Software*, 28:5 (2002), 243–249.

6. A. Okhotin, "Automaton representation of linear conjunctive languages", *Developments in Language Theory* (Proceedings of DLT 2002), LNCS, to appear.

7. A. Okhotin, "Whale Calf, a parser generator for conjunctive grammars", *Implementation and Application of Automata* (Proc. of CIAA 2002), *LNCS*, to appear; the software is available at `http://www.cs.queensu.ca/home/okhotin/whalecalf/`, `http://www.cs.queensu.ca/home/okhotin/whalecalf/`.

8. A. Okhotin, "Decision problems for language equations with Boolean operations", *Proceedings of ICALP 2003, LNCS*, to appear.

# Syntactic Semiring and Universal Automaton

Libor Polák*

Department of Mathematics, Masaryk University
Janáčkovo nám 2a, 662 95 Brno, Czech Republic
polak@math.muni.cz
http://www.math.muni.cz/~polak

**Abstract.** We discuss the relationships between the minimal automaton, the universal automaton, the syntactic monoid and the syntactic semiring of a given regular language. We use certain completions and reductions of the transformation matrix of the minimal automaton to clarify those connections.

**Keywords:** syntactic semiring, universal automaton
**MSC 2000 Classification:** 68Q45 Formal languages and automata

## 1 Introduction

In [6] the author introduced the so-called syntactic semiring of a language under the name syntactic semilattice-ordered monoid. The main result of that paper is an Eilenberg-type theorem giving a one-to-one correspondence between the so-called conjunctive varieties of regular languages and pseudovarieties of idempotent semirings. The author's next contribution [7] studies the relationships between the (ordered) syntactic monoid and the syntactic semiring of a given language. We mention here only that the first one is finite if and only if the second one is finite and that these two structures are equationally independent. In [8] we study the inequalities $r(x_1, \ldots, x_m) \subseteq L$ and equations $r(x_1, \ldots, x_m) = L$, where $L$ is a given regular language over a finite alphabet $A$ and $r$ is a given regular expression over $A$ in variables $x_1, \ldots, x_m$. We show that the search for their maximal solutions can be translated into the (finite) syntactic semiring of the language $L$. Moreover, connections between three models of the syntactic semiring are explored there.

The concept of the universal automaton of a regular language $L$ was implicitly introduced by Conway [1]; see Sakarovitch and Lombardy [3,4] and Lombardy [2] for historical remarks. The latter authors are using this concept for determining the so-called star height of a given language and for finding a reversible automaton for a reversible language. A crucial role is played there by the graph structure of this automaton and by its strongly connected components (so-called balls).

---

* Supported by the Ministry of Education of the Czech Republic under the project MSM 143100009

Z. Ésik and Z. Fülöp (Eds.): DLT 2003, LNCS 2710, pp. 411–422, 2003.

The aim of this note is to show that the syntactic semiring $S$ is an appropriate playground for considering together the (ordered) syntactic monoid, the minimal automaton and the universal automaton – all correspond to naturally described subsets of $S$.

We present preliminaries in the next section. Section 3 recalls the basics about the universal automata. We characterize there the languages with reversible balls. In Section 4 we present various tricks with the transformation matrix of the minimal automaton – it leads to methods for computing the universal automaton and its balls. Section 5 deals with variants of ball completeness. Finally, most of the discussed issues are illustrated by the examples in the last section.

## 2    Preliminaries

Let $A$ be a fixed finite alphabet. As usual, we denote the free semigroup and the free monoid over $A$ by $A^+$ and $A^* = A^+ \cup \{1\}$, respectively.

An *idempotent semiring* is a structure $(S, \cdot, \vee)$ where
$(S, \cdot)$ is a monoid with the neutral element 1,
$(S, \vee)$ is a semilattice with the smallest element 0,
$(\forall a, b, c \in S)(a(b \vee c) = ab \vee ac$ and $(a \vee b)c = ac \vee bc)$,
and $(\forall a \in S) a0 = 0a = 0$.
Such a structure becomes an ordered monoid with respect to the relation $\leq$ defined by $a \leq b \Leftrightarrow a \vee b = b$, $a, b \in S$.

If the type and the names of the operations of a considered structure are clear we usually do not write the operations explicitly.

Let $A^{\square}$ denote the set of all finite subsets of $A^*$. Note that this set with the operations $U \cdot V = \{uv \mid u \in U, v \in V\}$ and usual union form a free idempotent semiring over the set $A$.

Let $L \subseteq A^*$ be a regular language $u, v, u_1, \ldots, u_k, v_1, \ldots, v_l \in A^*$. We define the following relations :

$u \overset{\approx}{\to} v$ if and only if $(\forall y \in A^*)(uy \in L \Leftrightarrow vy \in L)$ that is iff $u^{-1}L = v^{-1}L$,

$u \overset{\approx}{\leftarrow} v$ if and only if $(\forall x \in A^*)(xu \in L \Leftrightarrow xv \in L)$ that is iff $Lu^{-1} = Lv^{-1}$,

$u \approx v$ if and only if $(\forall x, y \in A^*)(xuy \in L \Leftrightarrow xvy \in L)$,

$\{u_1, \ldots, u_k\} \overset{\approx}{\to} \{v_1, \ldots, v_l\}$ if and only if $(\forall y \in A^*)(u_1y, \ldots, u_ky \in L \Leftrightarrow v_1y, \ldots, v_ly \in L)$, that is iff $u_1^{-1}L \cap \cdots \cap u_k^{-1}L = v_1^{-1}L \cap \cdots \cap v_l^{-1}L$,

$\{u_1, \ldots, u_k\} \overset{\approx}{\leftarrow} \{v_1, \ldots, v_l\}$ if and only if $(\forall x \in A^*)(xu_1, \ldots, xu_k \in L \Leftrightarrow xv_1, \ldots, xv_l \in L)$, that is iff $Lu_1^{-1} \cap \cdots \cap Lu_k^{-1} = Lv_1^{-1} \cap \cdots \cap Lv_l^{-1}$,

$\{u_1, \ldots, u_k\} \sim \{v_1, \ldots, v_l\}$ if and only if $(\forall x, y \in A^*)(xu_1y, \ldots, xu_ky \in L \Leftrightarrow xv_1y, \ldots, xv_ly \in L)$.

It is immediate that the relation :
$\overset{\approx}{\to}$ is a right congruence,
$\overset{\approx}{\leftarrow}$ is a left congruence,
$\approx$ is a congruence of the monoid $(A^*, \cdot, 1)$, and similarly

$\rightsquigarrow$ is a right congruence,

$\leftsquigarrow$ is a left congruence,

$\sim$ is a congruence of the semiring $(A^\square, \cdot, \cup)$.

The factor-structures $(O, \cdot, 1 \approx) = (A^*, \cdot, 1)/\approx$ and $(S, \cdot, \vee) = (A^\square, \cdot, \cup)/\sim$ are called the *syntactic monoid* and the *syntactic semiring* of the language $L$. The monoid $O$ is ordered by the relation

$v \approx \, \le \, u \approx$   if and only if   $(\forall x, y \in A^*)(xuy \in L \Rightarrow xvy \in L)$

and we speak about the *ordered syntactic monoid*.

We further put :

$D = \{ u^{-1}L \mid u \in A^* \}$,

$U = \{ u_1^{-1}L \cap \cdots \cap u_k^{-1}L \mid k \ge 0,\ u_1, \ldots, u_k \in A^* \}$ (we get $A^*$ for $k = 0$),

$C = \{ u_1^{-1}Lv_1^{-1} \cap \cdots \cap u_k^{-1}Lv_k^{-1} \mid k \ge 0,\ u_1, \ldots, u_k, v_1, \ldots, v_k \in A^* \}$ (again we get $A^*$ for $k = 0$).

Notice that the mapping $\alpha : u \rightsquigarrow \, \mapsto u^{-1}L$ $(u \in A^*)$ is a "natural" bijection of the set $A^*/\rightsquigarrow$ and onto the set $D$, and

$\beta : \{u_1, \ldots, u_k\} \rightsquigarrow \, \mapsto u_1^{-1}L \cap \cdots \cap u_k^{-1}L$ $(u_1, \ldots, u_k \in A^*)$ is a "natural" bijection of the set $A^\square/\rightsquigarrow$ onto the set $U$.

For the sake of completeness we present now the links from the sets $U$ and $C$ to the Conway's factors and to the concepts used in [8]; see that paper for the proofs. Recall that a subset $X$ of $A^*$ is called *L-closed* (see [8]) if

$(\forall u_1, \ldots, u_k \in X)(\forall v \in A^*)(\{u_1, \ldots, u_k, v\} \sim \{u_1, \ldots, u_k\}$ implies $v \in X)$.

**Lemma 1.** *(i) Let $(P_1, Q_1), \ldots, (P_l, Q_l)$ be all maximal solutions of the inequality $xy \subseteq L$. Then $\{Q_1, \ldots, Q_l\} = U$.*

*(ii) The elements of $C$ are exactly the L-closed sets and the closure of a set $X \subseteq A^*$ is $X^L = \bigcap \{ x^{-1}Ly^{-1} \mid X \subseteq x^{-1}Ly^{-1},\ x, y \in A^* \}$.*

*(iii) The mapping $\{u_1, \ldots, u_k\} \sim \, \mapsto \{u_1, \ldots, u_k\}^L$ defines an isomorphism of $(S, \le)$ onto $(C, \subseteq)$.*

Classically, one assigns to $L$ its "canonical" *minimal automaton* $\mathcal{D} = (D, A, \cdot, L, F)$ using left derivatives; namely :

$D$ is the (finite) set of states,

$a \in A$ acts on $u^{-1}L$ by $(u^{-1}L) \cdot a = a^{-1}(u^{-1}L)$,

$L$ is the initial state and $Q \in D$ is a final state (i.e., element of $F$) if and only if $1 \in Q$.

For an arbitrary complete deterministic automaton $\mathcal{A} = (S, A, \cdot, i, G)$ we consider the transformation monoid $(T(\mathcal{A}), \circ) = (\{ \lceil u \rceil \mid u \in A^* \}, \circ)$ of $\mathcal{A}$ where $\lceil u \rceil : s \mapsto s \cdot u$ $(s \in S)$ is the transformation of the set $S$ induced by $u \in A^*$, and the composition $\lceil u \rceil \circ \lceil v \rceil$ is equal to $\lceil uv \rceil$. We have a quasiorder on $S$; namely : $s \le t$ iff $(\forall u \in A^*)(t \cdot u \in G \Rightarrow s \cdot u \in G)$. Note that $\mathcal{A}$ is minimal if and only if it is an order and each state is reachable. In such a case there is also an order $\le$ on $T(\mathcal{A})$ defined by : $\lceil u \rceil \le \lceil v \rceil$ iff $(\forall s \in S)\, s \cdot u \le s \cdot v$. For the automaton $\mathcal{D}$ we have $P \le Q$ iff $Q \subseteq P$.

Let $a \in A$ act on the whole $U$ by

$$(u_1^{-1}L \cap \cdots \cap u_k^{-1}L) \cdot a = (u_1^{-1}L) \cdot a \cap \cdots \cap (u_k^{-1}L) \cdot a$$

(which equals $a^{-1}(u_1^{-1}L) \cap \cdots \cap a^{-1}(u_k^{-1}L)$ ).

Define also an action of $\{u_1, \ldots, u_k\} \in A^\square$ on $\mathsf{U}$ by

$$Q \cdot \{u_1, \ldots, u_k\} = Q \cdot u_1 \cap \cdots \cap Q \cdot u_k$$

We denote this transformation of the set $\mathsf{U}$ by $[u_1, \ldots, u_k]$. Again,
$[u_1, \ldots, u_k] \circ [v_1, \ldots, v_l] = [u_1 v_1, \ldots, u_1 v_l, \ldots, u_k v_1, \ldots, u_k v_l]$ and put
$[u_1, \ldots, u_k] \vee [v_1, \ldots, v_l] = [u_1, \ldots, u_k, v_1, \ldots, v_l]$.

By the *extended automaton* of $L$ we mean a (generalized) automaton
$\mathcal{E} = (\mathsf{U}, A^\square, \cdot, L, T)$ where $Q \in \mathsf{U}$ is an element of $T$ if and only if $1 \in Q$. Let
$\mathsf{T}(\mathcal{E}) = \{ [u_1, \ldots, u_k] \mid k \geq 0, u_1, \ldots, u_k \in A^* \}$ (for $k = 0$ we get the constant
map onto $A^*$).

The first part of the following result is well-known and the second part is
from [8].

**Proposition 1.** *(i) The mapping* $\phi : u_\approx \mapsto [u]$, $u \in A^*$ *is an isomorphism of*
$(\mathsf{O}, \cdot, \leq)$ *onto* $(\mathsf{T}(\mathcal{D}), \circ, \leq)$.

*(ii) The mapping* $\psi : \{u_1, \ldots, u_k\}_\sim \mapsto [u_1, \ldots, u_k]$, $u_1, \ldots, u_k \in A^*$ *is an
isomorphism of* $(\mathsf{S}, \cdot, \vee)$ *onto* $(\mathsf{T}(\mathcal{E}), \circ, \vee)$.

## 3    Universal Automaton

The *universal automaton* of a language $L$ is a (non-deterministic) automaton
$\mathcal{U} = (\mathsf{U}, A, E, I, T)$ where $(P, a, Q) \in E$ if and only if $P \cdot a \supseteq Q$ and $Q \in \mathsf{U}$ is an
element of $I$ if and only if $Q \subseteq L$. Again, $Q \in T$ if and only if $1 \in Q$. In some
approaches (not here) the set $\emptyset$ is not considered as a state and the same for $A^*$
in case it is not an element from $\mathsf{D}$.

For the proofs of the following results consult [2].

**Proposition 2.** *Let* $\mathcal{U} = (\mathsf{U}, A, E, I, T)$ *be the universal automaton of a regular
language* $L$ *over an alphabet* $A$. *Then*

*(i)* $\mathcal{U}$ *accepts* $L$,

*(ii) for each non-deterministic automaton* $\mathcal{V} = (V, A, F, J, W)$ *accepting a
subset of* $L$, *the mapping*

$$\mu : q \mapsto \{ v \in A^* \mid (\forall u \in A^*)(q \in J \cdot u \Rightarrow uv \in L) \}, \quad q \in V$$

*is an automaton homomorphism of* $\mathcal{V}$ *into* $\mathcal{U}$.

*(iii) for each complete deterministic automaton* $\mathcal{A} = (S, A, \cdot, i, G)$ *accepting
the language* $L$, *we have* $\mu(s) = \{v \mid s \cdot v \in G\}$; *in particular,* $\mu$ *is the identity
on states of the automaton* $\mathcal{D}$.

Notice that there is a path from $P$ to $Q$ in $\mathcal{U}$ labeled by $u \in A^*$ if and only
if $P \cdot u \supseteq Q$, that is iff $u^{-1} P \supseteq Q$. We write in such a case also $P \xrightarrow{u} Q$. The set
$\mathsf{U}$ is ordered by the set-theoretical inclusion and strictly ordered by the proper
inclusion $\subset$. We define further relations on $\mathsf{U}$ :
$Q \sqsubseteq P$ if and only if there exists $u \in A^*$ such that $P \cdot u \supseteq Q$,
$P \equiv Q$ if and only if $P \sqsubseteq Q$ and $Q \sqsubseteq P$.

Clearly, $Q \subseteq P$ implies $Q \sqsubseteq P$, $\sqsubseteq$ is a quasiorder on $\mathsf{U}$, $\equiv$ is an equivalence relation on $\mathsf{U}$, and $\mathsf{U}/\equiv$ is ordered by $Q \equiv \preceq P \equiv$ iff $Q \sqsubseteq P$.

The classes of $\mathsf{U}/\equiv$ with the actions of letters are called the *balls* of $\mathcal{U}$. We do not take care of the initial and terminal states here; so the *determinism* means the absence of a configurations $P \xrightarrow{a} Q$, $P \xrightarrow{a} R$, $Q \neq R$. Similarly the *codeterminism* does not allow $P \xrightarrow{a} R$, $Q \xrightarrow{a} R$, $P \neq Q$. An automaton deterministic and codeterministic in this sense is called *reversible* (it can have several initial states). A language is reversible if it is recognized by a reversible automaton.

**Lemma 2.** *Let $B \in \mathsf{U}/\equiv$ and let $\mathcal{B}$ be the corresponding ball. The following conditions are equivalent :*
*(i) every two different states of $B$ are incomparable,*
*(ii) $\mathcal{B}$ is deterministic,*
*(iii) $\mathcal{B}$ is codeterministic.*

*Proof.* (i) implies (ii) : Suppose that there exist $P, Q, R \in B$, $a \in A$, $u, v \in A^*$ such that $Q \neq R$, $P \underset{u}{\overset{a}{\rightleftarrows}} Q$ and $P \underset{v}{\overset{a}{\rightleftarrows}} R$. The state $a^{-1}P$ is the greatest one with $P \xrightarrow{a} a^{-1}P$ and thus at least one of $Q, R$ is different from $a^{-1}P$; say $Q$. Then $Q \subset a^{-1}P$, $P \underset{u}{\overset{a}{\rightleftarrows}} a^{-1}P$.

(i) implies (iii) : Suppose that there exist $P, Q, R \in B$, $a \in A$, $u, v \in A^*$ such that $P \neq Q$, $P \underset{u}{\overset{a}{\rightleftarrows}} R$ and $Q \underset{v}{\overset{a}{\rightleftarrows}} R$. If $R \subset a^{-1}P$ then also $P \underset{u}{\overset{a}{\rightleftarrows}} a^{-1}P$ and similarly for $R \subset a^{-1}Q$. Finally if $a^{-1}P = a^{-1}Q = R$ then $P \cap Q \underset{u}{\overset{a}{\rightleftarrows}} R$.

not (i) implies not (ii), not (iii) : So let $P, Q \in B$, $P \subset Q$, $P \xrightarrow{u} Q$. Then both $P \xrightarrow{u} P, Q$ and $P, Q \xrightarrow{u} Q$.    □

Recall that for an element $a$ of a monoid $M$ the only idempotent in a sub-semigroup of $M$ generated by $a$ is denoted by $a^\omega$. The class of languages whose ordered syntactic monoids satisfy the inequality $x^\omega \geq 1$ is a quite prominent one – see Pin [5], p. 729. The reversible languages form its subclass. In [4] the authors proved the composition of Lemma 2 with the "if" part of the following result for the reversible languages.

**Proposition 3.** *The balls of $\mathcal{U}$ do not contain comparable different elements if and only if the ordered syntactic monoid $\mathsf{O}$ of $L$ satisfies the inequality $x^\omega \geq 1$.*

*Proof.* $\Leftarrow$ : Let $(\mathsf{O}, \cdot, \leq)$ satisfy $x^\omega \geq 1$ and suppose that there exist $P \in \mathsf{U}$, $u \in A^*$ such that $P \subset u^{-1}P$.
Let $(u \approx)^\omega = (u^n) \approx$. Then $P \subset u^{-1}P \subseteq (u^2)^{-1}P \subseteq \ldots \subseteq (u^n)^{-1}P \subseteq P$ since $\lceil u \rceil^\omega \geq \lceil 1 \rceil$.

$\Rightarrow$ : Now suppose an existence of an idempotent $\lceil e \rceil$ in the transformation monoid of $\mathcal{D}$ such that $\lceil e \rceil \not\geq \lceil 1 \rceil$. There exists $u \in A^*$ such that $e^{-1}(u^{-1}L) \not\subseteq u^{-1}L$. Denoting $P = (ue)^{-1}L \cap u^{-1}L$ we have $e^{-1}P = (ue)^{-1}L$ and thus $P \subset e^{-1}P$.    □

# 4   Playing with the Transformation Matrices

Our alphabet is usually $A = \{a_1, ..., a_n\}$ or $\{a, b, c, ...\}$ – so we can suppose that $A$ is linearly ordered. This gives also a strict linear order $\ll$ on the whole $A^*$ – we order words according their lengths first and the words of equal length lexicographically.

When calculating the transformation monoid $(\mathsf{T}(\mathcal{A}), \circ)$ of a complete deterministic automaton $\mathcal{A} = (S, A, \cdot, i, G)$, $S = \{s_1, ..., s_m\}$, we usually form so-called *transformation matrix* $\mathsf{M}(\mathcal{A})$. It is determined by the following conditions :

(i) rows are labeled by words $1 = u_1 \ll u_2 \ll \cdots \ll u_d$,

(ii) columns are labeled by states $s_1, ..., s_m$,

(iii) the $(i, j)$-entry is $s_j \cdot u_i$ – so the $i$-th row represents the transformation $\lceil u_i \rceil$,

(iv) for each $u \in A^*$ there exists exactly one $i$ such that $\lceil u \rceil = \lceil u_i \rceil$.

For any matrix $M$ over a set of states $S$ we define its *shadow* $\overline{M}$: put $\overline{M_{i,j}}$ instead of each entry $M_{i,j}$ where $\overline{s} = 1$ for a terminal state $s$ and $\overline{s} = 0$ otherwise. We define also the *left reduction* $\mathsf{lr}(M)$ of $M$ leaving from each set of identical columns only the left-most one. The *upwards reduction* $\mathsf{ur}(M)$ of $M$ is defined analogously.

Let $S$ consist either of sets and it is closed with respect to binary $\cap$ or $S = \{0, 1\}$ and $\cap$ is the semilattice operation on $S$ with $0 \cap 1 = 0$. We define the *right completion* $\mathsf{rc}(M)$ of $M$ by adding columns which are componentwise intersections of columns in $M$. Similarly we define the *downwards completion* $\mathsf{dc}(M)$ of $M$. The intersections of families with the empty indexed sets are also considered.

Notice that the automaton $\mathcal{A}$ is minimal iff every two columns of the shadow of its transformation matrix $\overline{\mathsf{M}(\mathcal{A})}$ with different indices are different. If not, $\mathsf{lr}(\overline{\mathsf{M}(\mathcal{A})})$ with necessary identifications lead to a minimalization of the set of the states. After possible renaming of states we can suppose from now on that the set of states of the minimal automaton is $\mathsf{D}$ and the columns of $\mathsf{M}(\mathcal{D})$ are labeled by $v_1^{-1}L, ..., v_m^{-1}L$. We know that the rows of $\mathsf{M}(\mathcal{D})$ correspond to elements of the syntactic monoid $\mathsf{O}$; in fact, the labels of the rows are the smallest elements with respect to $\ll$ in the corresponding $\approx$-classes. Observe that $v_j^{-1}L \cdot u_i = 1$ if and only if $u_i \in v_j^{-1}L$.

The following result shows what we can get using reductions and completions of the transformation matrix $\mathsf{M}(\mathcal{D})$. Notice that Lombardy in [2] identifies the rows of $\mathsf{dc}(\mathsf{ur}(\overline{\mathsf{M}(\mathcal{D})}))$ with the states of the universal automaton.

**Proposition 4.** *(i) The rule $\pi : j \mapsto \bigcup \{ u_i \approx \mid (i, j)\text{-entry is } 1 \}$ gives a one-to-one correspondence between the columns of $\mathsf{rc}(\overline{\mathsf{M}(\mathcal{D})})$ and elements of $\mathsf{U}$.*

*(ii) Considering the matrix $\mathsf{M}(\mathcal{D})$ over $\mathsf{U}$, the rows of $\mathsf{dc}(\mathsf{M}(\mathcal{D}))$ represent the elements of the syntactic semiring $\mathsf{S}$.*

*(iii) The elements of $A^*/\overset{\approx}{\approx}$ correspond to the rows of $\mathsf{ur}(\overline{\mathsf{M}(\mathcal{D})})$; a bijection is given by $w \overset{\approx}{\approx} \mapsto (\overline{v_1^{-1}L \cdot w}, ..., \overline{v_m^{-1}L \cdot w})$.*

*(iv) The elements of $A^\square/\underleftarrow{\approx}$ correspond to the rows of $\mathsf{dc}(\mathsf{ur}(\overline{\mathsf{M}\,(\mathcal{D})}))$; the mapping $\{w_1,\ldots,w_k\}\underleftarrow{\approx} \mapsto (v_1^{-1}L\cdot\{w_1,\ldots,w_k\},\ldots,v_m^{-1}L\cdot\{w_1,\ldots,w_k\})$ is a bijection.*

*(v) The rule $\rho : i \mapsto \bigcap\{\,v_j^{-1}L \mid (i,j)\text{-entry is 1}\,\}$ determines a one-to-one correspondence between the rows of $\mathsf{dc}(\mathsf{ur}(\overline{\mathsf{M}\,(\mathcal{D})}))$ and the elements of $\mathsf{U}$.*

*Proof.* (i) : Notice that $\pi(j) = v_j^{-1}L$ for $j \in \{1,\ldots,m\}$ and so the intersections of columns correspond to the intersections of their labels.

(ii) : This is a reformulation of Proposition 1 (ii).

(iii) : Notice that for each $w \in A^*$ there is a row $(v_1^{-1}L\cdot w,\ldots,v_m^{-1}L\cdot w)$ in $\mathsf{ur}(\overline{\mathsf{M}\,(\mathcal{D})})$ and observe that $w \underleftarrow{\approx} w'$ if and only if
$(\forall v \in A^*)(w \in v^{-1}L \Leftrightarrow w' \in v^{-1}L)$.

(iv) : Again, for each $\{w_1,\ldots,w_k\} \in A^\square$ there is a row
$(v_1^{-1}L\cdot\{w_1,\ldots,w_k\},\ldots,v_m^{-1}L\cdot\{w_1,\ldots,w_k\})$ in $\mathsf{dc}(\mathsf{ur}(\overline{\mathsf{M}\,(\mathcal{D})}))$. Notice that $\{w_1,\ldots,w_k\} \underleftarrow{\approx} \{w'_1,\ldots,w'_l\}$ if and only if
$(\forall v \in A^*)(w_1,\ldots,w_k \in v^{-1}L \Leftrightarrow w'_1,\ldots,w'_l \in v^{-1}L)$.
Further, $w_1,\ldots,w_k \in v^{-1}L$ iff $v^{-1}L\cdot\{w_1,\ldots,w_k\} = 1$ and similarly for the words $w'_1,\ldots,w'_l$.

(v) : Let $\sigma$ assign to $P \in \mathsf{U}$ a vector from $(\{0,1\})^m$ where the $j$-th entry is 1 iff $v_j^{-1}L \supseteq P$. We have $\rho(\sigma(P)) = P$ since $\rho(\sigma(P))$ is the intersection of all left derivatives of $L$ which are over the intersection $P$ of a concrete family of left derivatives.

Conversely, consider the $i$-th row of the matrix $\mathsf{dc}(\mathsf{ur}(\overline{\mathsf{M}\,(\mathcal{D})}))$. By (iv) there are $w_1,\ldots,w_k \in A^*$ such that $j$-th entry of this row is 1 iff $w_1,\ldots,w_k \in v_j^{-1}L$. Let $w_1,\ldots,w_k \in v_{j_1}^{-1}L,\ldots,v_{j_l}^{-1}L$ but not $w_1,\ldots,w_k \in v_j^{-1}L$ for other indices $j$. Then $P = v_{j_1}^{-1}L\cap\cdots\cap v_{j_l}^{-1}L = \rho(i\text{-th row})$ and $\sigma(P)$ has entries 1 at positions $j_1,\ldots,j_l$. Finally, $v_j^{-1}L \supseteq P$ gives $w_1,\ldots,w_k \in v_j^{-1}L$ and thus $\sigma(\rho(i\text{-th row})) = i$-th row. $\qquad\square$

All is illustrated in the last section.

# 5   Ball Completeness

**Proposition 5.** *The following conditions define strictly increasing families of languages :*

(G) *the syntactic monoid of $L$ is a group,*

(I) *the ordered syntactic monoid of $L$ satisfies ( $\forall p \in \mathsf{O}$ ) $p^\omega \leq 1$,*

(E) *the ordered syntactic monoid of $L$ satisfies ( $\forall p \in \mathsf{O}$ )( $\exists q \in \mathsf{O}$ ) $pq \leq 1$,*

(BC) *each ball of the universal automaton of $L$ is complete.*

*Proof.* (G) implies (I) is true since in a finite group $G$ we have even $p^\omega = 1$ for each $p \in G$.

(I) does not imply (G) : Consider, for instance, the complement of the language from Example 1.

(I) implies (E) : Observe that $p^\omega$ can be written as $p^n$ for some $n \geq 2$. Take $q = p^{n-1}$.

(E) does not imply (I) : Consider the language from Example 2.

(E) implies (BC) : Let $B \in \mathsf{U}/\equiv$, $P \in B$, $a \in A$. Let $u \approx$ $(u \in A^*)$ satisfy $a \approx \cdot u \approx \leq 1 \approx$ in $(\mathsf{O}, \cdot, \leq)$. Then for each $Q \in \mathsf{D}$ we have $Q \cdot au \supseteq Q$ and the same is even true for each $Q \in \mathsf{U}$. Now $(P \cdot a) \cdot u = P \cdot au \supseteq P$ and thus $P \cdot a \overset{u}{\rightarrow} P$.

(BC) does not imply (E) : Consider the language from Example 3.

## 6    Examples

**Example 1.** Let $A = \{a, b\}$ and $L = (ab)^* + b$. Denoting $K = b(ab)^*$, $M = 1$, $N = (ab)^*$, $O = \emptyset$, we see that the canonical minimal automaton $\mathcal{D}$ of $L$ looks as follows.

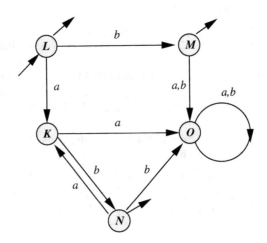

Let us calculate the matrix $\mathsf{M}(\mathcal{D})$ first.

|       | L | K | M | N | O |
|-------|---|---|---|---|---|
| 1     | L | K | M | N | O |
| $a$   | K | O | O | K | O |
| $b$   | M | N | O | O | O |
| $a^2$ | O | O | O | O | O |
| $ab$  | N | O | O | N | O |
| $ba$  | O | K | O | O | O |
| $bab$ | O | N | O | O | O |

The syntactic monoid $\mathsf{O}$ has the presentation

$$< a, b \mid a^2 = b^2 = 0, \ aba = a > .$$

Now consider consecutively the matrices $\overline{\mathsf{M}(\mathcal{D})}$ and $\mathsf{rc}(\overline{\mathsf{M}(\mathcal{D})})$.

|      | L | K | M | N | O | A* | K∩L |
|------|---|---|---|---|---|----|-----|
| 1    | 1 | 0 | 1 | 1 | 0 | 1  | 0   |
| a    | 0 | 0 | 0 | 0 | 0 | 1  | 0   |
| b    | 1 | 1 | 0 | 0 | 0 | 1  | 1   |
| a²   | 0 | 0 | 0 | 0 | 0 | 1  | 0   |
| ab   | 1 | 0 | 0 | 1 | 0 | 1  | 0   |
| ba   | 0 | 0 | 0 | 0 | 0 | 1  | 0   |
| bab  | 0 | 1 | 0 | 0 | 0 | 1  | 0   |

We see that $U = \{L, K, M, N, O, A^*, K \cap L\}$. Denoting $P = K \cap L$, the order on the states of the extended automaton (i.e. the reverse inclusion) is

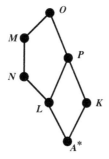

We can proceed by calculating the matrices $dc(M(\mathcal{D}))$ and $rc(M(\mathcal{D}))$ placing them into a single schema.

|       | L | K | M | N | O | A* | P |
|-------|---|---|---|---|---|----|---|
| 1     | L | K | M | N | O | A* | P |
| a     | K | O | O | K | O | A* | O |
| b     | M | N | O | O | O | A* | M |
| a²    | O | O | O | O | O | A* | O |
| ab    | N | O | O | N | O | A* | O |
| ba    | O | K | O | O | O | A* | O |
| bab   | O | N | O | O | O | A* | O |
| ∅     | A* | A* | A* | A* | A* |  |   |
| 1, a  | P | O | O | O | O |  |   |
| 1, b  | M | O | O | O | O |  |   |

The matrix $dc(M(\mathcal{D}))$ gives us the syntactic semiring $S$ of $L$; its order reduct is depicted below. The parts of labels after semicolons correspond to the representation of the syntactic semiring by $L$-closed subsets (see Lemma 1 (iii) ). Here $Q = Lb^{-1} = (ab)^*a + 1$, $R = (ab)^{-1}Lb^{-1} = (ab)^*a$, $S = a^{-1}Lb^{-1} = b(ab)^*a + 1$.

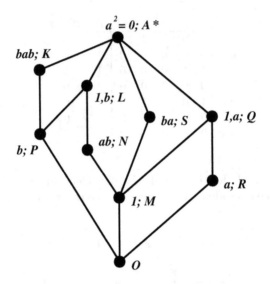

The rows of rc(M (𝒟)) indexed by $a$ and $b$ together with the inclusion order on the set U lead to a diagram of the universal automaton of $L$. The covering relation of this order on U is depicted by broken lines. For every state $T$ and a letter $c \in A$ we indicate only the largest state reachable by reading $c$ at $T$ – all the others are exactly its subsets.

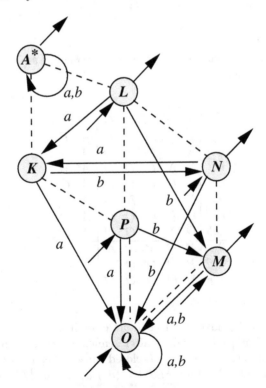

We can use the matrix rc($M(\mathcal{D})$) for a computation of the balls. Looking at the columns we see that exactly
$L, K, M, N, O, A^*, P \sqsubseteq A^*$,  $L, K, M, N, O, P \sqsubseteq L$,  $K, M, N, O, P \sqsubseteq K$,  $M, O \sqsubseteq M$,  $K, M, N, O, P \sqsubseteq N$,  $O \sqsubseteq O$,  $M, N, O \sqsubseteq P$. Thus the balls look as follows.

We see that all the balls are deterministic and codeterministic. Notice that $1^\omega = 1$, $a^\omega = b^\omega = 0^\omega = 0$, $(ab)^\omega = ab$, $(ba)^\omega = ba$, $(bab)^\omega = 0$ in $O$ and so $x^\omega \geq 1$ in $(O, \cdot, \leq)$ – compare it with Proposition 3.

Observe also that the only complete balls are the first one and the last one.

Finally we can calculate the matrices ur($\overline{M(\mathcal{D})}$) and dc(ur($\overline{M(\mathcal{D})}$)) presenting below the second one.

|       | L | K | M | N | O |
|-------|---|---|---|---|---|
| 1     | 1 | 0 | 1 | 1 | 0 |
| a     | 0 | 0 | 0 | 0 | 0 |
| b     | 1 | 1 | 0 | 0 | 0 |
| ab    | 1 | 0 | 0 | 1 | 0 |
| bab   | 0 | 1 | 0 | 0 | 0 |
| ∅     | 1 | 1 | 1 | 1 | 1 |
| 1, b  | 1 | 0 | 0 | 0 | 0 |

The relationship from Proposition 4 (v) is the following
$1 \mapsto M$, $a \mapsto A^*$, $b \mapsto P$, $ab \mapsto N$, $bab \mapsto K$, $\emptyset \mapsto O$, $\{1, b\} \mapsto L$.

**Example 2.** Let $A = \{a, b\}$, $L = A^*abA^* + A^*baA^* + 1$. We see that

$$O = \{1, a, b, ab\} = <a, b \mid a^2 = a, \ b^2 = b, \ ab = ba = 0 >$$

and the order is given by $ab < 1, a, b$. We have $1 \cdot 1$, $a \cdot b$, $b \cdot a$, $ab \cdot 1 \leq 1$ but $1^\omega = 1$, $a^\omega = a$, $b^\omega = b$, $(ab)^\omega = ab$ and so the identity $x^\omega \leq 1$ does not hold there.

**Example 3.** Consider the language $P = (A^*b + a)(a^2)^*$ over the alphabet $A = \{a, b\}$. Denote $Q = a^{-1}P = (A^*b + 1)(a^2)^*$, $R = P \cap Q$. We have

$$O = \{1, a, b, ba\} = < a, b \mid a^2 = 1, \ ab = b, \ b^2 = b > .$$

The elements of $O$ are incomparable and for $b$ there is no $q$ such that $bq \leq 1$. On the other hand, $U / \equiv = \{ \{A^*\}, \{P, Q, R\} \}$ and both balls are complete.

# References

1. J.H. Conway, *Regular Algebra and Finite Machines*, Chapman and Hall, Ltd., 1971
2. S. Lombardy, On the construction of reversible automata for reversible languages, ICALP, Springer Lecture Notes in Computer Science, Vol. 2380, pp. 170–182, 2002

3. S. Lombardy and J. Sakarovitch, On the star height of rational languages. A new presentation for two old results, *Proc. 3rd Int. Col. on Words, Languages and Combinatorics*, World Scientific, to appear

4. S. Lombardy and J. Sakarovitch, Star height of reversible languages and universal automata *LATIN*, Springer Lecture Notes in Computer Science, Vol. 2286, pp. 76–90, 2002

5. J.-E. Pin, Syntactic semigroups, Chapter 10 in *Handbook of Formal Languages*, G. Rozenberg and A. Salomaa eds, Springer, 1997

6. L. Polák, A classification of rational languages by semilattice-ordered monoids, submitted, see also `http://www.math.muni.cz/~polak`

7. L. Polák, Syntactic semiring of a language, *Proc. Mathematical Foundations of Computer Science 2001*, Springer Lecture Notes in Computer Science, Vol. 2136, 2001, pages 611–620

8. L. Polák, Syntactic Semiring and Language Equations, in *Proc. of the Seventh International Conference on Implementation and Application of Automata, Tours 2002*, Springer Lecture Notes in Computer Science, to appear

# Alphabetic Pushdown Tree Transducers

George Rahonis

Department of Mathematics
Aristotle University of Thessaloniki
54124 Thessaloniki, Greece
grahonis@math.auth.gr

**Abstract.** We introduce the concept of an alphabetic pushdown tree transducer, by adding a stack to an alphabetic tree transducer in the same way as a pushdown tree automaton is obtained from a top-down tree automaton. The stack of the general model contains trees, however, we also consider a restricted model of which the stack contains only unary trees. We give a characterization of the tree transformation induced by a restricted alphabetic pushdown tree transducer in terms of an algebraic forest over a suitable ranked alphabet and a bimorphism. We compare the class of tree relations induced by the alphabetic pushdown tree transducers with known classes of tree transformations. Finally, a new hierarchy of tree relations is established.

## 1 Introduction

Alphabetic transformations [2] constitute the natural generalization of rational relations, in the framework of trees. They have nice properties such as closure under composition and inversion and they can describe the most important operations on trees [2]. They have been also used as the basic tool, in order to build the AFL theory in the tree case [3]. In [17], we constructed a tree transducer model the so-called alphabetic, that realizes alphabetic transformations.

In this paper, we introduce the notion of an alphabetic pushdown tree transducer, by adding a stack to an alphabetic tree transducer [17]. This approach is similar to that of Guessarian [15], where a pushdown tree automaton is obtained by a top-down tree automaton. The stack of an alphabetic pushdown tree transducer may contain trees, however we also consider a restricted model of which the stack contains only unary trees. The restricted model is proved to be computationally equivalent to the general one. The class of tree relations induced by all alphabetic pushdown transducers is called algebraic and is denoted by $AGT$. We give a characterization of an algebraic tree relation in terms of an algebraic forest over a suitable ranked alphabet and a bimorphism. Precicely, we prove that each algebraic tree relation admits a factorization of an inverse alphabetic homomorphism an intersection with an algebraic forest and an alphabetic homomorphism. From this viewpoint, algebraic relations constitute a generalization of two concepts: pushdown translations on words [8], and alphabetic tree relations [2]. The underlying tree automaton of an (restricted) alphabetic pushdown

Z. Ésik and Z. Fülöp (Eds.): DLT 2003, LNCS 2710, pp. 423–436, 2003.

tree transducer is a (restricted) pushdown tree automaton [15]. Thus alphabetic pushdown tree transducers extend also in a natural way, the notion of pushdown tree automata.

We consider composition and inclusion properties for the class $AGT$. In fact, we prove that $AGT$ is not closed under composition whereas it is closed under composition with alphabetic relations. $AGT$ is incomparable with the class of tree relations induced by classical top-down, bottom-up, generalized [5] and macro tree transducers [7]. Moreover, $AGT$ is incomparable with the classes of tree relations induced by pushdown tree transducers introduced by Yamasaki [20,21]. Finally, we establish a new hierarchy of tree translations. The significance according to this hierarchy is that each component admits an algebraic characterization.

The organization of the paper is the following: in Sect. 2, we give some preliminary notions and results that will be used in the sequel. In Sect. 3, we introduce the general and the restricted model of alphabetic pushdown tree transducers and we prove the main results concerning them. Due to lack of space most of the proofs are omitted. The interested reader is referred to [18].

## 2   Preliminaries

As usual, in this section, we refer to notions and results that will be used in the sequel.

The power set of a set $V$, is denoted by $P(V)$. We write $V^*$ for the free monoid generated by $V$, whereas $\lambda$ denotes the empty string. For a natural number $n \geq 1$, $[n] = \{1, ..., n\}$, and $[0] = \emptyset$.

A *ranked alphabet* is a finite set $\Sigma$, such that to each symbol $\sigma \in \Sigma$, we associate a natural number denoted by $rank(\sigma)$ and called the *rank* of $\sigma$. $\Sigma_n$ denotes the set of symbols of $\Sigma$ with rank $n$. A symbol $\sigma \in \Sigma$ might have more than one ranks. The *degree* of $\Sigma$, $\deg(\Sigma)$ is the maximal number $n$ such that $\Sigma_n \neq \emptyset$ and $\Sigma_k = \emptyset$ for each $k > n$.

Let $X = \{x_1, x_2, ...\}$ be a countably infinite set of variables and $X_m = \{x_1, ..., x_m\}$, for $m \geq 0$. The set of all trees over $\Sigma$, indexed by the variables $x_1, ..., x_m$ is denoted by $T_\Sigma(X_m)$. A *forest* $F$ is a set of trees over an alphabet $\Sigma$, possibly with variables, that is $F \subseteq T_\Sigma(X_m)$, $m \geq 0$.

The *height* $hg(t)$ of a tree $t \in T_\Sigma(X_m)$ is defined by $hg(t) = 0$ if $t \in \Sigma_0 \cup X_m$, and $hg(t) = 1 + \max\{hg(t_1), ..., hg(t_n)\}$ if $t = \sigma(t_1, ..., t_n)$.

For $t \in T_\Sigma(X_n)$ and $t_1, ..., t_n \in T_\Sigma(X_m)$, we write $t(x_1/t_1, ..., x_n/t_n)$ or simply $t(t_1, ..., t_n)$ for the result of substituting $t_i$ for $x_i$ in $t$.

Let $\Sigma$, $\Gamma$ be ranked alphabets with $\deg(\Sigma) = n$. Assume that for each $k \in [n]$, there is a mapping $h_k : \Sigma_k \to T_\Gamma(X_k)$. Then the mappings $h_k$, $k \in [n]$ define a *tree homomorphism* $h : T_\Sigma \to T_\Gamma$ which is determined inductively: if $k \geq 0$, $\sigma \in \Sigma_k$ and $t_1, ..., t_k \in T_\Sigma$, then
$$h(\sigma(t_1, ..., t_k)) = h_k(\sigma)(h(t_1), ..., h(t_k)).$$

A tree homomorphism $h : T_\Sigma \to T_\Gamma$ is *linear*, if for each $\sigma \in \Sigma_k$, $k \geq 0$, $h_k(\sigma)$ is a linear tree, that is each variable from $X_k$ appears at most once in $h_k(\sigma)$.

A linear tree homomorphism $h : T_\Sigma \to T_\Gamma$ is called *alphabetic*, if for each $\sigma \in \Sigma_k$, $k \geq 0$, either
$h_k(\sigma) = \gamma(x_{i_1}, ..., x_{i_m})$, $\gamma \in \Gamma_m$, or
$h_k(\sigma) = x_n$, $1 \leq n \leq k$.

We denote by $REC$ the family of *recognizable forests* [13,14], by $OCF$ the family of *one counter forests* [3], by $NST$ the family of *synchronized forests* [19], and by $GST$ the family of *generalized synchronized forests* [16].

The notion of a pushdown tree automaton given below, introduced in [15].

**Definition 1.** A pushdown tree automaton (pdta) *is a six-tuple* $\mathcal{A} = (Q, \Sigma, \Pi, Q_0,$ $Z_0, R)$ *where $Q$ is the finite set of states, $\Sigma$ is the ranked alphabet of input symbols, $\Pi$ is the pushdown ranked alphabet with $\Pi \cap \Sigma = \emptyset$, $Q_0 \subseteq Q$ is the initial state set, $Z_0 \in \Pi$ is the initial pushdown symbol of rank 0, and $R$ is the finite set of rules of the form:*
*(i)* $q(\sigma(x_1, ..., x_n), \Theta(u_1, ..., u_s)) \to \sigma(q_1(x_1, \pi_1), ..., q_n(x_n, \pi_n))$,
*(ii)* $q(x, \Theta(u_1, ..., u_s)) \to q'(x, \pi)$,
*with $\sigma \in \Sigma_n$, $\Theta \in \Pi_s$, $\pi \in T_\Pi(u_1, ..., u_s)$, $\pi_i \in T_\Pi(u_1, ..., u_s)$, for $i \in [n]$, $x, x_i$ are variables ranging over $T_\Sigma$, and $u_1, ..., u_s$ are variables ranging over $T_\Pi$.* □

The rules of $R$ determine the computation steps of $\mathcal{A}$.

An input tree $t \in T_\Sigma$ is accepted by $\mathcal{A}$, if it can be consumed by a computation of $\mathcal{A}$, starting in an initial state and the initial stack symbol $Z_0$. The forest of all accepted trees is denoted by $L(\mathcal{A})$.

It has been proved in [15], that the family of forests accepted by all pdta's is the family $ALG$ of *algebraic forests*.

Guessarian in [15] considerd also a restricted model of a pushdown tree automaton where the alphabet $\Pi$ is unary. Precisely,

**Definition 2.** A restricted pushdown tree automaton (rpdta) *is a pdta, with* $\Pi = \Pi_1 \cup \{Z_0\}$. *The set of its rules is defined as:*
*(i)* $q(\sigma(x_1, ..., x_n), \Theta u) \to \sigma(q_1(x_1, \pi_1 u), ..., q_n(x_n, \pi_n u))$,
*(ii)* $q(x, \Theta u) \to q'(x, \pi u)$,
*(iii)* $q(\sigma(x_1, ..., x_n), u) \to \sigma(q_1(x_1, \pi_1 u), ..., q_n(x_n, \pi_n u))$,
*(iv)* $q(x, u) \to q'(x, \pi u)$
*with $\sigma \in \Sigma_n$, $\Theta \in \Pi_1$, $\pi \in \Pi_1^*$, $\pi_i \in \Pi_1^*$, for $i \in [n]$, $x, x_i$ are variables ranging over $T_\Sigma$, for $i \in [n]$ and $u$ is a variable ranging over $\Pi_1^* Z_0$.* □

The class of rpdta's is computationally equivalent to the class of pdta's [15].

A rpdta $\mathcal{A}$ is called *real-time* if it has no $\lambda$-moves, that is moves of type (ii) and (iv), and it is called *quasireal-time* if there is a constant $C$ such that in any computation of $\mathcal{A}$ there are at most $C$ consecutive $\lambda$-moves. A forest accepted by a real-time (quasireal-time) rpdta will be called *real-time* (*quasireal-time*) and the family of all real-time (quasireal-time) forests will be denoted by $RT^1$

$(QRT^1)$, where the superscript indicates that the above classes of forests are obtained using the restricted models [16] (it is not known if these models are equivalent to the corresponding general ones).

A *rewriting rule* over a ranked alphabet $\Sigma$, is a couple $(u, w)$ of trees of $T_\Sigma(X)$, such that: $var(w) \subseteq var(u)$. We usually denote a rule as $u \to w$.

A *rewriting system* $S$ over $\Sigma$, is a finite set of rewriting rules over $\Sigma$. For $t, p \in T_\Sigma$, we write $t \Rightarrow_S p$, if there exists a tree $\tau \in T_\Sigma(x)$ with just one occurrence of the variable $x$, a rule $u \to w$, $u \in T_\Sigma(X_n)$ and trees $v_1, ... v_n \in T_\Sigma$, such that $t = \tau(u(v_1, ... v_n))$, and $p = \tau(w(v_1, ... v_n))$. As it is usual, $\Rightarrow_S^*$ denotes the reflexive and transitive closure of $\Rightarrow_S$ .

A *rewriting system* $S$ over $\Sigma$ is called *confluent*, if for each $t, \tau_1, \tau_2 \in T_\Sigma$, $(t \Rightarrow_S^* \tau_1$ and $t \Rightarrow_S^* \tau_2)$ then there exists $p \in T_\Sigma$, such that $(\tau_1 \Rightarrow_S^* p$ and $\tau_2 \Rightarrow_S^* p)$. Finally, $S$ is called *Noetherian*, if there is no infinite sequence $t_1 \Rightarrow_S t_2 \Rightarrow_S ... t_k \Rightarrow_S ....$

## 3     Alphabetic Pushdown Tree Transducers

In this section, we introduce the general and the restricted model of alphabetic pushdown tree transducers. Then we give an algebraic characterization for the transformation induced by a restricted transducer. Using this result we establish the computationally equivalence between the two classes of models (the general and the restricted). Next, we state composition and inclusion properties for the class $AGT$ of tree relations induced by all alphabetic pushdown tree transducers. Then, we compare the class $AGT$ with other known families of tree transformations. Finally, we investigate the subclasses of $AGT$ induced by restricted alphabetic pushdown tree transducers with a certain number of stack symbols.

We shall need the notion of the *supremum alphabet* derived by two alphabets [2].

Let $\Sigma$ be a ranked alphabet and $k \geq \deg(\Sigma)$. We construct the ranked alphabet $\Sigma^{[k]}$ [2], in the following way:

$\Sigma_0^{[k]} = \Sigma_0$,

$\Sigma_n^{[k]} = \{\sigma_{i_1...i_p} \; / \; \sigma \in \Sigma_p, i_1, ..., i_p$ are distinct elements of $[n]$ and $\max\{i_1, ..., i_p\} = n\} \cup \{n\}$, for $1 \leq n \leq k$, and

$\Sigma_n^{[k]} = \emptyset$, for $n > k$.

Consider now, two ranked alphabets $\Sigma, \Gamma$ and a natural number $k \geq \max\{\deg(\Sigma), \deg(\Gamma)\}$. Then their *k-supremum alphabet* $\Sigma \vee_k \Gamma$ is

$(\Sigma \vee_k \Gamma)_0 = \Sigma_0 \times \Gamma_0$

$(\Sigma \vee_k \Gamma)_n = \bigcup_{\max(i,j)=n} \Sigma_i^{[k]} \times \Gamma_j^{[k]}, n \geq 1$.

If $k = \max\{\deg(\Sigma), \deg(\Gamma)\}$, we write $\Sigma \vee \Gamma$ and we call it the *supremum alphabet* of $\Sigma$ and $\Gamma$.

*Example 1.* Let $\Sigma$, $\Gamma$ be ranked alphabets, with $\Sigma_0 = \{a\}$, $\Sigma_1 = \{\sigma\}$ and $\Gamma_0 = \{b\}$, $\Gamma_2 = \{\gamma\}$. Then
- $\Sigma_0^{[2]} = \{a\}$, $\Sigma_1^{[2]} = \{\sigma_1, 1\}$, $\Sigma_2^{[2]} = \{\sigma_2, 2\}$, and
- $\Gamma_0^{[2]} = \{b\}$, $\Gamma_1^{[2]} = \{1\}$, $\Gamma_2^{[2]} = \{\gamma_{12}, \gamma_{21}, 2\}$.

The supremum alphabet of $\Sigma$ and $\Gamma$ is defined now as:
- $(\Sigma \vee \Gamma)_0 = \{<a, b>\}$,
- $(\Sigma \vee \Gamma)_1 = \{<a, 1>, <\sigma_1, b>, <1, b>, <\sigma_1, 1>, <1, 1>\}$, and
- $(\Sigma \vee \Gamma)_2 = \{<a, \gamma_{12}>, <a, \gamma_{21}>, <a, 2>, <\sigma_1, \gamma_{12}>, <\sigma_1, \gamma_{21}>, <\sigma_1, 2>, <1, \gamma_{12}>, <1, \gamma_{21}>, <1, 2>, <\sigma_2, b>, <2, b>, <\sigma_2, 1>, <2, 1>, <\sigma_2, \gamma_{12}>, <\sigma_2, \gamma_{21}>, <\sigma_2, 2>, <2, \gamma_{12}>, <2, \gamma_{21}>, <2, 2>\}$. $\square$

There are two naturally defined, alphabetic homomorphisms
$$\varphi_\Sigma : T_{\Sigma \vee \Gamma} \to T_\Sigma, \quad \varphi_\Gamma : T_{\Sigma \vee \Gamma} \to T_\Gamma,$$
with
$$\varphi_\Sigma(<\sigma_{i_1...i_p}, u>) = \sigma(x_{i_1}, ..., x_{i_p}), \quad \varphi_\Sigma(<n, u>) = x_n, \quad u \in \Gamma^{[k]},$$
$$\varphi_\Gamma(<w, \gamma_{j_1...j_m}>) = \gamma(x_{j_1}, ..., x_{j_m}), \quad \varphi_\Gamma(<w, n>) = x_n, \quad w \in \Sigma^{[k]},$$
where $k = \max\{\deg(\Sigma), \deg(\Gamma)\}$.

**Definition 3.** *[2] A tree transduction $\tau : T_\Sigma \to P(T_\Gamma)$ is called* alphabetic, *if there exists a recognizable forest $R \subseteq T_{\Sigma \vee \Gamma}$, such that $\#\tau = \{(\varphi_\Sigma(t), \varphi_\Gamma(t)) / t \in R\}$, where $\#\tau$ denotes the graph of $\tau$.* $\square$

We denote by *AT* the class of all alphabetic transductions.

### 3.1   The General and the Restricted Version

**Definition 4.** *An* alphabetic pushdown tree transducer (aptt) *is a system $\mathcal{T} = (Q, \Sigma, \Gamma, \Pi, d, \bar{q}, Q_0, Z_0, R)$, where*
*$Q \cup \{\bar{q}\}$ is the finite set of states,*
*$\Sigma$ is the ranked alphabet of input symbols,*
*$\Gamma$ is the ranked alphabet of output symbols,*
*$\Pi$ is the pushdown ranked alphabet,*
*$d$ is a symbol of rank 0, $d \notin \Sigma \cup \Gamma$,*
*$Q_0 \subseteq Q$ is the set of initial states,*
*$Z_0 \in \Pi_0$ is the initial pushdown symbol which appears initially in the pushdown store, and*
*$R$ is a finite set of rules of the form:*
*(I)* $q(\sigma(x_1, ..., x_n), \Theta(u_1, ..., u_s)) \to \bar{q}(<\sigma_{i_1...i_n}, v>$
$((q_1(d, \pi_1), ..., q_{i_{\rho(1)}}(x_{\rho(1)}, \pi_{i_{\rho(1)}}), ..., q_{i_{\rho(n)}}(x_{\rho(n)}, \pi_{i_{\rho(n)}}), ..., q_k(d, \pi_k)))$
*with $q, q_1, ..., q_k \in Q$, $\sigma \in \Sigma_n$, $n \geq 0$, $<\sigma_{i_1...i_n}, v> \in (\Sigma \vee \Gamma)_k$, $k \geq 0$, $v = \gamma_{j_1...j_m}$ or $v = j$, and $\rho$ is the unique permutation of $\{1, ..., n\}$, such that $i_{\rho(1)} < i_{\rho(2)} < ... < i_{\rho(n)}$,*

*(II)* $q(\sigma(x_1, ..., x_n), \Theta(u_1, ..., u_s)) \to$
$\bar{q}(<i, v> (q_1(d, \pi_1), ..., \underbrace{q(\sigma(x_1, ..., x_n), \Theta(u_1, ..., u_s))}_{ith-place}, ..., q_k(d, \pi_k)))$

*with $q, q_1, ..., q_k \in Q$, $\sigma \in \Sigma_n$, $n \geq 0$, $< i, v >\in (\Sigma \vee \Gamma)_k$, $k > 0$, $v = \gamma_{j_1...j_m}$ or*
*$v = j$,*

*(III) $q(d, \Theta(u_1, ..., u_s)) \to \overline{q}(< f, w > (q_1(d, \pi_1), ..., q_k(d, \pi_k)))$*
*with $q, q_1, ..., q_k \in Q$, $< f, w >\in (\Sigma \vee \Gamma)_k$, $k \geq 0$,*

*(IV) $q(x, \Theta(u_1, ..., u_s)) \to q'(x, \pi)$*
*with $q, q' \in Q$,*

*(V) $\overline{q}(< a, \gamma_{j_1...j_m} > (x_1, ..., x_k)) \to \gamma(x_{j_1}, ..., x_{j_m})$*
*with $< a, \gamma_{j_1...j_m} >\in (\Sigma \vee \Gamma)_k, k \geq 0$, and*

*(VI) $\overline{q}(< a, j > (x_1, ..., x_k)) \to x_j$*
*with $< a, j >\in (\Sigma \vee \Gamma)_k, k > 0$,*

*where in all the above rules, $\Theta \in \Pi_s$, $\pi, \pi_1, ..., \pi_k \in T_\Pi(u_1, ..., u_s)$ and $u_1, ..., u_s$ are variables ranging over $T_\Pi$.* $\square$

We denote by $E$ the ranked alphabet $(\Sigma \vee \Gamma) \cup \Sigma \cup \Gamma \cup Q \cup \{\overline{q}, d\} \cup \Pi$ (we associate the rank 2 to each state in $Q$ and the rank 1 to $\overline{q}$). *The one step computation of $\mathcal{T}$ is a binary relation on $T_E$* which is defined in the following way: for $t, p \in T_E$, $t \Rightarrow_\mathcal{T} p$ if and only if
(i) there is a rule $q(\sigma(x_1, ..., x_n), \Theta(u_1, ..., u_s)) \to v$ (resp. $q(d, \Theta(u_1, ..., u_s)) \to v$), of type (I) or (II) (resp. (III)), and
(ii) $t$ has a terminal subtree $t' = q(\sigma(t_1, ..., t_n), \Theta(r_1, ..., r_s))$, $t_1, ..., t_n \in T_\Sigma$, $r_1, ..., r_s \in T_\Pi$ (resp. $t' = q(d, \Theta(r_1, ..., r_s))$, $r_1, ..., r_s \in T_\Pi$), and $p$ is obtained by substituting $v(x_1/t_1, ..., x_n/t_n, u_1/r_1, ..., u_s/r_s)$ (resp.$v(u_1/r_1, ..., u_s/r_s)$) for an occurrence of $t'$ in $t$,
or
(i)' there is a rule $q(x, \Theta(u_1, ..., u_s)) \to q'(x, \pi)$ of type (IV), and
(ii)' $t$ has a terminal subtree $t' = q(w, \Theta(r_1, ..., r_s))$, $w \in T_\Sigma$, $r_1, ..., r_s \in T_\Pi$, and $p$ is obtained by substituting $q'(w, \pi(u_1/r_1, ..., u_s/r_s))$ for an occurrence of $t'$ in $t$,
or
(i)'' there is a rule $\overline{q}(< a, \gamma_{j_1...j_m} > (x_1, ..., x_k)) \to \gamma(x_{j_1}, ..., x_{j_m})$ (resp. $\overline{q}(< a, j > (x_1, ..., x_k)) \to x_j$), of type (V) (resp. of type (VI)), and
(ii)'' $t$ has a terminal subtree $t' = \overline{q}(< a, \gamma_{j_1...j_m} > (t_1, ..., t_k))$, $t_1, ..., t_k \in T_{\Sigma \vee \Gamma \cup \{\overline{q}\}}$, (resp. $t' = \overline{q}(< a, j > (t_1, ..., t_k))$, $t_1, ..., t_k \in T_{\Sigma \vee \Gamma \cup \{\overline{q}\}}$), and $p$ is obtained by substituting $\gamma(t_{j_1}, ..., t_{j_m})$ (resp. $t_j$), for an occurrence of $t'$ in $t$.

We denote by $\Rightarrow_\mathcal{T}^*$ the reflexive and transitive closure of $\Rightarrow_\mathcal{T}$.

Intuitively speaking, the transducer starts to process an input tree $t \in T_\Sigma$ in an initial state $q$ and the initial pushdown symbol $Z_0$. At the first computation step $\mathcal{T}$ applies a rule of type (I) or (II). Then $\mathcal{T}$ proceeds by using the rules of the form (I)-(III) transforming the tree $t$ to a tree $\tau$ over the alphabet $(\Sigma \vee \Gamma) \cup \{\overline{q}\}$. The rules of type (I)-(III) act on the tree $t \in T_\Sigma$ as the inverse of the alphabetic homomorphism $\phi_\Sigma : T_{(\Sigma \vee \Gamma) \cup \{\bar{q}\}} \to T_\Sigma$. Simultaneously, the rules of type (I)-(IV) simulate a pushdown tree automaton over the alphabet $\Sigma \vee \Gamma$. The next computation steps of $\mathcal{T}$ are realized by rules of the form (V) and (VI). In fact, with these rules the tree $\tau$ is projected to a tree $p \in T_\Gamma$. Thus $(\Sigma \vee \Gamma) \cup \{\bar{q}\}$ is used as an intermediate alphabet. On the other hand, the pushdown store gets empty when $\mathcal{T}$ finishes with the application of type (I)-(IV) rules. Next we do not need the pushdown automaton and we just have to simulate the alphabetic homomorphism $\phi_\Gamma : T_{\Sigma \vee \Gamma} \to T_\Gamma$. The state $\bar{q}$ just helps to the application of rules of the form (V) and (VI).

Notice that the rules of type (V) and (VI) are applied in a *look-ahead* manner [6,9,10,11]; the tree in process must belong to $T_{(\Sigma \vee \Gamma) \cup \{\bar{q}\}}$.

*The transformation realized* by an aptt $\mathcal{T}$ is the relation

$$| \mathcal{T} |= \{(t,p) \in T_\Sigma \times T_\Gamma \;/\; \exists q \in Q_0, \text{ such that } q(t, Z_0) \Rightarrow^*_\mathcal{T} p\}.$$

A relation computed by an aptt is called *algebraic*. The class of all algebraic relations is denoted by *AGT*.

For $t \in T_\Sigma$ we denote by $\mathcal{T}(t)$ the forest

$$\mathcal{T}(t) = \{p \in T_\Gamma \;/\; (t,p) \in| \mathcal{T} |\}$$

and

$$\mathcal{T}(F) = \bigcup_{t \in F} \mathcal{T}(t), \text{ where } F \subseteq T_\Sigma.$$

The *domain* and the *range* of $\mathcal{T}$ are respectively the forests

$$dom(\mathcal{T}) = \{t \in T_\Sigma \;/\; (t,p) \in| \mathcal{T} |, \text{ for some } p \in T_\Gamma\}$$
$$im(\mathcal{T}) = \{p \in T_\Gamma \;/\; (t,p) \in| \mathcal{T} |, \text{ for some } t \in T_\Sigma \}.$$

Two aptt's $\mathcal{T}, \mathcal{S}$ are called *equivalent* if $| \mathcal{T} |=| \mathcal{S} |$.
Below we give an example of an aptt.

*Example 2.* Let $\Sigma, \Gamma$ be ranked alphabets, such that $\Sigma_0 = \{a\}$, and $\Gamma_0 = \{b\}$, $\Gamma_1 = \{h, \gamma, \sigma\}$, $\Gamma_2 = \{f\}$. Consider the aptt $\mathcal{T} = (Q, \Sigma, \Gamma, \Pi, d, \bar{q}, Q_0, Z_0, R)$, with $Q = \{q_0, q_1, q_2, q_3, q_4\}$, $Q_0 = \{q_0\}$, $\Pi_1 = \{Z_1, Z_2\}$, $\Pi_0 = \{Z_0\}$. The set R of its rules is the following:

- $q_0(a, Z_0) \to \bar{q}(< a, f_{12} > (q_1(d, Z_1 Z_0), q_4(d, Z_1 Z_0)))$, (of type (I)),
- $q_1(d, Z_1 u) \to \bar{q}(< a, h_1 > (q_1(d, Z_1^2 u)))$, (of type (III)),
- $q_1(x, Z_1 u) \to q_2(x, u)$, (of type (IV)),
- $q_2(d, Z_1 u) \to \bar{q}(< a, \gamma_1 > (q_3(d, u)))$, (of type (III)),
- $q_3(d, Z_1 u) \to \bar{q}(< a, \gamma_1 > (q_3(d, u)))$, (of type (III)),
- $q_3(d, Z_0) \to \bar{q}(< a, b >)$, (of type (III)),
- $q_4(d, Z_1 u) \to \bar{q}(< a, \sigma_1 > (q_4(d, Z_1 u)))$, (of type (III)),
- $q_4(d, Z_1 u) \to \bar{q}(< a, b >)$, (of type (III)),
- $\bar{q}(< a, f_{12} > (x_1, x_2)) \to f_{12}(x_1, x_2)$, (of type (V)),
- $\bar{q}(< a, h_1 > (x_1)) \to h_1(x_1)$, (of type (V)),

- $\overline{q}(<a,\gamma_1>(x_1)) \to \gamma_1(x_1)$,    (of type (V)),
- $\overline{q}(<a,\sigma_1>(x_1)) \to \sigma_1(x_1)$,    (of type (V)),
- $\overline{q}(<a,b>) \to b$.

It is not difficult to see that the above transducer computes the transformation
$\{(a, f(h^n\gamma^n(b), \sigma^m(b)) \ / \ n, m \geq 1\}$. $\square$

The reader may notice in the above example, the use of the rules of type
(III). They simulate the part of the inverse alphabetic homomorphism $\phi_\Sigma$ with
which a letter is projected to a variable, e.g. $\gamma_1(x_1) \longmapsto x_1$.

The reader may also notice, that the alphabet $\Pi$ is unary. In fact, the transducer $\mathcal{T}$ is restricted in the sence of the following definition:

**Definition 5.** *A restricted alphabetic pushdown tree transducer (raptt) is an
aptt* $\mathcal{T} = (Q, \Sigma, \Gamma, \Pi, d, \overline{q}, Q_0, Z_0, R)$, *with* $\Pi = \Pi_1 \cup \{Z_0\}$ *that is all symbols in
$\Pi$ have rank 1 except the constant $Z_0$. In this case the rules of $\mathcal{T}$ have the form:*

*(I)* $q(\sigma(x_1, ..., x_n), \Theta u) \to \overline{q}(<\sigma_{i_1...i_n}, v >$
$((q_1(d, \pi_1 u), ..., q_{i_{\rho(1)}}(x_{\rho(1)}, \pi_{i_{\rho(1)}} u), ..., q_{i_{\rho(n)}}(x_{\rho(n)}, \pi_{i_{\rho(n)}} u), ..., q_k(d, \pi_k u)))$
*with* $<\sigma_{i_1...i_n}, v >\in(\Sigma \vee \Gamma)_k$, $k \geq 0$, $v = \gamma_{j_1...j_m}$ *or* $v = j$,

*(II)* $q(\sigma(x_1, ..., x_n), \Theta u) \to$
$\overline{q}(<i, v > (q_1(d, \pi_1 u), ..., \underbrace{q(\sigma(x_1, ..., x_n), \Theta u)}_{ith-place}, ..., q_k(d, \pi_k u)))$
*with* $<i, v >\in(\Sigma \vee \Gamma)_k$, $k > 0$, $v = \gamma_{j_1...j_m}$ *or* $v = j$,

*(III)* $q(d, \Theta u) \to \overline{q}(<f, w > (q_1(d, \pi_1 u), ..., q_k(d, \pi_k u)))$
*with* $<f, w >\in (\Sigma \vee \Gamma)_k$, $k \geq 0$,

*(IV)* $q(x, \Theta u) \to q'(x, \pi u)$,

*(V)* $\overline{q}(<a, \gamma_{j_1...j_m} > (x_1, ..., x_k)) \to \gamma(x_{j_1}, ..., x_{j_m})$
*with* $<a, \gamma_{j_1...j_m} >\in (\Sigma \vee \Gamma)_k$, $k \geq 0$, *and*

*(VI)* $\overline{q}(<a, j > (x_1, ..., x_k)) \to x_j$
*with* $<a, j >\in (\Sigma \vee \Gamma)_k$, $k > 0$,

*where in all the above rules,* $q, q', q_1, ..., q_k \in Q$, $\Theta \in \Pi_1 \cup \{\lambda\}$, $\pi, \pi_1, ..., \pi_k \in \Pi_1^*$,
*and $u$ is a variable ranging over* $\Pi_1^* Z_0$. $\square$

The pushdown store which belongs to $\Pi_1^* Z_0$ never gets empty, except the
case when $\mathcal{T}$ finishes the application of rules of type (I)-(IV).

In the next subsection we show that the class of tree relations computed by
all raptt's coincide with the class *AGT*. Thus the restricted version of aptt's do
not decrease their computational power.

## 3.2    The Main Results Concerning Alphabetic Pushdown Tree Transducers

In all the constructions below, we use the restricted model. The equivalence of the computational power of the classes of all aptt's and all raptt's will be clear from Theorems 1,2.

We introduce the notion of *the parallel computation* of a raptt $\mathcal{T} = (Q, \Sigma, \Gamma, \Pi, d,$ $\overline{q}, Q_0, Z_0, R)$. The one step of *a parallel computation of* $\mathcal{T}$ is a binary relation $\overset{par}{\Rightarrow}_{\mathcal{T}}$ on $T_E$ (recall that $E = (\Sigma \vee \Gamma) \cup \Sigma \cup \Gamma \cup Q \cup \{\overline{q}, d\} \cup \Pi$), such that for $t, p \in T_E$, $t \overset{par}{\Rightarrow}_{\mathcal{T}} p$ if and only if $\mathcal{T}$ does simultaneously all the possible one step (simple) computations. $\overset{par*}{\Rightarrow}_{\mathcal{T}}$ denotes the reflexive and transitive closure of $\overset{par}{\Rightarrow}_{\mathcal{T}}$. Since $\mathcal{T}$ is nondeterministic a computation $t \overset{par*}{\Rightarrow}_{\mathcal{T}} p$ may be done by different ways. In other words, writing down $t \overset{par*}{\Rightarrow}_{\mathcal{T}} p$ means that there exist trees $t_1, ..., t_{n-1} \in T_E$ such that

$$t \overset{par}{\Rightarrow}_{\mathcal{T}} t_1 \overset{par}{\Rightarrow}_{\mathcal{T}} ... \overset{par}{\Rightarrow}_{\mathcal{T}} t_{n-1} \overset{par}{\Rightarrow}_{\mathcal{T}} p \qquad (PAR)$$

but there may be also trees $r_1, ..., r_{m-1} \in T_E$ such that

$$t \overset{par}{\Rightarrow}_{\mathcal{T}} r_1 \overset{par}{\Rightarrow}_{\mathcal{T}} ... \overset{par}{\Rightarrow}_{\mathcal{T}} r_{m-1} \overset{par}{\Rightarrow}_{\mathcal{T}} p.$$

We define the *length* of the computation $t \overset{par*}{\Rightarrow}_{\mathcal{T}} p$ as the minimal number $n$ such that $(PAR)$ holds. In this case, we write $t \overset{par(n)}{\Rightarrow}_{\mathcal{T}} p$.

The transformation computed by $\mathcal{T}$ using the parallel computation mode is again a relation $|\mathcal{T}|_{par} \subseteq T_\Sigma \times T_\Gamma$:

$$|\mathcal{T}|_{par} = \{(t, p) \in T_\Sigma \times T_\Gamma \ / \ \exists q \in Q_0, \text{ such that } q(t, Z_0) \overset{par*}{\Rightarrow}_{\mathcal{T}} p\}.$$

Obviously $|\mathcal{T}|_{par} = |\mathcal{T}|$.

**Theorem 1.** *Let* $\Sigma, \Gamma$ *be ranked alphabets. A relation* $L \subseteq T_\Sigma \times T_\Gamma$ *is realized by a raptt if and only if there exists an algebraic forest* $F \subseteq T_{\Sigma \vee \Gamma}$, *such that* $L = \{(\varphi_\Sigma(t), \varphi_\Gamma(t)) \ / \ t \in F\}$. $\square$

The above result is also valid for aptt's. Thus we get:

**Theorem 2.** *A tree relation* $L \subseteq T_\Sigma \times T_\Gamma$ *is algebraic if and only if there exists an algebraic forest* $F$ *over* $\Sigma \vee \Gamma$, *such that* $L = \{(\varphi_\Sigma(t), \varphi_\Gamma(t)) \ / \ t \in F\}$. $\square$

The last Theorem shows that the model of an aptt is a natural generalization of pushdown transducers on words [8].

In [15] it is proved that for each pushdown tree automaton, we can effectively construct an equivalent restricted pushdown tree automaton. So taking into account Theorems 1,2, we obtain

**Corollary 1.** *The class of all raptt's have the same computational power with the class of all aptt's.* $\square$

**Corollary 2.** *For each aptt we can effectively construct an equivalent raptt.* $\square$

In the following examples the raptt's compute two important operations on trees. The first one realizes *the intersection with an algebraic forest*, and the latter *the $\alpha$-product with an algebraic forest*.

*Example 3.* Assume $F \subseteq T_\Sigma$ to be an algebraic forest and $\mathcal{A} = (Q, \Sigma, \Pi, Q_0, Z_0, R)$ a rpdta such that $L(\mathcal{A}) = F$. Consider the raptt $\mathcal{T} = (Q, \Sigma, \Sigma, \Pi, \bar{q}, Q_0, Z_0, R')$ (we do not need the symbol $d$ here, since we do not need type (III) rules). The set $R'$ of its rules is defined in the following manner:

(I) $q(\sigma(x_1, ..., x_n), \Theta u) \rightarrow \bar{q}(< \sigma_{1...n}, \sigma_{1...n} > (q_1(x_1, \pi_1 u), ..., q_n(x_n, \pi_n u)))$, with $\sigma \in \Sigma_n$, $n \geq 0$, $\Theta \in \Pi_1 \cup \{\lambda\}$, for each rule $q(\sigma(x_1, ..., x_n), \Theta u) \rightarrow \sigma(q_1(x_1, \pi_1 u), ..., q_n(x_n, \pi_n u))$ in $R$,

(IV) $q(x, \Theta u) \rightarrow q'(x, \pi u)$, with $\Theta \in \Pi_1 \cup \{\lambda\}$, for each rule $q(x, \Theta u) \rightarrow q'(x, \pi u)$, in $R$,

(V) $\bar{q}(< \sigma_{1...n}, \sigma_{1...n} > (x_1, ..., x_n)) \rightarrow \sigma(x_1, ..., x_n)$, for each $\sigma \in \Sigma_n$, $n \geq 0$.
It is an easy task to see that for each $t \in T_\Sigma$, $\mathcal{T}(t) = \{t\} \cap F$. $\square$

*Example 4.* Let $\mathcal{A} = (Q, \Sigma, \Pi, Q_0, Z_0, R)$ be an rpdta and $\alpha \in \Sigma_0$. We construct the transducer $\mathcal{T} = (Q, \Sigma, \Sigma, \Pi, d, \bar{q}, Q_0 \cup \{q'\}, Z_0, R')$, where $q'$ is a new state, and the set $R'$ has the following types of rules:

(I) $q'(\sigma(x_1, ..., x_n), Z_0) \rightarrow \bar{q}(< \sigma_{1...n}, \sigma_{1...n} > (q_1(x_1, Z_0), ..., q_n(x_n, Z_0)))$, for each $\sigma \in \Sigma_n - \{\alpha\}$, $n \geq 0$, and each $q_1, ..., q_n \in Q_0 \cup \{q'\}$,

(I) $q(\alpha, Z_0) \rightarrow \bar{q}(< \alpha, \sigma_{1...n} > (q_1(d, \pi_1 Z_0), ..., q_n(d, \pi_n Z_0)))$, for each $\sigma \in \Sigma_n - \{\alpha\}$, $n \geq 0$, $q \in Q_0$, $q_1, ..., q_n \in Q$, and $q(\sigma(x_1, ..., x_n), Z_0) \rightarrow \sigma(q_1(x_1, \pi_1 Z_0), ..., q_n(x_n, \pi_n Z_0))$, is a rule in $R$,

(I) $q(\alpha, \Theta u) \rightarrow \bar{q}(< \alpha, \alpha >)$, with $\Theta \in \Pi_1 \cup \{\lambda\}$, and $q(\alpha, \Theta u) \rightarrow \alpha$, is a rule in $R$,

(III) $q(d, \Theta u) \rightarrow \bar{q}(< \sigma_{1...n}, \sigma_{1...n} > (q_1(d, \pi_1 u), ..., q_n(d, \pi_n u)))$, with $\sigma \in \Sigma_n - \{\alpha\}$, $n \geq 0$, $q \in Q_0$, $q_1, ..., q_n \in Q$, $\Theta \in \Pi_1 \cup \{\lambda\}$, and $q(\sigma(x_1, ..., x_n), \Theta u) \rightarrow \sigma(q_1(x_1, \pi_1 u), ..., q_n(x_n, \pi_n u))$, is a rule in $R$,

(IV) $q(x, \Theta u) \rightarrow q'(x, \pi u)$, with $\Theta \in \Pi_1 \cup \{\lambda\}$, and $q(x, \Theta u) \rightarrow q'(x, \pi u)$, is a rule in $R$,

(V) $\bar{q}(< s, \sigma_{1...n} > (x_1, ..., x_n)) \rightarrow \sigma(x_1, ..., x_n)$, for each $\sigma \in \Sigma_n$, $n \geq 0$, and $s = \alpha$ or $s = \sigma_{1...n}$. $\square$

Next, we investigate subclasses of the class of raptt's.

A raptt $\mathcal{T}$ is called *real-time* if it has no $\lambda$-rules, that is rules of type (IV), and it is called *quasireal-time* if there is a constant $C$ such that in any computation of $\mathcal{T}$ there are at most $C$ consecutive $\lambda$-moves. The class of relations computed by all real-time raptt's (resp. quasireal-time raptt's) is denoted by $RAGT^1$ (resp. $QRAGT^1$) where the superscript indicates that we use the restricted version of the transducer. In a simiar way as in Theorem 1, we can prove the next

**Corollary 3.** *Let $\Sigma, \Gamma$ be ranked alphabets and $L \subseteq T_\Sigma \times T_\Gamma$. Then $L \in RAGT^1$ (resp. $L \in QRAGT^1$) if and only if there exists a real-time (resp. quasireal-time) algebraic forest $F \subseteq T_{\Sigma \vee \Gamma}$, such that $L = \{(\varphi_\Sigma(t), \varphi_\Gamma(t)) \, / \, t \in F\}$.* $\square$

**Proposition 1.** *$RAGT^1 \subsetneqq QRAGT^1 \subsetneqq AGT$.*

**Proof.** Take into account Theorem 1, Corollary 3 and the inclusions $RT^1 \subsetneqq QRT^1 \subsetneqq ALG$ [15,16]. $\square$

Our next task will be the investigation of the main properties of the class $AGT$.

Using the algebraic characterization of the class $AGT$, we derive in an easy way results for the classes of all aptt's and all raptt's. Indeed

**Proposition 2.** *The domain and the range of an (restricted) alphabetic pushdown tree transducer are algebraic forests.*

**Proof.** The image of an algebraic forest under an alphabetic homomorphism is still algebraic. $\square$

On the other hand

**Proposition 3.** *Let $\mathcal{T} = (Q, \Sigma, \Gamma, \Pi, d, \bar{q}, Q_0, Z_0, R)$ be an (restricted) alphabetic pushdown tree transducer and $F \subseteq T_\Sigma$ a recognizable forest. Then $\mathcal{T}(F) \in ALG$.*

**Proof.** The intersection of a recognizable and an algebraic forest is algebraic. $\square$

Recall that $AT$ denotes the class of alphabetic relations [2,17]. Since $REC \subseteq ALG$, from Theorem 7 in [17] and Theorem 2, we obtain that $AT \subseteq AGT$. The following results hold for alphabetic transducers and the class $AT$ [17]:
  - Equivalence is undecidable for the class $AT$.
  - Alphabetic tree transducers are not in general confluent.
  - Alphabetic tree transducers are not in general Noetherian.
  - Alphabetic tree transducers do not have in general bounded difference.
(A transducer $\mathcal{T}$ has bounded difference if there is a natural number $k$ such that $\mid hg(t) - hg(p) \mid \leq k$ for each $(t, p) \in \mid \mathcal{T} \mid$).
Thus

**Corollary 4.** *Equivalence is undecidable for the class $AGT$.* $\square$

**Corollary 5.** *The following statements hold:*
- *Alphabetic pushdown tree transducers are not in general confluent.*
- *Alphabetic pushdown tree transducers are not in general Noetherian.*
- *Alphabetic pushdown tree transducers do not have in general bounded difference.* $\square$

The composition operation of relations will be denoted by $\circ$.

**Proposition 4.** *It holds* $AGT \circ AT = AGT$, *and* $AT \circ AGT = AGT$. $\square$

**Proof.** We prove the relation $AT \circ AGT = AGT$. The other one is established similarly. Let $\Sigma, \Gamma, \Delta$ be finite ranked alphabets and $f : T_\Sigma \to P(T_\Gamma)$, $g : T_\Gamma \to P(T_\Delta)$ be algebraic and alphabetic transductions, respectively. By Theorem 7 in [17] and Theorem 2, we have that $\#f = \#\varphi_\Gamma(L \cap \varphi_\Sigma^{-1})$ and $\#g = \#h_\Delta(F \cap h_\Gamma^{-1})$, where $L \subseteq T_{\Sigma \vee \Gamma}$ is algebraic and $F \subseteq T_{\Gamma \vee \Delta}$ recognizable forests. By Lemma 5.1 of [2], there exists a ranked alphabet $\Theta$, a recognizable forest $R$ over $\Theta$ and two alphabetic homomorphisms $\alpha : T_\Theta \to T_{\Sigma \vee \Gamma}$, $\beta : T_\Theta \to T_{\Gamma \vee \Delta}$, such that $\{\alpha(t), \beta(t) \ / \ t \in R\} = \#(h_\Gamma^{-1} \circ \varphi_\Gamma)$.

Then for the transduction $g \circ f$, it holds

$$\#(g \circ f) = \#((h_\Delta \circ \beta)((\beta^{-1}(F) \cap R \cap \alpha^{-1}(L)) \cap (\varphi_\Sigma \circ \alpha)^{-1})).$$

The homomorphisms $\varphi_\Sigma \circ \alpha$, $h_\Delta \circ \beta$ are alphabetic and the forest $\beta^{-1}(F) \cap R \cap \alpha^{-1}(L)$ is algebraic, hence $g \circ f \in AGT$. Thus $AT \circ AGT \subseteq AGT$. The inverse inclusion is obvious. $\square$

**Theorem 3.** *The class* $AGT$ *is not closed under composition.* $\square$

Following the proof of Proposition 4, it is not difficult to see that if $f_1 : T_{\Sigma_1} \to P(T_\Gamma)$, $f_2 : T_\Gamma \to P(T_{\Sigma_2})$ are two algebraic transformations, then there exist a ranked alphabet $\Theta$, two algebraic forests $F_1, F_2$ over $\Theta$, and two alphabetic homomorphisms $h_1 : T_\Theta \to T_{\Sigma_1}$, $h_2 : T_\Theta \to T_{\Sigma_2}$, such that $\#(f_2 \circ f_1) = \#h_2((F_1 \cap F_2) \cap h_1^{-1})$. Thus the composition problem of algebraic relations reduces to the intersection problem of algebraic forests.

**Proposition 5.** *Let* $f_1 : T_\Sigma \to P(T_{\Gamma_1})$, $f_2 : T_{\Gamma_1} \to P(T_{\Gamma_2})$, ..., $f_n : T_{\Gamma_{n-1}} \to P(T_\Delta)$, *be* $n$ *algebraic tree transformations. Then, there exist a ranked alphabet* $\Theta$, $n$ *algebraic forests* $F_1, F_2, ..., F_n$ *over* $\Theta$, *and two alphabetic homomorphisms* $h : T_\Theta \to T_\Sigma$, $g : T_\Theta \to T_\Delta$, *such that* $\#(f_n \circ (... \circ (f_2 \circ f_1)...)) = \#g((F_1 \cap F_2... \cap F_n) \cap h^{-1})$. $\square$

In the following, we establish comparison results of the class $AGT$ with other known families of tree relations.

We denote by $SAT$ the class of synchronized tree relations [17], and by $T - FST, B-FST, GFST, MT$, the classes of tree transformations induced by top-down, bottom-up, generalized [5], and macro tree transducers [7], respectively. Finally, $t - PDTT$ denotes the class of tree translations induced by Yamasaki's pushdown tree transducers [20,21].

**Proposition 6.** $AT \subsetneq SAT \subsetneq AGT$.

**Proof.** It holds $REC \subsetneq NST \subsetneq ALG$ [19]. Hence the proof comes by Theorem 2 and Theorems 7, 22 in [17]. $\square$

**Theorem 4.** *The class* $AGT$ *is incomparable with all the classes* $T - FST$, $B - FST, GFST, MT$, *and* $t - PDTT$. $\square$

Below we investigate subclasses of $AGT$, induced by raptt's with a certain number of stack symbols. Explicitly, we establish a new hierarchy of tree relations.

We write $AGT_n$ for the family of algebraic transformations induced by all raptt's such that the cardinal number of their $\Pi_1$'s (denoted by $card(\Pi_1)$) equals to $n$. Obviously, $AGT_0 = AT$.

One counter tree automata were introduced in [3]. Here in a similar manner we define one counter tree transducers.

**Definition 6.** *A raptt* $\mathcal{T} = (Q, \Sigma, \Gamma, \Pi, d, \bar{q}, Q_0, Z_0, R)$ *is called* one counter *if* $card(\Pi_1) = 1$. $\square$

A tree relation induced by an one counter raptt is called *one counter*.

**Theorem 5.** *Let* $\Sigma, \Gamma$ *be ranked alphabets. A tree transformation* $L \subseteq T_\Sigma \times T_\Gamma$ *is one counter if and only if there exists an one counter forest* $F \subseteq T_{\Sigma \vee \Gamma}$, *such that* $L = \{(\varphi_\Sigma(t), \varphi_\Gamma(t)) \ / \ t \in F\}$. $\square$

We denote by $OCT$ the class of all one counter tree relations. By definition $OCT = AGT_1$.

**Proposition 7.** $AGT_n = AGT_2$, *for each* $n > 2$.

**Proof.** We can use the same technique as in Theorem 4.7. of [16]. $\square$

Since $REC \subsetneqq OCF \subsetneqq ALG$ [3], we obtain

$$AGT_0 = AT \subsetneqq AGT_1 = OCT \subsetneqq AGT_n = AGT_2, \text{ for each } n > 2.$$

On the other hand, $REC \subsetneqq OCF \subsetneqq NST$ [16], and thus

$$AT \subsetneqq OCT \subsetneqq SAT \subsetneqq AGT.$$

## 4   Conclusion

In the present paper we introduced the model of alphabetic pushdown tree transducers. We established an algebraic characterization for the class $AGT$ of the induced transformations. Further, we investigated the realionship of $AGT$ with well-known classes of tree relations, as well as compotition results.

**Acknowledgment.** I thank an anonymous referee for insightful suggestions.

## References

1. A.V. Aho and J.D. Ullman, *The Theory of Parsing, Translation and Compiling, Vol1*, Prentice-Hall, Englewood Cliffs, NJ, 1972.
2. S. Bozapalidis, Alphabetic tree relations, *Theoret. Comput. Sci.* **99** (1992) 177–211.

3. S. Bozapalidis and G. Rahonis, On two families of forests, *Acta Inform.* **31** (1994) 235–260.

4. C. Choffrut and K. Culik II, Properties of finite and pushdown transducers, *Siam J. Comput.* **12, 2,** (1983) 300–315.

5. J. Engelfriet, Bottom-up and top-down tree transformations. A comparison, *Math. Systems Theory* **9** (1975) 198–231.

6. J. Engelfriet, Top-down tree transducers with regular look-ahead, *Math. Systems Theory* **10** (1976/1977) 289–303.

7. J. Engelfriet and H. Vogler, Macro Tree Transducers, *J. Comput. System Sci.* **31** (1985) 71–146.

8. M. Fliess, Transductions algébriques, *R.A.I.R.O.* R1 (1970) 109–125.

9. Z. Fülöp and S. Vágvölgyi, Iterated deterministic top-down look-ahead, *Lecture Notes in Computer Science* **380** (1989) 175–184.

10. Z. Fülöp and S. Vágvölgyi, Top-down tree transducers with regular look-ahead, *Inform. Process. Lett.* **33** (1989/1990) 3–5.

11. Z. Fülöp and S. Vágvölgyi, Variants of top-down tree transducers with look-ahead, *Math. Systems Theory* **21** (1989) 125–145.

12. Z. Fülöp and H. Vogler, *Syntax-Directed Semantics*, Springer-Verlag, 1998.

13. F. Gécseg and M. Steinby, *Tree Automata*, Akadémiai Kiadó, Budapest 1984.

14. F. Gécseg and M. Steinby, Tree Languages, in: *Handbook of Formal Languages, Vol III*, G. Rozenberg and A. Salomaa, eds., Springer-Verlag, 1997, pp.1–68.

15. I. Guessarian, Pushdown tree automata, *Math. Systems Theory* **16** (1983) 237–263.

16. G. Rahonis and K. Salomaa, On the size of stack and synchronization alphabets of tree automata, *Fund. Inform.* **36** (1998) 57–69.

17. G. Rahonis, Alphabetic and synchronized tree transducers, *Theoret. Comput. Sci.* **255** (2001) 377–399.

18. G. Rahonis, Alphabetic pushdown tree transducers, *Dept. of Mathematics, University of Thessaloniki, Greece*, preprint.

19. K. Salomaa, Synchronized tree automata, *Theoret. Comput. Sci.* **127** (1994) 25–51.

20. K. Yamasaki, Fundamental Properties of Pushdown Tree Transducers (PDTT) – A Top-Down Case, *IEICE Trans. Inf. & Syst.* **E76-D, 10** (1993) 1234–1242.

21. K. Yamasaki, Note on Domain/Surface Tree Languages of t-PDTT's, *IEICE Trans. Inf. & Syst.* **E79-D, 6** (1996) 829–839.

# Author Index